commonsense CELESTIAL NAVIGATION

commonsense CELESTIAL NAVIGATION

HEWITT SCHLERETH

Henry Regnery Company · Chicago

Library of Congress Cataloging in Publication Data

Schlereth, Hewitt.
Common sense celestial navigation.

Includes index.
1. Nautical astronomy. 2. Navigation. I. Title.
VK555.S34 527 74-27829
ISBN 0-8092-8279-8

Published by Henry Regnery Company
180 North Michigan Avenue, Chicago, Illinois 60601
Manufactured in the United States of America
Library of Congress Catalog Card Number: 74-27829
International Standard Book Number: 0-8092-8279-8

Published simultaneously in Canada by
Fitzhenry & Whiteside Limited
150 Lesmill Road
Don Mills, Ontario M3B 2T5
Canada

To My Father and Mother

Contents

1
Broadly Speaking

Celestial navigation is the application of two sciences—astronomy and mathematics—to the problem of determining location at sea without reference to a stationary landmass or a fixed object such as a lighthouse on shore or a buoy in the water.

Basically, in lieu of a stationary physical object, astonomy supplies the navigator with another kind of known point. The sextant and the application of mathematics then enable him to deduce his position from that point. Astronomers call this known point the *geographical position*, or GP, of a celestial body. The GP is the point on the earth that is directly underneath the body at any instant, and astronomers have assembled a book—the *Nautical Almanac*—in which you can look up the geographical position of sun, moon, planet, or star at any second of any day of the year.

The particular branch of mathematics that applies to celestial navigation is spherical trigonometry. Fortunately, it is not necessary to understand spherical trigonometry in order to understand celestial navigation, nor is it necessary to know how to solve the formulas of spherical trigonometry. You simply have to accept the fact that the formulas exist and the fact that they can produce the information needed to carry out the process of celestial navigation.

Chapter 3 presents a fictional situation for the purpose of showing the logic underlying celestial navigation. Rest assured; you will not actually practice celestial navigation in the manner described. In practice, 99.9 percent of the time spent in celestial navigation is spent looking up GPs in the almanac and taking sextant shots. No time is spent on computations of spherical trigonometry. For that very reason, Chapter 4 of this book is called "Shoot! Don't Compute!"

As for Chapter 3, you will find that once you see the logic of celestial navigation, the details that are discussed in later chapters will not confuse you but will expand and refine what you already know.

2
The Sextant

You may already know that the sextant is a precision instrument for measuring angles very accurately. Specifically, when you use a sextant at sea, you measure the angle between the horizon, your eye, and the sun or another celestial body (Figure 2.1).

Because of the particular geometric relationships among you, the horizon, and the celestial body, what you actually do is measure the distance from the geographical position of the body to your location (Figure 2.2). Again, due to the geometry involved, this distance is an arc of what is called a *great circle* (a term that simply means a circle the same size as the equator), and one degree on a great circle is the same length as one degree along the equator—that is, sixty nautical miles.

This arc is called the *zenith distance* and is found by subtracting the sextant reading from 90° (Figure 2.3). Thus, if you observed the sun and the sextant said the angle between the horizon and the sun was

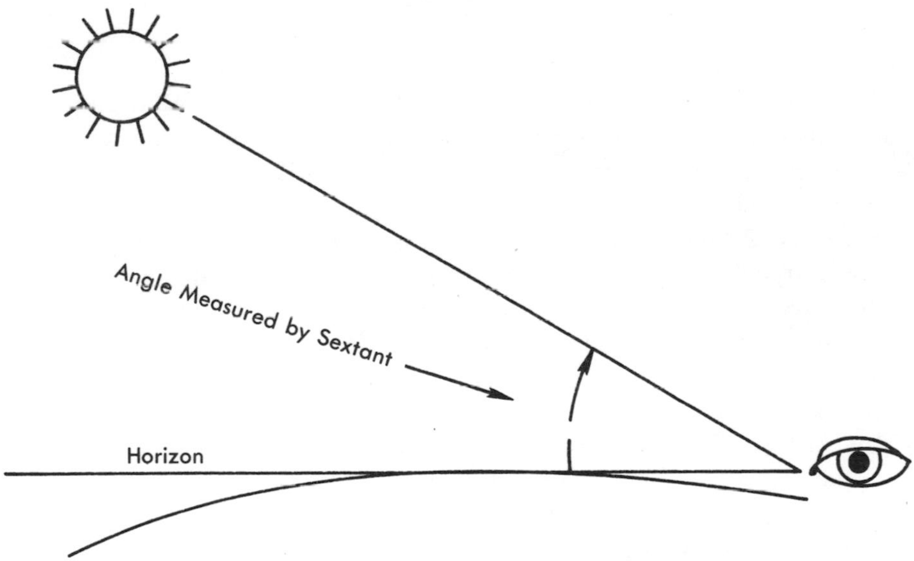

Fig. 2.1

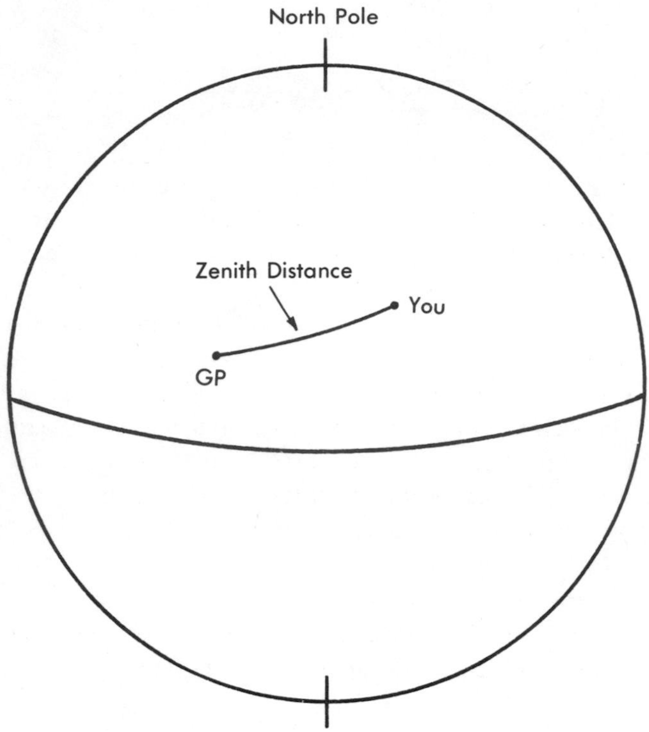

Fig. 2.2

31°, the zenith distance would be 59° (90° − 31°), or 60 × 59 nautical miles. This is the most important fact in celestial navigation. It is so basic that if you remember it, you will find you will not have to memorize a lot of rules later on.

Figure 2.4 emphasizes the point that the zenith distance is the length of the arc between you and the geographical position of a celestial body regardless of the direction in which the body lies.

To summarize, the sextant gives you the link between a known point (the geographical position of a celestial body) and your position by enabling you to calculate the zenith distance—the great circle arc on the surface of the earth between you and the GP of the body.

Obviously, if you know *how far* you are from a known point, you have gone a long way toward finding out *where* you are. The next chapter explains the final steps toward that end.

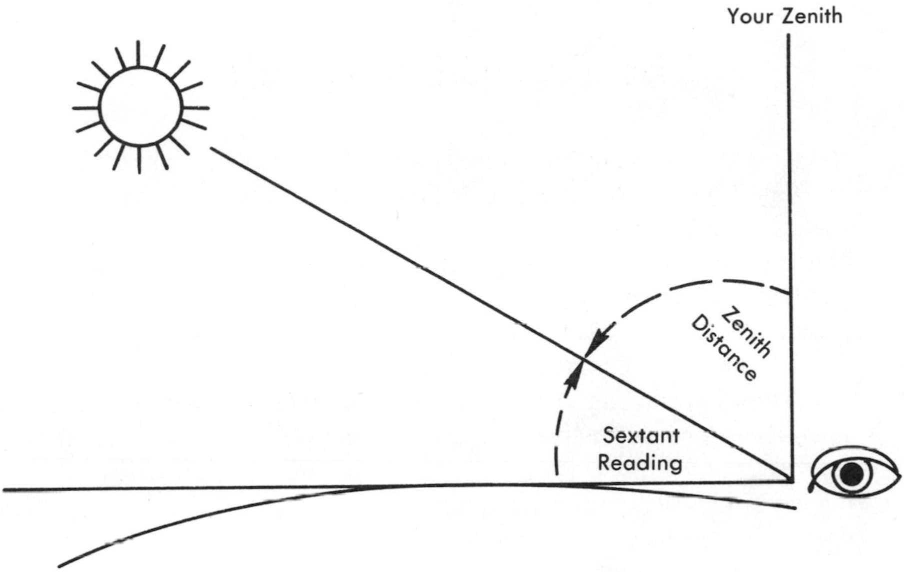
Your Zenith
Zenith Distance
Sextant Reading

Fig. 2.3

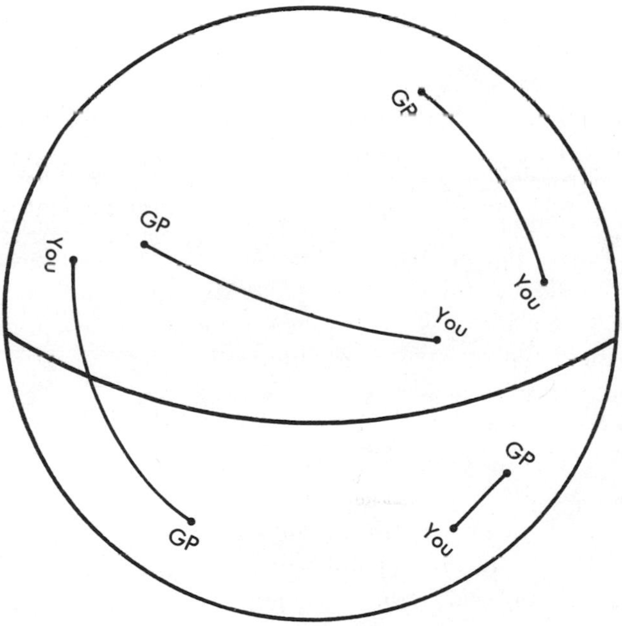
GP
You
GP
You
You
GP
GP
You

Fig. 2.4

3
The Eternal Triangle

Imagine the following situation. You are three days outward bound from New York for Bermuda. It is about 5 P.M., and, after two days of heavy overcast, the sky clears and you see the sun.

You pick up your sextant and begin using the procedure you read up on during the winter. You measure the angle between the horizon, yourself, and the sun. You note that the time is exactly 5 P.M. and the sextant reads 31°. The zenith distance is therefore 59°.

Unfortunately, during the winter you had a lot of other things to do, and you never got around to finding out what you do *after* you take the sextant sight, note the precise time, and figure out the zenith distance. You have, however, an exceptionally well-equipped boat, and you were at the top of your class in logic.

Logically, you decide to begin with what you know, so, from a well-stocked locker on your well-equipped yacht, you haul out a globe of the world. After looking at the globe for a while, you decide that it will be necessary to make one assumption: You will assume that you are at latitude 37° north and longitude 67° west. This is fairly close to your actual position, you believe, because you have kept track of your speed and heading during the days of overcast. You mark this spot on the globe and call it *Y*—for *you* (Figure 3.1).

After considering for a while, you decide it would advance things if you knew where the sun was on this day at 5 P.M. You look through your bookshelves and come across the *Nautical Almanac*. In it you find that at 5 P.M. the sun was directly over a spot on the earth at latitude 24° north and longitude 134° west (Figure 3.2). You mark this spot on your globe and draw a line between the two dots—one representing your assumed geographical position and the other representing that of the sun. You label the position of the sun *GP*.

You realize that the line you have just drawn is the zenith distance between your assumed position and the known position of the sun. You realize also that if you could figure out the length of this hypothetical zenith distance you could compare it with the zenith distance you ac-

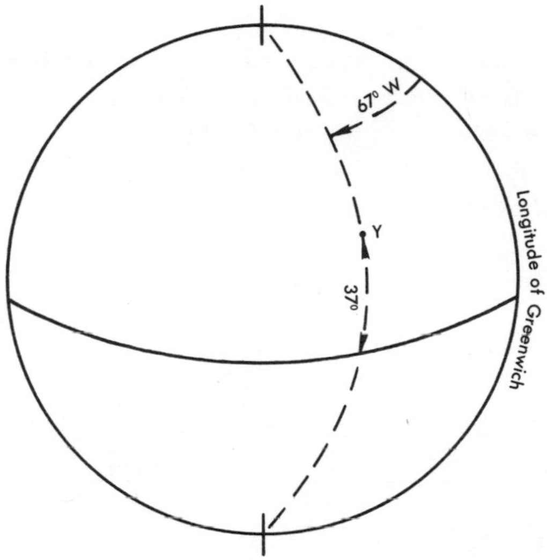
North Pole
67° W
Y
37°
Longitude of Greenwich

Fig. 3.1

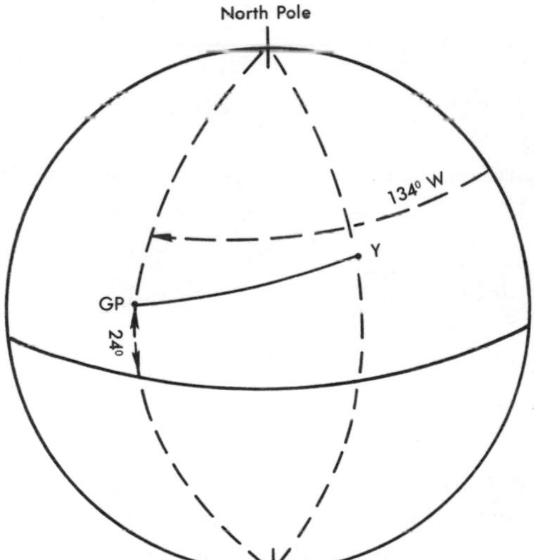
North Pole
134° W
Y
GP
24°

Fig. 3.2

tually measured with your sextant and find out where you were in relation to your assumed position.

Looking at what you have, you see that by drawing along the longitude line from you to the North Pole and from *GP* to the North Pole, you will have made an odd-looking triangle (Figure 3.3). You decide to see what you can figure out about this triangle.

First of all, you know that there are 90° of latitude between the equator and the North Pole. Since you are on 37° north latitude, there are 53° between you and the pole. This means that you know the length of one side of the triangle. By subtracting the latitude of the GP of the sun (24°) from 90° you find that a second side of the triangle is 66°.

You also conclude that since the sun is on longitude line 134° and you are on longitude 67°, the angle at the top of the triangle is 67° (134° − 67°; see Figure 3.4). From the almanac you learn that this angle is commonly labeled *LHA,* which stands for *local hour angle.*

You realize that you now know two sides of the triangle and the angle between them. Surely there is a way to figure out the third side—the zenith distance between your assumed position and the known position of the sun.

After rummaging again through your library, you find the right book (*Spherical Trigonometry*), and, by putting what you have figured out into a formula in the book, you find that the unknown side is 58° (Figure 3.5). In other words, if you actually were at *Y*, your zenith distance would be 58°.

Your sextant observation told you that your zenith distance was 59°. You differ by 1°. Obviously, you are not at point *Y.* But you did not expect to be; point *Y* was the position you *assumed* in order to construct your triangle. You do see, however, that you are within 1° of point *Y*, and 1°, you have learned, is 60 nautical miles on the surface of the earth. So now you know that you are within 60 miles of your assumed position.

You note further that the zenith distance you measured with your sextant is greater than it would have been had you actually been at point *Y.* Obviously, then, you are 60 miles farther from the GP of the sun than point *Y.* You tentatively mark spot *O* a little to the right of *Y* and in line with side *C* (Figure 3.6).

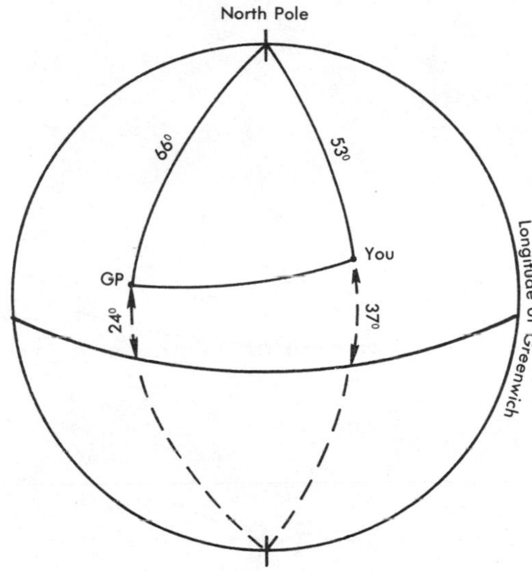
North Pole
66°
53°
You
GP
24°
37°
Longitude of Greenwich

Fig. 3.3

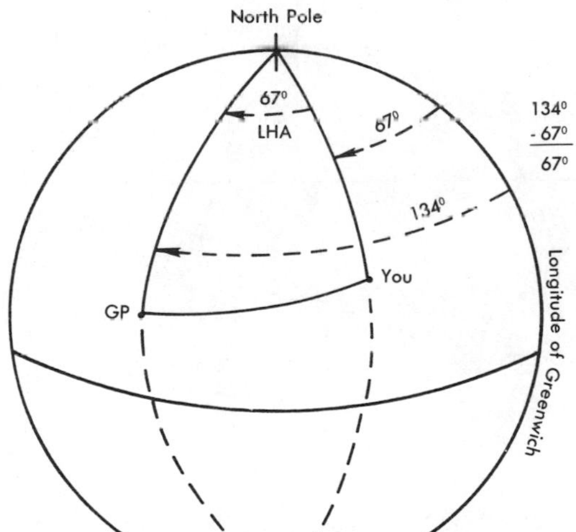
North Pole
67°
LHA
67°
134°
134°
- 67°
67°
You
GP
Longitude of Greenwich

Fig. 3.4

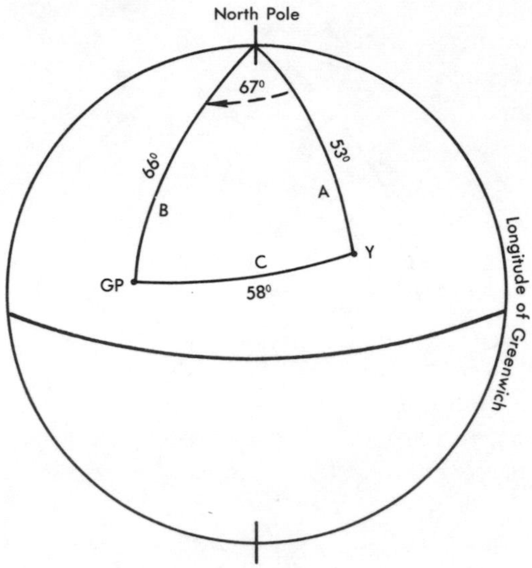
North Pole
67°
66°
53°
B
A
C
Y
GP
58°
Longitude of Greenwich

Fig. 3.5

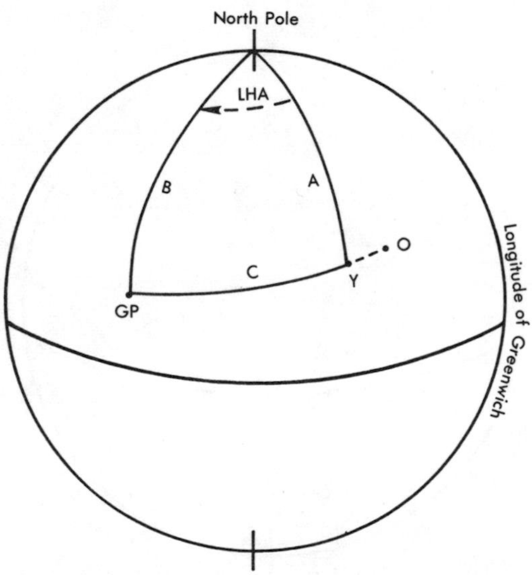
North Pole
LHA
B
A
C
Y
O
GP
Longitude of Greenwich

Fig. 3.6

Stepping back from the globe, you see that your actual position must be somewhere on the perimeter of a circle having a radius of side *C* plus 60 miles (Figure 3.7). This perimeter line would pass through point *O*, and since the radius is so large—over 3,000 miles—the part of the perimeter within a reasonable distance north or south of point *Y* would be a straight line for all practical purposes (Figure 3.8). You are, you realize, somewhere on that line. You label it *LOP*—for *line of position.*

You decide now to plot this information on your chart. To do this, you will need to know the angle in the lower right-hand corner of the triangle (Figure 3.9). You call it angle *Z* because it is apparently the last thing you are going to have to label. By going back to your book with the formulas, you discover that angle *Z* is 83°. This makes sense, because you could see that the sun was pretty nearly west of you at the time you took your sight.

The actual plotting procedure is simple (Figure 3.10). You locate point *Y* by its latitude and longitude. Through point *Y* you draw a line in the direction of the sun by marking off angle *Z*. Sixty miles from point *Y*, away from the sun, you mark point *O*. Through point *O* you draw a line at right angles to line *OY*. The line you just drew is your line of position. You are somewhere on that line.

You note with interest that the line from *Y* toward the GP of the sun is actually side *C* of your triangle; the longitude line is side *A*. And isn't it fortunate that you don't have to lay out the whole triangle on your chart? All you need is the lower right-hand corner—your assumed position—and from that you plot everything.

While you have been doing all this figuring and thinking, the sun has been moving on. Looking out the companionway, you can see that its altitude and bearing, or direction, have changed. This gives you an idea. You grab your sextant and take another sight. This time you don't have to start from scratch, so you are able to figure out the information you need much more quickly and, in a relative jiffy, you have a new LOP plotted on your chart (Figure 3.11).

Because angle *Z* has changed as the sun has moved west, your second LOP crosses your first. Since you know your approximate speed and course for the time between your two sights, you move your first

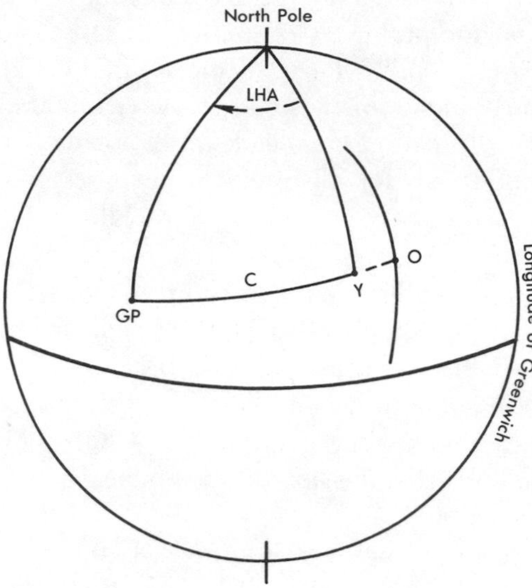
North Pole
LHA
C
Y
O
GP
Longitude of Greenwich

Fig. 3.7

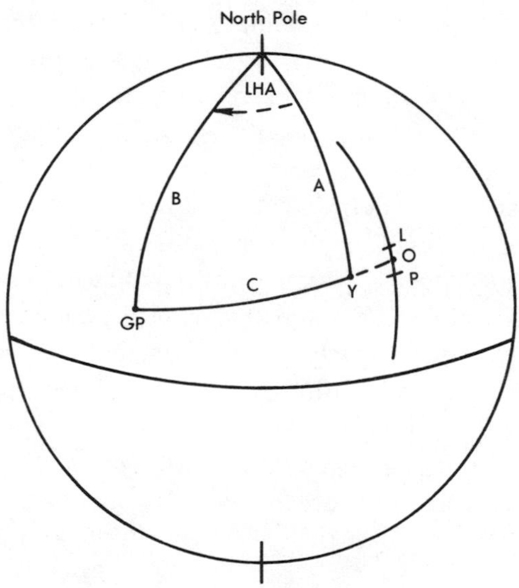
North Pole
LHA
B
A
L
O
P
C
Y
GP

Fig. 3.8

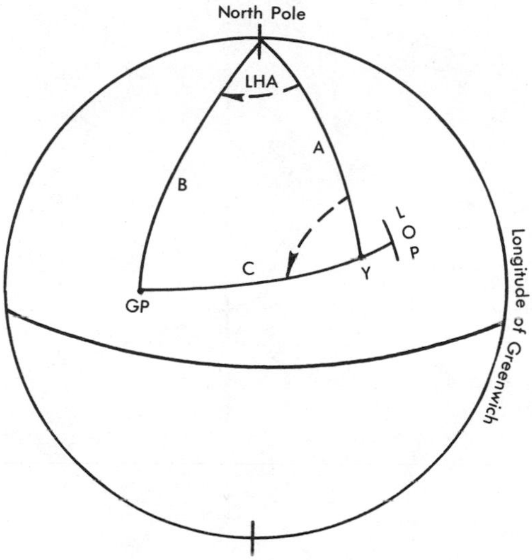

Fig. 3.9

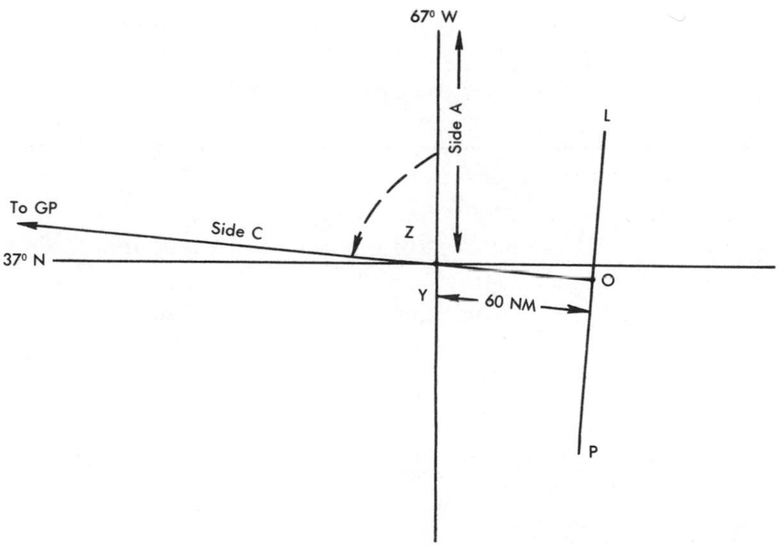

Fig. 3.10

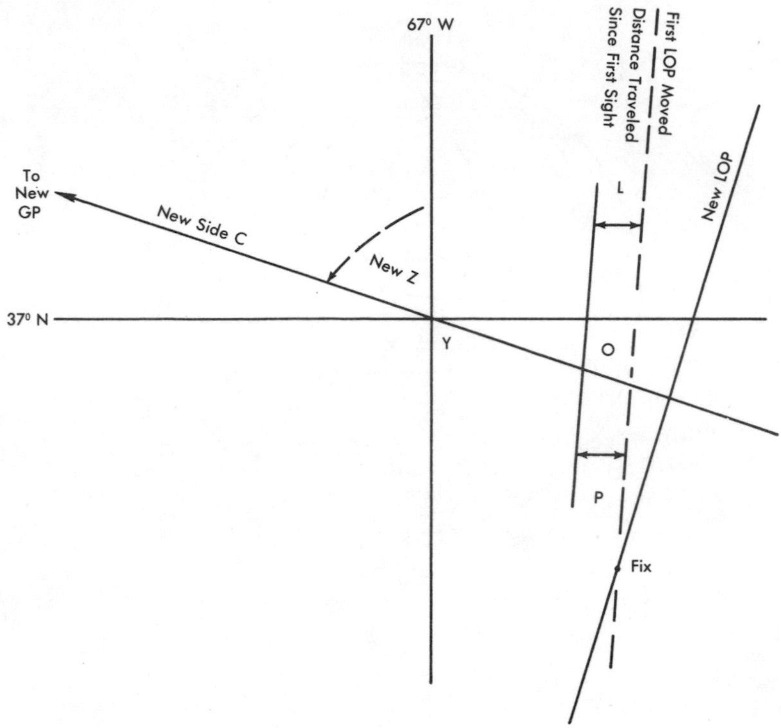

Fig. 3.11

LOP accordingly. The place where the two lines cross is your location, and you have fixed your position.

For the next few days you repeat this process morning and afternoon, adjusting your course according to your location, and—wonder of wonders—Bermuda appears on the horizon early one morning within twenty minutes of the time you expected to see it!

4
Shoot! Don't Compute!

Logically, the process of celestial navigation breaks down in the following way:

1. Find a known point (GP).
2. Using the sextant, determine your distance from that known point.
3. Assume a position and, using the known point, construct a navigational triangle.
4. Solve the triangle for your (theoretical) zenith distance from the known point.
5. Compare your actual distance from the known point as measured by the sextant with the theoretical or computed distance.
6. Based on whether your measured (or actual) distance is greater or less than the computed distance, lay out the appropriate line of position relative to your assumed position.

In practice, celestial navigation does not follow these steps in this order. Rather, the first step is to use a sextant and a watch or chronometer to take a sight. (This is step 2 above.) Then you follow step 1 and steps 3 through 6. There are other minor differences, too. These remarks are intended first as a review of the major contents of the first three chapters and second as a condensed presentation of the entire process of celestial navigation.

While you read Chapter 3, it may have occurred to you that had you known you would be at latitude 37° north and longitude 67° west, you could have set up the navigational triangle in the comfort of your den at home, written out the answer on a piece of paper, and had it all ready when you needed it. No need then to sweat through tables and formulas at the chart table of your bouncing boat.

Taking this thought a step further, suppose you could have figured out the third side of all triangles between latitudes 40° north and 30° north? In this day of computers, would that be impossible? Well, probably, because obviously there is an infinite number of triangles between those latitudes.

But suppose you figured out only those triangles for every whole degree of your latitude, whole degree of latitude of the GP of the sun, and whole degree of LHA? Not to belabor the point, this approach reduces the number of triangles to manageable proportions. As a matter of fact, the solutions for all such triangles between latitudes 0° and 89° north and south can be contained in two one-inch-thick volumes. Table 1 is a reproduction of a table from such a volume. This particular one is from *Sight Reduction Tables for Air Navigation* (H.O. 249). You will need to hold the book sideways to be properly oriented.

In the lower right-hand corner of Table 1 you see the designation "LAT 37°," or latitude 37°. This is the latitude of your assumed position in Chapter 3. At the left-hand edge of the page you see a column headed "LHA," and if you look down the column, you will find the LHA of the navigational triangle in Chapter 3—the third from the bottom, 67°.

Now you have to find the remaining factor of the triangle—the latitude of the sun. You may recall that the latitude was 24° north. The top of the table reads "Declination (15°-29°) Same Name as Latitude," and underneath are columns labeled "15°" through "29°." If you look down the column headed "24°" until you intersect the line across from LHA 67°, you will see 32°. Subtracting 32° from 90° gives you the zenith distance of 58°. You may remember that this is what you had to figure out in the fictional exercise in Chapter 3. Also, to the right side of the column you see angle *Z*, 83°.

Looking back at the heading "Declination (15°-29°) Same Name as Latitude" and translating, you can see that *declination* is the astronomers' word for the latitude of the GP of the celestial body (the sun in this case). That word is used to avoid confusion with *your* latitude. "Same Name" simply means that the sun's latitude is north and is therefore the same as your assumed latitude, which was north.

If you are in south latitude and the sun's declination is also south, you would use this same page. The only difference is that now the navigational triangle is turned upside down, and the South Pole is the top of the triangle. Obviously, the triangles are equivalent and the zenith distance is the same (Figure 4.1).

Had the sun's latitude been south and your latitude north, you

LHA	15°			16°			17°			18°			19°		
	Hc	d	Z	Hc	d	Z	Hc	d	Z	Hc	d	Z	Hc	d	Z
	° ′	′	°	° ′	′	°	° ′	′	°	° ′	′	°	° ′	′	°
0	68 00	+60	180	69 00	+60	180	70 00	+60	180	71 00	+60	180	72 00	+60	180
1	67 59	60	177	68 59	60	177	69 59	60	177	70 59	60	177	71 59	60	177
2	67 56	60	175	68 56	59	175	69 55	60	174	70 55	60	174	71 55	60	174
3	67 50	60	172	68 50	60	172	69 50	59	172	70 49	60	171	71 49	59	171
4	67 43	59	170	68 42	59	169	69 41	60	169	70 41	59	168	71 40	59	168
5	67 33	+59	167	68 32	+59	167	69 31	+59	166	70 30	+59	166	71 29	+58	165
6	67 22	58	165	68 20	59	164	69 19	58	164	70 17	58	163	71 15	58	162
7	67 08	58	162	68 06	58	162	69 04	58	161	70 02	57	160	70 59	57	159
8	66 53	57	160	67 50	57	159	68 47	57	158	69 44	57	158	70 41	56	157
9	66 35	57	158	67 32	57	157	68 29	56	156	69 25	56	155	70 21	55	154
10	66 16	+57	155	67 13	+55	155	68 08	+56	154	69 04	+55	153	69 59	+55	151
11	65 56	55	153	66 51	55	152	67 46	55	151	68 41	54	150	69 35	54	149
12	65 33	55	151	66 28	54	150	67 22	54	149	68 15	53	148	69 09	53	147
13	65 09	54	149	66 03	54	148	66 57	53	147	67 50	52	146	68 42	52	144
14	64 44	53	147	65 37	53	146	66 30	52	145	67 22	51	143	68 13	51	142
15	64 17	+52	145	65 09	+52	144	66 01	+51	143	66 52	+51	141	67 43	+50	140
16	63 48	52	143	64 40	51	142	65 31	51	141	66 22	50	139	67 12	48	138
17	63 19	51	141	64 10	50	140	65 00	50	139	65 50	49	137	66 39	48	136
18	62 48	50	139	63 38	50	138	64 28	49	137	65 17	48	135	66 05	47	134
19	62 16	50	138	63 06	48	136	63 54	48	135	64 42	48	134	65 30	46	132
20	61 43	+49	136	62 32	+48	135	63 20	+47	133	64 07	+47	132	64 54	+45	130
21	61 09	48	134	61 57	48	133	62 45	46	132	63 31	46	130	64 17	44	129
22	60 35	47	133	61 22	46	131	62 08	46	130	62 54	45	129	63 39	44	127
23	59 59	46	131	60 45	46	130	61 31	45	128	62 15	44	127	63 00	43	125
24	59 22	46	130	60 08	45	128	60 53	45	127	61 38	43	126	62 21	42	124
25	58 45	+45	128	59 30	+45	127	60 15	+43	126	60 53	+43	124	61 41	+41	123
26	58 07	45	127	58 52	43	125	59 35	43	124	60 18	42	123	61 00	41	121
27	57 28	44	125	58 12	43	124	58 55	43	123	59 38	41	121	60 19	40	120
28	56 49	43	124	57 32	43	123	58 15	41	122	58 56	41	120	59 37	40	119
29	56 09	43	123	56 52	42	122	57 34	41	120	58 15	40	119	58 55	39	117
30	55 28	+42	122	56 10	+42	120	56 52	+40	119	57 32	+40	118	58 12	+39	116
31	54 47	42	120	55 29	41	119	56 10	40	118	56 50	39	117	57 29	38	115
32	54 06	41	119	54 47	40	118	55 27	40	117	56 07	38	115	56 45	38	114
33	53 24	40	118	54 04	40	117	54 44	39	116	55 23	38	114	56 01	37	113
34	52 41	40	117	53 21	40	116	54 01	38	115	54 39	38	113	55 17	37	112
35	51 58	+40	116	52 38	+39	115	53 17	+38	114	53 55	+37	112	54 32	+37	111
36	51 15	39	115	51 54	39	114	52 33	37	112	53 10	37	111	53 47	36	110
37	50 31	39	114	51 10	38	113	51 48	38	112	52 26	36	110	53 02	36	109
38	49 47	39	113	50 26	38	112	51 04	37	111	51 41	36	109	52 17	35	108
39	49 03	38	112	49 41	38	111	50 19	36	110	50 55	36	108	51 31	35	107
40	48 18	+38	111	48 56	+37	110	49 33	+37	109	50 10	+35	107	50 45	+35	106
41	47 34	37	110	48 11	37	109	48 48	36	108	49 24	35	107	49 59	34	105
42	46 48	38	109	47 26	36	108	48 02	36	107	48 38	35	106	49 13	34	104
43	46 03	37	108	46 40	36	107	47 16	35	106	47 51	35	105	48 26	34	104
44	45 17	37	108	45 54	36	106	46 30	35	105	47 05	34	104	47 39	34	103
45	44 32	+36	107	45 08	+36	106	45 44	+34	104	46 18	+35	103	46 53	+33	102
46	43 46	36	106	44 22	35	105	44 57	35	104	45 32	34	102	46 06	33	101
47	42 59	36	105	43 35	35	104	44 10	35	103	44 45	34	102	45 19	33	101
48	42 13	36	104	42 49	35	103	43 24	34	102	43 58	33	101	44 31	33	100
49	41 27	35	104	42 02	35	102	42 37	34	101	43 11	33	100	43 44	33	99
50	40 40	+35	103	41 15	+35	102	41 50	+34	101	42 24	+33	99	42 57	+32	98
51	39 53	35	102	40 28	34	101	41 02	34	100	41 36	33	99	42 09	33	98
52	39 06	35	101	39 41	34	100	40 15	34	99	40 49	33	98	41 22	32	97
53	38 19	35	101	38 54	34	100	39 28	33	98	40 01	33	97	40 34	33	96
54	37 32	34	100	38 06	34	99	38 40	34	98	39 14	33	97	39 47	32	96
55	36 45	+34	99	37 19	+34	98	37 53	+33	97	38 26	+33	96	38 59	+32	95
56	35 57	35	98	36 32	33	97	37 05	33	96	37 38	33	95	38 11	32	94
57	35 10	34	98	35 44	34	97	36 18	33	96	36 51	32	95	37 23	32	94
58	34 22	34	97	34 56	34	96	35 30	33	95	36 03	32	94	36 35	32	93
59	33 35	34	96	34 09	33	95	34 42	33	94	35 15	33	93	35 48	32	92
60	32 47	+34	96	33 21	+33	95	33 54	+33	94	34 27	+33	93	35 00	+32	92
61	31 59	34	95	32 33	34	94	33 07	32	93	33 39	33	92	34 12	32	91
62	31 12	33	94	31 45	34	94	32 19	33	93	32 52	32	92	33 24	32	91
63	30 24	34	94	30 58	33	93	31 31	33	92	32 04	32	91	32 36	32	90
64	29 36	34	93	30 10	33	92	30 43	33	91	31 16	32	90	31 48	32	89
65	28 48	+34	93	29 22	+33	92	29 55	+33	91	30 28	+32	90	31 00	+32	89
66	28 00	34	92	28 34	33	91	29 07	33	90	29 40	32	89	30 12	32	88
67	27 12	34	91	27 46	33	90	28 19	33	89	28 52	32	89	29 24	32	88
68	26 24	34	91	26 58	33	90	27 31	33	89	28 04	32	88	28 36	33	87
69	25 37	33	90	26 10	33	89	26 43	33	88	27 16	33	87	27 49	32	86

LHA	20°			21°			22°			23°			24°		
	Hc	d	Z	Hc	d	Z	Hc	d	Z	Hc	d	Z	Hc	d	Z
	° ′	′	°	° ′	′	°	° ′	′	°	° ′	′	°	° ′	′	°
0	73 00	+60	180	74 00	+60	180	75 00	+60	180	76 00	+60	180	77 00	+60	180
1	72 59	60	177	73 59	60	177	74 59	59	176	75 58	60	176	76 58	60	176
2	72 55	59	174	73 54	60	173	74 54	60	173	75 54	59	172	76 53	60	172
3	72 48	59	170	73 47	60	170	74 47	59	169	75 46	59	169	76 45	59	168
4	72 39	59	167	73 38	58	167	74 36	59	166	75 35	58	165	76 33	58	164
5	72 27	+58	164	73 25	+58	163	74 23	+58	163	75 21	+58	162	76 19	+57	160
6	72 13	57	161	73 10	58	160	74 08	57	159	75 05	56	158	76 01	56	157
7	71 56	57	158	72 53	56	157	73 49	56	156	74 45	56	155	75 41	55	153
8	71 37	56	156	72 33	56	154	73 29	55	153	74 24	54	152	75 18	54	150
9	71 16	56	153	72 12	54	152	73 06	54	150	74 00	53	149	74 53	52	147
10	70 54	+54	150	71 48	+53	149	72 41	+53	147	73 34	+52	146	74 26	+50	144
11	70 29	53	148	71 22	52	146	72 14	52	145	73 06	50	143	73 56	50	141
12	70 02	52	145	70 54	52	144	71 46	50	142	72 36	49	140	73 25	48	138
13	69 34	51	143	70 25	50	141	71 15	49	140	72 04	49	138	72 53	46	136
14	69 04	50	141	69 54	49	139	70 43	49	137	71 32	46	135	72 18	46	133
15	68 33	+49	138	69 22	+48	137	70 10	+47	135	70 57	+46	133	71 43	+44	131
16	68 00	49	136	68 49	47	135	69 36	46	133	70 22	44	131	71 06	44	129
17	67 27	47	134	68 14	46	133	69 00	45	131	69 45	43	129	70 28	43	127
18	66 52	46	132	67 38	45	131	68 23	44	129	69 07	43	127	69 50	41	125
19	66 16	45	131	67 01	44	129	67 45	43	127	68 28	42	125	69 10	40	123
20	65 39	+44	129	66 23	+44	127	67 07	+42	125	67 49	+41	124	68 30	+39	122
21	65 01	44	127	65 45	42	126	66 27	42	124	67 09	39	122	67 48	39	120
22	64 23	42	126	65 05	42	124	65 47	41	122	66 28	38	120	67 07	37	118
23	63 43	42	124	64 25	41	122	65 06	40	121	65 46	38	119	66 24	37	117
24	63 03	41	123	63 44	41	121	64 25	39	119	65 04	37	117	65 41	36	116
25	62 22	+41	121	63 03	+40	120	63 43	+38	118	64 21	+37	116	64 58	+35	114
26	61 41	40	120	62 21	39	118	63 00	37	117	63 37	37	115	64 14	35	113
27	60 59	40	118	61 39	38	117	62 17	37	115	62 54	35	114	63 29	35	112
28	60 17	39	117	60 56	37	116	61 33	37	114	62 10	35	112	62 45	33	111
29	59 34	38	116	60 12	37	114	60 49	36	113	61 25	35	111	62 00	33	109
30	58 51	+37	115	59 28	+37	113	60 05	+35	112	60 40	+34	110	61 14	+33	108
31	58 07	37	114	58 44	36	112	59 20	35	111	59 55	34	109	60 29	32	107
32	57 23	37	113	58 00	35	111	58 35	35	110	59 10	33	108	59 43	32	106
33	56 38	37	112	57 15	35	110	57 50	34	109	58 24	33	107	58 57	31	105
34	55 54	35	110	56 29	35	109	57 04	34	108	57 38	32	106	58 10	32	104
35	55 09	+35	109	55 44	+34	108	56 18	+34	107	56 52	+32	105	57 24	+31	104
36	54 23	35	109	54 58	34	107	55 32	33	106	56 05	32	104	56 37	31	103
37	53 38	34	108	54 12	34	106	54 46	33	105	55 19	31	103	55 50	31	102
38	52 52	34	107	53 26	34	105	54 00	32	104	54 32	31	102	55 03	31	101
39	52 06	34	106	52 40	33	104	53 13	32	103	53 45	31	102	54 16	30	100
40	51 20	+33	105	51 53	+33	104	52 26	+32	102	52 58	+31	101	53 29	+30	99
41	50 33	34	104	51 07	32	103	51 39	32	101	52 11	31	100	52 42	29	99
42	49 47	33	103	50 20	32	102	50 52	32	101	51 24	30	99	51 54	30	98
43	49 00	33	102	49 33	32	101	50 05	31	100	50 36	31	98	51 07	29	97
44	48 13	33	102	48 46	32	100	49 18	31	99	49 49	30	98	50 19	29	96
45	47 26	+33	101	47 59	+32	100	48 31	+30	98	49 01	+30	97	49 31	+30	96
46	46 39	32	100	47 11	32	99	47 43	31	98	48 14	30	96	48 44	29	95
47	45 52	32	99	46 24	32	98	46 56	30	97	47 26	30	96	47 56	29	94
48	45 04	33	99	45 37	31	97	46 08	30	96	46 38	30	95	47 08	29	94
49	44 17	32	98	44 49	31	97	45 20	31	96	45 51	29	94	46 20	29	93
50	43 29	+32	97	44 01	+31	96	44 32	+31	95	45 03	+30	94	45 33	+28	92
51	42 42	32	97	43 14	31	95	43 45	30	94	44 15	30	93	44 45	28	92
52	41 54	32	96	42 26	31	95	42 57	30	94	43 27	30	92	43 57	28	91
53	41 07	31	95	41 38	31	94	42 09	30	93	42 39	30	92	43 09	29	91
54	40 19	31	95	40 50	31	93	41 21	30	92	41 51	30	91	42 21	29	90
55	39 31	+31	94	40 02	+31	93	40 33	+30	92	41 03	+30	90	41 33	+29	89
56	38 43	32	93	39 15	30	92	39 45	31	91	40 16	29	90	40 45	29	89
57	37 55	32	93	38 27	30	91	38 57	31	90	39 28	29	89	39 57	29	88
58	37 07	32	92	37 39	31	91	38 10	30	90	38 40	29	89	39 09	29	88
59	36 20	31	91	36 51	31	90	37 22	30	89	37 52	29	88	38 21	29	87
60	35 32	+31	91	36 03	+31	90	36 34	+30	89	37 04	+30	88	37 34	+29	86
61	34 44	31	90	35 15	31	89	35 46	30	88	36 16	30	87	36 46	29	86
62	33 56	31	90	34 27	31	88	34 58	30	87	35 28	30	86	35 58	29	85
63	33 08	31	89	33 39	31	88	34 10	30	87	34 40	30	86	35 10	30	85
64	32 20	31	88	32 51	31	87	33 22	31	86	33 53	30	85	34 23	29	84
65	31 32	+32	88	32 04	+30	87	32 34	+31	86	33 05	+30	85	33 35	+29	84
66	30 44	32	87	31 16	31	86	31 47	30	85	32 17	30	84	32 47	30	83
67	29 56	32	87	30 28	31	86	30 59	31	85	31 30	30	84	32 00	29	83
68	29 09	31	86	29 40	31	85	30 11	31	84	30 42	30	83	31 12	30	82
69	28 21	31	85	28 52	32	84	29 24	31	84	29 55	30	83	30 25	30	82

25°			26°			27°			28°			29°			LHA
Hc	d	Z	Hc	d	Z	Hc	d	Z	Hc	d	Z	Hc	d	Z	
° ′	′	°	° ′	′	°	° ′	′	°	° ′	′	°	° ′	′	°	
78 00	+60	180	79 00	+60	180	80 00	+60	180	81 00	+60	180	82 00	+60	180	360
77 58	60	176	78 58	60	175	79 58	60	175	80 58	59	174	81 57	60	174	359
77 53	59	171	78 52	60	171	79 52	59	170	80 51	59	169	81 50	58	168	358
77 44	59	167	78 43	58	166	79 41	58	165	80 39	58	164	81 37	57	162	357
77 31	58	163	78 29	58	162	79 27	57	160	80 24	56	158	81 20	55	156	356
77 16	+57	159	78 13	+56	158	79 09	+55	156	80 04	+54	154	80 58	+53	151	355
76 57	56	155	77 53	54	153	78 47	54	151	79 41	52	149	80 33	51	146	354
76 36	54	152	77 30	53	150	78 23	52	147	79 15	50	145	80 05	49	142	353
76 12	52	148	77 04	52	146	77 56	50	144	78 46	48	141	79 34	46	138	352
75 45	51	145	76 36	50	143	77 26	48	140	78 14	47	137	79 01	44	134	351
75 16	+50	142	76 06	+48	140	76 54	+47	137	77 41	+44	134	78 25	+43	131	350
74 46	48	139	75 34	47	137	76 21	45	134	77 06	42	131	77 48	41	128	349
74 13	47	136	75 00	45	134	75 45	44	131	76 29	41	128	77 10	38	125	348
73 39	46	134	74 25	44	131	75 09	41	129	75 50	40	126	76 30	37	123	347
73 04	44	131	73 48	43	129	74 31	40	126	75 11	38	123	75 49	36	120	346
72 27	+43	129	73 10	+41	127	73 51	+39	124	74 30	+37	121	75 07	+35	118	345
71 50	41	127	72 31	40	124	73 11	38	122	73 49	36	119	74 25	33	116	344
71 11	40	125	71 51	39	123	72 30	37	120	73 07	34	117	73 41	32	114	343
70 31	39	123	71 10	38	121	71 48	36	118	72 24	33	116	72 57	32	113	342
69 50	39	121	70 29	37	119	71 06	34	117	71 40	33	114	72 13	30	111	341
69 09	+38	119	69 47	+35	117	70 22	+34	115	70 56	+32	112	71 28	+30	110	340
68 27	37	118	69 04	35	116	69 39	33	113	70 12	31	111	70 43	29	108	339
67 44	36	116	68 20	34	114	68 54	33	112	69 27	30	110	69 57	28	107	338
67 01	35	115	67 36	34	113	68 10	31	111	68 41	30	108	69 11	28	106	337
66 17	35	114	66 52	33	112	67 25	31	109	67 56	29	107	68 25	27	105	336
65 33	+34	112	66 07	+32	110	66 39	+31	108	67 10	+29	106	67 39	+26	104	335
64 49	33	111	65 22	32	109	65 54	30	107	66 24	28	105	66 52	26	103	334
64 04	32	110	64 36	32	108	65 08	29	106	65 37	28	104	66 05	26	102	333
63 18	33	109	63 51	31	107	64 22	29	105	64 51	27	103	65 18	26	101	332
62 33	32	108	63 05	30	106	63 35	29	104	64 04	27	102	64 31	25	100	331
61 47	+32	107	62 19	+29	105	62 48	+29	103	63 17	+27	101	63 44	+25	99	330
61 01	31	106	61 32	30	104	62 02	28	102	62 30	26	100	62 56	25	98	329
60 15	30	105	60 45	30	103	61 15	27	101	61 42	27	99	62 09	24	97	328
59 28	31	104	59 59	29	102	60 28	27	100	60 55	26	98	61 21	25	97	327
58 42	30	103	59 12	28	101	59 40	28	99	60 08	26	98	60 34	24	96	326
57 55	+30	102	58 25	+28	100	58 53	+27	99	59 20	+26	97	59 46	+24	95	325
57 08	29	101	57 37	29	99	58 06	27	98	58 33	25	96	58 58	24	94	324
56 21	29	100	56 50	28	99	57 18	27	97	57 45	25	95	58 10	24	94	323
55 34	29	99	56 03	28	98	56 31	26	96	56 57	25	95	57 22	24	93	322
54 46	29	99	55 15	28	97	55 43	26	96	56 09	26	94	56 35	23	92	321
53 59	+29	98	54 28	+27	96	54 55	+26	95	55 21	+26	93	55 47	+23	92	320
53 11	29	97	53 40	27	96	54 07	27	94	54 34	25	93	54 59	24	91	319
52 24	28	96	52 52	28	95	53 20	26	93	53 46	25	92	54 11	24	90	318
51 36	28	96	52 04	28	94	52 32	26	93	52 58	25	91	53 23	24	90	317
50 48	29	95	51 17	27	94	51 44	26	92	52 10	25	91	52 35	24	89	316
50 01	+28	94	50 29	+27	93	50 56	+26	92	51 22	+25	90	51 47	+24	89	315
49 13	28	94	49 41	27	92	50 08	26	91	50 34	25	90	50 59	24	88	314
48 25	28	93	48 53	27	92	49 20	26	90	49 46	25	89	50 11	24	88	313
47 37	28	92	48 05	27	91	48 32	26	90	48 58	25	88	49 23	25	87	312
46 49	28	92	47 17	27	90	47 44	26	89	48 10	26	88	48 36	24	86	311
46 01	+28	91	46 29	+27	90	46 56	+27	89	47 23	+25	87	47 48	+24	86	310
45 13	28	91	45 41	27	89	46 08	27	88	46 35	25	87	47 00	24	85	309
44 25	28	90	44 53	28	89	45 21	26	87	45 47	25	86	46 12	25	85	308
43 38	28	89	44 06	27	88	44 33	26	87	44 59	26	86	45 25	24	84	307
42 50	28	89	43 18	27	88	43 45	26	86	44 11	26	85	44 37	25	84	306
42 02	+28	88	42 30	+27	87	42 57	+27	86	43 24	+25	85	43 49	+25	83	305
41 14	28	88	41 42	27	86	42 09	27	85	42 36	26	84	43 02	25	83	304
40 26	28	87	40 54	28	86	41 22	26	85	41 48	26	83	42 14	25	82	303
39 38	28	86	40 06	28	85	40 34	27	84	41 01	26	83	41 27	25	82	302
38 50	29	86	39 19	27	85	39 46	27	84	40 13	26	82	40 39	26	81	301
38 03	+28	85	38 31	+28	84	38 59	+27	83	39 26	+26	82	39 52	+26	81	300
37 15	28	85	37 43	28	84	38 11	27	83	38 38	27	81	39 05	26	80	299
36 27	29	84	36 56	28	83	37 24	27	82	37 51	27	81	38 18	26	80	298
35 40	28	84	36 08	28	83	36 36	28	82	37 04	27	80	37 31	26	79	297
34 52	29	83	35 21	28	82	35 49	28	81	36 17	27	80	36 44	26	79	296
34 04	+29	83	34 33	+29	82	35 02	+27	80	35 29	+28	79	35 57	+26	78	295
33 17	29	82	33 46	28	81	34 14	28	80	34 42	28	79	35 10	26	78	294
32 29	30	82	32 59	28	81	33 27	28	79	33 55	28	78	34 23	27	77	293
31 42	29	81	32 11	29	80	32 40	28	79	33 08	28	78	33 36	27	77	292
30 55	29	81	31 24	29	80	31 53	29	78	32 22	28	77	32 50	27	76	291

S. Lat. {LHA greater than 180°........Zn=180—Z
{LHA less than 180°............Zn=180+Z

DECLINATION (15°-29°) **SAME** NAME AS LATITUDE

LAT 37°

Table 1

LAT 37°

N. Lat. {LHA greater than 180°....... Zn=Z
{LHA less than 180°.......... Zn=360−Z

DECLINATION (15°-29°) SAME NAME AS LATITUDE

LAT 37°

LHA	15° Hc	d	Z	16° Hc	d	Z	17° Hc	d	Z	18° Hc	d	Z	19° Hc	d	Z	LHA
	° ′	′	°	° ′	′	°	° ′	′	°	° ′	′	°	° ′	′	°	
70	24 49	+33	90	25 22	+33	89	25 55	+33	88	26 28	+33	87	27 01	+32	86	290
71	24 01	33	89	24 34	34	88	25 08	33	87	25 41	32	86	26 13	32	85	289
72	23 13	33	88	23 46	34	87	24 20	33	87	24 53	32	86	25 25	33	85	288
73	22 25	34	88	22 59	33	87	23 32	33	86	24 05	33	85	24 38	32	84	287
74	21 37	34	87	22 11	33	86	22 44	33	85	23 17	33	84	23 50	32	84	286
75	20 49	+34	87	21 23	+33	86	21 56	+34	85	22 30	+32	84	23 02	+33	83	285
76	20 01	34	86	20 35	34	85	21 09	33	84	21 42	33	83	22 15	33	82	284
77	19 14	33	85	19 47	34	85	20 21	33	84	20 54	33	83	21 27	33	82	283
78	18 26	34	85	19 00	33	84	19 33	34	83	20 07	33	82	20 40	33	81	282
79	17 38	34	84	18 12	34	83	18 46	33	83	19 19	34	82	19 53	33	81	281
80	16 51	+34	84	17 25	+33	83	17 58	+34	82	18 32	+33	81	19 05	+34	80	280
81	16 03	34	83	16 37	34	82	17 11	34	81	17 45	33	81	18 18	34	80	279
82	15 15	35	83	15 50	34	82	16 24	34	81	16 58	33	80	17 31	34	79	278
83	14 28	34	82	15 02	34	81	15 36	34	80	16 10	34	79	16 44	34	79	277
84	13 40	35	81	14 15	34	81	14 49	34	80	15 23	34	79	15 57	34	78	276
85	12 53	+35	81	13 28	+34	80	14 02	+34	79	14 36	+34	78	15 10	+34	77	275
86	12 06	35	80	12 41	34	79	13 15	35	79	13 50	34	78	14 24	34	77	274
87	11 19	35	80	11 54	34	79	12 28	35	78	13 03	34	77	13 37	34	76	273
88	10 32	35	79	11 07	34	78	11 41	35	77	12 16	35	77	12 51	34	76	272
89	09 45	35	79	10 20	35	78	10 55	35	77	11 30	34	76	12 04	35	75	271
90	08 58	+35	78	09 33	+35	77	10 08	+35	76	10 43	+35	76	11 18	+35	75	270
91	08 11	35	77	08 46	36	77	09 22	35	76	09 57	35	75	10 32	35	74	269
92	07 24	36	77	08 00	35	76	08 35	36	75	09 11	35	74	09 46	35	74	268
93	06 38	35	76	07 13	36	75	07 49	36	75	08 25	35	74	09 00	35	73	267
94	05 51	36	76	06 27	36	75	07 03	36	74	07 39	35	73	08 14	36	72	266
95	05 05	+36	75	05 41	+36	74	06 17	+36	73	06 53	+36	73	07 29	+35	72	265
96	04 19	36	74	04 55	36	74	05 31	36	73	06 07	36	72	06 43	36	71	264
97	03 32	37	74	04 09	36	73	04 45	37	72	05 22	36	72	05 58	36	71	263
98	02 46	37	73	03 23	37	72	04 00	36	72	04 36	37	71	05 13	36	70	262
99	02 01	36	73	02 37	37	72	03 14	37	71	03 51	37	70	04 28	36	70	261
100	01 15	+37	72	01 52	+37	71	02 29	+37	71	03 06	+37	70	03 43	+37	69	260
101	00 29	38	71	01 07	37	71	01 44	37	70	02 21	37	69	02 58	37	68	259
102	−0 16	38	71	00 22	37	70	00 59	37	69	01 36	38	69	02 14	37	68	258
103	−1 01	38	70	−0 23	37	69	00 14	38	69	00 52	38	68	01 30	37	67	257
104	−1 46	38	70	−1 08	38	69	−0 30	38	68	00 08	38	67	00 46	37	67	256
105	−2 31	+38	69	−1 53	+38	68	−1 15	+39	68	−0 36	+38	67	00 02	+38	66	255
106	−3 16	39	68	−2 37	38	68	−1 59	39	67	−1 20	38	66	−0 42	38	65	254
107	−4 00	39	68	−3 21	38	67	−2 43	39	66	−2 04	39	66	−1 25	38	65	253
108	−4 44	39	67	−4 05	38	66	−3 27	39	66	−2 48	39	65	−2 09	39	64	252
109	−5 28	39	67	−4 49	39	66	−4 10	39	65	−3 31	39	64	−2 52	40	64	251
110	−6 12	+39	66	−5 33	+40	65	−4 53	+39	64	−4 14	+40	64	−3 34	+39	63	250
111				−6 16	39	64	−5 37	40	64	−4 57	40	63	−4 17	40	62	249
112							−6 19	40	63	−5 39	40	62	−4 59	40	62	248
113										−6 22	41	62	−5 41	40	61	247
114													−6 23	41	60	246

LHA	20° Hc	d	Z	21° Hc	d	Z	22° Hc	d	Z	23° Hc	d	Z	24° Hc	d	Z	LHA
	° ′	′	°	° ′	′	°	° ′	′	°	° ′	′	°	° ′	′	°	
70	27 33	+32	85	28 05	+31	84	28 36	+31	83	29 07	+31	82	29 38	+30	81	290
71	26 45	32	84	27 17	32	83	27 49	31	82	28 20	30	81	28 50	30	80	289
72	25 58	32	84	26 30	31	83	27 01	31	82	27 32	31	81	28 03	30	80	288
73	25 10	32	83	25 42	32	82	26 14	31	81	26 45	31	80	27 16	30	79	287
74	24 22	33	83	24 55	31	82	25 26	32	81	25 58	31	80	26 29	30	79	286
75	23 35	+32	82	24 07	+32	81	24 39	+32	80	25 11	+31	79	25 42	+31	78	285
76	22 48	32	82	23 20	32	81	23 52	32	80	24 24	31	79	24 55	31	78	284
77	22 00	33	81	22 33	32	80	23 05	32	79	23 37	31	78	24 08	31	77	283
78	21 13	33	80	21 46	32	80	22 18	32	79	22 50	32	78	23 22	31	77	282
79	20 26	32	80	20 58	33	79	21 31	32	78	22 03	32	77	2? 35	31	76	281
80	19 39	+32	79	20 11	+33	78	20 44	+32	78	21 16	+32	77	21 48	+32	76	280
81	18 52	33	79	19 25	32	78	19 57	33	77	20 30	32	76	21 02	32	75	279
82	18 05	33	78	18 38	33	77	19 11	32	76	19 43	33	76	20 16	32	75	278
83	17 18	33	78	17 51	33	77	18 24	33	76	18 57	33	75	19 30	32	74	277
84	16 31	33	77	17 04	34	76	17 38	33	75	18 11	33	75	18 44	32	74	276
85	15 44	+34	77	16 18	+33	76	16 51	+34	75	17 25	+33	74	17 58	+32	73	275
86	14 58	34	76	15 32	33	75	16 05	34	74	16 39	33	73	17 12	33	73	274
87	14 11	34	76	14 45	34	75	15 19	34	74	15 53	33	73	16 26	34	72	273
88	13 25	34	75	13 59	34	74	14 33	34	73	15 07	34	72	15 41	33	72	272
89	12 39	34	74	13 13	34	74	13 47	35	73	14 22	33	72	14 55	34	71	271
90	11 53	+34	74	12 27	+35	73	13 02	+34	72	13 36	+34	71	14 10	+34	70	270
91	11 07	35	73	11 42	34	72	12 16	35	72	12 51	34	71	13 25	34	70	269
92	10 21	35	73	10 56	35	72	11 31	35	71	12 06	34	70	12 40	35	69	268
93	09 35	36	72	10 11	35	71	10 46	35	71	11 21	34	70	11 55	35	69	267
94	08 50	35	72	09 25	36	71	10 01	35	70	10 36	35	69	11 11	35	68	266
95	08 04	+36	71	08 40	+36	70	09 16	+35	69	09 51	+35	69	10 26	+36	68	265
96	07 19	36	70	07 55	36	70	08 31	36	69	09 07	35	68	09 42	36	67	264
97	06 34	36	70	07 10	36	69	07 46	36	68	08 22	36	67	08 58	36	67	263
98	05 49	37	69	06 26	36	69	07 02	36	68	07 38	36	67	08 14	36	66	262
99	05 04	37	69	05 41	37	68	06 18	36	67	06 54	36	66	07 30	37	66	261
100	04 20	+37	68	04 57	+37	67	05 34	+36	67	06 10	+37	66	06 47	+37	65	260
101	03 35	38	68	04 13	37	67	04 50	37	66	05 27	37	65	06 04	37	64	259
102	02 51	38	67	03 29	37	66	04 06	37	65	04 43	38	65	05 21	37	64	258
103	02 07	38	66	02 45	38	66	03 23	37	65	04 00	38	64	04 38	37	63	257
104	01 23	38	66	02 01	38	65	02 39	38	64	03 17	38	63	03 55	38	63	256
105	00 40	+38	65	01 18	+38	64	01 56	+38	64	02 34	+38	63	03 12	+39	62	255
106	−0 04	39	65	00 35	38	64	01 13	39	63	01 52	38	62	02 30	39	62	254
107	−0 47	39	64	−0 08	39	63	00 31	38	62	01 09	39	62	01 48	39	61	253
108	−1 30	39	63	−0 51	39	63	−0 12	39	62	00 27	39	61	01 06	39	60	252
109	−2 12	39	63	−1 33	39	62	−0 54	40	61	−0 14	39	61	00 25	39	60	251
110	−2 55	+40	62	−2 15	+39	61	−1 36	+40	61	−0 56	+40	60	−0 16	+39	59	250
111	−3 37	40	62	−2 57	40	61	−2 17	40	60	−1 37	40	59	−0 57	40	59	249
112	−4 19	40	61	−3 39	40	60	−2 59	41	59	−2 18	40	59	−1 38	40	58	248
113	−5 01	41	60	−4 20	40	60	−3 40	41	59	−2 59	40	58	−2 19	41	57	247
114	−5 42	41	60	−5 01	40	59	−4 21	41	58	−3 40	41	57	−2 59	41	57	246
115				−5 42	+41	58	−5 01	+41	58	−4 20	+41	57	−3 39	+42	56	245
116							−5 41	41	57	−5 00	42	56	−4 18	41	55	244
117										−5 40	42	56	−4 58	42	55	243
118													−5 37	43	54	242
119													−6 15	42	54	241

LHA	25° Hc	d	Z	26° Hc	d	Z	27° Hc	d	Z	28° Hc	d	Z	29° Hc	d	Z	LHA
	° ′	′	°	° ′	′	°	° ′	′	°	° ′	′	°	° ′	′	°	
70	30 08	+29	80	30 37	+29	79	31 06	+29	78	31 35	+28	77	32 03	+28	76	290
71	29 20	30	79	29 50	29	78	30 19	29	77	30 48	29	76	31 17	28	75	289
72	28 33	30	79	29 03	30	78	29 33	29	77	30 02	28	76	30 30	28	75	288
73	27 46	30	78	28 16	30	77	28 46	29	76	29 15	29	75	29 44	28	74	287
74	26 59	31	78	27 30	30	77	28 00	29	76	28 29	29	75	28 58	29	74	286
75	26 13	+30	77	26 43	+30	76	27 13	+30	75	27 43	+29	74	28 12	+29	74	285
76	25 26	31	77	25 57	30	76	26 27	30	75	26 57	29	74	27 26	29	73	284
77	24 39	31	76	25 10	31	75	25 41	30	74	26 11	29	74	26 40	30	73	283
78	23 53	31	76	24 24	31	75	24 55	30	74	25 25	30	73	25 55	29	72	282
79	23 06	32	75	23 38	31	74	24 09	30	73	24 39	30	73	25 09	30	72	281
80	22 20	+32	75	22 52	+31	74	23 23	+30	73	23 53	+31	72	24 24	+30	71	280
81	21 34	32	74	22 06	31	73	22 37	31	72	23 08	31	72	23 39	30	71	279
82	20 48	32	74	21 20	31	73	21 51	32	72	22 23	31	71	22 54	30	70	278
83	20 02	32	73	20 34	32	72	21 06	31	71	21 37	32	71	22 09	30	70	277
84	19 16	32	73	19 48	32	72	20 20	32	71	20 52	32	70	21 24	31	69	276
85	18 30	+33	72	19 03	+32	71	19 35	+32	70	20 07	+32	70	20 39	+31	69	275
86	17 45	33	72	18 18	32	71	18 50	32	70	19 22	32	69	19 54	32	68	274
87	17 00	33	71	17 33	32	70	18 05	33	69	18 38	32	69	19 10	32	68	273
88	16 14	33	71	16 47	33	70	17 20	33	69	17 53	33	68	18 26	32	67	272
89	15 29	34	70	16 03	33	69	16 36	33	68	17 09	33	68	17 42	32	67	271
90	14 44	+34	70	15 18	+33	69	15 51	+34	68	16 25	+33	67	16 58	+33	66	270
91	13 59	34	69	14 33	34	68	15 07	34	67	15 41	33	67	16 14	33	66	269
92	13 15	34	69	13 49	34	68	14 23	34	67	14 57	34	66	15 31	33	65	268
93	12 30	35	68	13 05	34	67	13 39	34	66	14 13	34	66	14 47	34	65	267
94	11 46	35	67	12 21	34	67	12 55	35	66	13 30	34	65	14 04	34	64	266
95	11 02	+35	67	11 37	+35	66	12 12	+34	65	12 46	+35	64	13 21	+34	64	265
96	10 18	35	66	10 53	35	66	11 28	35	65	12 03	35	64	12 38	35	63	264
97	09 34	35	66	10 09	36	65	10 45	35	64	11 20	36	63	11 56	35	63	263
98	08 50	36	65	09 26	36	65	10 02	36	64	10 38	35	63	11 13	36	62	262
99	08 07	36	65	08 43	36	64	09 19	36	63	09 55	36	62	10 31	36	62	261
100	07 24	+36	64	08 00	+36	63	08 36	+37	63	09 13	+36	62	09 49	+36	61	260
101	06 41	36	64	07 17	37	63	07 54	37	62	08 31	36	61	09 07	37	60	259
102	05 58	37	63	06 35	37	62	07 12	37	62	07 49	37	61	08 26	36	60	258
103	05 15	38	63	05 53	37	62	06 30	37	61	07 07	37	60	07 44	37	59	257
104	04 33	37	62	05 10	38	61	05 48	38	60	06 26	37	60	07 03	38	59	256
105	03 51	+38	61	04 29	+38	61	05 07	+37	60	05 44	+38	59	06 22	+38	58	255
106	03 09	38	61	03 47	38	60	04 25	39	59	05 04	38	58	05 42	38	58	254
107	02 27	39	60	03 06	38	59	03 44	39	59	04 23	38	58	05 01	39	57	253
108	01 45	39	60	02 24	39	59	03 03	39	58	03 42	39	57	04 21	39	57	252
109	01 04	40	59	01 44	39	58	02 23	39	57	03 02	39	57	03 41	39	56	251
110	00 23	+40	58	01 03	+40	58	01 43	+39	57	02 22	+40	56	03 02	+39	55	250
111	−0 17	40	58	00 23	40	57	01 03	40	56	01 43	40	56	02 23	39	55	249
112	−0 58	41	57	−0 17	40	56	00 23	40	56	01 03	40	55	01 43	41	54	248
113	−1 38	41	57	−0 57	40	56	−0 17	41	55	00 24	41	54	01 05	40	54	247
114	−2 18	41	56	−1 37	41	55	−0 56	41	54	−0 15	41	54	00 26	41	53	246
115	−2 57	+41	55	−2 16	+41	55	−1 35	+42	54	−0 53	+41	53	−0 12	+42	52	245
116	−3 37	42	55	−2 55	42	54	−2 13	42	53	−1 31	41	53	−0 50	42	52	244
117	−4 16	43	54	−3 33	42	53	−2 51	42	53	−2 09	42	52	−1 27	42	51	243
118	−4 54	42	53	−4 12	43	53	−3 29	42	52	−2 47	43	51	−2 04	42	51	242
119	−5 33	43	53	−4 50	43	52	−4 07	43	51	−3 24	43	51	−2 41	43	50	241
120	−6 11	+44	52	−5 27	+43	51	−4 44	+43	51	−4 01	+43	50	−3 18	+44	49	240
121				−6 05	44	51	−5 21	43	50	−4 38	44	49	−3 54	44	49	239
122							−5 58	44	49	−5 14	44	49	−4 30	44	48	238
123										−5 50	45	48	−5 05	44	47	237
124													−5 40	44	47	236

LHA	15° Hc	d	Z	16° Hc	d	Z	17° Hc	d	Z	18° Hc	d	Z	19° Hc	d	Z	LHA
86	−5 51	36	104	−6 27	36	105										274
85	−5 05	−36	105	−5 41	−36	106	−6 17	−36	107							275
84	−4 19	36	106	−4 55	36	106	−5 31	36	107	−6 07	36	108				276
83	−3 32	37	106	−4 09	36	107	−4 45	37	108	−5 22	36	109	−5 58	36	109	277
82	−2 46	37	107	−3 23	37	108	−4 00	36	108	−4 36	37	109	−5 13	36	110	
81	−2 01	36	107	−2 37	37	108	−3 14	37	109	−3 51	37	110	−4 28	36	110	
80	−1 15	−37	108	−1 52	−37	109	−2 29	−37	109	−3 06	−37	110	−3 43	−37	111	
79	−0 29	38	109	−1 07	37	109	−1 44	37	110	−2 21	37	111	−2 58	37	112	
78	00 16	38	109	−0 22	37	110	−0 59	37	111	−1 36	38	111	−2 14	37	112	
77	01 01	38	110	00 23	37	111	−0 14	38	111	−0 52	38	112	−1 30	37	113	
76	01 46	38	110	01 08	38	111	00 30	38	112	−0 08	38	113	−0 46	37	113	
75	02 31	−38	111	01 53	−38	112	01 15	−39	112	00 36	−38	113	−0 02	−38	114	
74	03 16	39	112	02 37	38	112	01 59	39	113	01 20	38	114	00 42	38	115	
73	04 00	39	112	03 21	38	113	02 43	39	114	02 04	39	114	01 25	38	115	
72	04 44	39	113	04 05	38	114	03 27	39	114	02 48	39	115	02 09	39	116	
71	05 28	39	113	04 49	39	114	04 10	39	115	03 31	39	116	02 52	40	116	
70	06 12	−39	114	05 33	−40	115	04 53	−39	116	04 14	−40	116	03 34	−39	117	
	15°			16°			17°			18°			19°			

LHA	20° Hc	d	Z	21° Hc	d	Z	22° Hc	d	Z	23° Hc	d	Z	24° Hc	d	Z	LHA
82	−5 49	37	111	−6 26	36	112										278
81	−5 04	37	111	−5 41	37	112	−6 18	36	113							279
80	−4 20	−37	112	−4 57	−37	113	−5 34	−36	113	−6 10	−37	114				280
79	−3 35	38	112	−4 13	37	113	−4 50	37	114	−5 27	37	115	−6 04	37	116	281
78	−2 51	38	113	−3 29	37	114	−4 06	37	115	−4 43	38	115	−5 21	37	116	
77	−2 07	38	114	−2 45	38	114	−3 23	37	115	−4 00	38	116	−4 38	37	117	
76	−1 23	38	114	−2 01	38	115	−2 39	38	116	−3 17	38	117	−3 55	38	117	
75	−0 40	−38	115	−1 18	−38	116	−1 56	−38	116	−2 34	−38	117	−3 12	−39	118	
74	00 04	39	115	−0 35	38	116	−1 13	39	117	−1 52	38	118	−2 30	39	118	
73	00 47	39	116	00 08	39	117	−0 31	38	118	−1 09	39	118	−1 48	39	119	
72	01 30	39	117	00 51	39	117	00 12	39	118	−0 27	39	119	−1.06	39	120	
71	02 12	39	117	01 33	39	118	00 54	40	119	00 14	39	119	−0 25	39	120	
70	02 55	−40	118	02 15	−39	119	01 36	−40	119	00 56	−40	120	00 16	−39	121	
	20°			21°			22°			23°			24°			

LHA	25° Hc	d	Z	26° Hc	d	Z	27° Hc	d	Z	28° Hc	d	Z	29° Hc	d	Z	LHA
78	−5 58	37	117													282
77	−5 15	38	118	−5 53	37	118										283
76	−4 33	37	118	−5 10	38	119	−5 48	38	120							284
75	−3 51	−38	119	−4 29	−38	119	−5 07	−37	120	−5 44	−38	121				285
74	−3 09	38	119	−3 47	38	120	−4 25	39	121	−5 04	38	122	−5 42	38	122	286
73	−2 27	39	120	−3 06	38	121	−3 44	39	121	−4 23	38	122	−5 01	39	123	287
72	−1 45	39	120	−2 24	39	121	−3 03	39	122	−3 42	39	123	−4 21	39	123	288
71	−1 04	40	121	−1 44	39	122	−2 23	39	123	−3 02	39	123	−3 41	39	124	289
70	−0 23	−40	122	−1 03	−40	122	−1 43	−39	123	−2 22	−40	124	−3 02	−39	125	290
	25°			26°			27°			28°			29°			

Table 1, continued

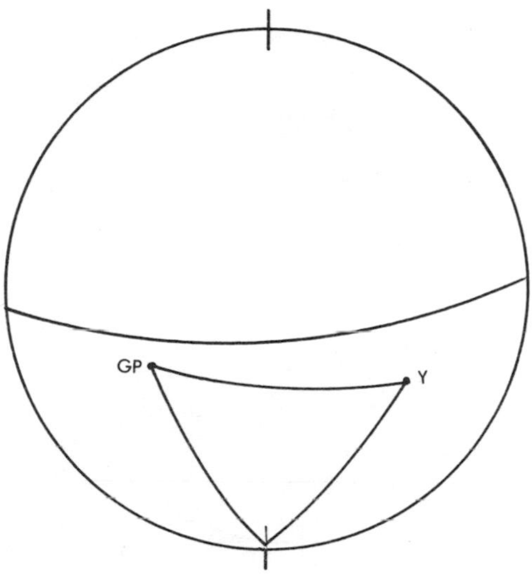

Fig. 4.1

would turn to another table headed "Declination (15°-29°) Contrary Name to Latitude." The reason for this is that there is another possible navigational triangle (Figure 4.2). As you can see, in this case, side *B* of the triangle is 90° *plus* the sun's latitude south, and, obviously, the zenith distance is different. Therefore, it is necessary to compute another set of solutions.

Suppose you were in south latitude and the sun was in north. Again, that's as though you turned the figure upside down and, mathematically, it's the same thing. The zenith distance remains the same.

Now turn back to Table 1 and look at the top of the column headed "24° " As you can see, the top row is for LHA of 0°, which means that there is no triangle because the sun is on the same longitude as you are and, in common terms, it is noon. The column is divided into three parts: "Hc," "d," and "Z." Hc (called *computed altitude*) is the angle that your sextant would have measured had you been at latitude 37° north at noon on the day when the declination of the sun was 24° north.

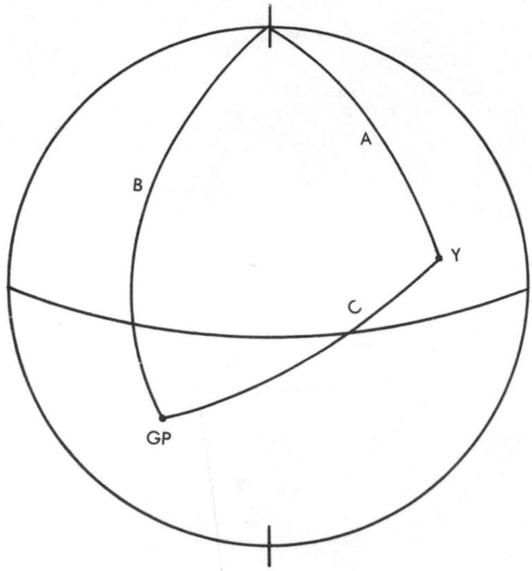

Fig. 4.2

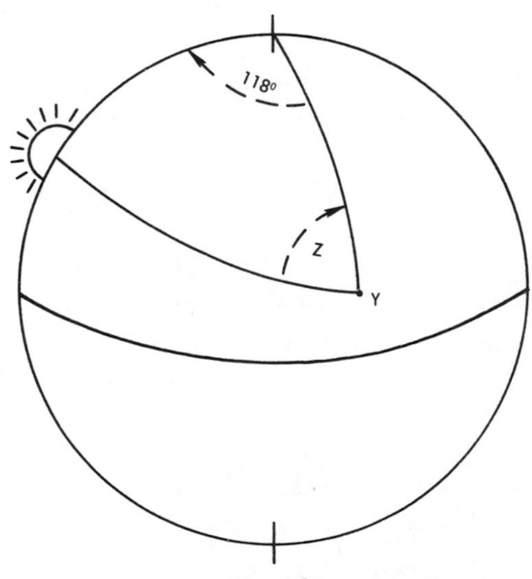

Fig. 4.3

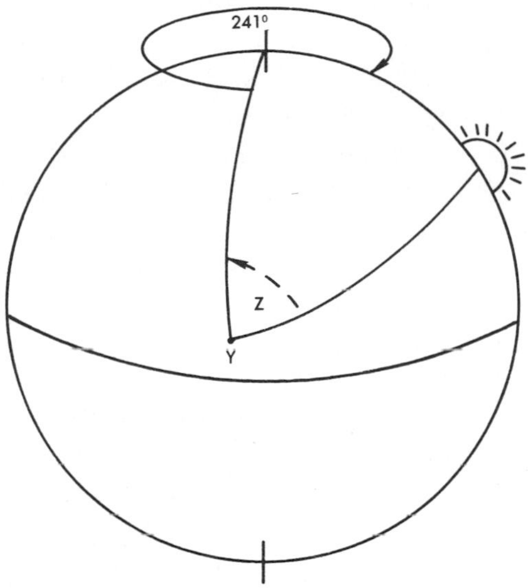

Fig. 4.4

Up to now you have been subtracting your sextant reading from 90° to find the zenith distance, because this is conceptually easier. You can save a step, however, if you compare your *actual* sextant reading with a *computed* sextant reading, so that is the way the tables are set up.

If you look down the column, you will see that as the LHA gets larger, the altitude (sextant reading) and angle *Z* get smaller. This is reasonable, because you know very well that as the sun appears to move west it gets lower in the sky and its direction becomes more and more westerly. As a matter of fact, when LHA is 54°, the sun is west of you—that is, $Z = 90°$.

As more time passes, the sun sinks lower and moves north of west. If you continue reading the 24° column, you can see that when LHA is 109°, the sun is just about at the horizon. The Hc here is read as "zero degrees 25 minutes." These minutes are minutes of arc, and there are 60 of them in 1°.

You will notice that as LHA increases further, Hc becomes negative. This is because the atmosphere bends rays of light, and as a result you can still see the sun when it is actually below the horizon (Figure 4.3).

At LHA of 119° it is no longer possible to see the sun, so the table stops there. The sun, however, is still moving, (or, rather, it appears to), and even though it is of no practical use, LHA is still increasing. Finally, when LHA is 241°, the sun is close to the horizon again (Figure 4.4). This LHA is on the right-hand side of the page, and if you follow the numbers up the page, you will see that as LHA increases, the sun gets higher in the sky until noon, when LHA is 360°—it has come full circle and is back where it started.

What you have done is to follow through the complete course of the sun in one day. Now you can see that for every moment of the day this volume of tables has a tabulated solution of the navigational triangle for you to look up. So instead of having to compute, you can devote your time to using your sextant, and that, as you will see, is really the most important part of celestial navigation.

5
Celestial Navigation in Your Backyard

The best way to learn to use a sextant is in the comfort of your own armchair or your backyard at home. This cuts down on the number of things you have to do to adapt to the process of celestial navigation on a boat. By practicing at home, all the procedures and techniques of taking a sight will become second nature to you, and, when you move to the boat, you will only have to accommodate your technique to the motion of the boat. In preparing for celestial navigation in an airplane, Sir Francis Chichester used to take practice sextant shots while roaring around in a convertible. You are not trying to become an air navigator, however, so such exertions will not be demanded of you.

What I suggest you do at this point is read the sections of Chapter 16 that deal with sextants and get one that seems suitable for your purpose. Having done this, sit down some evening in a comfortable chair and read the pamphlet that comes with the sextant so that you will be familiar with the names of its various parts. To help you orient yourself in the course of this chapter if you have not yet purchased your sextant, Figure 5.1 is a labeled illustration.

The marine sextant is my kind of machine because it has only one moving part—the index arm. If you are using the EBBCO sextant—which I strongly suggest—you will find that the index arm has a squeeze fitting at the bottom for making the arm move several inches at a time and also a "fine-tune knob" that can move the arm very slowly and almost imperceptibly.

Along the curved bottom part of the sextant is the scale; each line is 1°. One entire turn of the "fine-tune" knob (micrometer drum) is also 1°. This drum is divided into 60 parts, so each little line is one minute (1′).

As a beginning exercise, position the arm somewhere about the middle of the bottom arc, using the squeeze grip, and try to determine the sextant reading (See Figure 5.2). Chances are you will not land right on an exact degree but will be in between. The micrometer drum will tell you where you are between the two degree marks. Thus, in the ex-

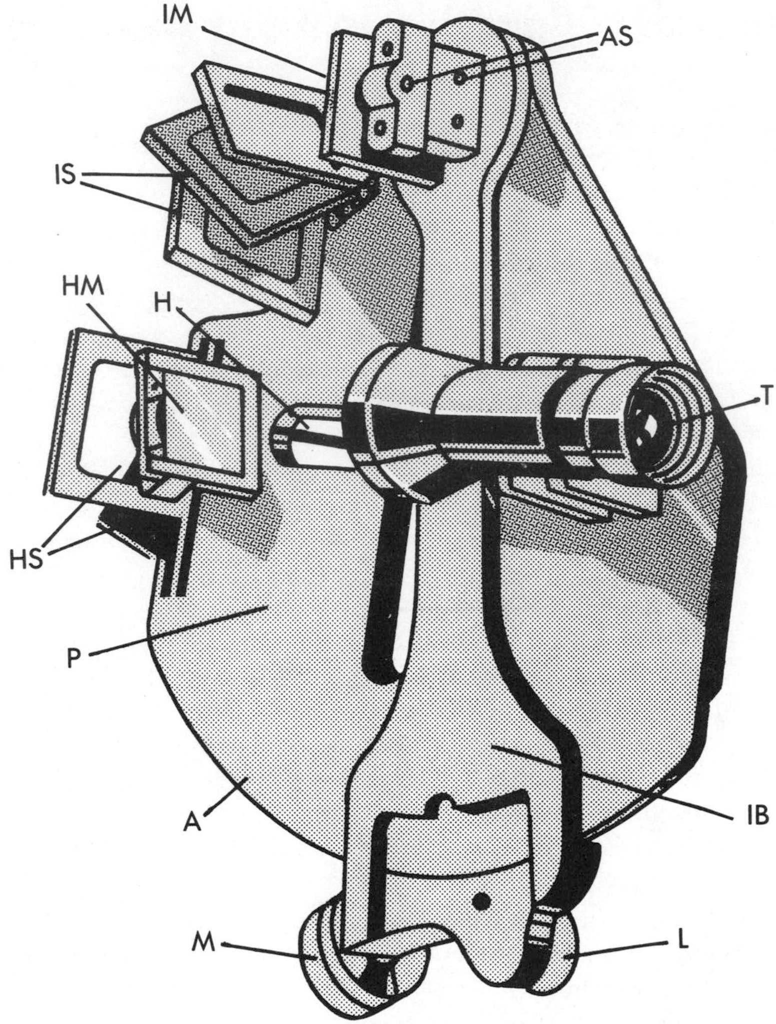

A: Arc
L: Squeeze Fitting
H: Handles
AS: Adjusting Screws
IB: Index Arm
T: Telescope
P: Plate of Sextant
IM: Index Mirror
IS: Index Shades
HM: Horizon Mirror
HS: Horizon Shades
M: Micrometer Drum

Fig. 5.1

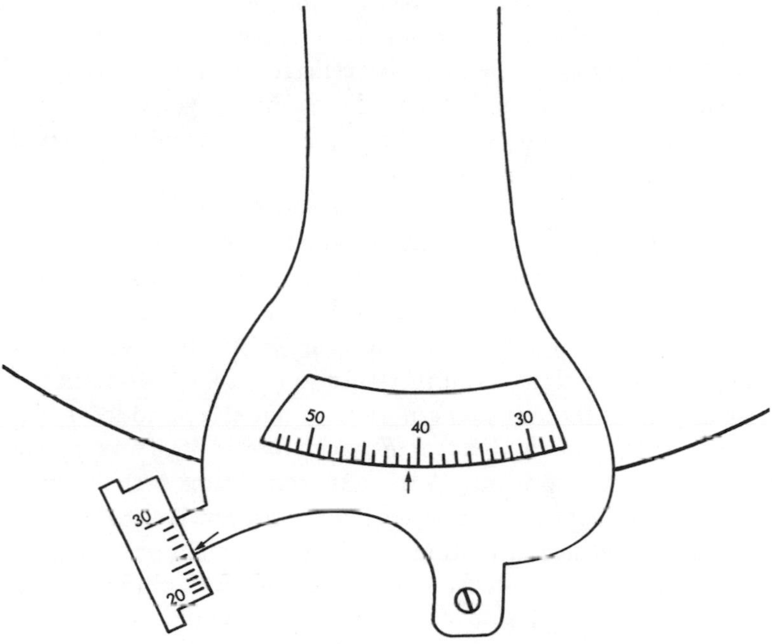

Fig. 5.2

ample, the sextant reads an angle of 41°26′. Turn the drum through a full revolution, and you will see that when the drum reads 0, the index arm is on an exact whole degree.

Now, sit down in a chair where you can face a window across the room. Set the index arm and the micrometer drum both to zero. Fold the shades away from the mirrors. Hold the sextant in your right hand by the handles and look through the telescope at the windowsill. Focus the scope if necessary. You will see the right-hand side of the sill reflected in the mirror in front of the scope and the left-hand side of the sill directly through the unsilvered part of the mirror (horizon mirror) in front of the telescope. With your left hand, turn the micrometer drum this way and that and observe what happens.

Now, turn your wristwatch around so that it is on the inside of your left wrist. Put the telescope back to your eye and turn the micrometer drum so that the sextant reading increases. Continue to do this until you see the top of the window reflected in the silvered side of the horizon mirror and in line with the sill. When you have lined them up exactly, look at your watch. Find the second hand and note the time. Read the sextant.

You have just measured the angle between the sill, your eye, and the top of the window, and you have taken the time at which you measured this angle. In carrying out this exercise you have just gone through all the motions of taking an actual celestial observation. The reason for having the watch on the inside of your wrist is that is simplifies taking the time of the sight—you just have to twist your·wrist, in fact. I personally don't think it takes more than one second to locate the second hand of the watch, but some navigators subtract as much as five seconds for this process. Work out what seems best for you.

The next logical exercise with your sextant is to take it out in the backyard and actually observe the sun. Some afternoon or morning when the sun is visible from your yard, go out with your sextant, put your watch on the inside of your left wrist, and proceed as follows. Pick out some horizontal line directly below the sun to use as a practice horizon. This can be the top edge of a fence, a neighbor's roof, or a convenient windowsill. When I first learned celestial navigation, I lived in an apartment complex, and I used to go out in the courtyard between the buildings and use a windowsill at approximately eye level. If you live in an apartment and have access to the roof, use the roof line of another apartment. For this exercise, it is not necessary that your reference line be at eye level; it can be considerably above or below.

The next step is to pick up the sun's image in the sextant. There are two ways of doing this. If you are good at estimating angles, estimate the angle between the sun and your reference line and set the sextant to that angle. Then put the darkest shade down between the index mirror and the horizon mirror, put the sextant to your eye, and move it slowly from side to side and up and down until you pick up the sun. Through the telescope it will appear as a disk about the size of a dime.

If you are not practiced at estimating angles, try this procedure: Make a fist with one hand and hold it out in front of you. Align the bottom edge of your fist with your reference line and estimate how many "fists" there are between there and the sun. A fist at arm's length covers about 8°, so if there is space for five fists between your reference line and the sun, the angle is about 40°. Whatever it is, set this angle on your sextant and try as before to pick up the image of the sun.

If you are unable to get the sun with either method, do this: set your sextant to zero. Put down the darkest horizon-mirror shade in addition to the darkest index-mirror shade and aim the scope right at the sun. With your left hand, release the index arm and move the index arm forward as you move the sextant down. Once you get the image close to your reference line, flip away the horizon shade.

Now proceed with the micrometer drum to set the image of the sun on the reference line so that the lower edge of the disk just touches the line (Figure 5.3). Look at your watch and note the time and sextant reading. I use a pocket-size pad for this and stick a pencil behind my left ear (this procedure works at sea, too).

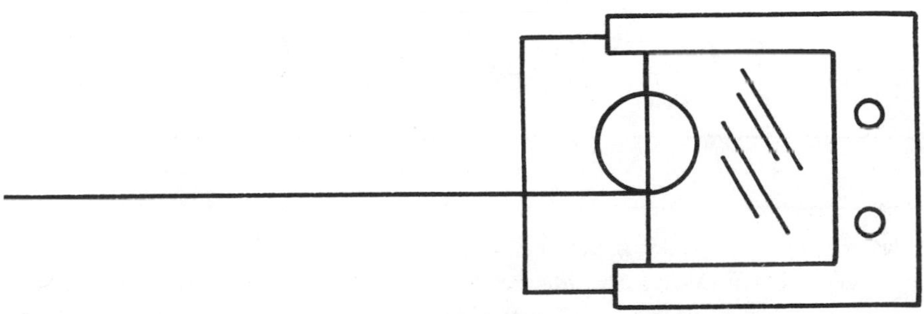

Fig. 5.3

Now look again through your telescope. You will notice that in the time it took you to write down the time and sextant reading, the sun has moved so that it is no longer resting on your roof line, windowsill, or whatever. With the micrometer drum, put the disk of the sun back in position so that the lower edge (called the *lower limb* by astronomers) is exactly touching. Again note the time and reading.

If you compare the two readings, you will see that the angle of the sun has changed by only a few minutes of arc, and you should have a pretty good idea of how accurately a sextant is able to measure angles. As a further exercise, you might want to try lining everything up correctly and then finding out how small a change in the sun's altitude you can detect. The results will probably amaze you. Since 1′ of arc equals 1 nautical mile on the surface of the earth, I think you can see that the sextant is capable of fixing a position very accurately.

There is one more thing that you should be doing while practicing this way with your sextant. When you have the disk balanced on the reference line, rotate the sextant slightly to the left and right around the axis of the telescope. You will see the image of the sun rise off the reference line as though it were a pendulum at the bottom of a long string. What you want to do is be sure that when the disk is on the line it is also at the deepest part of the pendulum's swing (Figure 5.4). This procedure ensures that at the moment you take the time and make the sight, the sun is in a vertical line and you are measuring the proper angle and not a slightly greater angle.

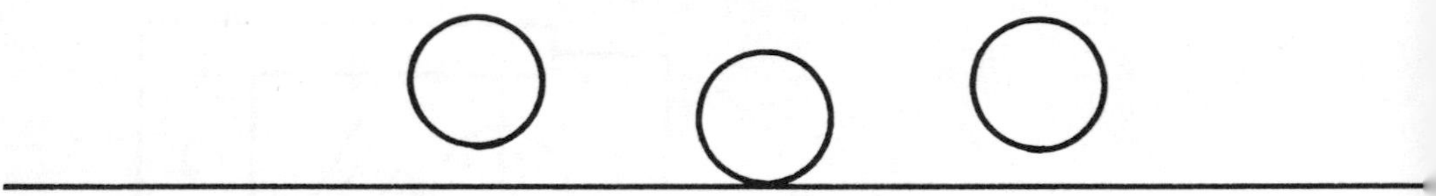

Fig. 5.4

Once you have practiced a bit and become familiar with the mechanics of taking a sight, the next step is to take actual usable sights without being on a boat. To do this you will need to find someplace in your area where you can see the natural horizon. If you live in a building with a view of the sea, you are in good shape; at my home in Larchmont, New York, I use the opposite shore of Long Island Sound, where Long Island meets the water. Possibilities other than the sea horizon are the line where a distant mountain meets a plain or the shoreline of a lake. In the Great Plains, the vast grasslands have their own horizon just like the sea.

In connection with any possible horizon, you must consider another factor, called *dip*. Celestial navigation is based on the theory that the spot from which you measure the angle of the sun is located at eye level. This would be true only if you were right down in the water or if you dug a hole in the ground up to eye level and stood in it to take your sights. In practice, the horizon you look at is normally several miles farther away than it would be in that case, and the higher you are, the farther it is. If you are standing on the deck of a medium-sized sailboat, the horizon you see is about three miles farther away than the theory and mathematics call for, and you have to compensate. This is very simple to do and, in the case above, if the horizon is, say, 3 miles away, since 1′ of arc = 1 nautical mile, you simply subtract 3′ from the angle your sextant measures. Do not worry about this, because the table giving the correction to apply is printed in the *Nautical Almanac*, which you will learn to use later.

To return to practicing with the sextant without going to sea: Suppose you live near the shore of a lake that is 3 miles wide and your back porch is 35 feet above the surface of the lake. Some morning or afternoon when the sun is over the lake, get out your sextant and take a sight, using the line where the shore opposite you meets the water and being careful to "rock" the sextant to establish the low point of the pendulum. Let's say the sextant reads 35°16′. Go to the table at the end of this chapter (Table 2) and look down the left-hand column to 3.0 and then across to the column headed 35 under "Height of eye above the sea, in feet" and find the correction—7.8′. Round this to 8′ and subtract from 35°16′ to give you a sextant altitude of 35°08′.

This observation, with one further possible correction, is a usable sight that can be used to establish a line of position. The other possible correction is called *index error* and is a part of the sextant. To check for index error, set the sextant at 0°0′ and look through the telescope at your horizon or a horizontal line at least two miles away. You should see a straight line in both sides of the horizon mirror. If you do, there is no index error. If the line is broken, turn the micrometer drum until you see a continuous straight line and read the drum. Assume that the drum shows 5′. Look at the arc of the sextant to determine whether this 5′ is to the left or right side of the 0° mark. If it is to the

Dip of the Sea Short of the Horizon

Distance	Height of eye above the sea, in feet										Distance
	5	10	15	20	25	30	35	40	45	50	
Miles	′	′	′	′	′	′	′	′	′	′	*Miles*
0.1	28.3	56.6	84.9	113.2	141.5	169.8	198.0	226.3	254.6	282.9	0.1
0.2	14.2	28.4	42.5	56.7	70.8	84.9	99.1	113.2	127.4	141.5	0.2
0.3	9.6	19.0	28.4	37.8	47.3	56.7	66.1	75.6	85.0	94.4	0.3
0.4	7.2	14.3	21.4	28.5	35.5	42.6	49.7	56.7	63.8	70.9	0.4
0.5	5.9	11.5	17.2	22.8	28.5	34.2	39.8	45.5	51.1	56.8	0.5
0.6	5.0	9.7	14.4	19.1	23.8	28.5	33.3	38.0	42.7	47.4	0.6
0.7	4.3	8.4	12.4	16.5	20.5	24.5	28.6	32.6	36.7	40.7	0.7
0.8	3.9	7.4	10.9	14.5	18.0	21.5	25.1	28.6	32.2	35.7	0.8
0.9	3.5	6.7	9.8	12.9	16.1	19.2	22.4	25.5	28.7	31.8	0.9
1.0	3.2	6.1	8.9	11.7	14.6	17.4	20.2	23.0	25.9	28.7	1.0
1.1	3.0	5.6	8.2	10.7	13.3	15.9	18.5	21.0	23.6	26.2	1.1
1.2	2.9	5.2	7.6	9.9	12.3	14.6	17.0	19.4	21.7	24.1	1.2
1.3	2.7	4.9	7.1	9.2	11.4	13.6	15.8	17.9	20.1	22.3	1.3
1.4	2.6	4.6	6.6	8.7	10.7	12.7	14.7	16.7	18.8	20.8	1.4
1.5	2.5	4.4	6.3	8.2	10.0	11.9	13.8	15.7	17.6	19.5	1.5
1.6	2.4	4.2	6.0	7.7	9.5	11.3	13.0	14.8	16.6	18.3	1.6
1.7	2.4	4.0	5.7	7.4	9.0	10.7	12.4	14.0	15.7	17.3	1.7
1.8	2.3	3.9	5.5	7.0	8.6	10.2	11.7	13.3	14.9	16.5	1.8
1.9	2.3	3.8	5.3	6.7	8.2	9.7	11.2	12.7	14.2	15.7	1.9
2.0	2.2	3.7	5.1	6.5	7.9	9.3	10.7	12.1	13.6	15.0	2.0
2.1	2.2	3.6	4.9	6.3	7.6	9.0	10.3	11.6	13.0	14.3	2.1
2.2	2.2	3.5	4.8	6.1	7.3	8.6	9.9	11.2	12.5	13.8	2.2
2.3	2.2	3.4	4.6	5.9	7.1	8.3	9.6	10.8	12.0	13.3	2.3
2.4	2.2	3.4	4.5	5.7	6.9	8.1	9.2	10.4	11.6	12.8	2.4
2.5	2.2	3.3	4.4	5.6	6.7	7.8	9.0	10.1	11.2	12.4	2.5
2.6	2.2	3.3	4.3	5.4	6.5	7.6	8.7	9.8	10.9	12.0	2.6
2.7	2.2	3.2	4.3	5.3	6.4	7.4	8.4	9.5	10.6	11.6	2.7
2.8	2.2	3.2	4.2	5.2	6.2	7.2	8.2	9.2	10.3	11.3	2.8
2.9	2.2	3.2	4.1	5.1	6.1	7.1	8.0	9.0	10.0	11.0	2.9
3.0	2.2	3.1	4.1	5.0	6.0	6.9	7.8	8.8	9.7	10.7	3.0
3.1	2.2	3.1	4.0	4.9	5.9	6.8	7.7	8.6	9.5	10.4	3.1
3.2	2.2	3.1	4.0	4.9	5.7	6.6	7.5	8.4	9.3	10.2	3.2
3.3	2.2	3.1	3.9	4.8	5.7	6.5	7.4	8.2	9.1	9.9	3.3
3.4	2.2	3.1	3.9	4.7	5.6	6.4	7.2	8.1	8.9	9.7	3.4
3.5	2.2	3.1	3.9	4.7	5.5	6.3	7.1	7.9	8.7	9.5	3.5
3.6	2.2	3.1	3.8	4.6	5.4	6.2	7.0	7.8	8.6	9.4	3.6
3.7	2.2	3.1	3.8	4.6	5.4	6.1	6.9	7.7	8.4	9.2	3.7
3.8	2.2	3.1	3.8	4.6	5.3	6.0	6.8	7.5	8.3	9.0	3.8
3.9	2.2	3.1	3.8	4.5	5.2	6.0	6.7	7.4	8.1	8.9	3.9
4.0	2.2	3.1	3.8	4.5	5.2	5.9	6.6	7.3	8.0	8.7	4.0
4.1	2.2	3.1	3.8	4.5	5.1	5.8	6.5	7.2	7.9	8.6	4.1
4.2	2.2	3.1	3.8	4.4	5.1	5.8	6.5	7.1	7.8	8.5	4.2
4.3	2.2	3.1	3.8	4.4	5.1	5.7	6.4	7.0	7.7	8.4	4.3
4.4	2.2	3.1	3.8	4.4	5.0	5.7	6.3	7.0	7.6	8.3	4.4
4.5	2.2	3.1	3.8	4.4	5.0	5.6	6.3	6.9	7.5	8.2	4.5
4.6	2.2	3.1	3.8	4.4	5.0	5.6	6.2	6.8	7.4	8.1	4.6
4.7	2.2	3.1	3.8	4.4	5.0	5.6	6.2	6.8	7.4	8.0	4.7
4.8	2.2	3.1	3.8	4.4	4.9	5.5	6.1	6.7	7.3	7.9	4.8
4.9	2.2	3.1	3.8	4.3	4.9	5.5	6.1	6.7	7.2	7.8	4.9
5.0	2.2	3.1	3.8	4.3	4.9	5.5	6.0	6.6	7.2	7.7	5.0
5.5	2.2	3.1	3.8	4.3	4.9	5.4	5.9	6.4	6.9	7.4	5.5
6.0	2.2	3.1	3.8	4.3	4.9	5.3	5.8	6.3	6.7	7.2	6.0
6.5	2.2	3.1	3.8	4.3	4.9	5.3	5.7	6.2	6.6	7.1	6.5
7.0	2.2	3.1	3.8	4.3	4.9	5.3	5.7	6.1	6.5	6.9	7.0
7.5	2.2	3.1	3.8	4.3	4.9	5.3	5.7	6.1	6.5	6.9	7.5
8.0	2.2	3.1	3.8	4.3	4.9	5.3	5.7	6.1	6.5	6.9	8.0
8.5	2.2	3.1	3.8	4.3	4.9	5.3	5.7	6.1	6.5	6.9	8.5
9.0	2.2	3.1	3.8	4.3	4.9	5.3	5.7	6.1	6.5	6.9	9.0
9.5	2.2	3.1	3.8	4.3	4.9	5.3	5.7	6.1	6.5	6.9	9.5
10.0	2.2	3.1	3.8	4.3	4.9	5.3	5.7	6.1	6.5	6.9	10.0

Table 2

Dip of the Sea Short of the Horizon

Distance	Height of eye above the sea, in feet										Distance
	55	60	65	70	75	80	85	90	95	100	
Miles	′	′	′	′	′	′	′	′	′	′	*Miles*
0. 1	311. 2	339. 5	367. 8	396. 1	424. 4	452. 6	480. 9	509. 2	537. 5	565. 8	0. 1
0. 2	155. 6	169. 8	184. 0	198. 1	212. 2	226. 4	240. 5	254. 7	268. 8	283. 0	0. 2
0. 3	103. 8	113. 3	122. 7	132. 1	141. 6	151. 0	160. 4	169. 9	179. 3	188. 7	0. 3
0. 4	78. 0	85. 0	92. 1	99. 2	106. 2	113. 3	120. 4	127. 5	134. 5	141. 6	0. 4
0. 5	62. 4	68. 1	73. 8	79. 4	85. 1	90. 7	96. 4	102. 0	107. 7	113. 4	0. 5
0. 6	52. 1	56. 8	61. 5	66. 3	71. 0	75. 7	80. 4	85. 1	89. 8	94. 5	0. 6
0. 7	44. 7	48. 8	52. 8	56. 9	60. 9	64. 9	69. 0	73. 0	77. 1	81. 1	0. 7
0. 8	39. 2	42. 8	46. 3	49. 8	53. 4	56. 9	60. 4	64. 0	67. 5	71. 1	0. 8
0. 9	34. 9	38. 1	41. 2	44. 4	47. 5	50. 7	53. 8	56. 9	60. 1	63. 2	0. 9
1. 0	31. 5	34. 4	37. 2	40. 0	42. 8	45. 7	48. 5	51. 3	54. 2	57. 0	1. 0
1. 1	28. 7	31. 3	33. 9	36. 5	39. 0	41. 6	44. 2	46. 7	49. 3	51. 9	1. 1
1. 2	26. 4	28. 8	31. 1	33. 5	35. 9	38. 2	40. 6	42. 9	45. 3	47. 6	1. 2
1. 3	24. 5	26. 7	28. 8	31. 0	33. 2	35. 4	37. 5	39. 7	41. 9	44. 1	1. 3
1. 4	22. 8	24. 8	26. 8	28. 9	30. 9	32. 9	34. 9	37. 0	39. 0	41. 0	1. 4
1. 5	21. 4	23. 3	25. 1	27. 0	28. 9	30. 8	32. 7	34. 6	36. 5	38. 3	1. 5
1. 6	20. 1	21. 9	23. 6	25. 4	27. 2	29. 0	30. 7	32. 5	34. 3	36. 0	1. 6
1. 7	19. 0	20. 7	22. 3	24. 0	25. 7	27. 3	29. 0	30. 7	32. 3	34. 0	1. 7
1. 8	18. 0	19. 6	21. 2	22. 8	24. 3	25. 9	27. 5	29. 0	30. 6	32. 2	1. 8
1. 9	17. 2	18. 7	20. 1	21. 6	23. 1	24. 6	26. 1	27. 6	29. 1	30. 6	1. 9
2. 0	16. 4	17. 8	19. 2	20. 6	22. 0	23. 5	24. 9	26. 3	27. 7	29. 1	2. 0
2. 1	15. 7	17. 0	18. 4	19. 7	21. 1	22. 4	23. 8	25. 1	26. 5	27. 8	2. 1
2. 2	15. 1	16. 3	17. 6	18. 9	20. 2	21. 5	22. 7	24. 1	25. 3	26. 6	2. 2
2. 3	14. 5	15. 7	16. 9	18. 2	19. 4	20. 6	21. 9	23. 1	24. 3	25. 6	2. 3
2. 4	14. 0	15. 1	16. 3	17. 5	18. 7	19. 9	21. 0	22. 2	23. 4	24. 6	2. 4
2. 5	13. 5	14. 6	15. 7	16. 9	18. 0	19. 1	20. 3	21. 4	22. 5	23. 7	2. 5
2. 6	13. 0	14. 1	15. 2	16. 3	17. 4	18. 5	19. 6	20. 7	21. 8	22. 8	2. 6
2. 7	12. 6	13. 7	14. 7	15. 8	16. 8	17. 9	18. 9	20. 0	21. 0	22. 1	2. 7
2. 8	12. 3	13. 3	14. 3	15. 3	16. 3	17. 3	18. 3	19. 3	20. 4	21. 4	2. 8
2. 9	11. 9	12. 9	13. 9	14. 9	15. 8	16. 8	17. 8	18. 8	19. 7	20. 7	2. 9
3. 0	11. 6	12. 6	13. 5	14. 4	15. 4	16. 3	17. 3	18. 2	19. 2	20. 1	3. 0
3. 1	11. 3	12. 2	13. 2	14. 1	15. 0	15. 9	16. 8	17. 7	18. 6	10. 5	3. 1
3. 2	11. 1	11. 9	12. 8	13. 7	14. 6	15. 5	16. 4	17. 2	18. 1	19. 0	3. 2
3. 3	10. 8	11. 7	12. 5	13. 4	14. 2	15. 1	15. 9	16. 8	17. 7	18. 5	3. 3
3. 4	10. 6	11. 4	12. 2	13. 1	13. 0	14. 7	15. 6	16. 4	17. 2	18. 1	3. 4
3. 5	10. 3	11. 2	12. 0	12. 8	13. 6	14. 4	15. 2	16. 0	16. 8	17. 6	3. 5
3. 6	10. 1	10. 9	11. 7	12. 4	13. 3	14. 1	14. 9	15. 6	16. 4	17. 2	3. 6
3. 7	9. 9	10. 7	11. 5	12. 2	13. 0	13. 8	14. 5	15. 3	16. 1	16. 8	3. 7
3. 8	9. 8	10. 5	11. 3	12. 0	12. 7	13. 5	14. 2	15. 0	15. 7	16. 5	3. 8
3. 9	9. 6	10. 3	11. 1	11. 8	12. 5	13. 2	14. 0	14. 7	15. 4	16. 1	3. 9
4. 0	9. 4	10. 1	10. 9	11. 6	12. 3	13. 0	13. 7	14. 4	15. 1	15. 8	4. 0
4. 1	9. 3	10. 0	10. 7	11. 4	12. 1	12. 7	13. 4	14. 1	14. 8	15. 5	4. 1
4. 2	9. 2	9. 8	10. 5	11. 2	11. 8	12. 5	13. 2	13. 9	14. 5	15. 2	4. 2
4. 3	9. 0	9. 7	10. 3	11. 0	11. 7	12. 3	13. 0	13. 6	14. 3	14. 9	4. 3
4. 4	8. 9	9. 5	10. 2	10. 8	11. 5	12. 1	12. 8	13. 4	14. 0	14. 7	4. 4
4. 5	8. 8	9. 4	10. 0	10. 7	11. 3	11. 9	12. 6	13. 2	13. 8	14. 4	4. 5
4. 6	8. 7	9. 3	9. 9	10. 5	11. 1	11. 8	12. 4	13. 0	13. 6	14. 2	4. 6
4. 7	8. 6	9. 2	9. 8	10. 4	11. 0	11. 6	12. 2	12. 8	13. 4	14. 0	4. 7
4. 8	8. 5	9. 1	9. 7	10. 2	10. 8	11. 4	12. 0	12. 6	13. 2	13. 8	4. 8
4. 9	8. 4	9. 0	9. 5	10. 1	10. 7	11. 3	11. 9	12. 4	13. 0	13. 6	4. 9
5. 0	8. 3	8. 9	9. 4	10. 0	10. 6	11. 1	11. 7	12. 3	12. 8	13. 4	5. 0
5. 5	7. 9	8. 5	9. 0	9. 5	10. 0	10. 5	11. 0	11. 5	12. 1	12. 6	5. 5
6. 0	7. 7	8. 2	8. 6	9. 1	9. 6	10. 0	10. 5	11. 0	11. 5	11. 9	6. 0
6. 5	7. 5	7. 9	8. 4	8. 8	9. 2	9. 7	10. 1	10. 5	11. 0	11. 4	6. 5
7. 0	7. 4	7. 8	8. 2	8. 6	9. 0	9. 4	9. 8	10. 2	10. 6	11. 0	7. 0
7. 5	7. 3	7. 6	8. 0	8. 4	8. 8	9. 2	9. 5	9. 9	10. 3	10. 7	7. 5
8. 0	7. 2	7. 6	7. 9	8. 3	8. 6	9. 0	9. 3	9. 7	10. 0	10. 4	8. 0
8. 5	7. 2	7. 5	7. 9	8. 2	8. 5	8. 9	9. 2	9. 5	9. 9	10. 2	8. 5
9. 0	7. 2	7. 5	7. 8	8. 1	8. 5	8. 8	9. 1	9. 4	9. 7	10. 0	9. 0
9. 5	7. 2	7. 5	7. 8	8. 1	8. 4	8. 7	9. 0	9. 3	9. 6	9. 9	9. 5
10. 0	7. 2	7. 5	7. 8	8. 1	8. 4	8. 7	9. 0	9. 2	9. 5	9. 8	10. 0

Table 2, continued

right, the error is said to be "off the arc." This means that the sextant is measuring 5′ too low and that you will have to *add* 5′ to your sextant reading. If it is to the left of the 0° mark, the error is said to be "on the arc"; your sextant is measuring 5′ too much and you will have to *subtract* 5′ from your sextant altitude. My EBBCO has 5′ "off the arc," so let's say yours does too. This means that you add 5′ to your reading of 35°08′ to give a reading of 35°13′. This is called *apparent altitude* (ha) and could now be used with the almanac and tables to obtain a line of position.

Index Error:		
Minutes off Arc	=	Plus (+) Correction
Minutes on Arc	=	Minus (−) Correction
Sextant Reading (Hs)	=	35° 16′
Dip (always minus)	=	− 08′
Index Correction (IC)	=	+ 05′
Apparent Altitude (ha)	=	35° 13′

Try practicing as much as you conveniently can, because this is really the most important part of celestial navigation. The more you practice ashore, the easier it will be afloat.

I think you can see by now that the modern "no math or trig" methods of solving the navigational triangle mean that you can spend the bulk of your time shooting, not computing.

One of the most difficult things to determine when you are learning celestial navigation is how much progress you are making. After all, you probably can't stand alongside an experienced navigator and compare your sights with his. Nor are there any fixed standards as there are in such sports as sprinting, in which you can compare yourself against a time standard.

When I was learning celestial navigation, I plotted my backyard sights on graph paper. I would put a series of sights down and draw a straight line through them as nearly as I could. Later, when I actually started navigating from a boat at sea, I continued this practice and recommend it highly. You can tell how your sights are falling, and you will get a "feel" for how consistent you are. After a little practice, you will find that your sights on land will fall very nearly on a straight line.

If you have done this on land and have an idea of how close your fixes are to your known position (more of this in Chapter 8) you will be able to evaluate very well the probable accuracy of your fixes derived from sextant observations at sea.

Figures 5.5 and 5.6 are several samples of this technique together with comments. One thing needs to be mentioned here—you should not expect a straight line to lie through a series of sights taken within one-half hour or so of noon, because at that time the path of the sun or of any other celestial body is a rather marked curve. In fact, the path is always a curve; but at most times the curve is so slight that it can be ignored for practical purposes.

Now that you have seen the graphs, let me make a few suggestions on practical procedure. When I draw a graph such as one of these, I normally pick the sight closest to the line to work out. Thus, in Figure 5.6, I took the sight circled. If the sights are spread more than 2 or 3 minutes from the line, I throw out the worst and average the rest.

There comes a time in a small boat when it becomes impossible to plot little points on graph paper. When it is rough, the thing to do is take as many sights as you can—preferably five or more—and average them. You will know when this time has arrived because you will no longer be able to write your sights down yourself. You will have to content yourself with taking and timing; then you will have to shout the time and angle to a crew member, who will write them down for you. To the greatest extent possible, however, get in the habit of taking and timing your sights yourself. Don't rely on someone else for this. You are the person least likely to make errors because *you* are a *very* interested party! This is an additional reason for taking more than one sight. You are likely to catch yourself if you are misreading the time by a minute or more or being a degree out in your sextant reading. Believe me, these things happen, especially at dawn and dusk when you are out after planets and stars and other exotic game.

Here are a few more observations from my experience. As a rule, the rougher the sea, the higher on the boat you want to be. This is because you want to get your line of sight above the waves, so that you bring the sun down to the horizon and not to the top of a wave a mile or so away. When it is rough on *Paper Tiger*, a forty-foot yawl owned by

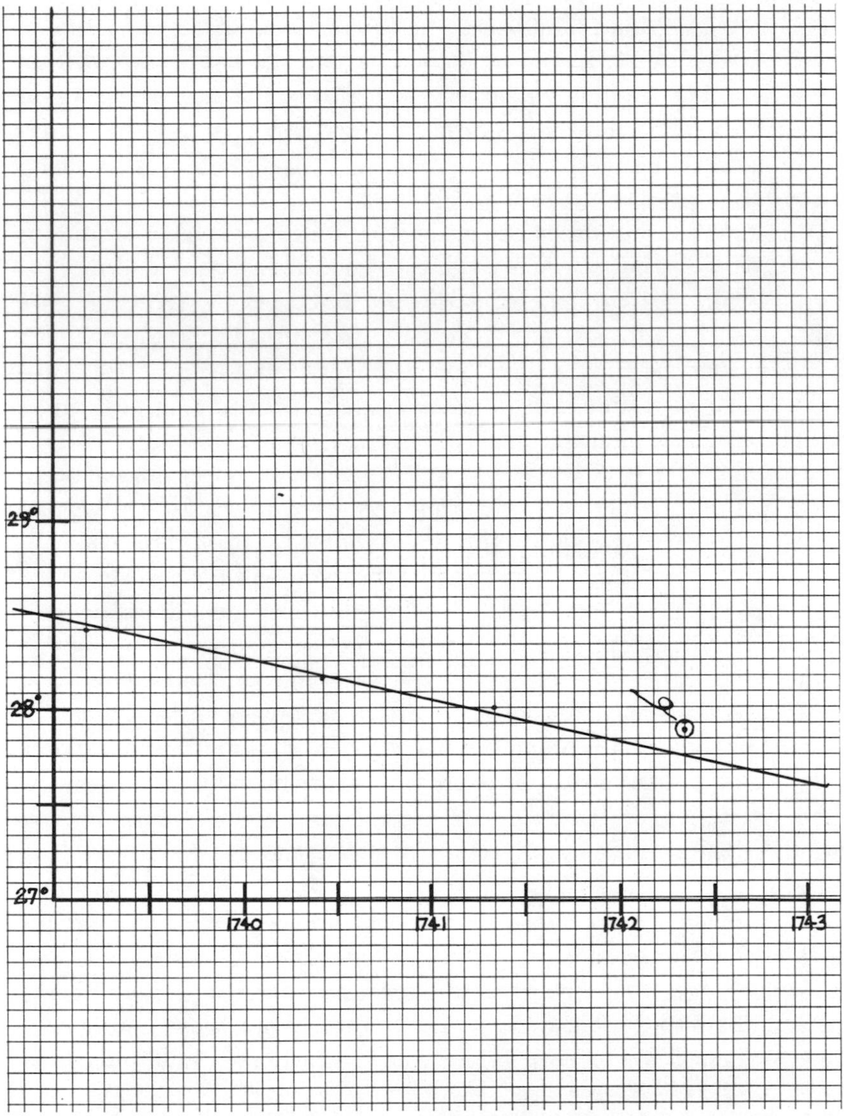

Fig. 5.5. When I am seasick, as I was when I prepared this graph, I have a tendency to rush my last shot. As you see, it is nearly 7′ from the line-of-best-fit, and I discarded it and used the middle sight of the remaining three.

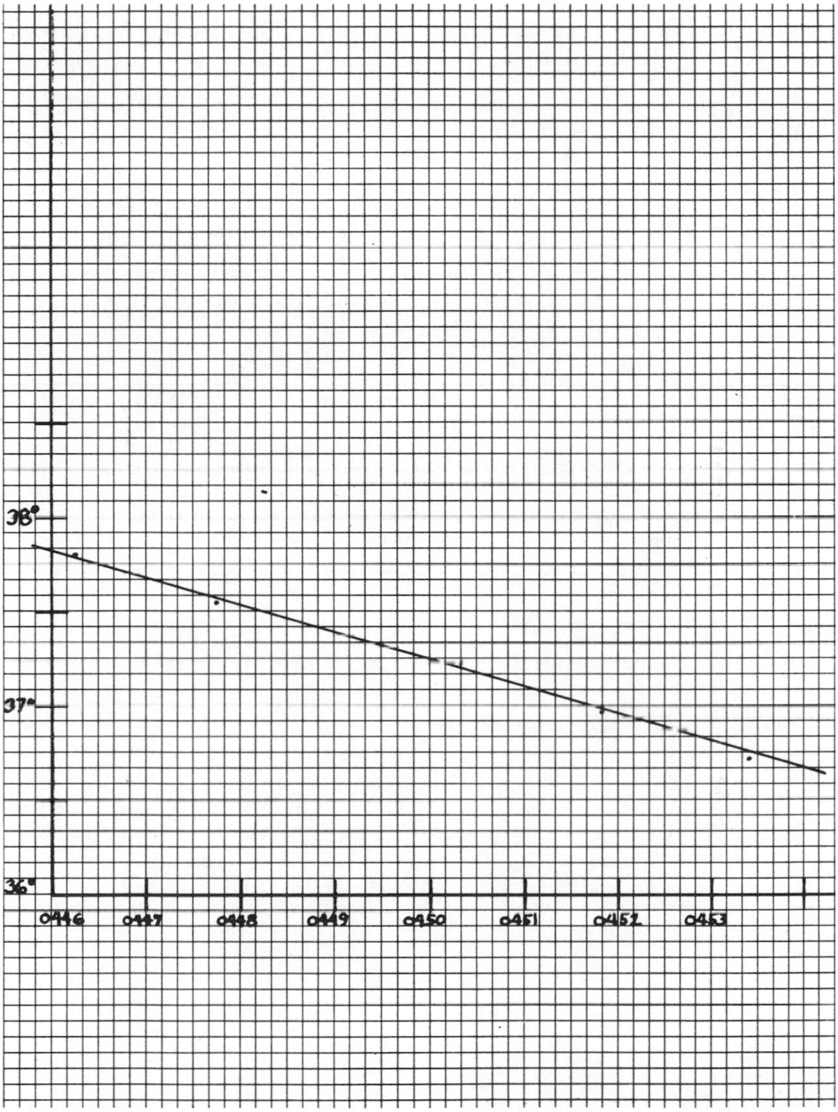

Fig. 5.6. This series was made at dawn. I had been concentrating on making observations of Venus and then Jupiter, which were beautifully situated for shooting. Then I happened to look aft and saw a star just above the mizzen. Since it was nearly sunrise, the background stars had faded, and I could not identify it. I made these five observations, took a compass bearing of the star, and identified it, as you will learn to do in Chapter 11.

my friend Phil Strenger, I lean against the mizzen, bracing myself well with my back and legs. This leaves my arms and shoulders reasonably free and keeps me out of the worst of the spray. In calmer weather, I sit on the cabin amidships with my feet on deck. This is the point of least motion on the boat, and I have great confidence in sights taken from this position and usually take only three. From the mizzen in rough weather, I try to take at least five.

When you have read this book through Chapter 8, come back to this chapter. Take your practice location as your dead reckoning (DR) position, and actually take sights and plot them. Since you are not moving between observations, you do not advance the LOPs. If you use the same plotting sheet for a while, you will find the lines crossing at, or very near, your location. Try it; it's fun.

6
Time

As you undoubtedly realize by now, time is a very important part of celestial navigation. You were asked to practice timing sights in Chapter 5, and in Chapter 1 you learned that astonomers have put together an almanac in which the geographical positions of many celestial bodies are set down for every second of every hour of every day. You need to know the time in order to use the data in the almanac properly.

Of course, the information in the almanac could not be recorded for all the different times that are kept in every country or state around the world. The standard or reference time on which the book is based is the time at the Royal Observatory near Greenwich, England. Just as longitude begins at Greenwich, so does time for navigational purposes. The time is called *Greenwich Mean Time*, abbreviated GMT; it is also called Coordinated Universal Time (UTC) and Zulu Time.

The main thing a navigator needs to know is how to convert his watch time to GMT. Of course, for several hundred dollars you can buy a chronometer, set it to GMT, and forget the problem. There are other cheaper ways, however.

If you have a short-wave receiver, you can pick up the time signals broadcast by the United States Bureau of Standards from Fort Collins, Colorado. These signals are broadcast continuously, and every minute a voice announces the exact GMT. These signals are broadcast on 2.5, 5, 10, 15, and 25 Megahertz, abbreviated MHz or MC. Station CHU in Canada broadcasts on 3.3 and 7.7 MHz.[1] Reception is usually best at night.

I have found in practice that by checking my watch against the time signal nightly, figuring the gain or loss from the previous signal, and then interpolating this gain or loss as I take observations, things work out just fine. The watch I use is a twenty-year-old Omega that was given to me when I graduated from high school. Oddly enough, it does not have a sweep second hand, just a little one in a small circle above

[1]See appendix for schedule of BBC time broadcasts.

the 6. Although a sweep would be better, I find the small dial works adequately.

Basically, I think it is better to use the equipment you already have until you have developed your own style and abilities. Then you can decide what suits you best and get it. So, if you have a short-wave that will pick up time signals, use it. For practice purposes you can simply dial the telephone company time number and get the time that way. It is said by the telephone company to be within one second of the Greenwich Observatory or Fort Collins.

After many years of using various short-wave receivers, I recently discovered a specialized one designed solely to pick up time signals. It is called Time Kube and is manufactured by Radio Shack, Inc., a company with a chain of retail stores located throughout the United States (100 stores in England). This wizard device costs $50 and is quite compact, and from now on it goes with me on every trip.

One thing you should know is that the system of time notation used in navigation is the 24-hour system. This simply means that the hours from midnight to noon are 0000, 0100, 0200, 0300, and so on, to 1200 (noon); 1 P.M. is 1300; 2 P.M. is 1400; and so on, right up to midnight, which is 2400 (or 0000). Then the cycle starts again. As you will see in the next chapters, this will all become second nature after a few practical examples and exercises.

As I said at the beginning of the chapter, what you really need to know is how to convert your watch time to GMT. In the back of the *Nautical Almanac* or the *Air Almanac* is a list of all the countries of the world and the hours and minutes to add to or subtract from your standard time to get GMT. Therefore, all you have to do is find a source for accurate local time, and you will be able to figure Greenwich time from your watch. The easier way, of course, is to listen to the short-wave time signals and work from there, using the following form:

		Hr.,	Min.,	Sec.
Watch Time	=			
Fast (−) or Slow (+)	=			
Conversion to GMT	=			
GMT	=			

To use this form, write down the GMT and your watch time. Then figure what it takes to convert your watch to GMT. The factor to convert from your standard time is usually a whole number of hours; in some countries it is a whole number of hours and 30 minutes. It is never an odd number of minutes such as 5 hours and 36 minutes or 6 hours and 47 minutes.

As an example, I live in New York, and I hear on the short-wave that it is 21-06-00 GMT. In other words, it is exactly 9:06 P.M. in Greenwich. My watch says 17-06-38. The filled-out form looks like this:

Watch Time	=	17-	06	-38
Fast (−) or Slow (+)	=			−38
Conversion to GMT	=	+4		
GMT	=	21-	06	-00

If I take a sight later and my watch, at the instant when the sun's edge just touches the horizon, says 19-12-46, I use the same form to figure out GMT:

Watch Time	=	19-	12	-46
Fast (−) or Slow (+)	=			−38
Conversion to GMT	=	+4		
GMT	=	23-	12	-08

It is 12 minutes and 8 seconds past 11 in the evening in Greenwich, England. As I said, the 24-hour system will become second nature as you gain practice.

7
Almanacs

In this book so far you have seen several references to the *Nautical Almanac*. You read in Chapter 1 that the *Nautical Almanac* is the source for the geographical position (GP) or known point on which celestial navigation depends. It is now time to learn how to use this almanac.

Actually, there is not just one almanac in general use for celestial navigation but two—the *Nautical Almanac* and the *Air Almanac.* Both contain the same information in slightly different forms, and since each has its advantages, you will learn how to use both. These two books are published in identical form in England, where they are also entitled *Nautical Almanac* (or *Nautical Publication 314*) and *Air Almanac* (or *Air Publication 1602.*) Sources for obtaining these volumes are listed in the last chapter, and appropriate extracts are reproduced in this chapter as needed.

In Chapter 5 we discussed a hypothetical practice session in which the sextant reading was 35°16′. Correcting for dip and index error, we had the following:

Sextant Reading (Hs)	=	35°	16′
Dip	=	−	08′
Index Correction (IC)	=	+	05′
Apparent Altitude (ha)	=	35°	13′

Now, if you turn to the inside front cover of the *Nautical Almanac*, you will see the table that is reproduced here as Table 3. This table contains the additional correction necessary in order to determine the true or "observed" altitude of the sun. If you look down the column for the sun under "Apr.-Sept.," or April-September, you will see that the apparent altitude of 35°13′ falls between 33°20′ and 35°17′ on the table and that the correction to be applied to an observation of the lower limb of the sun is +14.6′. Round this to 15′ and complete the observation:

ALTITUDE CORRECTION TABLES 10°–90°—SUN, STARS, PLANETS

SUN

OCT.–MAR. App. Alt. ° ′	Lower Limb ′	Upper Limb ′	APR.–SEPT. App. Alt. ° ′	Lower Limb ′	Upper Limb ′
9 34	+10.8	−22.7	9 39	+10.6	−22.4
9 45	+10.9	−22.6	9 51	+10.7	−22.3
9 56	+11.0	−22.5	10 03	+10.8	−22.2
10 08	+11.1	−22.4	10 15	+10.9	−22.1
10 21	+11.2	−22.3	10 27	+11.0	−22.0
10 34	+11.3	−22.2	10 40	+11.1	−21.9
10 47	+11.4	−22.1	10 54	+11.2	−21.8
11 01	+11.5	−22.0	11 08	+11.3	−21.7
11 15	+11.6	−21.9	11 23	+11.4	−21.6
11 30	+11.7	−21.8	11 38	+11.5	−21.5
11 46	+11.8	−21.7	11 54	+11.6	−21.4
12 02	+11.9	−21.6	12 10	+11.7	−21.3
12 19	+12.0	−21.5	12 28	+11.8	−21.2
12 37	+12.1	−21.4	12 46	+11.9	−21.1
12 55	+12.2	−21.3	13 05	+12.0	−21.0
13 14	+12.3	−21.2	13 24	+12.1	−20.9
13 35	+12.4	−21.1	13 45	+12.2	−20.8
13 56	+12.5	−21.0	14 07	+12.3	−20.7
14 18	+12.6	−20.9	14 30	+12.4	−20.6
14 42	+12.7	−20.8	14 54	+12.5	−20.5
15 06	+12.8	−20.7	15 19	+12.6	−20.4
15 32	+12.9	−20.6	15 46	+12.7	−20.3
15 59	+13.0	−20.5	16 14	+12.8	−20.2
16 28	+13.1	−20.4	16 44	+12.9	−20.1
16 59	+13.2	−20.3	17 15	+13.0	−20.0
17 32	+13.3	−20.2	17 48	+13.1	−19.9
18 06	+13.4	−20.1	18 24	+13.2	−19.8
18 42	+13.5	−20.0	19 01	+13.3	−19.7
19 21	+13.6	−19.9	19 42	+13.4	−19.6
20 03	+13.7	−19.8	20 25	+13.5	−19.5
20 48	+13.8	−19.7	21 11	+13.6	−19.4
21 35	+13.9	−19.6	22 00	+13.7	−19.3
22 26	+14.0	−19.5	22 54	+13.8	−19.2
23 22	+14.1	−19.4	23 51	+13.9	−19.1
24 21	+14.2	−19.3	24 53	+14.0	−19.0
25 26	+14.3	−19.2	26 00	+14.1	−18.9
26 36	+14.4	−19.1	27 13	+14.2	−18.8
27 52	+14.5	−19.0	28 33	+14.3	−18.7
29 15	+14.6	−18.9	30 00	+14.4	−18.6
30 46	+14.7	−18.8	31 35	+14.5	−18.5
32 26	+14.8	−18.7	33 20	+14.6	−18.4
34 17	+14.9	−18.6	35 17	+14.7	−18.3
36 20	+15.0	−18.5	37 26	+14.8	−18.2
38 36	+15.1	−18.4	39 50	+14.9	−18.1
41 08	+15.2	−18.3	42 31	+15.0	−18.0
43 59	+15.3	−18.2	45 31	+15.1	−17.9
47 10	+15.4	−18.1	48 55	+15.2	−17.8
50 46	+15.5	−18.0	52 44	+15.3	−17.7
54 49	+15.6	−17.9	57 02	+15.4	−17.6
59 23	+15.7	−17.8	61 51	+15.5	−17.5
64 30	+15.8	−17.7	67 17	+15.6	−17.4
70 12	+15.9	−17.6	73 16	+15.7	−17.3
76 26	+16.0	−17.5	79 43	+15.8	−17.2
83 05	+16.1	−17.4	86 32	+15.9	−17.1
90 00			90 00		

STARS AND PLANETS

App. Alt. ° ′	Corrⁿ ′
9 56	−5.3
10 08	−5.2
10 20	−5.1
10 33	−5.0
10 46	−4.9
11 00	−4.8
11 14	−4.7
11 29	−4.6
11 45	−4.5
12 01	−4.4
12 18	−4.3
12 35	−4.2
12 54	−4.1
13 13	−4.0
13 33	−3.9
13 54	−3.8
14 16	−3.7
14 40	−3.6
15 04	−3.5
15 30	−3.4
15 57	−3.3
16 26	−3.2
16 56	−3.1
17 28	−3.0
18 02	−2.9
18 38	−2.8
19 17	−2.7
19 58	−2.6
20 42	−2.5
21 28	−2.4
22 19	−2.3
23 13	−2.2
24 11	−2.1
25 14	−2.0
26 22	−1.9
27 36	−1.8
28 56	−1.7
30 24	−1.6
32 00	−1.5
33 45	−1.4
35 40	−1.3
37 48	−1.2
40 08	−1.1
42 44	−1.0
45 36	−0.9
48 47	−0.8
52 18	−0.7
56 11	−0.6
60 28	−0.5
65 08	−0.4
70 11	−0.3
75 34	−0.2
81 13	−0.1
87 03	0.0
90 00	

App. Alt. — Additional Corrⁿ

1965

VENUS

App. Alt.	Additional Corrⁿ
Jan. 1–Oct. 2	
0°	′
42	+0.1
Oct. 3–Nov. 18	
0°	′
47	+0.2
Nov. 19–Dec. 15	
0°	′
46	+0.3
Dec. 16–Dec. 31	
0°	′
11	+0.4
41	+0.5

MARS

App. Alt.	Additional Corrⁿ
Jan. 1–Jan. 10	
0°	′
60	+0.1
Jan. 11–May 16	
0°	′
41	+0.2
75	+0.1
May 17–Dec. 31	
0°	′
60	+0.1

DIP

Ht. of Eye ft.	Corrⁿ ′	Ht. of Eye ft.	Corrⁿ ′
1.1	−1.1	44	−6.5
1.4	−1.2	45	−6.6
1.6	−1.3	47	−6.7
1.9	−1.4	48	−6.8
2.2	−1.5	49	−6.9
2.5	−1.6	51	−7.0
2.8	−1.7	52	−7.1
3.2	−1.8	54	−7.2
3.6	−1.9	55	−7.3
4.0	−2.0	57	−7.4
4.4	−2.1	58	−7.5
4.9	−2.2	60	−7.6
5.3	−2.3	62	−7.7
5.8	−2.4	63	−7.8
6.3	−2.5	65	−7.9
6.9	−2.6	67	−8.0
7.4	−2.7	68	−8.1
8.0	−2.8	70	−8.2
8.6	−2.9	72	−8.3
9.2	−3.0	74	−8.4
9.8	−3.1	75	−8.5
10.5	−3.2	77	−8.6
11.2	−3.3	79	−8.7
11.9	−3.4	81	−8.8
12.6	−3.5	83	−8.9
13.3	−3.6	85	−9.0
14.1	−3.7	87	−9.1
14.9	−3.8	88	−9.2
15.7	−3.9	90	−9.3
16.5	−4.0	92	−9.4
17.4	−4.1	94	−9.5
18.3	−4.2	96	−9.6
19.1	−4.3	98	−9.7
20.1	−4.4	101	−9.8
21.0	−4.5	103	−9.9
22.0	−4.6	105	−10.0
22.9	−4.7	107	−10.1
23.9	−4.8	109	−10.2
24.9	−4.9	111	−10.3
26.0	−5.0	113	−10.4
27.1	−5.1	116	−10.5
28.1	−5.2	118	−10.6
29.2	−5.3	120	−10.7
30.4	−5.4	122	−10.8
31.5	−5.5	125	−10.9
32.7	−5.6	127	−11.0
33.9	−5.7	129	−11.1
35.1	−5.8	132	−11.2
36.3	−5.9	134	−11.3
37.6	−6.0	136	−11.4
38.9	−6.1	139	−11.5
40.1	−6.2	141	−11.6
41.5	−6.3	144	−11.7
42.8	−6.4	146	−11.8
44.2		149	

App. Alt. = Apparent altitude = Sextant altitude corrected for index error and dip.

Table 3

Sextant Reading (Hs)	=	35°	16′
Dip	=	−	08′
Index Correction (IC)	=	+	05′
Apparent Altitude (ha)	=	35°	13′
Main Correction	=	+	15′
Observed Altitude (Ho)	=	35°	28′

This final correction is the result of two factors. The first arises from the fact that you have measured the angle between the horizon and the lower edge of the sun instead of the angle between the horizon and the center of the sun, which is the angle that is really required. Since the sun has an apparent diameter of 32′, the sextant reading is 16′ too small and 16′ should be added to the reading (Figure 7.1). If you have difficulty visualizing this, imagine that you were going to measure to the middle of the sun. You would have to bring the sun down farther by turning the micrometer drum, and you would therefore increase the angle measured. Since it is much easier visually, and therefore more accurate, to bring the edge of the sun to the horizon, it is preferable to make the correction instead of trying to measure to the middle of the sun.

The second factor in this correction is the phenomenon known as *refraction*. Refraction is caused by the fact that the atmosphere bends rays of light that enter it at an oblique angle. Figure 7.2 shows how refraction tends to make the sun appear higher in the sky than it actually is. Therefore, the angle between the sun, your eye, and the horizon is actually less than the sextant measures, and the correction is made by subtracting. The table in the almanac combines this correction and the correction for half the diameter (the semidiameter) of the sun and gives the net result in one main correction.

Note that the table is divided into two half-year segments. The reason is that refraction is slightly different during the winter and summer months. You can see, however, that the difference is so slight that it wouldn't throw you off very much if you happened to take your correction from the wrong half of this table.

The other columns in Table 3 contain corrections for use with stars

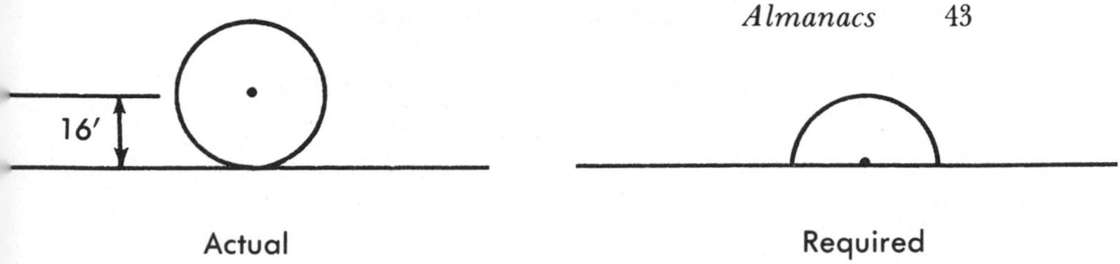

Fig. 7.1

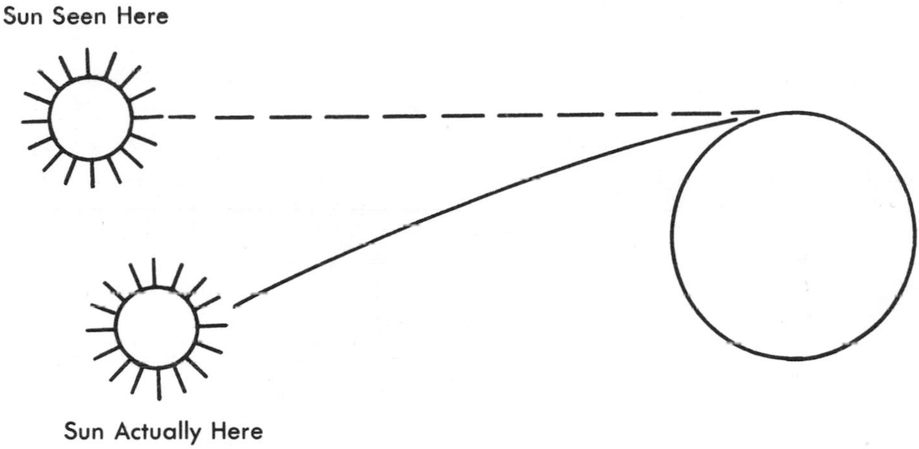

Fig. 7.2

and planets and the dip corrections for various heights of eye.

Note, too, that this table is for apparent altitudes between 9°34′ and 90°. For smaller apparent altitudes there is another page that you will study in a later chapter.

Now assume that you took a sight on June 28, 1974, at 23-12-08 Greenwich Mean Time (as per Chapter 6) and that you are going to proceed to look up the remaining information given in the *Nautical Almanac.*

Table 4 is a reproduction of the relevant page from the almanac. Looking at the column headed "Sun," you will see subcolumns headed "G.H.A." and "Dec." *Dec.* stands for *declination*, and you already know that this is the equivalent of latitude. *G.H.A.* (or *GHA*) stands for *Greenwich Hour Angle* and is the astronomer's equivalent of

1974 JUNE 27, 28, 29 (THURS., FRI., SAT.)

G.M.T.		SUN		MOON				
d	h	G.H.A. ° ′	Dec. ° ′	G.H.A. ° ′	v ′	Dec. ° ′	d ′	H.P. ′
27	00	179 17·6	N23 20·9	89 52·7	11·7	S 7 33·6	12·5	58·0
	01	194 17·5	20·8	104 23·4	11·7	7 46·1	12·6	58·0
	02	209 17·3	20·7	118 54·1	11·8	7 58·7	12·4	58·0
	03	224 17·2	· · 20·6	133 24·9	11·7	8 11·1	12·4	57·9
	04	239 17·1	20·5	147 55·6	11·7	8 23·5	12·4	57·9
	05	254 16·9	20·4	162 26·3	11·7	8 35·9	12·3	57·9
	06	269 16·8	N23 20·3	176 57·0	11·8	S 8 48·2	12·2	57·8
	07	284 16·7	20·2	191 27·8	11·7	9 00·4	12·2	57·8
T	08	299 16·6	20·1	205 58·5	11·7	9 12·6	12·1	57·8
H	09	314 16·4	· · 20·0	220 29·2	11·7	9 24·7	12·1	57·7
U	10	329 16·3	19·9	234 59·9	11·7	9 36·8	12·0	57·7
R	11	344 16·2	19·9	249 30·6	11·7	9 48·8	11·9	57·7
S	12	359 16·0	N23 19·8	264 01·3	11·7	S10 00·7	11·9	57·7
D	13	14 15·9	19·7	278 32·0	11·7	10 12·6	11·8	57·6
A	14	29 15·8	19·6	293 02·7	11·7	10 24·4	11·7	57·6
Y	15	44 15·6	· · 19·5	307 33·4	11·7	10 36·1	11·7	57·6
	16	59 15·5	19·4	322 04·1	11·6	10 47·8	11·6	57·5
	17	74 15·4	19·3	336 34·7	11·7	10 59·4	11·6	57·5
	18	89 15·3	N23 19·1	351 05·4	11·6	S11 11·0	11·5	57·5
	19	104 15·1	19·0	5 36·0	11·7	11 22·5	11·4	57·4
	20	119 15·0	18·9	20 06·7	11·6	11 33·9	11·3	57·4
	21	134 14·9	· · 18·8	34 37·3	11·6	11 45·2	11·3	57·4
	22	149 14·7	18·7	49 07·9	11·6	11 56·5	11·2	57·4
	23	164 14·6	18·6	63 38·5	11·6	12 07·7	11·1	57·3
28	00	179 14·5	N23 18·5	78 09·1	11·6	S12 18·8	11·1	57·3
	01	194 14·3	18·4	92 39·7	11·5	12 29·9	11·0	57·3
	02	209 14·2	18·3	107 10·2	11·6	12 40·9	10·9	57·2
	03	224 14·1	· · 18·2	121 40·8	11·5	12 51·8	10·8	57·2
	04	239 14·0	18·1	136 11·3	11·5	13 02·6	10·8	57·2
	05	254 13·8	18·0	150 41·8	11·5	13 13·4	10·7	57·2
	06	269 13·7	N23 17·9	165 12·3	11·5	S13 24·1	10·6	57·1
	07	284 13·6	17·8	179 42·8	11·5	13 34·7	10·5	57·1
	08	299 13·5	17·6	194 13·3	11·4	13 45·2	10·5	57·1
F	09	314 13·3	· · 17·5	208 43·7	11·4	13 55·7	10·4	57·0
R	10	329 13·2	17·4	223 14·1	11·4	14 06·1	10·2	57·0
I	11	344 13·1	17·3	237 44·5	11·4	14 16·3	10·3	57·0
D	12	359 12·9	N23 17·2	252 14·9	11·4	S14 26·6	10·1	57·0
A	13	14 12·8	17·1	266 45·3	11·4	14 36·7	10·0	56·9
Y	14	29 12·7	17·0	281 15·7	11·3	14 46·7	10·0	56·9
	15	44 12·6	· · 16·8	295 46·0	11·3	14 56·7	9·9	56·9
	16	59 12·4	16·7	310 16·3	11·3	15 06·6	9·8	56·8
	17	74 12·3	16·6	324 46·6	11·3	15 16·4	9·7	56·8
	18	89 12·2	N23 16·5	339 16·9	11·2	S15 26·1	9·6	56·8
	19	104 12·0	16·4	353 47·1	11·3	15 35·7	9·5	56·8
	20	119 11·9	16·2	8 17·4	11·2	15 45·2	9·5	56·7
	21	134 11·8	· · 16·1	22 47·6	11·2	15 54·7	9·4	56·7
	22	149 11·7	16·0	37 17·8	11·1	16 04·1	9·2	56·7
	23	164 11·5	15·9	51 47·9	11·2	16 13·3	9·2	56·7
29	00	179 11·4	N23 15·8	66 18·1	11·1	S16 22·5	9·1	56·6
	01	194 11·3	15·6	80 48·2	11·1	16 31·6	9·0	56·6
	02	209 11·2	15·5	95 18·3	11·1	16 40·6	8·9	56·6
	03	224 11·0	· · 15·4	109 48·4	11·0	16 49·5	8·8	56·5
	04	239 10·9	15·3	124 18·4	11·0	16 58·3	8·8	56·5
	05	254 10·8	15·1	138 48·4	11·0	17 07·1	8·6	56·5
	06	269 10·7	N23 15·0	153 18·4	11·0	S17 15·7	8·5	56·5
	07	284 10·5	14·9	167 48·4	11·0	17 24·2	8·5	56·4
S	08	299 10·4	14·7	182 18·4	10·9	17 32·7	8·3	56·4
A	09	314 10·3	· · 14·6	196 48·3	11·0	17 41·0	8·3	56·4
T	10	329 10·1	14·5	211 18·3	10·8	17 49·3	8·1	56·4
U	11	344 10·0	14·4	225 48·1	10·9	17 57·4	8·1	56·3
R	12	359 09·9	N23 14·2	240 18·0	10·9	S18 05·5	8·0	56·3
D	13	14 09·8	14·1	254 47·9	10·8	18 13·5	7·8	56·3
A	14	29 09·6	14·0	269 17·7	10·8	18 21·3	7·8	56·3
Y	15	44 09·5	· · 13·8	283 47·5	10·8	18 29·1	7·6	56·2
	16	59 09·4	13·7	298 17·3	10·7	18 36·7	7·6	56·2
	17	74 09·3	13·5	312 47·0	10·8	18 44·3	7·5	56·2
	18	89 09·1	N23 13·4	327 16·8	10·7	S18 51·8	7·3	56·2
	19	104 09·0	13·3	341 46·5	10·6	18 59·1	7·3	56·1
	20	119 08·9	13·1	356 16·1	10·7	19 06·4	7·2	56·1
	21	134 08·8	· · 13·0	10 45·8	10·7	19 13·6	7·0	56·1
	22	149 08·6	12·9	25 15·5	10·6	19 20·6	7·0	56·1
	23	164 08·5	12·7	39 45·1	10·6	19 27·6	6·8	56·1
		S.D. 15·8	*d* 0·1	S.D. 15·7		15·5		15·3

Lat.	Twilight Naut.	Twilight Civil	Sun-rise	Moonrise 27	Moonrise 28	Moonrise 29	Moonrise 30
°	h m	h m	h m	h m	h m	h m	h m
N 72	□	□	□	14 52	17 14	■	■
N 70	□	□	□	14 33	16 36	19 01	■
68	□	□	□	14 19	16 09	18 05	20 19
66	□	□	□	14 07	15 49	17 31	19 12
64	////	////	01 35	13 57	15 33	17 07	18 36
62	////	////	02 13	13 49	15 20	16 48	18 11
60	////	00 55	02 39	13 42	15 09	16 33	17 51
N 58	////	01 44	02 59	13 36	14 59	16 19	17 34
56	////	02 14	03 16	13 30	14 51	16 08	17 20
54	00 51	02 36	03 30	13 26	14 43	15 58	17 08
52	01 36	02 53	03 42	13 21	14 37	15 49	16 58
50	02 03	03 08	03 53	13 17	14 31	15 41	16 48
45	02 48	03 38	04 15	13 09	14 18	15 25	16 29
N 40	03 19	04 00	04 33	13 01	14 07	15 11	16 13
35	03 42	04 18	04 48	12 55	13 58	15 00	15 59
30	04 00	04 34	05 01	12 50	13 50	14 50	15 47
20	04 29	04 59	05 23	12 41	13 37	14 32	15 27
N 10	04 52	05 19	05 42	12 33	13 25	14 17	15 10
0	05 11	05 37	05 59	12 25	13 14	14 03	14 54
S 10	05 28	05 54	06 17	12 18	13 03	13 50	14 38
20	05 44	06 11	06 35	12 10	12 52	13 35	14 21
30	06 00	06 30	06 56	12 01	12 38	13 18	14 01
35	06 09	06 40	07 09	11 56	12 31	13 08	13 49
40	06 18	06 52	07 23	11 50	12 22	12 57	13 36
45	06 28	07 06	07 39	11 44	12 12	12 44	13 21
S 50	06 40	07 21	08 00	11 36	12 00	12 29	13 02
52	06 45	07 29	08 10	11 32	11 55	12 21	12 53
54	06 51	07 37	08 21	11 28	11 49	12 13	12 43
56	06 57	07 46	08 33	11 24	11 42	12 04	12 32
58	07 04	07 56	08 48	11 19	11 34	11 54	12 19
S 60	07 11	08 08	09 05	11 13	11 26	11 42	12 04

Lat.	Sun-set	Twilight Civil	Twilight Naut.	Moonset 27	Moonset 28	Moonset 29	Moonset 30
°	h m	h m	h m	h m	h m	h m	h m
N 72	□	□	□	21 59	21 19	■	■
N 70	□	□	□	22 19	21 59	21 17	■
68	□	□	□	22 35	22 27	22 15	21 46
66	□	□	□	22 49	22 48	22 49	22 54
64	22 29	////	////	23 00	23 05	23 14	23 30
62	21 53	////	////	23 09	23 19	23 33	23 56
60	21 27	23 09	////	23 17	23 31	23 50	24 16
N 58	21 07	22 21	////	23 24	23 41	24 03	00 03
56	20 50	21 52	////	23 31	23 50	24 15	00 15
54	20 36	21 30	23 14	23 36	23 58	24 25	00 25
52	20 24	21 12	22 30	23 41	24 06	00 06	00 35
50	20 13	20 57	22 02	23 46	24 12	00 12	00 43
45	19 51	20 28	21 18	23 56	24 26	00 26	01 00
N 40	19 33	20 06	20 47	24 05	00 05	00 38	01 15
35	19 18	19 48	20 24	24 12	00 12	00 48	01 27
30	19 05	19 32	20 06	24 18	00 18	00 57	01 38
20	18 43	19 08	19 37	24 29	00 29	01 12	01 56
N 10	18 24	18 47	19 14	24 39	00 39	01 25	02 12
0	18 07	18 29	18 55	00 00	00 48	01 38	02 28
S 10	17 49	18 12	18 39	00 05	00 58	01 50	02 43
20	17 31	17 55	18 22	00 11	01 08	02 03	02 59
30	17 10	17 36	18 06	00 18	01 19	02 19	03 17
35	16 58	17 26	17 57	00 22	01 25	02 28	03 28
40	16 44	17 14	17 48	00 27	01 33	02 38	03 41
45	16 27	17 01	17 38	00 32	01 42	02 50	03 56
S 50	16 06	16 45	17 26	00 38	01 52	03 05	04 14
52	15 56	16 37	17 21	00 41	01 57	03 12	04 22
54	15 45	16 29	17 15	00 44	02 03	03 19	04 32
56	15 33	16 20	17 09	00 47	02 09	03 28	04 43
58	15 18	16 10	17 02	00 51	02 16	03 37	04 55
S 60	15 01	15 58	16 55	00 56	02 23	03 49	05 10

Day	SUN Eqn. of Time 00h	SUN Eqn. of Time 12h	SUN Mer. Pass.	MOON Mer. Pass. Upper	MOON Mer. Pass. Lower	MOON Age	MOON Phase
	m s	m s	h m	h m	h m	d	
27	02 49	02 56	12 03	18 37	06 13	07	
28	03 02	03 08	12 03	19 26	07 01	08	
29	03 14	03 20	12 03	20 15	07 50	09	

Table 4

longitude. Look at the entry for midnight (00 hours) on June 28. You see that GHA is 179°14.5′, and declination is N 23°18.5′. Translated, this means that at midnight GMT the sun's geographical position was 179°14.5′ west longitude and 23°18.5′ north latitude. (That is a spot in the Pacific Ocean near Midway Island.)

If you look down at the next few rows of the GHA column, you will see that GHA continues to increase. As you know, longitude is measured east and west from Greenwich 180° each way. GHA, however, is only measured westward from Greenwich, so there is no exact equivalence between GHA and longitude except in the Western Hemisphere.

In Figures 7.3 and 7.4 you can see that a GHA of 270° is equal to 90° east longitude and that a GHA of 300° is equal to 60° east longitude. GHA is a navigational convention adopted to make celestial navigation easier. If you look down the GHA column, you will see that GHA increases until at 1200 hours (noon) the sun's angle has gone through nearly 360° and is back at Greenwich, at which time another cycle starts. Thus, GHA at 1300 hours (1 P.M.) is 14°12.8′.

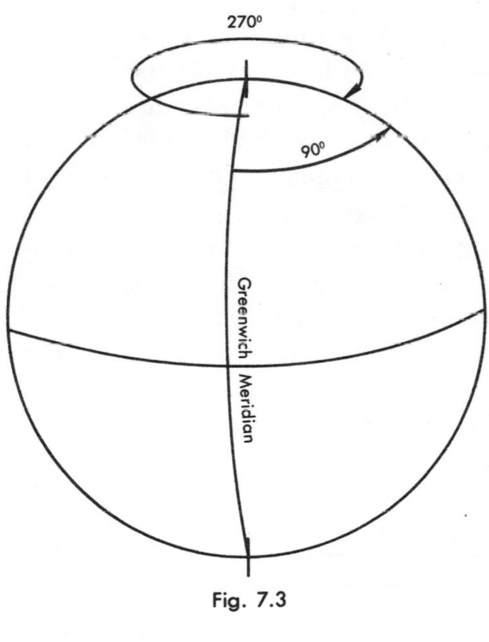

Fig. 7.3

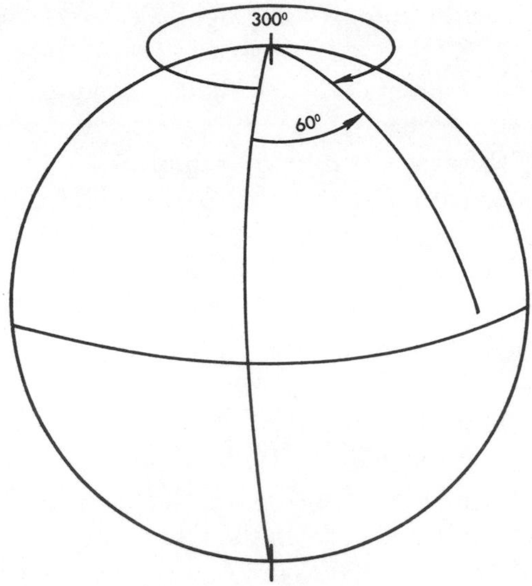

Fig. 7.4

Now proceed to use the almanac to get the data you need to continue working your sight. GMT is 23-12-08, June 28, 1974. From the almanac page for June 28 (Table 4), at 2300 hours the sun's GHA was 164°11.5′ and its declination was N 23°15.9′.

Now you must figure out how much farther west the sun moved in 12 minutes and 8 seconds. The sun moves through 15° of longitude (or GHA) in one hour, and in the back of the almanac is a table that gives the calculation for odd minutes and seconds. It is our Table 5.

In the top left-hand corner you will see that the table on the left covers increases in GHA for 12 minutes 00 seconds to 12 minutes 60 seconds. If you look down the left-hand column to 08 seconds and across to the first column on the right headed "Sun [and] Planets," you find that in 12 minutes and 8 seconds the GHA of the sun increases 3°02.0′.

Therefore, what you have is this:

SUN			
GHA 23 Hr.	=	164°11.5′	Dec. = N 23°15.9′
12 Min. 08 Sec.	=	3°02.0′	
GHA	=	167°13.5′	

12ᵐ INCREMENTS AND CORRECTIONS 13ᵐ

12ᵐ	SUN PLANETS	ARIES	MOON	v or d	Corrⁿ	v or d	Corrⁿ	v or d	Corrⁿ
s	° ′	° ′	° ′	′	′	′	′	′	′
00	3 00·0	3 00·5	2 51·8	0·0	0·0	6·0	1·3	12·0	2·5
01	3 00·3	3 00·7	2 52·0	0·1	0·0	6·1	1·3	12·1	2·5
02	3 00·5	3 01·0	2 52·3	0·2	0·0	6·2	1·3	12·2	2·5
03	3 00·8	3 01·2	2 52·5	0·3	0·1	6·3	1·3	12·3	2·6
04	3 01·0	3 01·5	2 52·8	0·4	0·1	6·4	1·3	12·4	2·6
05	3 01·3	3 01·7	2 53·0	0·5	0·1	6·5	1·4	12·5	2·6
06	3 01·5	3 02·0	2 53·2	0·6	0·1	6·6	1·4	12·6	2·6
07	3 01·8	3 02·2	2 53·5	0·7	0·1	6·7	1·4	12·7	2·6
08	3 02·0	3 02·5	2 53·7	0·8	0·2	6·8	1·4	12·8	2·7
09	3 02·3	3 02·7	2 53·9	0·9	0·2	6·9	1·4	12·9	2·7
10	3 02·5	3 03·0	2 54·2	1·0	0·2	7·0	1·5	13·0	2·7
11	3 02·8	3 03·3	2 54·4	1·1	0·2	7·1	1·5	13·1	2·7
12	3 03·0	3 03·5	2 54·7	1·2	0·3	7·2	1·5	13·2	2·8
13	3 03·3	3 03·8	2 54·9	1·3	0·3	7·3	1·5	13·3	2·8
14	3 03·5	3 04·0	2 55·1	1·4	0·3	7·4	1·5	13·4	2·8
15	3 03·8	3 04·3	2 55·4	1·5	0·3	7·5	1·6	13·5	2·8
16	3 04·0	3 04·5	2 55·6	1·6	0·3	7·6	1·6	13·6	2·8
17	3 04·3	3 04·8	2 55·9	1·7	0·4	7·7	1·6	13·7	2·9
18	3 04·5	3 05·0	2 56·1	1·8	0·4	7·8	1·6	13·8	2·9
19	3 04·8	3 05·3	2 56·3	1·9	0·4	7·9	1·6	13·9	2·9
20	3 05·0	3 05·5	2 56·6	2·0	0·4	8·0	1·7	14·0	2·9
21	3 05·3	3 05·8	2 56·8	2·1	0·4	8·1	1·7	14·1	2·9
22	3 05·5	3 06·0	2 57·0	2·2	0·5	8·2	1·7	14·2	3·0
23	3 05·8	3 06·3	2 57·3	2·3	0·5	8·3	1·7	14·3	3·0
24	3 06·0	3 06·5	2 57·5	2·4	0·5	8·4	1·8	14·4	3·0
25	3 06·3	3 06·8	2 57·8	2·5	0·5	8·5	1·8	14·5	3·0
26	3 06·5	3 07·0	2 58·0	2·6	0·5	8·6	1·8	14·6	3·0
27	3 06·8	3 07·3	2 58·2	2·7	0·6	8·7	1·8	14·7	3·1
28	3 07·0	3 07·5	2 58·5	2·8	0·6	8·8	1·8	14·8	3·1
29	3 07·3	3 07·8	2 58·7	2·9	0·6	8·9	1·9	14·9	3·1
30	3 07·5	3 08·0	2 59·0	3·0	0·6	9·0	1·9	15·0	3·1
31	3 07·8	3 08·3	2 59·2	3·1	0·6	9·1	1·9	15·1	3·1
32	3 08·0	3 08·5	2 59·4	3·2	0·7	9·2	1·9	15·2	3·2
33	3 08·3	3 08·8	2 59·7	3·3	0·7	9·3	1·9	15·3	3·2
34	3 08·5	3 09·0	2 59·9	3·4	0·7	9·4	2·0	15·4	3·2
35	3 08·8	3 09·3	3 00·2	3·5	0·7	9·5	2·0	15·5	3·2
36	3 09·0	3 09·5	3 00·4	3·6	0·8	9·6	2·0	15·6	3·3
37	3 09·3	3 09·8	3 00·6	3·7	0·8	9·7	2·0	15·7	3·3
38	3 09·5	3 10·0	3 00·9	3·8	0·8	9·8	2·0	15·8	3·3
39	3 09·8	3 10·3	3 01·1	3·9	0·8	9·9	2·1	15·9	3·3
40	3 10·0	3 10·5	3 01·3	4·0	0·8	10·0	2·1	16·0	3·3
41	3 10·3	3 10·8	3 01·6	4·1	0·9	10·1	2·1	16·1	3·4
42	3 10·5	3 11·0	3 01·8	4·2	0·9	10·2	2·1	16·2	3·4
43	3 10·8	3 11·3	3 02·1	4·3	0·9	10·3	2·1	16·3	3·4
44	3 11·0	3 11·5	3 02·3	4·4	0·9	10·4	2·2	16·4	3·4
45	3 11·3	3 11·8	3 02·5	4·5	0·9	10·5	2·2	16·5	3·4
46	3 11·5	3 12·0	3 02·8	4·6	1·0	10·6	2·2	16·6	3·5
47	3 11·8	3 12·3	3 03·0	4·7	1·0	10·7	2·2	16·7	3·5
48	3 12·0	3 12·5	3 03·3	4·8	1·0	10·8	2·3	16·8	3·5
49	3 12·3	3 12·8	3 03·5	4·9	1·0	10·9	2·3	16·9	3·5
50	3 12·5	3 13·0	3 03·7	5·0	1·0	11·0	2·3	17·0	3·5
51	3 12·8	3 13·3	3 04·0	5·1	1·1	11·1	2·3	17·1	3·6
52	3 13·0	3 13·5	3 04·2	5·2	1·1	11·2	2·3	17·2	3·6
53	3 13·3	3 13·8	3 04·4	5·3	1·1	11·3	2·4	17·3	3·6
54	3 13·5	3 14·0	3 04·7	5·4	1·1	11·4	2·4	17·4	3·6
55	3 13·8	3 14·3	3 04·9	5·5	1·1	11·5	2·4	17·5	3·6
56	3 14·0	3 14·5	3 05·2	5·6	1·2	11·6	2·4	17·6	3·7
57	3 14·3	3 14·8	3 05·4	5·7	1·2	11·7	2·4	17·7	3·7
58	3 14·5	3 15·0	3 05·6	5·8	1·2	11·8	2·5	17·8	3·7
59	3 14·8	3 15·3	3 05·9	5·9	1·2	11·9	2·5	17·9	3·7
60	3 15·0	3 15·5	3 06·1	6·0	1·3	12·0	2·5	18·0	3·8

13ᵐ	SUN PLANETS	ARIES	MOON	v or d	Corrⁿ	v or d	Corrⁿ	v or d	Corrⁿ
s	° ′	° ′	° ′	′	′	′	′	′	′
00	3 15·0	3 15·5	3 06·1	0·0	0·0	6·0	1·4	12·0	2·7
01	3 15·3	3 15·8	3 06·4	0·1	0·0	6·1	1·4	12·1	2·7
02	3 15·5	3 16·0	3 06·6	0·2	0·0	6·2	1·4	12·2	2·7
03	3 15·8	3 16·3	3 06·8	0·3	0·1	6·3	1·4	12·3	2·8
04	3 16·0	3 16·5	3 07·1	0·4	0·1	6·4	1·4	12·4	2·8
05	3 16·3	3 16·8	3 07·3	0·5	0·1	6·5	1·5	12·5	2·8
06	3 16·5	3 17·0	3 07·5	0·6	0·1	6·6	1·5	12·6	2·8
07	3 16·8	3 17·3	3 07·8	0·7	0·2	6·7	1·5	12·7	2·9
08	3 17·0	3 17·5	3 08·0	0·8	0·2	6·8	1·5	12·8	2·9
09	3 17·3	3 17·8	3 08·3	0·9	0·2	6·9	1·6	12·9	2·9
10	3 17·5	3 18·0	3 08·5	1·0	0·2	7·0	1·6	13·0	2·9
11	3 17·8	3 18·3	3 08·7	1·1	0·2	7·1	1·6	13·1	2·9
12	3 18·0	3 18·5	3 09·0	1·2	0·3	7·2	1·6	13·2	3·0
13	3 18·3	3 18·8	3 09·2	1·3	0·3	7·3	1·6	13·3	3·0
14	3 18·5	3 19·0	3 09·5	1·4	0·3	7·4	1·7	13·4	3·0
15	3 18·8	3 19·3	3 09·7	1·5	0·3	7·5	1·7	13·5	3·0
16	3 19·0	3 19·5	3 09·9	1·6	0·4	7·6	1·7	13·6	3·1
17	3 19·3	3 19·8	3 10·2	1·7	0·4	7·7	1·7	13·7	3·1
18	3 19·5	3 20·0	3 10·4	1·8	0·4	7·8	1·8	13·8	3·1
19	3 19·8	3 20·3	3 10·7	1·9	0·4	7·9	1·8	13·9	3·1
20	3 20·0	3 20·5	3 10·9	2·0	0·5	8·0	1·8	14·0	3·2
21	3 20·3	3 20·8	3 11·1	2·1	0·5	8·1	1·8	14·1	3·2
22	3 20·5	3 21·0	3 11·4	2·2	0·5	8·2	1·8	14·2	3·2
23	3 20·8	3 21·3	3 11·6	2·3	0·5	8·3	1·9	14·3	3·2
24	3 21·0	3 21·6	3 11·8	2·4	0·5	8·4	1·9	14·4	3·2
25	3 21·3	3 21·8	3 12·1	2·5	0·6	8·5	1·9	14·5	3·3
26	3 21·5	3 22·1	3 12·3	2·6	0·6	8·6	1·9	14·6	3·3
27	3 21·8	3 22·3	3 12·6	2·7	0·6	8·7	2·0	14·7	3·3
28	3 22·0	3 22·6	3 12·8	2·8	0·6	8·8	2·0	14·8	3·3
29	3 22·3	3 22·8	3 13·0	2·9	0·7	8·9	2·0	14·9	3·4
30	3 22·5	3 23·1	3 13·3	3·0	0·7	9·0	2·0	15·0	3·4
31	3 22·8	3 23·3	3 13·5	3·1	0·7	9·1	2·0	15·1	3·4
32	3 23·0	3 23·6	3 13·8	3·2	0·7	9·2	2·1	15·2	3·4
33	3 23·3	3 23·8	3 14·0	3·3	0·7	9·3	2·1	15·3	3·4
34	3 23·5	3 24·1	3 14·2	3·4	0·8	9·4	2·1	15·4	3·5
35	3 23·8	3 24·3	3 14·5	3·5	0·8	9·5	2·1	15·5	3·5
36	3 24·0	3 24·6	3 14·7	3·6	0·8	9·6	2·2	15·6	3·5
37	3 24·3	3 24·8	3 14·9	3·7	0·8	9·7	2·2	15·7	3·5
38	3 24·5	3 25·1	3 15·2	3·8	0·9	9·8	2·2	15·8	3·6
39	3 24·8	3 25·3	3 15·4	3·9	0·9	9·9	2·2	15·9	3·6
40	3 25·0	3 25·6	3 15·7	4·0	0·9	10·0	2·3	16·0	3·6
41	3 25·3	3 25·8	3 15·9	4·1	0·9	10·1	2·3	16·1	3·6
42	3 25·5	3 26·1	3 16·1	4·2	0·9	10·2	2·3	16·2	3·6
43	3 25·8	3 26·3	3 16·4	4·3	1·0	10·3	2·3	16·3	3·7
44	3 26·0	3 26·6	3 16·6	4·4	1·0	10·4	2·3	16·4	3·7
45	3 26·3	3 26·8	3 16·9	4·5	1·0	10·5	2·4	16·5	3·7
46	3 26·5	3 27·1	3 17·1	4·6	1·0	10·6	2·4	16·6	3·7
47	3 26·8	3 27·3	3 17·3	4·7	1·1	10·7	2·4	16·7	3·8
48	3 27·0	3 27·6	3 17·6	4·8	1·1	10·8	2·4	16·8	3·8
49	3 27·3	3 27·8	3 17·8	4·9	1·1	10·9	2·5	16·9	3·8
50	3 27·5	3 28·1	3 18·0	5·0	1·1	11·0	2·5	17·0	3·8
51	3 27·8	3 28·3	3 18·3	5·1	1·1	11·1	2·5	17·1	3·8
52	3 28·0	3 28·6	3 18·5	5·2	1·2	11·2	2·5	17·2	3·9
53	3 28·3	3 28·8	3 18·8	5·3	1·2	11·3	2·5	17·3	3·9
54	3 28·5	3 29·1	3 19·0	5·4	1·2	11·4	2·6	17·4	3·9
55	3 28·8	3 29·3	3 19·2	5·5	1·2	11·5	2·6	17·5	3·9
56	3 29·0	3 29·6	3 19·5	5·6	1·3	11·6	2·6	17·6	4·0
57	3 29·3	3 29·8	3 19·7	5·7	1·3	11·7	2·6	17·7	4·0
58	3 29·5	3 30·1	3 20·0	5·8	1·3	11·8	2·7	17·8	4·0
59	3 29·8	3 30·3	3 20·2	5·9	1·3	11·9	2·7	17·9	4·0
60	3 30·0	3 30·6	3 20·4	6·0	1·4	12·0	2·7	18·0	4·1

Table 5

You will notice that the declination of the sun doesn't change very fast—only one-tenth of a minute in an hour, so I don't bother figuring out the part of the one-tenth that is covered in 12 minutes. In fact, for practical purposes on a small boat, I would round off the numbers and fill out the form this way:

SUN			
GHA 23 Hr.	=	164°12′	Dec. N 23°16′
12 Min. 08 Sec.	=	3°02′	
GHA	=	167°14′	

I have found in practice that this is sufficiently accurate, and not having to mess around with decimal points reduces the chances of an arithmetical blunder—to which I am prone, unfortunately.

Now let's look at the *Air Almanac* and use the same hypothetical sight so that you will appreciate the differences between the air and nautical almanacs. The principal difference is that the *Air Almanac* gives the GHA for each 10-minute segment of every day, so the interpolation for odd minutes and seconds is quicker. Also, instead of giving one main correction for the sun, the *Air Almanac* tabulates refraction and semidiameter separately, and the corrections must be applied separately.

Turning, then, to the *Air Almanac* tables, you will find the correction for dip on the back cover (Table 6) and the correction for refrac-

CORRECTIONS TO BE APPLIED TO MARINE SEXTANT ALTITUDES

MARINE SEXTANT ERROR	CORRECTIONS
Sextant No. Index Error	In addition to sextant error and dip, corrections are to be applied for: Refraction Semi-diameter (for the Sun and Moon) Parallax (for the Moon) Dome refraction (if applicable)

CORRECTION FOR DIP OF THE HORIZON
To be subtracted from sextant altitude

Ht.	Dip	Ht.	Dip	Ht.	Dip	Ht.	Dip	Ht.	Dip
Ft.	′	Ft.	′	Ft.	′	Ft.	′	Ft.	′
0		114		437		968		1,707	
	1		11		21		31		41
2		137		481		1,033		1,792	
	2		12		22		32		42
6		162		527		1,099		1,880	
	3		13		23		33		43
12		189		575		1,168		1,970	
	4		14		24		34		44
21		218		625		1,239		2,061	
	5		15		25		35		45
31		250		677		1,311		2,155	
	6		16		26		36		46
43		283		731		1,386		2,251	
	7		17		27		37		47
58		318		787		1,463		2,349	
	8		18		28		38		48
75		356		845		1,543		2,449	
	9		19		29		39		49
93		395		906		1,624		2,551	
	10		20		30		40		50
114		437		968		1,707		2,655	

Table 6

tion on the inside back cover (Table 7). From Table 7, the refraction (Ro) at sea level (height of 0) for a sextant altitude between 33° and 63° is 1′. Refraction always makes the body appear higher, so you correct by subtracting.

Turning to the page for June 28, 1974 (Table 8), in the bottom right-hand corner is the sun's semidiameter—15.8′—which you can round to 16′. This correction is added. Therefore, the form would look like this:

Sextant Reading (Hs)	=	35°	16′
Dip	=	−	08′
Index Correction (IC)	=	+	05′
Apparent Altitude (ha)	=	35°	13′
Semidiameter	=	+	16′
Refraction	=	−	01′
Observed Altitude (Ho)	=	35°	28′

This is the same result that you got with the *Nautical Almanac.* Now if you look at the column labeled "Sun," you find that at 23-10-00 hours GMT the sun's GHA was 166°41.6′ and declination was 23°15.9′ N. The time of the sight was 23-12-08, so you need the sun's increase in GHA for 2 minutes and 8 seconds. The table for this is on the inside front cover, Table 9. The increment, located between 0205 and 0209 is 32′. Therefore, the form we are using would look like this:

GHA 23 Hr., 10 Min.	=	166°41.6′		
2 Min. 08 Sec.	=	32.0′		
GHA	=	166°73.6′	=	167°13.6′

The difference between this and the result using the *Nautical Almanac* is due to the differences in the GHA interpolation tables of the two books. It is negligible.

CORRECTIONS TO BE APPLIED TO SEXTANT ALTITUDE

REFRACTION

To be subtracted from sextant altitude (referred to as observed altitude in A.P. 3270)

R_o	Height above sea level in units of 1,000 ft. — Sextant Altitude: 0	5	10	15	20	25	30	35	40	45	50	55	R_o	$R = R_o \times f$; f: 0·9	1·0	1·1	1·2
′	° ′	° ′	° ′	° ′	° ′	° ′	° ′	° ′	° ′	° ′	° ′	° ′	′	R ′	′	′	′
	90	90	90	90	90	90	90	90	90	90	90	90					
0													0	0	0	0	0
	63	59	55	51	46	41	36	31	26	20	17	13					
1													1	1	1	1	1
	33	29	26	22	19	16	14	11	9	7	6	4					
2													2	2	2	2	2
	21	19	16	14	12	10	8	7	5	4	2 40	1 40					
3													3	3	3	3	4
	16	14	12	10	8	7	6	5	3 10	2 20	1 30	0 40					
4													4	4	4	4	5
	12	11	9	8	7	5	4 00	3 10	2 10	1 30	0 39	+0 05					
5													5	5	5	5	6
	10	9	7	5 50	4 50	3 50	3 10	2 20	1 30	0 49	+0 11	−0 19					
6													6	5	6	7	7
	8 10	6 50	5 50	4 50	4 00	3 00	2 20	1 50	1 10	0 24	−0 11	−0 38					
7													7	6	7	8	8
	6 50	5 50	5 00	4 00	3 10	2 30	1 50	1 20	0 38	+0 04	−0 28	−0 54					
8													8	7	8	9	10
	6 00	5 10	4 10	3 20	2 40	2 00	1 30	1 00	0 19	−0 13	−0 42	−1 08					
9													9	8	9	10	11
	5 20	4 30	3 40	2 50	2 10	1 40	1 10	0 35	+0 03	−0 27	−0 53	−1 18					
10													10	9	10	11	12
	4 30	3 40	2 50	2 20	1 40	1 10	0 37	+0 11	−0 16	−0 43	−1 08	−1 31					
12													12	11	12	13	14
	3 30	2 50	2 10	1 40	1 10	0 34	+0 09	−0 14	−0 37	−1 00	−1 23	−1 44					
14													14	13	14	15	17
	2 50	2 10	1 40	1 10	0 37	+0 10	−0 13	−0 34	−0 53	−1 14	−1 35	−1 56					
16													16	14	16	18	19
	2 20	1 40	1 20	0 43	+0 15	−0 08	−0 31	−0 52	−1 08	−1 27	−1 46	−2 05					
18													18	16	18	20	22
	1 50	1 20	0 49	+0 23	−0 02	−0 26	−0 46	−1 06	−1 22	−1 39	−1 57	−2 14					
20													20	18	20	22	24
	1 12	0 44	+0 19	−0 06	−0 28	−0 48	−1 09	−1 27	−1 42	−1 58	−2 14	−2 30					
25													25	22	25	28	30
	0 34	+0 10	−0 13	−0 36	−0 55	−1 14	−1 32	−1 51	−2 06	−2 21	−2 34	−2 49					
30													30	27	30	33	36
	+0 06	−0 16	−0 37	−0 59	−1 17	−1 33	−1 51	−2 07	−2 23	−2 37	−2 51	−3 04					
35													35	31	35	38	42
	−0 18	−0 37	−0 58	−1 16	−1 34	−1 49	−2 06	−2 22	−2 35	−2 49	−3 03	−3 16					
40													40	36	40	44	48
		−0 53	−1 14	−1 31	−1 47	−2 03	−2 18	−2 33	−2 47	−2 59	−3 13	−3 25					
45													45	40	45	50	54
		−1 10	−1 28	−1 44	−1 59	−2 15	−2 28	−2 43	−2 56	−3 08	−3 22	−3 33					
50													50	45	50	55	60
			−1 40	−1 53	−2 09	−2 24	−2 38	−2 52	−3 04	−3 17	−3 29	−3 41					
55													55	49	55	60	66
				−2 03	−2 18	−2 33	−2 46	−3 01	−3 12	−3 25	−3 37	−3 48					
60													60	54	60	66	72
							−2 53	−3 07	−3 19	−3 31	−3 42	−3 53					

f	Temperature in °C. — 0	5	10	15	20	25	30	35	40	45	50	55	f
	+47	+36	+27	+18	+10	+ 3	− 5	−13					
0·9													0·9
	+26	+16	+ 6	− 4	−13	−22	−31	−40					
1·0													1·0
	+ 5	− 5	−15	−25	−36	−46	−57	−68					
1·1													1·1
	−16	−25	−36	−46	−58	−71	−83	−95					
1·2													1·2
	−37	−45	−56	−67	−81	−95							

(Heights 45–55:) For these heights no temperature correction is necessary, so use $R = R_o$

(Right-hand table, f = 0·9, 1·0, 1·1, 1·2:) Where R_o is less than 10′ or the height greater than 35,000 ft. use $R = R_o$

Choose the column appropriate to height, in units of 1,000 ft., and find the range of altitude in which the sextant altitude lies; the corresponding value of R_o is the refraction, to be subtracted from sextant altitude, unless conditions are extreme. In that case find f from the lower table, with critical argument temperature. Use the table on the right to form the refraction, $R = R_o \times f$.

CORIOLIS (Z) CORRECTION

To be applied by moving the position line a distance Z to starboard (right) of the track in northern latitudes and to port (left) in southern latitudes.

G/S KNOTS	Latitude 0°	10°	20°	30°	40°	50°	60°	70°	80°	90°	G/S KNOTS	Latitude 0°	10°	20°	30°	40°	50°	60°	70°	80°	90°
	′	′	′	′	′	′	′	′	′	′		′	′	′	′	′	′	′	′	′	′
150	0	1	1	2	3	3	3	4	4	4	550	0	3	5	7	9	11	12	14	14	14
200	0	1	2	3	3	4	5	5	5	5	600	0	3	5	8	10	12	14	15	16	16
250	0	1	2	3	4	5	6	6	6	7	650	0	3	6	9	11	13	15	16	17	17
300	0	1	3	4	5	6	7	7	8	8	700	0	3	6	9	12	14	16	17	18	18
350	0	2	3	5	6	7	8	9	9	9	750	0	3	7	10	13	15	17	18	19	20
400	0	2	4	5	7	8	9	10	10	10	800	0	4	7	10	13	16	18	20	21	21
450	0	2	4	6	8	9	10	11	12	12	850	0	4	8	11	14	17	19	21	22	22
500	0	2	4	7	8	10	11	12	13	13	900	0	4	8	12	15	18	20	22	23	24

Table 7

(DAY 179) GREENWICH P. M. 1974 JUNE 28 (FRIDAY)

GMT	☉ SUN GHA	SUN Dec.	ARIES GHA ♈	VENUS −3.4 GHA	VENUS Dec.	MARS 2.0 GHA	MARS Dec.	JUPITER −2.2 GHA	JUPITER Dec.	☾ MOON GHA	MOON Dec.
h m	° ′	° ′	° ′	° ′	° ′	° ′	° ′	° ′	° ′	° ′	° ′
12 00	359 13.0	N23 17.2	96 12.2	34 56	N19 06	321 26	N18 25	107 01	S 5 57	252 15	S14 27
10	1 43.0	17.2	98 42.6	37 26		323 56		109 32		254 40	29
20	4 13.0	17.2	101 13.0	39 56		326 26		112 02		257 05	31
30	6 42.9	· 17.1	103 43.4	42 26	· ·	328 56	· ·	114 32	· ·	259 30	· 32
40	9 12.9	17.1	106 13.8	44 56		331 26		117 03		261 55	34
50	11 42.9	17.1	108 44.2	47 26		333 57		119 33		264 20	36
13 00	14 12.9	N23 17.1	111 14.6	49 56	N19 06	336 27	N18 24	122 04	S 5 57	266 45	S14 38
10	16 42.9	17.1	113 45.0	52 26		338 57		124 34		269 11	39
20	19 12.8	17.0	116 15.4	54 56		341 27		127 04		271 36	41
30	21 42.8	· 17.0	118 45.9	57 25	· ·	343 57	· ·	129 35	· ·	274 01	· 43
40	24 12.8	17.0	121 16.3	59 55		346 27		132 05		276 26	44
50	26 42.8	17.0	123 46.7	62 25		348 57		134 36		278 51	46
14 00	29 12.7	N23 17.0	126 17.1	64 55	N19 07	351 28	N18 24	137 06	S 5 57	281 16	S14 48
10	31 42.7	16.9	128 47.5	67 25		353 58		139 36		283 41	49
20	34 12.7	16.9	131 17.9	69 55		356 28		142 07		286 06	51
30	36 42.7	· 16.9	133 48.3	72 25	· ·	358 58	· ·	144 37	· ·	288 31	· 53
40	39 12.7	16.9	136 18.7	74 55		1 28		147 08		290 56	54
50	41 42.6	16.9	138 49.1	77 25		3 58		149 38		293 21	56
15 00	44 12.6	N23 16.8	141 19.5	79 55	N19 08	6 29	N18 24	152 08	S 5 57	295 46	S14 58
10	46 42.6	16.8	143 50.0	82 24		8 59		154 39		298 11	14 59
20	49 12.6	16.8	146 20.4	84 54		11 29		157 09		300 36	15 01
30	51 42.6	· 16.8	148 50.8	87 24	· ·	13 59	· ·	159 40	· ·	303 01	· 02
40	54 12.5	16.8	151 21.2	89 54		16 29		162 10		305 26	04
50	56 42.5	16.7	153 51.6	92 24		18 59		164 40		307 51	06
16 00	59 12.5	N23 16.7	156 22.0	94 54	N19 08	21 29	N18 23	167 11	S 5 57	310 16	S15 07
10	61 42.5	16.7	158 52.4	97 24		24 00		169 41		312 41	09
20	64 12.5	16.7	161 22.8	99 54		26 30		172 12		315 07	11
30	66 42.4	· 16.7	163 53.2	102 24	· ·	29 00	· ·	174 42	· ·	317 32	12
40	69 12.4	16.6	166 23.7	104 54		31 30		177 12		319 57	14
50	71 42.4	16.6	168 54.1	107 23		34 00		179 43		322 22	16
17 00	74 12.4	N23 16.6	171 24.5	109 53	N19 09	36 30	N18 23	182 13	S 5 57	324 47	S15 17
10	76 42.3	16.6	173 54.9	112 23		39 00		184 44		327 12	19
20	79 12.3	16.6	176 25.3	114 53		41 31		187 14		329 37	20
30	81 42.3	· 16.5	178 55.7	117 23	· ·	44 01	· ·	189 44	· ·	332 02	· 22
40	84 12.3	16.5	181 26.1	119 53		46 31		192 15		334 27	24
50	86 42.3	16.5	183 56.5	122 23		49 01		194 45		336 52	25
18 00	89 12.2	N23 16.5	186 26.9	124 53	N19 10	51 31	N18 22	197 16	S 5 56	339 17	S15 27
10	91 42.2	16.5	188 57.4	127 23		54 01		199 46		341 42	29
20	94 12.2	16.4	191 27.8	129 53		56 32		202 16		344 07	30
30	96 42.2	· 16.4	193 58.2	132 22	· ·	59 02	· ·	204 47	· ·	346 32	· 32
40	99 12.2	16.4	196 28.6	134 52		61 32		207 17		348 57	33
50	101 42.1	16.4	198 59.0	137 22		64 02		209 48		351 22	35
19 00	104 12.1	N23 16.4	201 29.4	139 52	N19 10	66 32	N18 22	212 18	S 5 56	353 47	S15 36
10	106 42.1	16.3	203 59.8	142 22		69 02		214 48		356 12	38
20	109 12.1	16.3	206 30.2	144 52		71 32		217 19		358 37	40
30	111 42.0	· 16.3	209 00.6	147 22	· ·	74 03	· ·	219 49	· ·	1 02	· 41
40	114 12.0	16.3	211 31.0	149 52		76 33		222 20		3 27	43
50	116 42.0	16.3	214 01.5	152 22		79 03		224 50		5 52	44
20 00	119 12.0	N23 16.2	216 31.9	154 52	N19 11	81 33	N18 21	227 20	S 5 56	8 18	S15 46
10	121 42.0	16.2	219 02.3	157 22		84 03		229 51		10 43	48
20	124 11.9	16.2	221 32.7	159 51		86 33		232 21		13 08	49
30	126 41.9	· 16.2	224 03.1	162 21	· ·	89 03	· ·	234 52	· ·	15 33	· 51
40	129 11.9	16.2	226 33.5	164 51		91 34		237 22		17 58	52
50	131 41.9	16.1	229 03.9	167 21		94 04		239 52		20 23	54
21 00	134 11.9	N23 16.1	231 34.3	169 51	N19 12	96 34	N18 21	242 23	S 5 56	22 48	S15 55
10	136 41.8	16.1	234 04.7	172 21		99 04		244 53		25 13	57
20	139 11.8	16.1	236 35.2	174 51		101 34		247 24		27 38	15 59
30	141 41.8	· 16.1	239 05.6	177 21	· ·	104 04	· ·	249 54	· ·	30 03	16 00
40	144 11.8	16.0	241 36.0	179 51		106 35		252 24		32 28	02
50	146 41.8	16.0	244 06.4	182 21		109 05		254 55		34 53	03
22 00	149 11.7	N23 16.0	246 36.8	184 50	N19 12	111 35	N18 20	257 25	S 5 56	37 18	S16 05
10	151 41.7	16.0	249 07.2	187 20		114 05		259 56		39 43	06
20	154 11.7	16.0	251 37.6	189 50		116 35		262 26		42 08	08
30	156 41.7	· 15.9	254 08.0	192 20	· ·	119 05	· ·	264 56	· ·	44 33	· 09
40	159 11.6	15.9	256 38.4	194 50		121 36		267 27		46 58	11
50	161 41.6	15.9	259 08.9	197 20		124 06		269 57		49 23	13
23 00	164 11.6	N23 15.9	261 39.3	199 50	N19 13	126 36	N18 20	272 28	S 5 56	51 48	S16 14
10	166 41.6	15.9	264 09.7	202 20		129 06		274 58		54 13	16
20	169 11.6	15.8	266 40.1	204 50		131 36		277 28		56 38	17
30	171 41.5	· 15.8	269 10.5	207 20	· ·	134 06	· ·	279 59	· ·	59 03	· 19
40	174 11.5	15.8	271 40.9	209 49		136 36		282 29		61 28	20
50	176 41.5	15.8	274 11.3	212 19		139 07		285 00		63 53	22
Rate	14 59.9	S0 00.1		14 59.4	N0 00.6	15 00.9	S0 00.4	15 02.4	0 00.0	14 30.3	S0 09.7

Lat.	Moon-set	Diff.
N		
°	h m	m
72	21 19	*
70	21 59	−16
68	22 27	−05
66	22 48	00
64	23 05	+04
62	23 19	06
60	23 31	08
58	23 41	10
56	23 50	11
54	23 58	12
52	24 06	14
50	24 12	14
45	24 26	16
40	00 05	16
35	00 12	18
30	00 18	19
20	00 29	21
10	00 39	23
0	00 48	25
10	00 58	26
20	01 08	28
30	01 19	30
35	01 25	32
40	01 33	33
45	01 42	35
50	01 52	37
52	01 57	38
54	02 03	39
56	02 09	40
58	02 16	42
60	02 23	43
S		

Moon's P. in A.

At.	Corr.	At.	Corr.
°	+ ′	°	+ ′
0		55	
	57		32
5		56	
	56		31
12		57	
	55		30
16		58	
	54		29
19		59	
	53		28
22		61	
	52		27
24		62	
	51		26
27		63	
	50		25
29		64	
	49		24
31		65	
	48		23
33		66	
	47		22
35		67	
	46		21
36		68	
	45		20
38		69	
	44		19
39		70	
	43		18
41		72	
	42		17
43		73	
	41		16
44		74	
	40		15
45		75	
	39		14
47		76	
	38		13
48		77	
	37		12
49		78	
	36		11
51		79	
	35		10
52		80	
	34		
53			
	33		
55			
	32		
56			

Sun SD 15′.8

Moon SD 15′

Age 9d

Table 8

STARS, MAY—AUG., 1968

No.	Name		Mag.	S.H.A.	Dec.
				° ′	° ′
7*	*Acamar*		3·1	315 44	S.40 26
5*	*Achernar*		0·6	335 52	S.57 24
30*	*Acrux*		1·1	173 48	S.62 56
19	*Adhara*	†	1·6	255 40	S.28 56
10*	*Aldebaran*	†	1·1	291 29	N.16 27
32*	*Alioth*		1·7	166 50	N.56 08
34*	*Alkaid*		1·9	153 25	N.49 28
55	*Al Na'ir*		2·2	28 26	S.47 07
15	*Alnilam*	†	1·8	276 21	S. 1 13
25*	*Alphard*	†	2·2	218 30	S. 8 31
41*	*Alphecca*	†	2·3	126 40	N.26 49
1*	*Alpheratz*	†	2·2	358 19	N.28 55
51*	*Altair*	†	0·9	62 41	N. 8 47
2	*Ankaa*		2·4	353 49	S.42 28
42*	*Antares*	†	1·2	113 08	S.26 22
37*	*Arcturus*	†	0·2	146 27	N.19 21
43	*Atria*		1·9	108 40	S.68 59
22	*Avior*		1·7	234 33	S.59 24
13	*Bellatrix*	†	1·7	279 09	N. 6 20
16*	*Betelgeuse*	†	0·1–1·2	271 38	N. 7 24
17*	*Canopus*		–0·9	264 12	S.52 41
12*	*Capella*		0·2	281 25	N.45 58
53*	*Deneb*		1·3	49 54	N.45 10
28*	*Denebola*	†	2·2	183 08	N.14 45
4*	*Diphda*	†	2·2	349 30	S.18 09
27*	*Dubhe*		2·0	194 33	N.61 56
14	*Elnath*	†	1·8	278 56	N.28 35
47	*Eltanin*		2·4	91 02	N.51 29
54*	*Enif*	†	2·5	34 20	N. 9 44
56*	*Fomalhaut*	†	1·3	16 01	S.29 47
31	*Gacrux*		1·6	172 39	S.56 56
29*	*Gienah*	†	2·8	176 27	S.17 22
35	*Hadar*		0·9	149 36	S.60 14
6*	*Hamal*	†	2·2	328 39	N.23 19
48	*Kaus Aust.*		2·0	84 29	S.34 24
40*	*Kochab*		2·2	137 17	N.74 17
57	*Markab*	†	2·6	14 12	N.15 02
8*	*Menkar*	†	2·8	314 51	N. 3 58
36	*Menkent*		2·3	148 48	S.36 13
24*	*Miaplacidus*		1·8	221 48	S.69 35
9*	*Mirfak*		1·9	309 30	N.49 45
50*	*Nunki*	†	2·1	76 40	S.26 20
52*	*Peacock*		2·1	54 12	S.56 50
21*	*Pollux*	†	1·2	244 10	N.28 06
20*	*Procyon*	†	0·5	245 36	N. 5 19
46*	*Rasalhague*	†	2·1	96 38	N.12 35
26*	*Regulus*	†	1·3	208 20	N.12 07
11*	*Rigel*	†	0·3	281 45	S. 8 14
38*	*Rigil Kent.*		0·1	140 38	S.60 43
44	*Sabik*	†	2·6	102 51	S.15 41
3*	*Schedar*		2·5	350 20	N.56 22
45*	*Shaula*		1·7	97 08	S.37 05
18*	*Sirius*	†	–1·6	259 04	S.16 40
33*	*Spica*	†	1·2	159 07	S.11 00
23*	*Suhail*		2·2	223 18	S.43 18
49*	*Vega*		0·1	81 02	N.38 45
39	*Zuben'ubi*	†	2·9	137 43	S.15 55

INTERPOLATION OF G.H.A.

Increment to be added for intervals of G.M.T. to G.H.A. of: Sun, Aries (♈) and planets; Moon

SUN, etc. m s	° ′	MOON m s	SUN, etc. m s	° ′	MOON m s	SUN, etc. m s	° ′	MOON m s
00 00		00 00	03 17		03 25	06 37		06 52
	0 00			0 50			1 40	
01		00 02	21		03 29	41		06 56
	0 01			0 51			1 41	
05		00 06	25		03 33	45		07 00
	0 02			0 52			1 42	
09		00 10	29		03 37	49		07 04
	0 03			0 53			1 43	
13		00 14	33		03 41	53		07 08
	0 04			0 54			1 44	
17		00 18	37		03 45	06 57		07 13
	0 05			0 55			1 45	
21		00 22	41		03 49	07 01		07 17
	0 06			0 56			1 46	
25		00 26	45		03 54	05		07 21
	0 07			0 57			1 47	
29		00 31	49		03 58	09		07 25
	0 08			0 58			1 48	
33		00 35	53		04 02	13		07 29
	0 09			0 59			1 49	
37		00 39	03 57		04 06	17		07 33
	0 10			1 00			1 50	
41		00 43	04 01		04 10	21		07 37
	0 11			1 01			1 51	
45		00 47	05		04 14	25		07 42
	0 12			1 02			1 52	
49		00 51	09		04 19	29		07 46
	0 13			1 03			1 53	
53		00 55	13		04 23	33		07 50
	0 14			1 04			1 54	
00 57		01 00	17		04 27	37		07 54
	0 15			1 05			1 55	
01 01		01 04	21		04 31	41		07 58
	0 16			1 06			1 56	
05		01 08	25		04 35	45		08 02
	0 17			1 07			1 57	
09		01 12	29		04 39	49		08 06
	0 18			1 08			1 58	
13		01 16	33		04 43	53		08 11
	0 19			1 09			1 59	
17		01 20	37		04 48	07 57		08 15
	0 20			1 10			2 00	
21		01 24	41		04 52	08 01		08 19
	0 21			1 11			2 01	
25		01 29	45		04 56	05		08 23
	0 22			1 12			2 02	
29		01 33	49		05 00	09		08 27
	0 23			1 13			2 03	
33		01 37	53		05 04	13		08 31
	0 24			1 14			2 04	
37		01 41	04 57		05 08	17		08 35
	0 25			1 15			2 05	
41		01 45	05 01		05 12	21		08 40
	0 26			1 16			2 06	
45		01 49	05		05 17	25		08 44
	0 27			1 17			2 07	
49		01 53	09		05 21	29		08 48
	0 28			1 18			2 08	
53		01 58	13		05 25	33		08 52
	0 29			1 19			2 09	
01 57		02 02	17		05 29	37		08 56
	0 30			1 20			2 10	
02 01		02 06	21		05 33	41		09 00
	0 31			1 21			2 11	
05		02 10	25		05 37	45		09 04
	0 32			1 22			2 12	
09		02 14	29		05 41	49		09 09
	0 33			1 23			2 13	
13		02 18	33		05 46	53		09 13
	0 34			1 24			2 14	
17		02 22	37		05 50	08 57		09 17
	0 35			1 25			2 15	
21		02 27	41		05 54	09 01		09 21
	0 36			1 26			2 16	
25		02 31	45		05 58	05		09 25
	0 37			1 27			2 17	
29		02 35	49		06 02	09		09 29
	0 38			1 28			2 18	
33		02 39	53		06 06	13		09 33
	0 39			1 29			2 19	
37		02 43	05 57		06 10	17		09 38
	0 40			1 30			2 20	
41		02 47	06 01		06 15	21		09 42
	0 41			1 31			2 21	
45		02 51	05		06 19	25		09 46
	0 42			1 32			2 22	
49		02 56	09		06 23	29		09 50
	0 43			1 33			2 23	
53		03 00	13		06 27	33		09 54
	0 44			1 34			2 24	
02 57		03 04	17		06 31	37		09 58
	0 45			1 35			2 25	
03 01		03 08	21		06 35	41		10 00
	0 46			1 36			2 26	
05		03 12	25		06 39	45		
	0 47			1 37			2 27	
09		03 16	29		06 44	49		
	0 48			1 38			2 28	
13		03 20	33		06 48	53		
	0 49			1 39			2 29	
17		03 25	37		06 52	09 57		
	0 50			1 40			2 30	
03 21		03 29	06 41		06 56	10 00		

* Stars used in H.O. 249 (A.P. 3270) Vol. 1.

† Stars that may be used with Vols. 2 and 3.

Table 9

8
A Day with the Sun

This chapter will explain the entire process of taking two sun sights at an interval of several hours and plotting a position from them as outlined in Chapter 3.

First, however, you should be familiar with the plotting sheet. Figure 8.1 is a reproduction of the Universal Plotting Sheet. The actual sheets measure 14 × 12 inches, and they are printed in pads of fifty. While it is possible to do celestial navigation on a chart, I find that the chart very quickly gets crowded with lines and that the scale of the chart is not always convenient. Therefore, I will use the plotting sheet and supply the necessary details of its use as you proceed.

Assume that it is about 10 A.M. on July 2, 1974. You read the log and then take the following series of sights:

10-09-00	54°47′
10-12-05	55°25′
10-13-05	55°36′
10-13-40	55°46′
10-14-40	56°19′

Plotted on graph paper, these sights result in the configuration in Figure 8.2. Obviously, something went wrong with the last sight. Possibly you were rushing, or probably you misread the time. In any case, throw that one out and choose the one at 10-12-05, since it lies quite close to the line.

Your watch, according to the time check the previous evening, was 35 seconds fast and gaining 15 seconds a day. You figure the day is about one-third over, so you add 5 seconds, making your watch 40 seconds fast. Then you complete the time form:

Watch Time	= 10 -12- 05
Fast (−) or Slow (+)	= − 40
Conversion to GMT	= +4
GMT	= 14 -11- 25

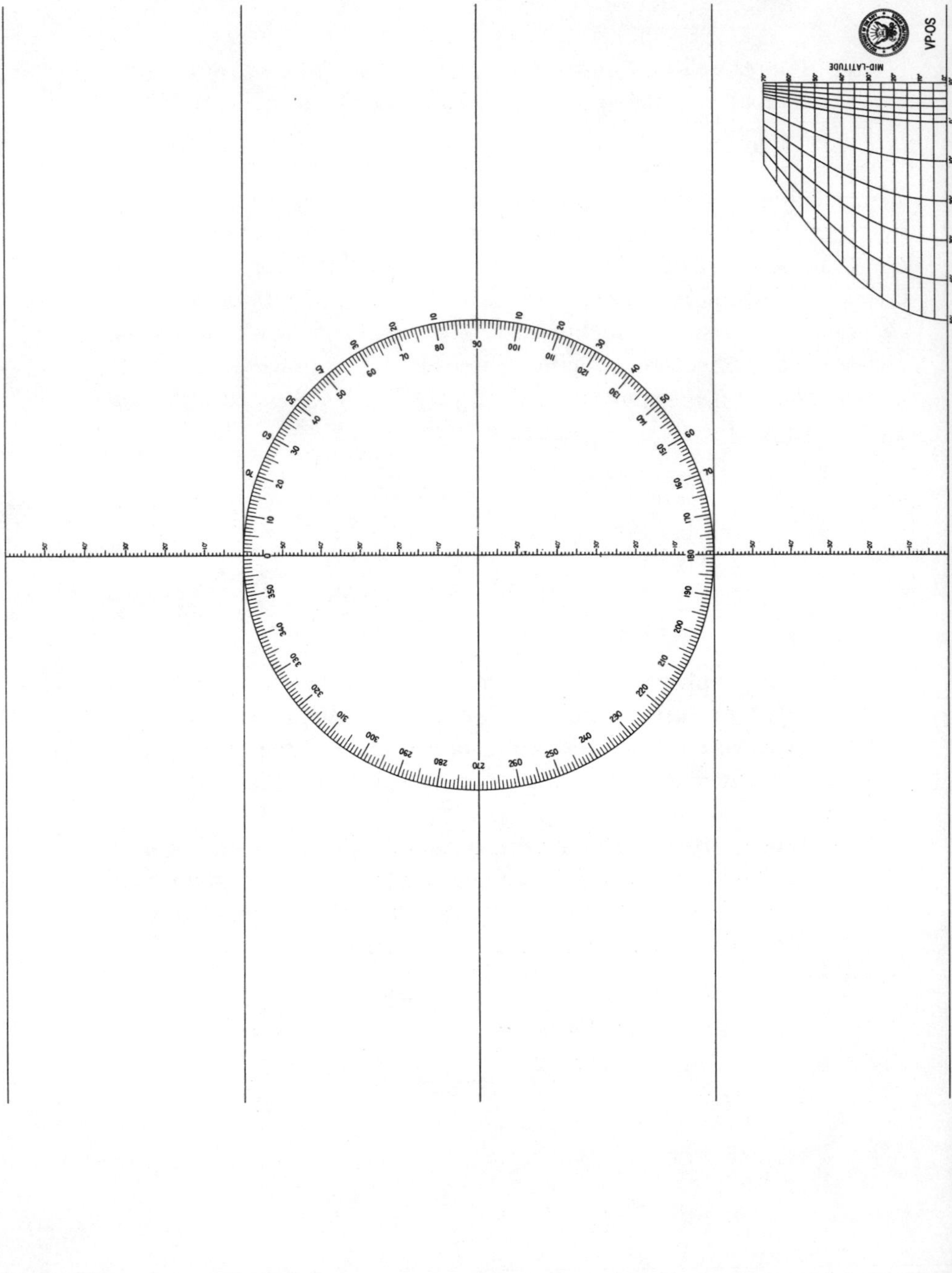
VP-OS
MID-LATITUDE
UNIVERSAL PLOTTING SHEET
1st EDITION: September 1942
Published by the Hydrographic Office, Washington, D.C. Sept. 1942

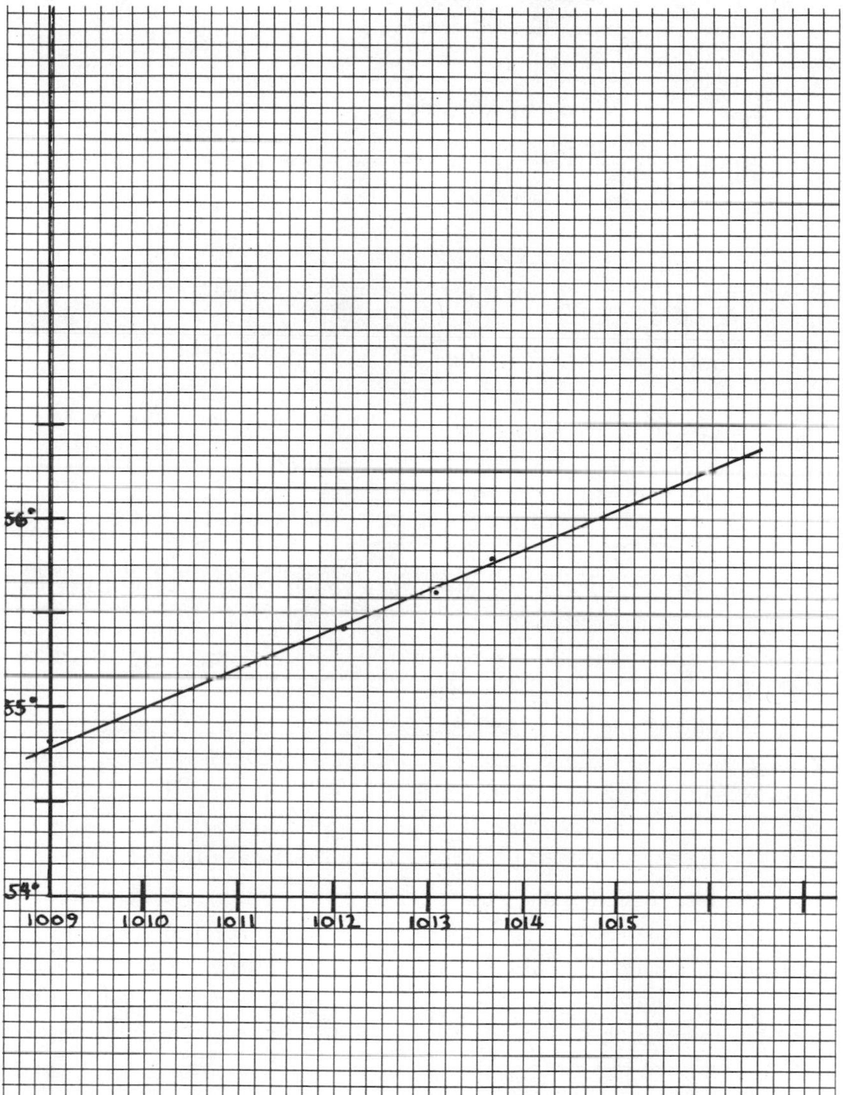

Fig. 8.2

The next step is to record the sextant data, using the format of the last chapter. Assume height of eye is 9½ feet (see Table 6 or 3) and index error is 5′ off the arc. The main correction comes from Table 3.

Sextant Reading (Hs)	= 55° 25′
Index Correction (IC)	= + 05′
Dip	= − 03′
Apparent Altitude (ha)	= 55° 27′
Main Correction	= + 15′
Observed Altitude (Ho)	= 55° 42′

Now open the *Nautical Almanac* to the page for July 2, 1974 (Table 10), and using that page and the "Increments and Corrections" page in the back of the almanac (Table 11) for the minutes and seconds of GHA, complete the following form:

SUN		
GHA 14 Hr.	= 29°01′	Dec. = N 23°03′
11 Min. 25 Sec.	= 2°51′	
GHA	= 31°52′	

You can now put away the almanac; you have all the information you need for this observation.

You now know the GHA of the sun, so the next step is to figure the LHA. Remember that the LHA is the angle at the top of the navigational triangle and is measured westward from your longitude to the longitude of the sun. This is a case like the one shown in Figure 4.4, in which the sun is east of you and LHA is greater than 180°. To find LHA, you add 360° to the GHA and subtract your assumed longitude. As a matter of fact, this is the rule: In *west* longitude, *subtract* your longitude from GHA (adding 360° to GHA if necessary to perform the subtraction), and in *east* longitude, *add* your longitude to GHA (subtracting 360° if the result of the addition is greater than 360°). Study Figures 8.3 and 8.4, and you will understand why. This is one case in which it is probably easier to remember a rule than to have to visualize the situation each time you sit down to work out a sight.

1974 JUNE 30, JULY 1, 2 (SUN., MON., TUES.)

G.M.T.	SUN G.H.A.	SUN Dec.	MOON G.H.A.	v	MOON Dec.	d	H.P.
d h	° ′	° ′	° ′	′	° ′	′	′
30 00	179 08·4	N23 12·6	54 14·7	10·6	S19 34·4	6·8	56·0
01	194 08·3	12·4	68 44·3	10·5	19 41·2	6·6	56·0
02	209 08·2	12·3	83 13·8	10·6	19 47·8	6·5	56·0
03	224 08·0	· · 12·2	97 43·4	10·5	19 54·3	6·5	56·0
04	239 07·9	12·0	112 12·9	10·5	20 00·8	6·3	55·9
05	254 07·8	11·9	126 42·4	10·4	20 07·1	6·2	55·9
06	269 07·7	N23 11·7	141 11·8	10·5	S20 13·3	6·1	55·9
07	284 07·5	11·6	155 41·3	10·4	20 19·4	6·0	55·9
08	299 07·4	11·4	170 10·7	10·5	20 25·4	5·9	55·8
S 09	314 07·3	· · 11·3	184 40·2	10·4	20 31·3	5·8	55·8
U 10	329 07·2	11·1	199 09·6	10·4	20 37·1	5·7	55·8
N 11	344 07·0	11·0	213 39·0	10·3	20 42·8	5·5	55·8
D 12	359 06·9	N23 10·8	228 08·3	10·4	S20 48·3	5·5	55·8
A 13	14 06·8	10·7	242 37·7	10·3	20 53·8	5·3	55·7
Y 14	29 06·7	10·5	257 07·0	10·3	20 59·1	5·3	55·7
15	44 06·5	· · 10·4	271 36·3	10·3	21 04·4	5·1	55·7
16	59 06·4	10·2	286 05·6	10·3	21 09·5	5·0	55·7
17	74 06·3	10·1	300 34·9	10·3	21 14·5	4·9	55·6
18	89 06·2	N23 09·9	315 04·2	10·2	S21 19·4	4·8	55·6
19	104 06·1	09·8	329 33·4	10·3	21 24·2	4·7	55·6
20	119 05·9	09·6	344 02·7	10·2	21 28·9	4·5	55·6
21	134 05·8	· · 09·5	358 31·9	10·2	21 33·4	4·5	55·6
22	149 05·7	09·3	13 01·1	10·2	21 37·9	4·3	55·5
23	164 05·6	09·2	27 30·3	10·2	21 42·2	4·3	55·5
1 00	179 05·4	N23 09·0	41 59·5	10·2	S21 46·5	4·1	55·5
01	194 05·3	08·8	56 28·7	10·2	21 50·6	4·0	55·5
02	209 05·2	08·7	70 57·9	10·1	21 54·6	3·9	55·5
03	224 05·1	· · 08·5	85 27·0	10·2	21 58·5	3·7	55·4
04	239 05·0	08·4	99 56·2	10·1	22 02·2	3·7	55·4
05	254 04·8	08·2	114 25·3	10·2	22 05·9	3·5	55·4
06	269 04·7	N23 08·0	128 54·5	10·1	S22 09·4	3·5	55·4
07	284 04·6	07·9	143 23·6	10·1	22 12·9	3·3	55·4
08	299 04·5	07·7	157 52·7	10·2	22 16·2	3·2	55·3
M 09	314 04·4	· · 07·6	172 21·9	10·1	22 19·4	3·1	55·3
O 10	329 04·2	07·4	186 51·0	10·1	22 22·5	2·9	55·3
N 11	344 04·1	07·2	201 20·1	10·1	22 25·4	2·9	55·3
D 12	359 04·0	N23 07·1	215 49·2	10·1	S22 28·3	2·7	55·3
A 13	14 03·9	06·9	230 18·3	10·1	22 31·0	2·6	55·3
Y 14	29 03·7	06·7	244 47·4	10·1	22 33·6	2·5	55·2
15	44 03·6	· · 06·6	259 16·5	10·1	22 36·1	2·4	55·2
16	59 03·5	06·4	273 45·6	10·1	22 38·5	2·3	55·2
17	74 03·4	06·2	288 14·7	10·1	22 40·8	2·1	55·2
18	89 03·3	N23 06·0	302 43·8	10·1	S22 42·9	2·1	55·2
19	104 03·1	05·9	317 12·9	10·2	22 45·0	1·9	55·1
20	119 03·0	05·7	331 42·1	10·1	22 46·9	1·8	55·1
21	134 02·9	· · 05·5	346 11·2	10·1	22 48·7	1·7	55·1
22	149 02·8	05·4	0 40·3	10·1	22 50·4	1·6	55·1
23	164 02·7	05·2	15 09·4	10·1	22 52·0	1·4	55·1
2 00	179 02·6	N23 05·0	29 38·5	10·2	S22 53·4	1·4	55·1
01	194 02·4	04·8	44 07·7	10·1	22 54·8	1·2	55·0
02	209 02·3	04·7	58 36·8	10·2	22 56·0	1·1	55·0
03	224 02·2	· · 04·5	73 06·0	10·1	22 57·1	1·0	55·0
04	239 02·1	04·3	87 35·1	10·2	22 58·1	0·9	55·0
05	254 02·0	04·1	102 04·3	10·2	22 59·0	0·7	55·0
06	269 01·8	N23 04·0	116 33·5	10·2	S22 59·7	0·7	55·0
07	284 01·7	03·8	131 02·7	10·2	23 00·4	0·5	54·9
08	299 01·6	03·6	145 31·9	10·2	23 00·9	0·4	54·9
T 09	314 01·5	· · 03·4	160 01·1	10·3	23 01·3	0·3	54·9
U 10	329 01·4	03·2	174 30·4	10·2	23 01·6	0·2	54·9
E 11	344 01·3	03·1	188 59·6	10·3	23 01·8	0·0	54·9
S 12	359 01·1	N23 02·9	203 28·9	10·3	S23 01·8	0·0	54·9
D 13	14 01·0	02·7	217 58·2	10·3	23 01·8	0·2	54·8
A 14	29 00·9	02·5	232 27·5	10·3	23 01·6	0·3	54·8
Y 15	44 00·8	· · 02·3	246 56·8	10·3	23 01·3	0·4	54·8
16	59 00·7	02·1	261 26·1	10·4	23 00·9	0·5	54·8
17	74 00·5	01·9	275 55·5	10·4	23 00·4	0·6	54·8
18	89 00·4	N23 01·8	290 24·9	10·4	S22 59·8	0·7	54·8
19	104 00·3	01·6	304 54·3	10·4	22 59·1	0·9	54·8
20	119 00·2	01·4	319 23·7	10·4	22 58·2	1·0	54·7
21	134 00·1	· · 01·2	333 53·1	10·5	22 57·2	1·0	54·7
22	149 00·0	01·0	348 22·6	10·5	22 56·2	1·2	54·7
23	163 59·8	00·8	2 52·1	10·5	22 55·0	1·3	54·7
	S.D. 15·8	*d* 0·2	S.D. 15·2		15·1		14·9

Lat.	Twilight Naut.	Twilight Civil	Sunrise	Moonrise 30	Moonrise 1	Moonrise 2	Moonrise 3
°	h m	h m	h m	h m	h m	h m	h m
N 72	□	□	□	■	■	■	■
N 70	□	□	□	■	■	■	■
68	□	□	□	20 19	■	■	23 23
66	////	////	00 13	19 12	20 41	21 37	21 57
64	////	////	01 40	18 36	19 53	20 47	21 19
62	////	////	02 16	18 11	19 22	20 16	20 52
60	////	01 02	02 42	17 51	18 59	19 52	20 31
N 58	////	01 48	03 01	17 34	18 40	19 34	20 14
56	////	02 16	03 18	17 20	18 25	19 18	19 59
54	00 57	02 38	03 32	17 08	18 11	19 04	19 47
52	01 39	02 56	03 44	16 58	18 00	18 52	19 36
50	02 06	03 10	03 54	16 48	17 49	18 42	19 26
45	02 50	03 39	04 17	16 29	17 28	18 20	19 06
N 40	03 20	04 02	04 34	16 13	17 10	18 02	18 49
35	03 43	04 20	04 49	15 59	16 55	17 48	18 35
30	04 02	04 35	05 02	15 47	16 43	17 35	18 22
20	04 30	04 59	05 24	15 27	16 21	17 13	18 01
N 10	04 53	05 20	05 43	15 10	16 02	16 54	17 43
0	05 11	05 38	06 00	14 54	15 45	16 36	17 26
S 10	05 28	05 54	06 17	14 38	15 27	16 18	17 09
20	05 44	06 11	06 35	14 21	15 09	15 59	16 51
30	06 00	06 30	06 56	14 01	14 47	15 37	16 30
35	06 09	06 41	07 08	13 49	14 35	15 24	16 17
40	06 18	06 52	07 23	13 36	14 20	15 10	16 03
45	06 28	07 05	07 39	13 21	14 03	14 52	15 46
S 50	06 40	07 21	07 59	13 02	13 42	14 30	15 25
52	06 45	07 29	08 09	12 53	13 32	14 20	15 15
54	06 50	07 37	08 20	12 43	13 21	14 08	15 04
56	06 56	07 46	08 33	12 32	13 08	13 55	14 51
58	07 03	07 56	08 47	12 19	12 53	13 39	14 36
S 60	07 10	08 07	09 04	12 04	12 36	13 20	14 19

Lat.	Sunset	Twilight Civil	Twilight Naut.	Moonset 30	Moonset 1	Moonset 2	Moonset 3
°	h m	h m	h m	h m	h m	h m	h m
N 72	□	□	□	■	■	■	■
N 70	□	□	□	■	■	■	■
68	□	□	□	21 46	■	■	23 59
66	23 44	////	////	22 54	23 11	24 01	00 01
64	22 26	////	////	23 30	24 00	00 00	00 51
62	21 50	////	////	23 56	24 30	00 30	01 22
60	21 25	23 03	////	24 16	00 16	00 54	01 45
N 58	21 05	22 19	////	00 03	00 33	01 12	02 04
56	20 49	21 50	////	00 15	00 47	01 28	02 20
54	20 35	21 29	23 08	00 25	00 59	01 42	02 33
52	20 23	21 11	22 27	00 35	01 10	01 53	02 45
50	20 12	20 57	22 01	00 43	01 20	02 04	02 55
45	19 50	20 28	21 17	01 00	01 40	02 26	03 17
N 40	19 33	20 05	20 47	01 15	01 57	02 43	03 35
35	19 18	19 47	20 24	01 27	02 10	02 58	03 49
30	19 05	19 32	20 06	01 38	02 23	03 11	04 02
20	18 43	19 08	19 37	01 56	02 43	03 33	04 24
N 10	18 25	18 48	19 15	02 12	03 01	03 52	04 43
0	18 07	18 30	18 56	02 28	03 18	04 10	05 00
S 10	17 50	18 13	18 39	02 43	03 35	04 27	05 18
20	17 32	17 56	18 23	02 59	03 53	04 46	05 37
30	17 11	17 37	18 07	03 17	04 14	05 08	05 58
35	16 59	17 27	17 59	03 28	04 27	05 21	06 11
40	16 45	17 15	17 49	03 41	04 41	05 36	06 25
45	16 28	17 02	17 39	03 56	04 57	05 53	06 42
S 50	16 08	16 46	17 28	04 14	05 18	06 15	07 04
52	15 58	16 39	17 23	04 22	05 28	06 25	07 14
54	15 47	16 31	17 17	04 32	05 39	06 37	07 25
56	15 35	16 22	17 11	04 43	05 52	06 51	07 38
58	15 21	16 12	17 04	04 55	06 06	07 06	07 53
S 60	15 04	16 01	16 57	05 10	06 24	07 25	08 11

Day	SUN Eqn. of Time 00ʰ	SUN Eqn. of Time 12ʰ	SUN Mer. Pass.	MOON Mer. Pass. Upper	MOON Mer. Pass. Lower	Age	Phase
	m s	m s	h m	h m	h m	d	
30	03 26	03 32	12 04	21 06	08 41	10	
1	03 38	03 44	12 04	21 57	09 32	11	
2	03 50	03 55	12 04	22 48	10 23	12	

Table 10

10ᵐ INCREMENTS AND CORRECTIONS 11ᵐ

10ᵐ	SUN PLANETS	ARIES	MOON	v or d	Corrⁿ	v or d	Corrⁿ	v or d	Corrⁿ
s	° ′	° ′	° ′	′	′	′	′	′	′
00	2 30·0	2 30·4	2 23·2	0·0	0·0	6·0	1·1	12·0	2·1
01	2 30·3	2 30·7	2 23·4	0·1	0·0	6·1	1·1	12·1	2·1
02	2 30·5	2 30·9	2 23·6	0·2	0·0	6·2	1·1	12·2	2·1
03	2 30·8	2 31·2	2 23·9	0·3	0·1	6·3	1·1	12·3	2·2
04	2 31·0	2 31·4	2 24·1	0·4	0·1	6·4	1·1	12·4	2·2
05	2 31·3	2 31·7	2 24·4	0·5	0·1	6·5	1·1	12·5	2·2
06	2 31·5	2 31·9	2 24·6	0·6	0·1	6·6	1·2	12·6	2·2
07	2 31·8	2 32·2	2 24·8	0·7	0·1	6·7	1·2	12·7	2·2
08	2 32·0	2 32·4	2 25·1	0·8	0·1	6·8	1·2	12·8	2·2
09	2 32·3	2 32·7	2 25·3	0·9	0·2	6·9	1·2	12·9	2·3
10	2 32·5	2 32·9	2 25·6	1·0	0·2	7·0	1·2	13·0	2·3
11	2 32·8	2 33·2	2 25·8	1·1	0·2	7·1	1·2	13·1	2·3
12	2 33·0	2 33·4	2 26·0	1·2	0·2	7·2	1·3	13·2	2·3
13	2 33·3	2 33·7	2 26·3	1·3	0·2	7·3	1·3	13·3	2·3
14	2 33·5	2 33·9	2 26·5	1·4	0·2	7·4	1·3	13·4	2·3
15	2 33·8	2 34·2	2 26·7	1·5	0·3	7·5	1·3	13·5	2·4
16	2 34·0	2 34·4	2 27·0	1·6	0·3	7·6	1·3	13·6	2·4
17	2 34·3	2 34·7	2 27·2	1·7	0·3	7·7	1·3	13·7	2·4
18	2 34·5	2 34·9	2 27·5	1·8	0·3	7·8	1·4	13·8	2·4
19	2 34·8	2 35·2	2 27·7	1·9	0·3	7·9	1·4	13·9	2·4
20	2 35·0	2 35·4	2 27·9	2·0	0·4	8·0	1·4	14·0	2·5
21	2 35·3	2 35·7	2 28·2	2·1	0·4	8·1	1·4	14·1	2·5
22	2 35·5	2 35·9	2 28·4	2·2	0·4	8·2	1·4	14·2	2·5
23	2 35·8	2 36·2	2 28·7	2·3	0·4	8·3	1·5	14·3	2·5
24	2 36·0	2 36·4	2 28·9	2·4	0·4	8·4	1·5	14·4	2·5
25	2 36·3	2 36·7	2 29·1	2·5	0·4	8·5	1·5	14·5	2·5
26	2 36·5	2 36·9	2 29·4	2·6	0·5	8·6	1·5	14·6	2·6
27	2 36·8	2 37·2	2 29·6	2·7	0·5	8·7	1·5	14·7	2·6
28	2 37·0	2 37·4	2 29·8	2·8	0·5	8·8	1·5	14·8	2·6
29	2 37·3	2 37·7	2 30·1	2·9	0·5	8·9	1·6	14·9	2·6
30	2 37·5	2 37·9	2 30·3	3·0	0·5	9·0	1·6	15·0	2·6
31	2 37·8	2 38·2	2 30·6	3·1	0·5	9·1	1·6	15·1	2·6
32	2 38·0	2 38·4	2 30·8	3·2	0·6	9·2	1·6	15·2	2·7
33	2 38·3	2 38·7	2 31·0	3·3	0·6	9·3	1·6	15·3	2·7
34	2 38·5	2 38·9	2 31·3	3·4	0·6	9·4	1·6	15·4	2·7
35	2 38·8	2 39·2	2 31·5	3·5	0·6	9·5	1·7	15·5	2·7
36	2 39·0	2 39·4	2 31·8	3·6	0·6	9·6	1·7	15·6	2·7
37	2 39·3	2 39·7	2 32·0	3·7	0·6	9·7	1·7	15·7	2·7
38	2 39·5	2 39·9	2 32·2	3·8	0·7	9·8	1·7	15·8	2·8
39	2 39·8	2 40·2	2 32·5	3·9	0·7	9·9	1·7	15·9	2·8
40	2 40·0	2 40·4	2 32·7	4·0	0·7	10·0	1·8	16·0	2·8
41	2 40·3	2 40·7	2 32·9	4·1	0·7	10·1	1·8	16·1	2·8
42	2 40·5	2 40·9	2 33·2	4·2	0·7	10·2	1·8	16·2	2·8
43	2 40·8	2 41·2	2 33·4	4·3	0·8	10·3	1·8	16·3	2·9
44	2 41·0	2 41·4	2 33·7	4·4	0·8	10·4	1·8	16·4	2·9
45	2 41·3	2 41·7	2 33·9	4·5	0·8	10·5	1·8	16·5	2·9
46	2 41·5	2 41·9	2 34·1	4·6	0·8	10·6	1·9	16·6	2·9
47	2 41·8	2 42·2	2 34·4	4·7	0·8	10·7	1·9	16·7	2·9
48	2 42·0	2 42·4	2 34·6	4·8	0·8	10·8	1·9	16·8	2·9
49	2 42·3	2 42·7	2 34·9	4·9	0·9	10·9	1·9	16·9	3·0
50	2 42·5	2 42·9	2 35·1	5·0	0·9	11·0	1·9	17·0	3·0
51	2 42·8	2 43·2	2 35·3	5·1	0·9	11·1	1·9	17·1	3·0
52	2 43·0	2 43·4	2 35·6	5·2	0·9	11·2	2·0	17·2	3·0
53	2 43·3	2 43·7	2 35·8	5·3	0·9	11·3	2·0	17·3	3·0
54	2 43·5	2 43·9	2 36·1	5·4	0·9	11·4	2·0	17·4	3·0
55	2 43·8	2 44·2	2 36·3	5·5	1·0	11·5	2·0	17·5	3·1
56	2 44·0	2 44·4	2 36·5	5·6	1·0	11·6	2·0	17·6	3·1
57	2 44·3	2 44·7	2 36·8	5·7	1·0	11·7	2·0	17·7	3·1
58	2 44·5	2 45·0	2 37·0	5·8	1·0	11·8	2·1	17·8	3·1
59	2 44·8	2 45·2	2 37·2	5·9	1·0	11·9	2·1	17·9	3·1
60	2 45·0	2 45·5	2 37·5	6·0	1·1	12·0	2·1	18·0	3·2

11ᵐ	SUN PLANETS	ARIES	MOON	v or d	Corrⁿ	v or d	Corrⁿ	v or d	Corrⁿ
s	° ′	° ′	° ′	′	′	′	′	′	′
00	2 45·0	2 45·5	2 37·5	0·0	0·0	6·0	1·2	12·0	2·3
01	2 45·3	2 45·7	2 37·7	0·1	0·0	6·1	1·2	12·1	2·3
02	2 45·5	2 46·0	2 38·0	0·2	0·0	6·2	1·2	12·2	2·3
03	2 45·8	2 46·2	2 38·2	0·3	0·1	6·3	1·2	12·3	2·4
04	2 46·0	2 46·5	2 38·4	0·4	0·1	6·4	1·2	12·4	2·4
05	2 46·3	2 46·7	2 38·7	0·5	0·1	6·5	1·2	12·5	2·4
06	2 46·5	2 47·0	2 38·9	0·6	0·1	6·6	1·3	12·6	2·4
07	2 46·8	2 47·2	2 39·2	0·7	0·1	6·7	1·3	12·7	2·4
08	2 47·0	2 47·5	2 39·4	0·8	0·2	6·8	1·3	12·8	2·5
09	2 47·3	2 47·7	2 39·6	0·9	0·2	6·9	1·3	12·9	2·5
10	2 47·5	2 48·0	2 39·9	1·0	0·2	7·0	1·3	13·0	2·5
11	2 47·8	2 48·2	2 40·1	1·1	0·2	7·1	1·4	13·1	2·5
12	2 48·0	2 48·5	2 40·3	1·2	0·2	7·2	1·4	13·2	2·5
13	2 48·3	2 48·7	2 40·6	1·3	0·2	7·3	1·4	13·3	2·5
14	2 48·5	2 49·0	2 40·8	1·4	0·3	7·4	1·4	13·4	2·6
15	2 48·8	2 49·2	2 41·1	1·5	0·3	7·5	1·4	13·5	2·6
16	2 49·0	2 49·5	2 41·3	1·6	0·3	7·6	1·5	13·6	2·6
17	2 49·3	2 49·7	2 41·5	1·7	0·3	7·7	1·5	13·7	2·6
18	2 49·5	2 50·0	2 41·8	1·8	0·3	7·8	1·5	13·8	2·6
19	2 49·8	2 50·2	2 42·0	1·9	0·4	7·9	1·5	13·9	2·7
20	2 50·0	2 50·5	2 42·3	2·0	0·4	8·0	1·5	14·0	2·7
21	2 50·3	2 50·7	2 42·5	2·1	0·4	8·1	1·6	14·1	2·7
22	2 50·5	2 51·0	2 42·7	2·2	0·4	8·2	1·6	14·2	2·7
23	2 50·8	2 51·2	2 43·0	2·3	0·4	8·3	1·6	14·3	2·7
24	2 51·0	2 51·5	2 43·2	2·4	0·5	8·4	1·6	14·4	2·8
25	2 51·3	2 51·7	2 43·4	2·5	0·5	8·5	1·6	14·5	2·8
26	2 51·5	2 52·0	2 43·7	2·6	0·5	8·6	1·6	14·6	2·8
27	2 51·8	2 52·2	2 43·9	2·7	0·5	8·7	1·7	14·7	2·8
28	2 52·0	2 52·5	2 44·2	2·8	0·5	8·8	1·7	14·8	2·8
29	2 52·3	2 52·7	2 44·4	2·9	0·6	8·9	1·7	14·9	2·9
30	2 52·5	2 53·0	2 44·6	3·0	0·6	9·0	1·7	15·0	2·9
31	2 52·8	2 53·2	2 44·9	3·1	0·6	9·1	1·7	15·1	2·9
32	2 53·0	2 53·5	2 45·1	3·2	0·6	9·2	1·8	15·2	2·9
33	2 53·3	2 53·7	2 45·4	3·3	0·6	9·3	1·8	15·3	2·9
34	2 53·5	2 54·0	2 45·6	3·4	0·7	9·4	1·8	15·4	3·0
35	2 53·8	2 54·2	2 45·8	3·5	0·7	9·5	1·8	15·5	3·0
36	2 54·0	2 54·5	2 46·1	3·6	0·7	9·6	1·8	15·6	3·0
37	2 54·3	2 54·7	2 46·3	3·7	0·7	9·7	1·9	15·7	3·0
38	2 54·5	2 55·0	2 46·6	3·8	0·7	9·8	1·9	15·8	3·0
39	2 54·8	2 55·2	2 46·8	3·9	0·7	9·9	1·9	15·9	3·0
40	2 55·0	2 55·5	2 47·0	4·0	0·8	10·0	1·9	16·0	3·1
41	2 55·3	2 55·7	2 47·3	4·1	0·8	10·1	1·9	16·1	3·1
42	2 55·5	2 56·0	2 47·5	4·2	0·8	10·2	2·0	16·2	3·1
43	2 55·8	2 56·2	2 47·7	4·3	0·8	10·3	2·0	16·3	3·1
44	2 56·0	2 56·5	2 48·0	4·4	0·8	10·4	2·0	16·4	3·1
45	2 56·3	2 56·7	2 48·2	4·5	0·9	10·5	2·0	16·5	3·2
46	2 56·5	2 57·0	2 48·5	4·6	0·9	10·6	2·0	16·6	3·2
47	2 56·8	2 57·2	2 48·7	4·7	0·9	10·7	2·1	16·7	3·2
48	2 57·0	2 57·5	2 48·9	4·8	0·9	10·8	2·1	16·8	3·2
49	2 57·3	2 57·7	2 49·2	4·9	0·9	10·9	2·1	16·9	3·2
50	2 57·5	2 58·0	2 49·4	5·0	1·0	11·0	2·1	17·0	3·3
51	2 57·8	2 58·2	2 49·7	5·1	1·0	11·1	2·1	17·1	3·3
52	2 58·0	2 58·5	2 49·9	5·2	1·0	11·2	2·1	17·2	3·3
53	2 58·3	2 58·7	2 50·1	5·3	1·0	11·3	2·2	17·3	3·3
54	2 58·5	2 59·0	2 50·4	5·4	1·0	11·4	2·2	17·4	3·3
55	2 58·8	2 59·2	2 50·6	5·5	1·1	11·5	2·2	17·5	3·4
56	2 59·0	2 59·5	2 50·8	5·6	1·1	11·6	2·2	17·6	3·4
57	2 59·3	2 59·7	2 51·1	5·7	1·1	11·7	2·2	17·7	3·4
58	2 59·5	3 00·0	2 51·3	5·8	1·1	11·8	2·3	17·8	3·4
59	2 59·8	3 00·2	2 51·6	5·9	1·1	11·9	2·3	17·9	3·4
60	3 00·0	3 00·5	2 51·8	6·0	1·2	12·0	2·3	18·0	3·5

Table 11

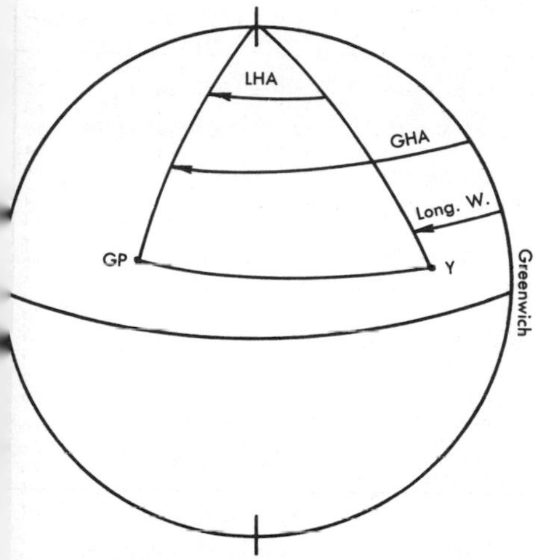

West Longitude, Body West
LHA = GHA - Long. W.

West Longitude, Body East
LHA + t = 360°
t = 360° - LHA
Also: t = Long. W. - GHA
Therefore:
360° - LHA = Long. W. - GHA
- LHA = Long. W. - GHA - 360°
LHA = - Long. W. + GHA + 360°
LHA = GHA + 360° - Long. W.

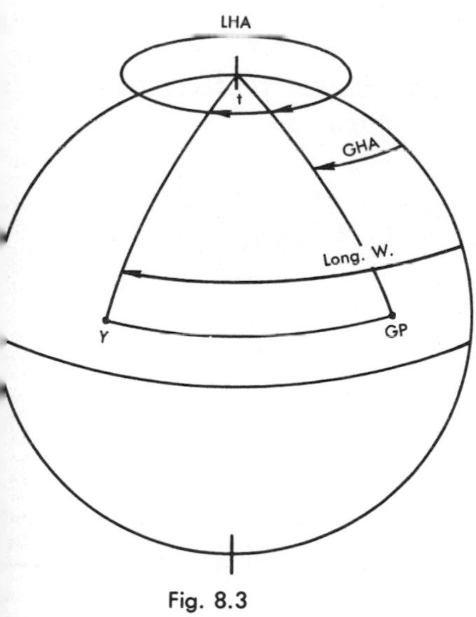

Fig. 8.3

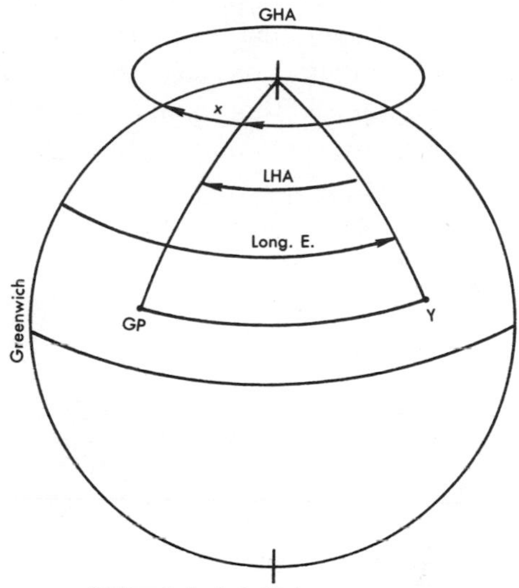

East Longitude, Body West
GHA + x = 360°
x = 360° - GHA
Also:
x = Long. E. - LHA
Therefore:
Long. E. - LHA = 360° - GHA
- LHA = 360° - GHA - Long. E.
LHA = -360° + GHA + Long. E.
LHA = Long. E. + GHA - 360°

East Longitude, Body East
LHA = GHA + Long. E.

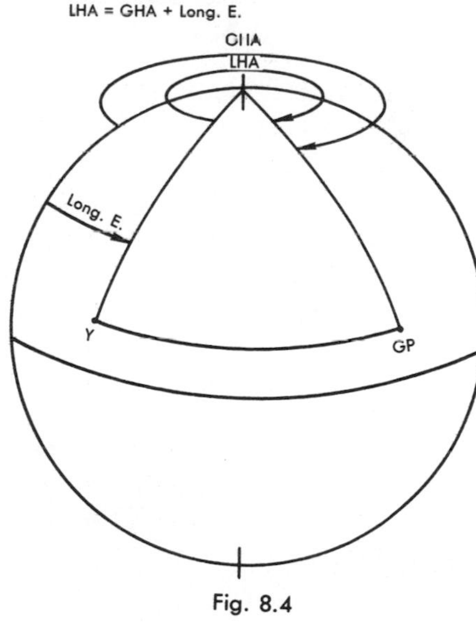

Fig. 8.4

You may remember that the solutions for the navigational triangles were computed for every whole degree of LHA. Therefore, you want the longitude of your assumed position to be of such a value that when subtracted from or added to GHA it will leave a whole number of degrees—no minutes. In practice, this is very easy to do. Here is an example.

You have learned that the GHA of the sun at the time you took your sight was 31°52′. Suppose you had been at longitude 20°25′ east. You would assume that you were at longitude 20°08′ east and do the following:

GHA	=	31°52′
Assumed Longitude	=	20°08′ E (add)
LHA	=	51°60′ = 52°

Suppose your record (DR) of course steered and distance run since your last fix indicates that you are at longitude 68°30′ west. You assume you are at 68°52′ west and carry out the following steps:

GHA	=	31°52′	
		360°	(add)
GHA	=	391°52′	
Assumed Longitude	=	68°52′ W	(subtract)
LHA	=	323°	

For the sake of the example, assume further that your estimated (DR) latitude is 36°40′ north when your estimated longitude is 68°30′ west as above.

Remember that the table of precomputed triangles (the sight reduction table) is set up on the basis of whole degrees of latitude. Therefore, since your DR indicates that you are at latitude 36°40′ N, assume that you are at latitude 36° N. Your assumed position, therefore, is 68°52′ W, 36° N.

So far, you have the following:

Assumed Latitude	=	36° N
Declination	=	23°03′ N
LHA	=	323°

With this information, turn to *Sight Reduction Tables for Air Navigation* (H.O. 249). Since declination and your latitude have the same name (north), you want the page for 36° latitude containing the column for 23° declination same name as latitude. It is reproduced as Table 12. LHA 323° is in the right-hand column and under 23° declination is the following information:

Hc	d	Z
55°32′	+30′	102°

Hc, computed altitude, is what your observed altitude (Ho) would have been at your assumed position had the declination of the sun been 23°. But the sun's declination was 23°03′, or 3′ more than the value used to solve the particular triangle listed. On the table, the +30′ (d) is the amount of change in altitude that occurs for each 1° change in declination. You can check this by looking in the 24° declination column for the same 323° LHA and noting that Hc there is 30′ more (56°02′).

The actual declination (23°03′) at the time of your sight was not a whole degree different; it was only 3′ or 1/20 of a degree larger. Therefore, you add 1/20 of 30′ to Hc (1/20 × 30′ = 1.5). So the correct Hc is 55°32′ plus 1.5′, or 55°33.5′. Actually, you don't have to do this arithmetic, because a handy table in the back of the sight reduction book, reproduced as Table 13, does it for you. If you follow 3′ on the top line down to 30′ in the left-hand column, you come up with 2′, which is 1.5 rounded off. This table works either way. You can start at 3 on the left-hand side and follow it to the right to 30, and you will find the same 2′. This feature is one of the main reasons I prefer H.O. 249. As you will see when we discuss H.O. 229, this interpolation table is by far the simplest and fastest to use.

To backtrack a bit, when you are using H.O. 249, write down the information this way:

Hc	= 55°32′	d	= +30′
Corr.	=	Dec. Inc.	= 3′
Hc	=	Corr.	= 2′

Then turn to the table at the back of H.O. 249, reproduced here as

LAT 36°

N. Lat. {LHA greater than 180°……. Zn=Z; LHA less than 180°……….. Zn=360−Z}

DECLINATION (15°-29°) SAME NAME AS LATITUDE

LHA	15°			16°			17°			18°			19°			20°			21°			22°		
	Hc	d	Z	Hc	d	Z	Hc	d	Z	Hc	d	Z	Hc	d	Z	Hc	d	Z	Hc	d	Z	Hc	d	Z
	° ′	′	°	° ′	′	°	° ′	′	°	° ′	′	°	° ′	′	°	° ′	′	°	° ′	′	°	° ′	′	°
0	69 00	+60	180	70 00	+60	180	71 00	+60	180	72 00	+60	180	73 00	+60	180	74 00	+60	180	75 00	+60	180	76 00	+60	180
1	68 59	60	177	69 59	60	177	70 59	60	177	71 59	60	177	72 59	60	177	73 59	60	177	74 59	59	176	75 58	60	176
2	68 55	60	175	69 55	60	174	70 55	60	174	71 55	60	174	72 55	59	174	73 54	60	173	74 54	60	173	75 54	59	172
3	68 50	59	172	69 49	60	172	70 49	59	171	71 48	60	171	72 48	59	170	73 47	59	170	74 46	60	169	75 46	59	169
4	68 42	59	169	69 41	59	169	70 40	59	168	71 39	59	168	72 38	59	167	73 37	59	167	74 36	58	166	75 34	59	165
5	68 32	+59	167	69 31	+58	166	70 29	+59	166	71 28	+58	165	72 26	+59	164	73 25	+58	163	74 23	+57	162	75 20	+58	161
6	68 20	58	164	69 18	58	164	70 16	58	163	71 14	58	162	72 12	57	161	73 09	58	160	74 07	57	159	75 04	56	158
7	68 05	58	162	69 03	58	161	70 01	57	160	70 58	57	159	71 55	57	158	72 52	56	157	73 48	56	156	74 44	55	155
8	67 49	57	159	68 46	57	158	69 43	57	157	70 40	56	156	71 36	56	155	72 32	55	154	73 27	55	153	74 22	54	151
9	67 31	56	157	68 27	56	156	69 23	56	155	70 19	56	154	71 15	54	153	72 09	55	151	73 04	54	150	73 58	52	148
10	67 11	+56	154	68 07	+55	153	69 02	+55	152	69 57	+54	151	70 51	+54	150	71 45	+53	149	72 38	+53	147	73 31	+52	145
11	66 49	55	152	67 44	54	151	68 38	54	150	69 32	54	149	70 26	53	147	71 19	52	146	72 11	52	144	73 03	50	143
12	66 25	54	150	67 19	54	149	68 13	53	148	69 06	53	146	69 59	52	145	70 51	51	144	71 42	50	142	72 32	49	140
13	66 00	54	148	66 54	52	147	67 46	53	145	68 39	51	144	69 30	51	143	70 21	50	141	71 11	49	139	72 00	48	138
14	65 33	53	146	66 26	52	144	67 18	51	143	68 09	51	142	69 00	50	140	69 50	49	139	70 39	48	137	71 27	46	135
15	65 05	+52	144	65 57	+51	142	66 48	+51	141	67 39	+49	140	68 28	+49	138	69 17	+48	137	70 05	+47	135	70 52	+45	133
16	64 36	51	142	65 27	50	140	66 17	50	139	67 07	48	138	67 55	48	136	68 43	47	135	69 30	46	133	70 16	44	131
17	64 05	50	140	64 55	50	139	65 45	48	137	66 33	48	136	67 21	47	134	68 08	46	133	68 54	44	131	69 38	44	129
18	63 33	49	138	64 22	49	137	65 11	48	135	65 59	47	134	66 46	46	132	67 32	44	131	68 16	44	129	69 00	42	127
19	63 00	49	136	63 49	47	135	64 36	47	134	65 23	46	132	66 09	45	130	66 54	44	129	67 38	43	127	68 21	41	125
20	62 26	+48	135	63 14	+47	133	64 01	+46	132	64 47	+45	130	65 32	+44	129	66 16	+43	127	66 59	+42	125	67 41	+40	123
21	61 51	47	133	62 38	46	132	63 24	45	130	64 09	44	129	64 53	44	127	65 37	42	125	66 19	41	124	67 00	39	122
22	61 15	46	131	62 01	45	130	62 46	45	129	63 31	43	127	64 14	43	125	64 57	41	124	65 38	40	122	66 18	39	120
23	60 38	45	130	61 23	45	128	62 08	44	127	62 52	42	126	63 34	42	124	64 16	41	122	64 57	39	121	65 36	38	119
24	60 00	45	128	60 45	44	127	61 29	43	126	62 12	42	124	62 54	41	122	63 35	39	121	64 14	39	119	64 53	37	117
25	59 22	+44	127	60 06	+43	125	60 49	+42	124	61 31	+41	123	62 12	+41	121	62 53	+39	119	63 32	+38	118	64 10	+36	116
26	58 42	44	125	59 26	42	124	60 08	42	123	60 50	41	121	61 31	39	120	62 10	39	118	62 49	37	116	63 26	36	115
27	58 02	43	124	58 45	42	123	59 27	41	121	60 08	40	120	60 48	39	118	61 27	38	117	62 05	36	115	62 41	36	113
28	57 22	42	123	58 04	41	121	58 45	41	120	59 26	39	119	60 05	38	117	60 43	38	116	61 21	36	114	61 57	34	112
29	56 41	41	122	57 22	41	120	58 03	40	119	58 43	39	117	59 22	37	116	59 59	37	114	60 36	36	113	61 12	34	111
30	55 59	+41	120	56 40	+40	119	57 20	+40	118	58 00	+38	116	58 38	+37	115	59 15	+36	113	59 51	+35	112	60 26	+34	110
31	55 17	40	119	55 57	40	118	56 37	39	117	57 16	38	115	57 54	36	114	58 30	36	112	59 06	34	111	59 40	34	109
32	54 34	40	118	55 14	39	117	55 53	39	115	56 32	37	114	57 09	36	113	57 45	35	111	58 20	34	110	58 54	33	108
33	53 51	40	117	54 31	38	116	55 09	38	114	55 47	37	113	56 24	36	112	57 00	34	110	57 34	34	109	58 08	33	107
34	53 08	39	116	53 47	38	115	54 25	37	113	55 02	37	112	55 39	35	111	56 14	34	109	56 48	34	108	57 22	32	106
35	52 24	+38	115	53 02	+38	114	53 40	+37	112	54 17	+36	111	54 53	+35	110	55 28	+34	108	56 02	+33	107	56 35	+32	105
36	51 40	38	114	52 18	37	113	52 55	37	111	53 32	35	110	54 07	35	109	54 42	33	107	55 15	33	106	55 48	31	104
37	50 55	38	113	51 33	37	112	52 10	36	110	52 46	35	109	53 21	34	108	53 55	33	106	54 28	33	105	55 01	31	103
38	50 10	37	112	50 47	37	111	51 24	36	109	52 00	35	108	52 35	33	107	53 08	33	105	53 41	32	104	54 13	31	103
39	49 25	37	111	50 02	36	110	50 38	35	108	51 13	35	107	51 48	34	106	52 22	32	105	52 54	32	103	53 26	30	102
40	48 39	+37	110	49 16	+36	109	49 52	+35	108	50 27	+34	106	51 01	+33	105	51 34	+33	104	52 07	+31	102	52 38	+31	101
41	47 54	36	109	48 30	35	108	49 05	35	107	49 40	34	105	50 14	33	104	50 47	32	103	51 19	32	102	51 51	30	100
42	47 08	36	108	47 44	35	107	48 19	34	106	48 53	34	105	49 27	33	103	50 00	32	102	50 32	31	101	51 03	30	99
43	46 21	36	107	46 57	35	106	47 32	34	105	48 06	34	104	48 40	32	103	49 12	32	101	49 44	31	100	50 15	30	99
44	45 35	35	107	46 10	35	105	46 45	34	104	47 19	33	103	47 52	33	102	48 25	31	101	48 56	31	99	49 27	29	98
45	44 48	+36	106	45 24	+34	105	45 58	+34	103	46 32	+33	102	47 05	+32	101	47 37	+31	100	48 08	+31	99	48 39	+29	97
46	44 02	34	105	44 36	35	104	45 11	33	103	45 44	33	101	46 17	32	100	46 49	31	99	47 20	30	98	47 50	30	97
47	43 15	34	104	43 49	34	103	44 23	34	102	44 57	32	101	45 29	32	100	46 01	31	98	46 32	30	97	47 02	29	96
48	42 27	35	103	43 02	34	102	43 36	33	101	44 09	32	100	44 41	32	99	45 13	31	98	45 44	30	96	46 14	29	95
49	41 40	34	103	42 14	34	102	42 48	33	100	43 21	32	99	43 53	32	98	44 25	30	97	44 55	30	96	45 25	30	95
50	40 53	+34	102	41 27	+33	101	42 00	+33	100	42 33	+32	99	43 05	+31	97	43 36	+31	96	44 07	+30	95	44 37	+29	94
51	40 05	34	101	40 39	33	100	41 12	33	99	41 45	32	98	42 17	31	97	42 48	31	96	43 19	30	94	43 49	29	93
52	39 17	34	100	39 51	33	99	40 24	33	98	40 57	32	97	41 29	31	96	42 00	30	95	42 30	30	94	43 00	29	93
53	38 30	33	100	39 03	33	99	39 36	33	98	40 09	31	97	40 40	31	95	41 11	31	94	41 42	30	93	42 12	29	92
54	37 42	33	99	38 15	33	98	38 48	32	97	39 20	32	96	39 52	31	95	40 23	30	94	40 53	30	93	41 23	29	91
55	36 54	+33	98	37 27	+33	97	38 00	+32	96	38 32	+32	95	39 04	+31	94	39 35	+30	93	40 05	+30	92	40 35	+28	91
56	36 06	33	98	36 39	33	97	37 12	32	96	37 44	31	95	38 15	31	94	38 46	30	92	39 16	30	91	39 46	29	90
57	35 18	33	97	35 51	32	96	36 23	32	95	36 55	32	94	37 27	31	93	37 58	30	92	38 28	29	91	38 57	29	90
58	34 29	33	96	35 02	33	95	35 35	32	94	36 07	31	93	36 38	31	92	37 09	30	91	37 39	30	90	38 09	29	89
59	33 41	33	96	34 14	32	95	34 46	32	94	35 18	32	93	35 50	30	92	36 20	31	91	36 51	29	90	37 20	29	88
60	32 53	+33	95	33 26	+32	94	33 58	+32	93	34 30	+31	92	35 01	+31	91	35 32	+30	90	36 02	+30	89	36 32	+29	88
61	32 04	33	95	32 37	32	94	33 09	32	93	33 41	32	91	34 13	30	90	34 43	31	89	35 14	29	88	35 43	29	87
62	31 16	33	94	31 49	32	93	32 21	32	92	32 53	31	91	33 24	31	90	33 55	30	89	34 25	30	88	34 55	29	87
63	30 28	32	93	31 00	32	92	31 32	32	91	32 04	32	90	32 36	30	89	33 06	31	88	33 37	29	87	34 06	30	86
64	29 39	33	93	30 12	32	92	30 44	32	91	31 16	31	90	31 47	31	89	32 18	30	88	32 48	30	87	33 18	29	86
65	28 51	+32	92	29 23	+32	91	29 55	+32	90	30 27	+31	89	30 58	+31	88	31 29	+31	87	32 00	+30	86	32 30	+29	85
66	28 02	33	91	28 35	32	90	29 07	32	90	29 39	31	89	30 10	31	88	30 41	30	87	31 11	30	86	31 41	30	85
67	27 13	33	91	27 46	32	90	28 18	32	89	28 50	31	88	29 21	31	87	29 52	31	86	30 23	30	85	30 53	30	84
68	26 25	33	90	26 58	32	89	27 30	32	88	28 02	31	87	28 33	31	86	29 04	31	85	29 35	30	84	30 05	29	83
69	25 36	33	90	26 09	32	89	26 41	32	88	27 13	32	87	27 45	31	86	28 16	30	85	28 46	31	84	29 17	29	83

LHA	23°			24°			25°			26°			27°			28°			29°			LHA
	Hc	d	Z	Hc	d	Z	Hc	d	Z	Hc	d	Z	Hc	d	Z	Hc	d	Z	Hc	d	Z	
	° ′	′	°	° ′	′	°	° ′	′	°	° ′	′	°	° ′	′	°	° ′	′	°	° ′	′	°	
0	77 00	+60	180	78 00	+60	180	79 00	+60	180	80 00	+60	180	81 00	+60	180	82 00	+60	180	83 00	+60	180	360
1	76 58	60	176	77 58	60	176	78 58	60	175	79 58	60	175	80 58	59	174	81 57	60	174	82 57	60	173	359
2	76 53	60	172	77 53	59	171	78 52	59	171	79 51	59	170	80 50	59	169	81 49	59	168	82 48	58	166	358
3	76 45	58	168	77 43	59	167	78 42	59	166	79 41	58	165	80 39	57	163	81 36	58	162	82 34	56	159	357
4	76 33	58	164	77 31	58	163	78 29	57	162	79 26	57	160	80 23	56	158	81 19	55	156	82 14	54	153	356
5	76 18	+57	160	77 15	+57	159	78 12	+56	157	79 08	+55	156	80 03	+54	153	80 57	+53	151	81 50	+51	148	355
6	76 00	56	157	76 56	55	155	77 51	55	153	78 46	53	151	79 39	52	149	80 31	51	146	81 22	48	143	354
7	75 39	55	153	76 34	54	151	77 28	53	149	78 21	52	147	79 13	50	145	80 03	48	142	80 51	46	138	353
8	75 16	53	150	76 09	53	148	77 02	51	146	77 53	50	143	78 43	48	141	79 31	46	138	80 17	43	134	352
9	74 50	53	147	75 43	50	145	76 33	50	142	77 23	48	140	78 11	46	137	78 57	44	134	79 41	41	130	351
10	74 23	+50	144	75 13	+50	142	76 03	+48	139	76 51	+46	137	77 37	+44	134	78 21	+42	131	79 03	+39	127	350
11	73 53	49	141	74 42	48	139	75 30	47	136	76 17	44	134	77 01	43	131	77 44	40	128	78 24	36	124	349
12	73 21	48	138	74 09	47	136	74 56	45	134	75 41	43	131	76 24	40	128	77 04	39	125	77 43	35	121	348
13	72 48	47	136	73 35	45	133	74 20	43	131	75 03	42	128	75 45	39	126	76 24	37	122	77 01	33	119	347
14	72 13	46	133	72 59	44	131	73 43	42	129	74 25	40	126	75 05	38	123	75 43	35	120	76 18	32	117	346
15	71 37	+44	131	72 21	+43	129	73 04	+41	126	73 45	+39	124	74 24	+36	121	75 00	+34	118	75 34	+31	115	345
16	71 00	43	129	71 43	41	127	72 24	40	124	73 04	38	122	73 42	35	119	74 17	33	116	74 50	30	113	344
17	70 22	42	127	71 04	40	125	71 44	38	122	72 22	37	120	72 59	34	117	73 33	32	114	74 05	29	111	343
18	69 42	41	125	70 23	39	123	71 02	38	121	71 40	35	118	72 15	33	115	72 48	31	113	73 19	28	110	342
19	69 02	40	123	69 42	38	121	70 20	37	119	70 57	34	116	71 31	32	114	72 03	30	111	72 33	28	108	341
20	68 21	+39	121	69 00	+37	119	69 37	+36	117	70 13	+33	115	70 46	+32	112	71 18	+29	110	71 47	+27	107	340
21	67 39	38	120	68 17	37	118	68 54	34	116	69 28	33	113	70 01	31	111	70 32	28	108	71 00	26	106	339
22	66 57	37	118	67 34	36	116	68 10	34	114	68 44	32	112	69 16	30	110	69 46	27	107	70 13	26	105	338
23	66 14	36	117	66 50	35	115	67 25	33	113	67 58	32	111	68 30	29	108	68 59	27	106	69 26	25	103	337
24	65 30	36	115	66 06	34	114	66 40	33	111	67 13	30	109	67 43	29	107	68 12	27	105	68 39	25	102	336
25	64 46	+35	114	65 21	+34	112	65 55	+32	110	66 27	+30	108	66 57	+28	106	67 25	+27	104	67 52	+24	101	335
26	64 02	34	113	64 36	33	111	65 09	31	109	65 40	30	107	66 10	28	105	66 38	26	103	67 04	24	100	334
27	63 17	34	112	63 51	32	110	64 23	31	108	64 54	29	106	65 23	27	104	65 50	26	102	66 16	24	99	333
28	62 31	34	111	63 05	32	109	63 37	30	107	64 07	29	105	64 36	27	103	65 03	25	101	65 28	23	99	332
29	61 46	33	109	62 19	31	108	62 50	30	106	63 20	28	104	63 48	27	102	64 15	25	100	64 40	23	98	331
30	61 00	+32	108	61 32	+31	107	62 03	+30	105	62 33	+28	103	63 01	+26	101	63 27	+25	99	63 52	+23	97	330
31	60 14	32	107	60 46	30	106	61 16	29	104	61 45	28	102	62 13	26	100	62 39	25	98	63 04	22	96	329
32	59 27	32	106	59 59	30	105	60 29	29	103	60 58	27	101	61 25	26	99	61 51	24	97	62 15	23	95	328
33	58 41	31	105	59 12	30	104	59 42	28	102	60 10	27	100	60 37	26	98	61 03	24	97	61 27	22	95	327
34	57 54	30	104	58 24	30	103	58 54	28	101	59 22	27	99	59 49	26	98	60 15	24	96	60 39	22	94	326
35	57 07	+30	104	57 37	+29	102	58 06	+28	100	58 34	+27	99	59 01	+25	97	59 26	+24	95	59 50	+22	93	325
36	56 19	31	103	56 50	29	101	57 19	27	100	57 46	27	98	58 13	25	96	58 38	24	94	59 02	22	93	324
37	55 32	30	102	56 02	29	100	56 31	27	99	56 58	27	97	57 25	25	95	57 50	23	94	58 13	22	92	323
38	54 44	30	101	55 14	29	100	55 43	27	98	56 10	26	96	56 36	25	95	57 01	24	93	57 25	22	91	322
39	53 56	30	100	54 26	29	99	54 55	27	97	55 22	26	96	55 48	25	94	56 13	23	92	56 36	22	91	321
40	53 09	+29	99	53 38	+28	98	54 06	+27	97	54 33	+26	95	54 59	+25	93	55 24	+24	92	55 48	+22	90	320
41	52 21	29	99	52 50	28	97	53 18	27	96	53 45	26	94	54 11	25	93	54 36	23	91	54 59	22	90	319
42	51 33	29	98	52 02	28	97	52 30	27	95	52 57	25	94	53 22	25	92	53 47	23	91	54 10	23	89	318
43	50 45	28	97	51 13	28	96	51 41	27	94	52 08	26	93	52 34	25	92	52 59	23	90	53 22	22	88	317
44	49 56	29	97	50 25	28	95	50 53	27	94	51 20	25	92	51 45	25	91	52 10	23	89	52 33	23	88	316
45	49 08	+29	96	49 37	+27	95	50 04	+27	93	50 31	+26	92	50 57	+24	90	51 21	+24	89	51 45	+22	87	315
46	48 20	28	95	48 48	28	94	49 16	27	93	49 43	25	91	50 08	25	90	50 33	23	88	50 56	23	87	314
47	47 31	29	95	48 00	27	93	48 27	27	92	48 54	26	91	49 20	24	89	49 44	24	88	50 08	23	86	313
48	46 43	28	94	47 11	28	93	47 39	27	91	48 06	25	90	48 31	25	89	48 56	24	87	49 20	22	86	312
49	45 55	28	93	46 23	27	92	46 50	27	91	47 17	26	89	47 43	24	88	48 07	24	87	48 31	23	85	311
50	45 06	+28	93	45 34	+28	91	46 02	+27	90	46 29	+25	89	46 54	+25	87	47 19	+24	86	47 43	+23	85	310
51	44 18	28	92	44 46	27	91	45 13	27	90	45 40	26	88	46 06	25	87	46 31	24	86	46 55	23	84	309
52	43 29	28	91	43 57	28	90	44 25	26	89	44 51	26	88	45 17	25	86	45 42	24	85	46 06	23	84	308
53	42 41	28	91	43 09	27	90	43 36	27	88	44 03	26	87	44 29	25	86	44 54	24	85	45 18	23	83	307
54	41 52	28	90	42 20	28	89	42 48	27	88	43 15	25	87	43 40	26	85	44 06	24	84	44 30	23	83	306
55	41 03	+29	90	41 32	+27	88	41 59	+27	87	42 26	+26	86	42 52	+25	85	43 17	+25	84	43 42	+23	82	305
56	40 15	28	89	40 43	28	88	41 11	27	87	41 38	26	86	42 04	25	84	42 29	25	83	42 54	23	82	304
57	39 26	29	88	39 55	27	87	40 22	27	86	40 49	27	85	41 16	25	84	41 41	25	83	42 06	24	81	303
58	38 38	28	88	39 06	28	87	39 34	27	86	40 01	26	84	40 27	26	83	40 53	25	82	41 18	24	81	302
59	37 49	29	87	38 18	28	86	38 46	27	85	39 13	26	84	39 39	26	83	40 05	25	82	40 30	24	80	301
60	37 01	+28	87	37 29	+28	86	37 57	+27	85	38 24	+27	83	38 51	+26	82	39 17	+25	81	39 42	+24	80	300
61	36 12	29	86	36 41	28	85	37 09	27	84	37 36	27	83	38 03	26	82	38 29	25	81	38 54	25	79	299
62	35 24	29	86	35 53	28	85	36 21	27	84	36 48	27	82	37 15	26	81	37 41	26	80	38 07	24	79	298
63	34 36	28	85	35 04	28	84	35 32	28	83	36 00	27	82	36 27	26	81	36 53	26	80	37 19	25	79	297
64	33 47	29	85	34 16	28	84	34 44	28	82	35 12	27	81	35 39	27	80	36 06	25	79	36 31	26	78	296
65	32 59	+29	84	33 28	+28	83	33 56	+28	82	34 24	+27	81	34 51	+27	80	35 18	+26	79	35 44	+25	78	295
66	32 11	29	84	32 40	28	82	33 08	28	81	33 36	28	80	34 04	26	79	34 30	27	78	34 57	25	77	294
67	31 23	29	83	31 52	28	82	32 20	28	81	32 48	28	80	33 16	27	79	33 43	26	78	34 09	26	77	293
68	30 34	30	82	31 04	28	81	31 32	29	80	32 01	27	79	32 28	28	78	32 56	26	77	33 22	26	76	292
69	29 46	30	82	30 16	29	81	30 45	28	80	31 13	28	79	31 41	27	78	32 08	27	77	32 35	26	76	291

LAT 36°

S. Lat. {LHA greater than 180°……. Zn=180−Z; LHA less than 180°……. Zn=180+Z}

DECLINATION (15°-29°) SAME NAME AS LATITUDE

LAT 36°

LAT 35°

LHA	15° Hc	d	Z	16° Hc	d	Z	17° Hc	d	Z	18° Hc	d	Z	19° Hc	d	Z
	° ′	′	°	° ′	′	°	° ′	′	°	° ′	′	°	° ′	′	°
0	70 00	+60	180	71 00	+60	180	72 00	+60	180	73 00	+60	180	74 00	+60	180
1	69 59	60	177	70 59	60	177	71 59	60	177	72 59	50	177	73 59	59	177
2	69 55	60	174	70 55	60	174	71 55	59	174	72 54	60	174	73 54	60	173
3	69 49	60	172	70 49	59	171	71 48	60	171	72 48	59	170	73 47	59	170
4	69 41	59	169	70 40	59	168	71 39	59	168	72 38	59	167	73 37	58	167
5	69 30	+59	166	70 29	+58	166	71 27	+59	165	72 26	+58	164	73 24	+58	163
6	69 17	58	163	70 15	58	163	71 13	58	162	72 11	57	161	73 08	58	160
7	69 02	58	161	70 00	57	160	70 57	57	159	71 54	56	158	72 50	57	157
8	68 45	57	158	69 42	56	157	70 38	56	156	71 34	56	155	72 30	55	154
9	68 26	56	156	69 22	55	155	70 17	56	154	71 13	55	153	72 08	54	151
10	68 05	+55	153	69 00	+55	152	69 55	+54	151	70 49	+54	150	71 43	+53	148
11	67 42	54	151	68 36	54	150	69 30	53	149	70 23	53	147	71 16	52	146
12	67 17	53	149	68 10	53	148	69 03	53	146	69 56	52	145	70 48	51	143
13	66 50	53	147	67 43	52	145	68 35	52	144	69 27	50	143	70 17	50	141
14	66 23	51	144	67 14	52	143	68 06	50	142	68 56	50	140	69 46	48	139
15	65 53	+51	142	66 44	+50	141	67 34	+50	140	68 24	+48	138	69 12	+48	136
16	65 22	51	140	66 13	49	139	67 02	48	138	67 50	48	136	68 38	47	134
17	64 50	50	138	65 40	48	137	66 28	48	136	67 16	46	134	68 02	46	132
18	64 17	49	137	65 06	47	135	65 53	47	134	66 40	45	132	67 25	45	130
19	63 43	47	135	64 30	47	133	65 17	46	132	66 03	44	130	66 47	44	129
20	63 07	+47	133	63 54	+46	132	64 40	+45	130	65 25	+44	129	66 09	+42	127
21	62 31	46	131	63 17	45	130	64 02	44	129	64 46	43	127	65 29	42	125
22	61 54	45	130	62 39	44	128	63 23	43	127	64 06	42	125	64 48	41	124
23	61 16	44	128	62 00	43	127	62 43	43	125	63 26	41	124	64 07	40	122
24	60 37	43	127	61 20	43	125	62 03	42	124	62 45	40	122	63 25	40	121
25	59 57	+43	125	60 40	+42	124	61 22	+41	123	62 03	+40	121	62 43	+38	119
26	59 16	43	124	59 59	41	123	60 40	40	121	61 20	40	120	62 00	38	118
27	58 35	42	123	59 17	41	121	59 58	39	120	60 37	39	118	61 16	37	117
28	57 54	41	121	58 35	40	120	59 15	39	119	59 54	38	117	60 32	37	116
29	57 12	40	120	57 52	39	119	58 31	39	117	59 10	37	116	59 47	36	114
30	56 29	+40	119	57 09	+38	118	57 47	+38	116	58 25	+37	115	59 02	+36	113
31	55 46	39	118	56 25	38	117	57 03	38	115	57 41	36	114	58 17	35	112
32	55 02	39	117	55 41	37	115	56 18	37	114	56 55	36	113	57 31	35	111
33	54 18	38	116	54 56	37	114	55 33	37	113	56 10	35	112	56 45	35	110
34	53 33	38	115	54 11	37	113	54 48	36	112	55 24	35	111	55 59	34	109
35	52 48	+38	114	53 26	+36	112	54 02	+36	111	54 38	+34	110	55 12	+34	108
36	52 03	37	113	52 40	36	111	53 16	35	110	53 51	34	109	54 25	34	107
37	51 18	36	112	51 54	36	110	52 30	35	109	53 05	33	108	53 38	33	106
38	50 32	36	111	51 08	35	109	51 43	35	108	52 18	33	107	52 51	33	105
39	49 46	35	110	50 21	35	109	50 56	34	107	51 30	34	106	52 04	32	105
40	48 59	+36	109	49 35	+34	108	50 09	+34	106	50 43	+33	105	51 16	+32	104
41	48 13	35	108	48 48	34	107	49 22	34	106	49 56	32	104	50 28	32	103
42	47 26	35	107	48 01	34	106	48 35	33	105	49 08	32	104	49 40	32	102
43	46 39	34	106	47 13	34	105	47 47	33	104	48 20	32	103	48 52	31	101
44	45 52	34	106	46 26	33	104	46 59	33	103	47 32	32	102	48 04	31	101
45	45 04	+34	105	45 38	+33	104	46 11	+33	102	46 44	+31	101	47 15	+31	100
46	44 17	33	104	44 50	33	103	45 23	33	102	45 56	31	100	46 27	31	99
47	43 29	33	103	44 02	33	102	44 35	32	101	45 07	31	100	45 38	31	99
48	42 41	33	103	43 14	33	101	43 47	32	100	44 19	31	99	44 50	30	98
49	41 53	33	102	42 26	32	101	42 58	32	100	43 30	31	98	44 01	30	97
50	41 05	+33	101	41 38	+32	100	42 10	+31	99	42 41	+31	98	43 12	+30	97
51	40 16	33	100	40 49	32	99	41 21	32	98	41 53	30	97	42 23	30	96
52	39 28	32	100	40 00	32	99	40 32	32	97	41 04	30	96	41 34	30	95
53	38 39	33	99	39 12	32	98	39 44	31	97	40 15	30	96	40 45	30	95
54	37 51	32	98	38 23	32	97	38 55	31	96	39 26	30	95	39 56	30	94
55	37 02	+32	98	37 34	+32	97	38 06	+31	96	38 37	+30	94	39 07	+30	93
56	36 13	32	97	36 45	32	96	37 17	31	95	37 48	30	94	38 18	30	93
57	35 25	32	96	35 57	31	95	36 28	31	94	36 59	30	93	37 29	30	92
58	34 36	32	96	35 08	31	95	35 39	31	94	36 10	30	93	36 40	30	92
59	33 47	32	95	34 19	31	94	34 50	31	93	35 21	30	92	35 51	30	91
60	32 58	+32	94	33 30	+31	93	34 01	+31	92	34 32	+30	91	35 02	+30	90
61	32 09	31	94	32 40	32	93	33 12	30	92	33 42	31	91	34 13	29	90
62	31 20	31	93	31 51	32	92	32 23	30	91	32 53	31	90	33 24	29	89
63	30 31	31	93	31 02	31	92	31 33	31	91	32 04	30	90	32 34	30	89
64	29 41	32	92	30 13	31	91	30 44	31	90	31 15	30	89	31 45	30	88
65	28 52	+32	91	29 24	+31	91	29 55	+31	90	30 26	+30	89	30 56	+30	88
66	28 03	32	91	28 35	31	90	29 06	31	89	29 37	30	88	30 07	30	87
67	27 14	32	90	27 46	31	89	28 17	31	88	28 48	30	87	29 18	30	86
68	26 25	32	90	26 57	31	89	27 28	31	88	27 59	30	87	28 29	30	86
69	25 36	31	89	26 07	32	88	26 39	30	87	27 09	31	86	27 40	30	85

LHA	20° Hc	d	Z	21° Hc	d	Z	22° Hc	d	Z	23° Hc	d	Z	24° Hc	d	Z
	° ′	′	°	° ′	′	°	° ′	′	°	° ′	′	°	° ′	′	°
0	75 00	+60	180	76 00	+60	180	77 00	+60	180	78 00	+60	180	79 00	+60	180
1	74 58	60	176	75 58	60	176	76 58	60	176	77 58	60	176	78 58	60	175
2	74 54	59	173	75 53	60	172	76 53	59	172	77 52	60	171	78 52	59	171
3	74 46	59	169	75 45	59	169	76 44	59	168	77 43	59	167	78 42	58	166
4	74 35	59	166	75 34	58	165	76 32	58	164	77 30	58	163	78 28	57	161
5	74 22	+58	162	75 20	+57	161	76 17	+57	160	77 14	+57	159	78 11	+55	157
6	74 06	56	159	75 02	57	158	75 59	56	156	76 55	55	155	77 50	54	153
7	73 47	56	156	74 43	55	154	75 38	54	153	76 32	54	151	77 26	53	149
8	73 25	55	153	74 20	54	151	75 14	53	150	76 07	53	148	77 00	51	146
9	73 02	53	150	73 55	53	148	74 48	52	146	75 40	51	144	76 31	49	142
10	72 36	+52	147	73 28	+52	145	74 20	+50	143	75 10	+50	141	76 00	+47	139
11	72 08	51	144	72 59	51	143	73 50	49	141	74 39	47	139	75 26	47	136
12	71 39	50	142	72 29	48	140	73 17	48	138	74 05	46	136	74 51	45	133
13	71 07	49	139	71 56	48	137	72 44	46	135	73 30	45	133	74 15	43	131
14	70 34	48	137	71 22	46	135	72 08	46	133	72 54	43	131	73 37	42	128
15	70 00	+47	135	70 47	+45	133	71 32	+44	131	72 16	+42	129	72 58	+40	126
16	69 25	45	133	70 10	44	131	70 54	43	129	71 37	41	127	72 18	39	124
17	68 48	44	131	69 32	43	129	70 15	42	127	70 57	40	125	71 37	38	122
18	68 10	43	129	68 53	42	127	69 35	41	125	70 16	39	123	70 55	37	120
19	67 31	42	127	68 13	41	125	68 54	40	123	69 34	38	121	70 12	36	119
20	66 51	+42	125	67 33	+40	123	68 13	+38	121	68 51	+37	119	69 28	+35	117
21	66 11	40	124	66 51	39	122	67 30	38	120	68 08	36	118	68 44	35	116
22	65 29	40	122	66 09	38	120	66 47	37	118	67 24	36	116	68 00	33	114
23	64 47	39	121	65 26	38	119	66 04	36	117	66 40	34	115	67 14	33	113
24	64 05	38	119	64 43	37	117	65 20	35	115	65 55	34	113	66 29	32	111
25	63 21	+38	118	63 59	+36	116	64 35	+35	114	65 10	+33	112	65 43	+32	110
26	62 38	36	116	63 14	36	115	63 50	34	113	64 24	33	111	64 57	31	109
27	61 53	37	115	62 30	34	113	63 04	34	112	63 38	32	110	64 10	31	108
28	61 09	35	114	61 44	35	112	62 19	33	111	62 52	31	109	63 23	30	107
29	60 23	36	113	60 59	33	111	61 32	33	109	62 05	31	108	62 36	29	106
30	59 38	+35	112	60 13	+33	110	60 46	+32	108	61 18	+31	107	61 49	+29	105
31	58 52	34	111	59 26	33	109	59 59	32	107	60 31	30	106	61 01	29	104
32	58 06	34	110	58 40	32	108	59 12	31	106	59 43	30	105	60 13	29	103
33	57 20	33	109	57 53	32	107	58 25	31	105	58 56	29	104	59 25	28	102
34	56 33	33	108	57 06	31	106	57 37	31	105	58 08	29	103	58 37	28	101
35	55 46	+32	107	56 18	+32	105	56 50	+30	104	57 20	+29	102	57 49	+27	100
36	54 59	32	106	55 31	31	104	56 02	30	103	56 32	28	101	57 00	28	100
37	54 11	32	105	54 43	31	103	55 14	29	102	55 43	29	100	56 12	27	99
38	53 24	31	104	53 55	31	103	54 26	29	101	54 55	28	100	55 23	27	98
39	52 36	31	103	53 07	30	102	53 37	29	100	54 06	28	99	54 34	27	97
40	51 48	+31	102	52 19	+30	101	52 49	+29	100	53 18	+28	98	53 46	+26	97
41	51 00	31	102	51 31	29	100	52 00	29	99	52 29	28	97	52 57	26	96
42	50 12	30	101	50 42	30	100	51 12	28	98	51 40	28	97	52 08	26	95
43	49 23	31	100	49 54	29	99	50 23	29	97	50 52	27	96	51 19	26	95
44	48 35	30	99	49 05	29	98	49 34	29	97	50 03	27	95	50 30	26	94
45	47 46	+30	99	48 16	+29	97	48 45	+29	96	49 14	+27	95	49 41	+26	93
46	46 58	30	98	47 28	29	97	47 57	28	95	48 25	27	94	48 52	26	93
47	46 09	30	97	46 39	29	96	47 08	28	95	47 36	27	93	48 03	26	92
48	45 20	30	97	45 50	29	95	46 19	28	94	46 47	27	93	47 14	26	92
49	44 31	30	96	45 01	29	95	45 30	27	93	45 57	27	92	46 24	27	91
50	43 42	+30	95	44 12	+28	94	44 40	+28	93	45 08	+27	92	45 35	+26	90
51	42 53	30	95	43 23	28	93	43 51	28	92	44 19	27	91	44 46	26	90
52	42 04	30	94	42 34	28	93	43 02	28	92	43 30	27	90	43 57	26	89
53	41 15	30	93	41 45	28	92	42 13	28	91	42 41	27	90	43 08	26	89
54	40 26	29	93	40 55	29	92	41 24	28	90	41 52	27	89	42 19	26	88
55	39 37	+29	92	40 06	+29	91	40 35	+28	90	41 03	+27	89	41 30	+26	88
56	38 48	29	92	39 17	29	90	39 46	27	89	40 13	28	88	40 41	26	87
57	37 59	29	91	38 28	29	90	38 57	27	89	39 24	27	88	39 51	27	86
58	37 10	29	90	37 39	28	89	38 07	28	88	38 35	27	87	39 02	27	86
59	36 21	29	90	36 50	28	89	37 18	28	88	37 46	27	87	38 13	27	85
60	35 32	+29	89	36 01	+28	88	36 29	+28	87	36 57	+27	86	37 24	+27	85
61	34 42	30	89	35 12	28	88	35 40	28	87	36 08	27	85	36 35	27	84
62	33 53	29	88	34 22	29	87	34 51	28	86	35 19	28	85	35 47	26	84
63	33 04	29	88	33 33	29	87	34 02	28	86	34 30	28	84	34 58	27	83
64	32 15	29	87	32 44	29	86	33 13	28	85	33 41	28	84	34 09	27	83
65	31 26	+29	87	31 55	+29	86	32 24	+28	84	32 52	+28	83	33 20	+27	82
66	30 37	29	86	31 06	29	85	31 35	29	84	32 04	28	83	32 32	27	82
67	29 48	29	85	30 17	29	84	30 46	29	83	31 15	28	82	31 43	27	81
68	28 59	29	85	29 28	30	84	29 58	28	83	30 26	28	82	30 54	28	81
69	28 10	30	84	28 40	29	83	29 09	29	82	29 38	28	81	30 06	28	80

25° Hc	d	Z	26° Hc	d	Z	27° Hc	d	Z	28° Hc	d	Z	29° Hc	d	Z	LHA
° ′	′	°	° ′	′	°	° ′	′	°	° ′	′	°	° ′	′	°	
80 00	+60	180	81 00	+60	130	82 00	+60	180	83 00	+60	180	84 00	+60	180	360
79 58	60	175	80 58	59	174	81 57	50	174	82 57	59	173	83 56	60	172	359
79 51	59	170	80 50	59	169	81 49	59	167	82 48	58	166	83 46	57	164	358
79 40	58	165	80 38	58	163	81 36	57	161	82 33	56	159	83 29	55	156	357
79 25	57	160	80 22	56	158	81 18	55	156	82 13	54	153	83 07	51	149	356
79 06	+56	155	80 02	+54	153	80 56	+52	151	81 48	+51	147	82 39	+49	143	355
78 44	54	151	79 38	52	149	80 30	50	146	81 20	48	142	82 08	45	138	354
78 19	51	147	79 10	50	144	80 00	48	141	80 48	46	138	81 34	42	133	353
77 51	49	143	78 40	48	140	79 28	46	137	80 14	43	134	80 57	39	129	352
77 20	48	140	78 08	46	137	78 54	43	134	79 37	41	130	80 18	37	126	351
76 47	+46	137	77 33	+44	134	78 17	+42	130	78 59	+38	127	79 37	+35	123	350
76 13	44	134	76 57	42	131	77 39	40	127	78 19	36	124	78 55	33	120	349
75 36	43	131	76 19	40	128	76 59	38	125	77 37	35	121	78 12	31	117	348
74 58	41	128	75 39	39	125	76 18	37	122	76 55	33	119	77 28	30	115	347
74 19	40	126	74 59	37	123	75 36	35	120	76 11	32	117	76 43	29	113	346
73 38	+39	124	74 17	+36	121	74 53	+34	118	75 27	+30	115	75 57	+28	111	345
72 57	37	122	73 34	35	119	74 09	33	116	74 42	29	113	75 11	27	109	344
72 15	36	120	72 51	34	117	73 25	31	114	73 56	29	111	74 25	25	108	343
71 32	35	118	72 07	33	115	72 40	30	113	73 10	28	110	73 38	25	107	342
70 48	34	116	71 22	32	114	71 54	29	111	72 23	27	108	72 50	25	105	341
70 03	+34	115	70 37	+31	112	71 08	+29	110	71 37	+26	107	72 03	+24	104	340
69 19	32	113	69 51	30	111	70 21	28	108	70 49	26	106	71 15	23	103	339
68 33	32	112	69 05	29	109	69 34	28	107	70 02	25	104	70 27	23	102	338
67 47	31	111	68 18	29	108	68 47	27	106	69 14	25	103	69 39	22	101	337
67 01	30	109	67 31	29	107	68 00	26	105	68 26	25	102	68 51	21	100	336
66 15	+29	108	66 44	+28	106	67 12	+26	104	67 38	+24	101	68 02	+22	99	335
65 28	29	107	65 57	27	105	66 24	26	103	66 50	23	100	67 13	22	98	334
64 41	28	106	65 09	27	104	65 36	25	102	66 01	24	99	66 25	21	97	333
63 53	28	105	64 21	27	103	64 48	25	101	65 13	23	99	65 36	21	96	332
63 05	28	104	63 33	27	102	64 00	24	100	64 24	23	98	64 47	21	96	331
62 18	+27	103	62 45	+26	101	63 11	+25	99	63 36	+22	97	63 58	+21	95	330
61 30	27	102	61 57	26	100	62 23	24	98	62 47	22	96	63 09	21	94	329
60 42	27	101	61 09	25	99	61 34	24	97	61 58	22	96	62 20	20	94	328
59 53	27	100	60 20	25	99	60 45	24	97	61 09	22	95	61 31	20	93	327
59 05	26	100	59 31	25	98	59 56	24	96	60 20	22	94	60 42	20	92	326
58 16	+27	99	58 43	+24	97	59 07	+24	95	59 31	+22	93	59 53	+20	92	325
57 28	26	98	57 54	24	96	58 18	24	95	58 42	22	93	59 04	20	91	324
56 39	26	97	57 05	24	96	57 29	24	94	57 53	21	92	58 14	21	90	323
55 50	26	97	56 16	24	95	56 40	24	93	57 04	21	92	57 25	21	90	322
55 01	26	96	55 27	24	94	55 51	23	93	56 14	22	91	56 36	20	89	321
54 12	+26	95	54 38	+24	94	55 02	+23	92	55 25	+22	90	55 47	+20	89	320
53 23	26	94	53 49	24	93	54 13	23	91	54 36	22	90	54 58	20	88	319
52 34	26	94	53 00	24	92	53 24	23	91	53 47	22	89	54 09	20	88	318
51 45	26	93	52 11	24	92	52 35	23	90	52 58	22	89	53 20	20	87	317
50 56	25	93	51 21	25	91	51 46	23	90	52 09	22	88	52 31	20	87	316
50 07	+25	92	50 32	+24	91	50 56	+24	89	51 20	+22	88	51 42	+20	86	315
49 18	25	91	49 43	24	90	50 07	23	89	50 30	22	87	50 52	21	86	314
48 29	25	91	48 54	24	89	49 18	23	88	49 41	23	87	50 04	21	85	313
47 40	25	90	48 05	24	89	48 29	23	87	48 52	23	86	49 15	21	85	312
46 51	25	90	47 16	24	88	47 40	23	87	48 03	23	86	48 26	21	84	311
46 01	+26	89	46 27	+24	88	46 51	+23	86	47 14	+23	85	47 37	+21	84	310
45 12	26	89	45 38	24	87	46 02	23	86	46 25	23	85	46 48	22	83	309
44 23	25	88	44 48	25	87	45 13	24	85	45 37	22	84	45 59	22	83	308
43 34	25	87	43 59	25	86	44 24	24	85	44 48	22	84	45 10	22	82	307
42 45	25	87	43 10	25	86	43 35	24	84	43 59	23	83	44 22	22	82	306
41 56	+25	86	42 21	+25	85	42 46	+24	84	43 10	+23	83	43 33	+22	81	305
41 07	25	86	41 32	25	85	41 57	24	83	42 21	24	82	42 45	22	81	304
40 18	26	85	40 44	25	84	41 09	24	83	41 33	23	82	41 56	23	80	303
39 29	26	85	39 55	25	84	40 20	24	82	40 44	24	81	41 08	22	80	302
38 40	26	84	39 06	25	83	39 31	25	82	39 56	23	81	40 19	23	80	301
37 51	+26	84	38 17	+25	83	38 42	+25	81	39 07	+24	80	39 31	+23	79	300
37 02	26	83	37 28	26	82	37 54	25	81	38 19	24	80	38 43	23	79	299
36 13	27	83	36 40	25	82	37 05	25	81	37 30	25	79	37 55	23	78	298
35 25	26	82	35 51	26	81	36 17	25	80	36 42	25	79	37 07	23	78	297
34 36	27	82	35 03	26	81	35 29	25	80	35 54	25	78	36 19	24	77	296
33 47	+27	81	34 14	+26	80	34 40	+26	79	35 06	+25	78	35 31	+24	77	295
32 59	27	81	33 26	26	80	33 52	26	79	34 18	25	78	34 43	24	76	294
32 10	27	80	32 37	27	79	33 04	26	78	33 30	25	77	33 55	25	76	293
31 22	27	80	31 49	27	79	32 16	26	78	32 42	26	77	33 08	25	76	292
30 34	27	79	31 01	27	78	31 28	26	77	31 54	26	76	32 20	25	75	291

S. Lat. {LHA greater than 180°........Zn=180−Z; LHA less than 180°...........Zn=180+Z}

DECLINATION (15°-29°) SAME NAME AS LATITUDE

LAT 35°

Table 12A

Correction to Tabulated Altitude for Minutes of Declination

d / ′	1	2	3	4	5	6	7	8	9	10	11	12	13	14	15	16	17	18	19	20	21	22	23	24	25	26	27	28	29	30
0	0	0	0	0	0	0	0	0	0	0	0	0	0	0	0	0	0	0	0	0	0	0	0	0	0	0	0	0	0	0
1	0	0	0	0	0	0	0	0	0	0	0	0	0	0	0	0	0	0	0	0	0	0	0	0	0	0	0	0	0	0
2	0	0	0	0	0	0	0	0	0	0	0	0	0	0	0	1	1	1	1	1	1	1	1	1	1	1	1	1	1	1
3	0	0	0	0	0	0	0	0	0	0	1	1	1	1	1	1	1	1	1	1	1	1	1	1	1	1	1	1	1	2
4	0	0	0	0	0	0	0	1	1	1	1	1	1	1	1	1	1	1	1	1	1	1	2	2	2	2	2	2	2	2
5	0	0	0	0	0	0	1	1	1	1	1	1	1	1	1	1	1	2	2	2	2	2	2	2	2	2	2	2	2	2
6	0	0	0	0	0	1	1	1	1	1	1	1	1	1	2	2	2	2	2	2	2	2	2	2	2	3	3	3	3	3
7	0	0	0	0	1	1	1	1	1	1	1	1	2	2	2	2	2	2	2	2	2	3	3	3	3	3	3	3	3	4
8	0	0	0	1	1	1	1	1	1	1	1	2	2	2	2	2	2	2	3	3	3	3	3	3	3	3	4	4	4	4
9	0	0	0	1	1	1	1	1	1	2	2	2	2	2	2	2	3	3	3	3	3	3	3	4	4	4	4	4	4	4
10	0	0	0	1	1	1	1	1	2	2	2	2	2	2	2	3	3	3	3	3	4	4	4	4	4	4	4	5	5	5
11	0	0	1	1	1	1	1	1	2	2	2	2	2	3	3	3	3	3	3	4	4	4	4	4	5	5	5	5	5	6
12	0	0	1	1	1	1	1	2	2	2	2	2	3	3	3	3	3	4	4	4	4	4	5	5	5	5	5	6	6	6
13	0	0	1	1	1	1	2	2	2	2	2	3	3	3	3	3	4	4	4	4	5	5	5	5	5	6	6	6	6	6
14	0	0	1	1	1	1	2	2	2	2	3	3	3	3	4	4	4	4	4	5	5	5	5	6	6	6	6	7	7	7
15	0	0	1	1	1	2	2	2	2	2	3	3	3	4	4	4	4	4	5	5	5	6	6	6	6	6	7	7	7	8
16	0	1	1	1	1	2	2	2	2	3	3	3	3	4	4	4	5	5	5	5	6	6	6	6	7	7	7	7	8	8
17	0	1	1	1	1	2	2	2	3	3	3	3	4	4	4	5	5	5	5	6	6	6	7	7	7	7	8	8	8	8
18	0	1	1	1	2	2	2	2	3	3	3	4	4	4	4	5	5	5	6	6	6	7	7	7	8	8	8	8	9	9
19	0	1	1	1	2	2	2	3	3	3	3	4	4	4	5	5	5	6	6	6	7	7	7	8	8	8	9	9	9	10
20	0	1	1	1	2	2	2	3	3	3	4	4	4	5	5	5	6	6	6	7	7	7	8	8	8	9	9	9	10	10
21	0	1	1	1	2	2	2	3	3	4	4	4	5	5	5	6	6	6	7	7	7	8	8	8	9	9	9	10	10	10
22	0	1	1	1	2	2	3	3	3	4	4	4	5	5	6	6	6	7	7	7	8	8	8	9	9	10	10	10	11	11
23	0	1	1	2	2	2	3	3	3	4	4	5	5	5	6	6	7	7	7	8	8	8	9	9	10	10	10	11	11	12
24	0	1	1	2	2	2	3	3	4	4	4	5	5	6	6	6	7	7	8	8	8	9	9	10	10	10	11	11	12	12
25	0	1	1	2	2	2	3	3	4	4	5	5	5	6	6	7	7	8	8	8	9	9	10	10	10	11	11	12	12	12
26	0	1	1	2	2	3	3	3	4	4	5	5	6	6	6	7	7	8	8	9	9	10	10	10	11	11	12	12	13	13
27	0	1	1	2	2	3	3	4	4	4	5	5	6	6	7	7	8	8	9	9	9	10	10	11	11	12	12	13	13	14
28	0	1	1	2	2	3	3	4	4	5	5	6	6	7	7	7	8	8	9	9	10	10	11	11	12	12	13	13	14	14
29	0	1	1	2	2	3	3	4	4	5	5	6	6	7	7	8	8	9	9	10	10	11	11	12	12	13	13	14	14	14
30	0	1	2	2	2	3	4	4	4	5	6	6	6	7	8	8	8	9	10	10	10	11	12	12	12	13	14	14	14	15
31	1	1	2	2	3	3	4	4	5	5	6	6	7	7	8	8	9	9	10	10	11	11	12	12	13	13	14	14	15	16
32	1	1	2	2	3	3	4	4	5	5	6	6	7	7	8	9	9	10	10	11	11	12	12	13	13	14	14	15	15	16
33	1	1	2	2	3	3	4	4	5	6	6	7	7	8	8	9	9	10	10	11	12	12	13	13	14	14	15	15	16	16
34	1	1	2	2	3	3	4	5	5	6	6	7	7	8	8	9	10	10	11	11	12	12	13	14	14	15	15	16	16	17
35	1	1	2	2	3	4	4	5	5	6	6	7	8	8	9	9	10	10	11	12	12	13	13	14	15	15	16	16	17	18
36	1	1	2	2	3	4	4	5	5	6	7	7	8	8	9	10	10	11	11	12	13	13	14	14	15	16	16	17	17	18
37	1	1	2	2	3	4	4	5	6	6	7	7	8	9	9	10	10	11	12	12	13	14	14	15	15	16	17	17	18	18
38	1	1	2	3	3	4	4	5	6	6	7	8	8	9	10	10	11	11	12	13	13	14	15	15	16	16	17	18	18	19
39	1	1	2	3	3	4	5	5	6	6	7	8	8	9	10	10	11	12	12	13	14	14	15	16	16	17	18	18	19	20
40	1	1	2	3	3	4	5	5	6	7	7	8	9	9	10	11	11	12	13	13	14	15	15	16	17	17	18	19	19	20
41	1	1	2	3	3	4	5	5	6	7	8	8	9	10	10	11	12	12	13	14	14	15	16	16	17	18	18	19	20	20
42	1	1	2	3	4	4	5	6	6	7	8	8	9	10	10	11	12	13	13	14	15	15	16	17	18	18	19	20	20	21
43	1	1	2	3	4	4	5	6	6	7	8	9	9	10	11	11	12	13	14	14	15	16	16	17	18	19	19	20	21	22
44	1	1	2	3	4	4	5	6	7	7	8	9	10	10	11	12	12	13	14	15	15	16	17	18	18	19	20	21	21	22
45	1	2	2	3	4	4	5	6	7	8	8	9	10	10	11	12	13	14	14	15	16	16	17	18	19	20	20	21	22	22
46	1	2	2	3	4	5	5	6	7	8	8	9	10	11	12	12	13	14	15	15	16	17	18	18	19	20	21	21	22	23
47	1	2	2	3	4	5	5	6	7	8	9	9	10	11	12	13	13	14	15	16	16	17	18	19	20	20	21	22	23	24
48	1	2	2	3	4	5	6	6	7	8	9	10	10	11	12	13	14	14	15	16	17	18	18	19	20	21	22	22	23	24
49	1	2	2	3	4	5	6	7	7	8	9	10	11	11	12	13	14	15	16	16	17	18	19	20	20	21	22	23	24	24
50	1	2	2	3	4	5	6	7	8	8	9	10	11	12	12	13	14	15	16	17	18	18	19	20	21	22	22	23	24	25
51	1	2	3	3	4	5	6	7	8	8	9	10	11	12	13	14	14	15	16	17	18	19	20	20	21	22	23	24	25	26
52	1	2	3	3	4	5	6	7	8	9	10	10	11	12	13	14	15	16	16	17	18	19	20	21	22	23	23	24	25	26
53	1	2	3	4	4	5	6	7	8	9	10	11	11	12	13	14	15	16	17	18	19	19	20	21	22	23	24	25	26	26
54	1	2	3	4	4	5	6	7	8	9	10	11	12	13	14	14	15	16	17	18	19	20	21	22	22	23	24	25	26	27
55	1	2	3	4	5	6	6	7	8	9	10	11	12	13	14	15	16	16	17	18	19	20	21	22	23	24	25	26	27	28
56	1	2	3	4	5	6	7	7	8	9	10	11	12	13	14	15	16	17	18	19	20	21	21	22	23	24	25	26	27	28
57	1	2	3	4	5	6	7	8	9	10	10	11	12	13	14	15	16	17	18	19	20	21	22	23	24	25	26	27	28	28
58	1	2	3	4	5	6	7	8	9	10	11	12	13	14	14	15	16	17	18	19	20	21	22	23	24	25	26	27	28	29
59	1	2	3	4	5	6	7	8	9	10	11	12	13	14	15	16	17	18	19	20	21	22	23	24	25	26	27	28	29	30

d / ′	31	32	33	34	35	36	37	38	39	40	41	42	43	44	45	46	47	48	49	50	51	52	53	54	55	56	57	58	59	60	d / ′
0	0	0	0	0	0	0	0	0	0	0	0	0	0	0	0	0	0	0	0	0	0	0	0	0	0	0	0	0	0	0	0
1	1	1	1	1	1	1	1	1	1	1	1	1	1	1	1	1	1	1	1	1	1	1	1	1	1	1	1	1	1	1	1
2	1	1	1	1	1	1	1	1	1	1	1	1	1	1	2	2	2	2	2	2	2	2	2	2	2	2	2	2	2	2	2
3	2	2	2	2	2	2	2	2	2	2	2	2	2	2	2	2	2	2	2	2	3	3	3	3	3	3	3	3	3	3	3
4	2	2	2	2	2	2	2	3	3	3	3	3	3	3	3	3	3	3	3	3	3	3	4	4	4	4	4	4	4	4	4
5	3	3	3	3	3	3	3	3	3	3	3	4	4	4	4	4	4	4	4	4	4	4	4	4	5	5	5	5	5	5	5
6	3	3	3	3	4	4	4	4	4	4	4	4	4	4	4	5	5	5	5	5	5	5	5	5	6	6	6	6	6	6	6
7	4	4	4	4	4	4	4	4	5	5	5	5	5	5	5	5	5	6	6	6	6	6	6	6	6	7	7	7	7	7	7
8	4	4	4	5	5	5	5	5	5	5	5	6	6	6	6	6	6	6	7	7	7	7	7	7	7	7	8	8	8	8	8
9	5	5	5	5	5	5	6	6	6	6	6	6	6	7	7	7	7	7	7	8	8	8	8	8	8	8	9	9	9	9	9
10	5	5	6	6	6	6	6	6	6	7	7	7	7	7	8	8	8	8	8	8	8	9	9	9	9	9	10	10	10	10	10
11	6	6	6	6	6	7	7	7	7	7	8	8	8	8	8	8	9	9	9	9	9	10	10	10	10	10	10	11	11	11	11
12	6	6	7	7	7	7	7	8	8	8	8	8	9	9	9	9	9	10	10	10	10	10	11	11	11	11	11	12	12	12	12
13	7	7	7	7	8	8	8	8	8	9	9	9	9	10	10	10	10	10	11	11	11	11	11	12	12	12	12	13	13	13	13
14	7	7	8	8	8	8	9	9	9	9	10	10	10	10	10	11	11	11	11	12	12	12	12	13	13	13	13	14	14	14	14
15	8	8	8	8	9	9	9	10	10	10	10	10	11	11	11	12	12	12	12	12	13	13	13	14	14	14	14	14	15	15	15
16	8	9	9	9	9	10	10	10	10	11	11	11	11	12	12	12	13	13	13	13	14	14	14	14	15	15	15	15	16	16	16
17	9	9	9	10	10	10	10	11	11	11	12	12	12	12	13	13	13	14	14	14	14	15	15	15	16	16	16	16	17	17	17
18	9	10	10	10	10	11	11	11	12	12	12	13	13	13	14	14	14	14	15	15	15	16	16	16	16	17	17	17	18	18	18
19	10	10	10	11	11	11	12	12	12	13	13	13	14	14	14	15	15	15	16	16	16	16	17	17	17	18	18	18	19	19	19
20	10	11	11	11	12	12	12	13	13	13	14	14	14	15	15	15	16	16	16	17	17	17	18	18	18	19	19	19	20	20	20
21	11	11	12	12	12	13	13	13	14	14	14	15	15	15	16	16	16	17	17	18	18	18	19	19	19	20	20	20	21	21	21
22	11	12	12	12	13	13	14	14	14	15	15	15	16	16	16	17	17	18	18	18	19	19	19	20	20	21	21	21	22	22	22
23	12	12	13	13	13	14	14	15	15	15	16	16	16	17	17	18	18	18	19	19	20	20	20	21	21	21	22	22	23	23	23
24	12	13	13	14	14	14	15	15	16	16	16	17	17	18	18	18	19	19	20	20	20	21	21	22	22	22	23	23	24	24	24
25	13	13	14	14	15	15	15	16	16	17	17	18	18	18	19	19	20	20	20	21	21	22	22	22	23	23	24	24	25	25	25
26	13	14	14	15	15	16	16	16	17	17	18	18	19	19	20	20	20	21	21	22	22	23	23	23	24	24	25	25	26	26	26
27	14	14	15	15	16	16	17	17	18	18	18	19	19	20	20	21	21	22	22	22	23	23	24	24	25	25	26	26	27	27	27
28	14	15	15	16	16	17	17	18	18	19	19	20	20	21	21	21	22	22	23	23	24	24	25	25	26	26	27	27	28	28	28
29	15	15	16	16	17	17	18	18	19	19	20	20	21	21	22	22	23	23	24	24	25	25	26	26	27	27	28	28	29	29	29
30	16	16	16	17	18	18	18	19	20	20	20	21	22	22	22	23	24	24	24	25	26	26	26	27	28	28	28	29	30	30	30
31	16	17	17	18	18	19	19	20	20	21	21	22	22	23	23	24	24	25	25	26	26	27	27	28	28	29	29	30	30	31	31
32	17	17	18	18	19	19	20	20	21	21	22	22	23	23	24	25	25	26	26	27	27	28	28	29	29	30	30	31	31	32	32
33	17	18	18	19	19	20	20	21	21	22	23	23	24	24	25	25	26	26	27	28	28	29	29	30	30	31	31	32	32	33	33
34	18	18	19	19	20	20	21	22	22	23	23	24	24	25	26	26	27	27	28	28	29	29	30	31	31	32	32	33	33	34	34
35	18	19	19	20	20	21	22	22	23	23	24	24	25	26	26	27	27	28	29	29	30	30	31	32	32	33	33	34	34	35	35
36	19	19	20	20	21	22	22	23	23	24	25	25	26	26	27	28	28	29	29	30	31	31	32	32	33	34	34	35	35	36	36
37	19	20	20	21	22	22	23	23	24	25	25	26	27	27	28	28	29	30	30	31	31	32	33	33	34	35	35	36	36	37	37
38	20	20	21	22	22	23	23	24	25	25	26	27	27	28	28	29	30	30	31	32	32	33	34	34	35	35	36	37	37	38	38
39	20	21	21	22	23	23	24	25	25	26	27	27	28	29	29	30	31	31	32	32	33	34	34	35	36	36	37	38	38	39	39
40	21	21	22	23	23	24	25	25	26	27	27	28	29	29	30	31	31	32	33	33	34	35	35	36	37	37	38	39	39	40	40
41	21	22	23	23	24	25	25	26	27	27	28	29	29	30	31	31	32	33	33	34	35	36	36	37	38	38	39	40	40	41	41
42	22	22	23	24	24	25	26	27	27	28	29	29	30	31	32	32	33	34	34	35	36	36	37	38	38	39	40	41	41	42	42
43	22	23	24	24	25	26	27	27	28	29	29	30	31	32	32	33	34	34	35	36	37	37	38	39	39	40	41	42	42	43	43
44	23	23	24	25	26	26	27	28	29	29	30	31	32	32	33	34	34	35	36	37	37	38	39	40	40	41	42	43	43	44	44
45	23	24	25	26	26	27	28	28	29	30	31	32	32	33	34	34	35	36	37	38	38	39	40	40	41	42	43	44	44	45	45
46	24	25	25	26	27	28	28	29	30	31	31	32	33	34	34	35	36	37	38	38	39	40	41	41	42	43	44	44	45	46	46
47	24	25	26	27	27	28	29	30	31	31	32	33	34	34	35	36	37	38	38	39	40	41	42	42	43	44	45	45	46	47	47
48	25	26	26	27	28	29	30	30	31	32	33	34	34	35	36	37	38	38	39	40	41	42	42	43	44	45	46	46	47	48	48
49	25	26	27	28	29	29	30	31	32	33	33	34	35	36	37	38	38	39	40	41	42	42	43	44	45	46	47	47	48	49	49
50	26	27	28	28	29	30	31	32	32	33	34	35	36	37	38	38	39	40	41	42	42	43	44	45	46	47	48	48	49	50	50
51	26	27	28	29	30	31	31	32	33	34	35	36	37	37	38	39	40	41	42	42	43	44	45	46	47	48	48	49	50	51	51
52	27	28	29	29	30	31	32	33	34	35	36	36	37	38	39	40	41	42	42	43	44	45	46	47	48	49	49	50	51	52	52
53	27	28	29	30	31	32	33	34	34	35	36	37	38	39	40	41	42	42	43	44	45	46	47	48	49	49	50	51	52	53	53
54	28	29	30	31	32	32	33	34	35	36	37	38	39	40	40	41	42	43	44	45	46	47	48	49	50	50	51	52	53	54	54
55	28	29	30	31	32	33	34	35	36	37	38	38	39	40	41	42	43	44	45	46	47	48	49	50	50	51	52	53	54	55	55
56	29	30	31	32	33	34	35	35	36	37	38	39	40	41	42	43	44	45	46	47	48	49	49	50	51	52	53	54	55	56	56
57	29	30	31	32	33	34	35	36	37	38	39	40	41	42	43	44	45	46	47	48	48	49	50	51	52	53	54	55	56	57	57
58	30	31	32	33	34	35	36	37	38	39	40	41	42	43	44	44	45	46	47	48	49	50	51	52	53	54	55	56	57	58	58
59	30	31	32	33	34	35	36	37	38	39	40	41	42	43	44	45	46	47	48	49	50	51	52	53	54	55	56	57	58	59	59

Table 13, find the correction, and fill in the final Hc. Now it looks like this:

Hc	= 55° 32′
Corr.	= + 02′
Hc	= 55° 34′

Finally, go back to your sextant sights and write down the observed altitude (Ho) underneath Hc:

Hc	=	55°34′
Ho	=	55°42′

Looking at it, you can see that you measured an angle larger than it should have been had you been at the assumed position. Remember, this same kind of thing happened in the fictional explanation in Chapter 3. What does it mean? Think of it this way: "If I had been farther away from the sun than my assumed position, the angle I measured would have been less than the computed or theoretical angle." You know this from everyday experience—as you move away from a lighthouse, an island, or whatever, it appears to get shorter. Conversely, as you move toward something, it appears to get taller—that is, the angle between its base, your eye, and its top gets larger.

In the calculations you made, your measured angle is 8′ larger than the theoretical angle, so you are 8 nautical miles closer to the sun than you assumed you were.

To fix the process in your mind and clarify it, look at all these calculations as they would be jotted down in a logical format on a worksheet:

Date: 7-2-74		
Log: 5807		
Watch Time	=	10 -12- 05
Fast (−) or Slow (+)	=	− 40
Conversion to GMT	=	+4
GMT	=	14 -11- 25
Sextant Reading (Hs)	=	55° 25′
Index Correction (IC)	=	+ 05′
Dip	=	− 03′
Apparent Altitude (ha)	=	55° 27′
Main Correction	=	+ 15′
Observed Altitude (Ho)	=	55° 42′

SUN			
GHA 14 Hr.	=	29° 01′	Dec. = 23°03′N
11 Min. 25 Sec.	=	2° 51′	
GHA	=	31° 52′	
(+360?)	=	360°	
GHA	=	391° 52′	
Assumed Long. W. (−)	=	−68° 52′	Assumed Lat. = 36° N
Assumed Long. E. (+)	=	(+)	
LHA	=	323°	
(− 360?)	=		
LHA	=	323°	

Using LHA, assumed latitude, and whole degree of declination, enter H.O. 249 and extract the following:

Hc	=	55° 32′	d	=	+30′
Corr.	=	+ 02′	Dec. Inc.	=	3′
Hc	=	55° 34′	Corr.	=	2′
Ho	=	55° 42′			
		8′ Toward Sun			
Z	=	102°			

For purposes of using the plotting sheet, angle *Z* (or *azimuth angle*) must be converted to azimuth (or "true" direction). The rule for this is in the upper left-hand corner of Table 12 if you are in north latitude and in the lower left-hand corner if you are in south latitude. If you remember the triangle from Chapter 3, you should be able to see the basis for these rules.

Get out a plotting sheet, mark the middle latitude line "36°," and draw in the longitude lines. You do this to the proper scale by putting a pencil mark on the compass rose at 36° from the horizontal and ruling a line through the pencil mark parallel to the vertical centerline. Put a second longitude line the same distance from the center on the other side, and mark appropriately (Figure 8.5).

Using the longitude scale in the lower right-hand corner of the sheet, mark off your DR longitude (68°30′), and using the scale on the vertical centerline, mark off your DR latitude (36°40′; Figure 8.6). Now mark your assumed position (*AP* in Figure 8.6)—68°52′ W, 36° N.

Using a pair of parallel rulers, put a line through *AP* in the direction 102°. This is the azimuth (Zn)—the bearing to the sun. Put an arrow on the end of the line and draw the symbol for the sun as shown. You know you are 8′ of arc or 8 nautical miles closer to the sun than your assumed position, so with a pair of dividers measure 8′ off the vertical center scale and mark a point on the sun's azimuth line that far from *AP* toward the sun. Through this point, draw a line perpendicular to the azimuth line. This is your line of position (LOP) from your 1012 observation. Label the line so that you will be able to identify it later.

At about 2 P.M. the same day, you decide to take another sight of the sun and your filled-out worksheet looks like this (note new assumed latitude and see Table 12a):

Date: 7-2-74
Log: 5827

Watch Time	=	14	-07- 30
Fast (−) or Slow (+)	=		− 40
Conversion to GMT	=	+4	
GMT	=	18	-06- 50

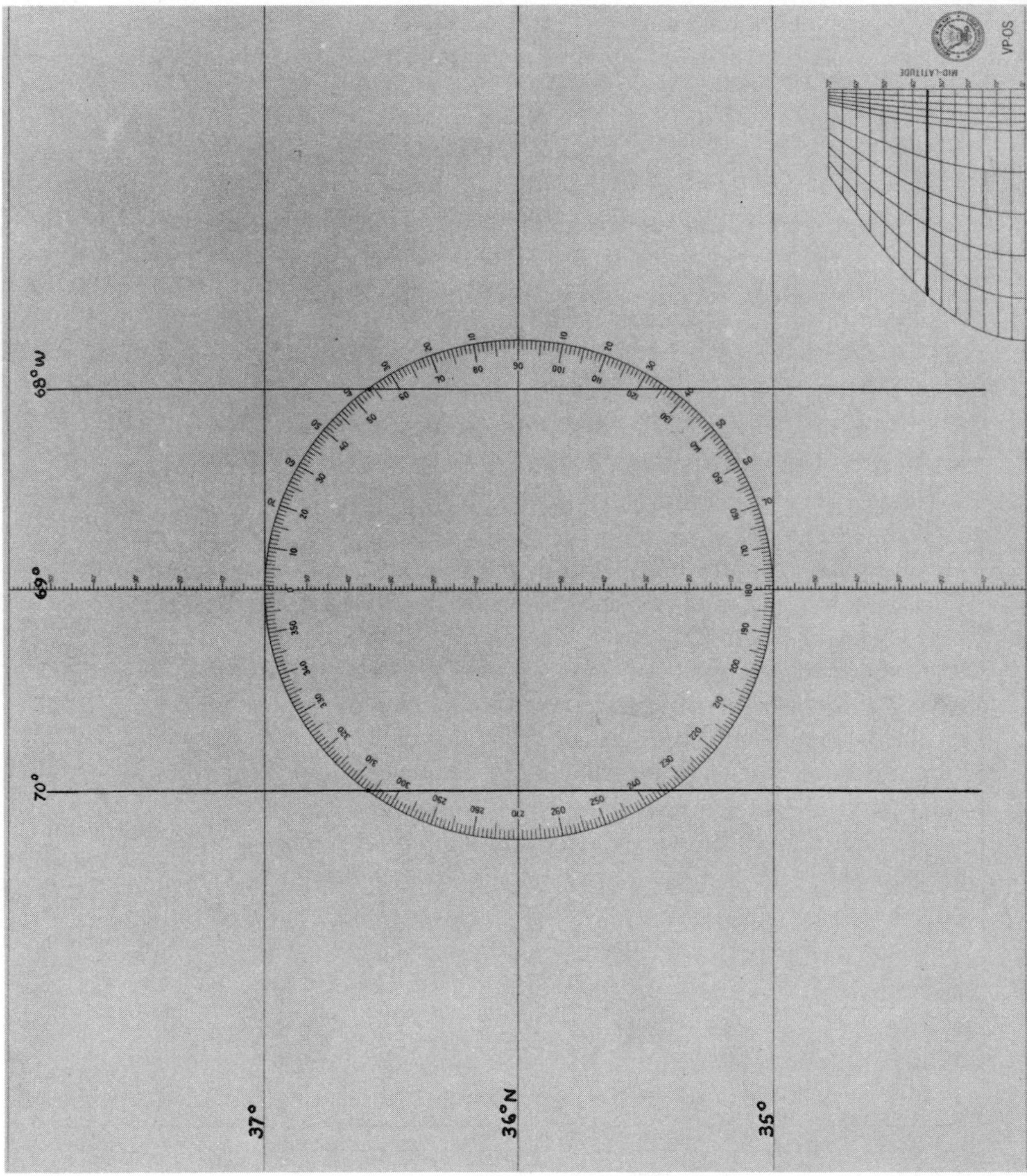

68° W
69°
70°
37°
36° N
35°
VP-OS
MID-LATITUDE
LONGITUDE SCALE
UNIVERSAL PLOTTING SHEET
1st. EDITION: September, 1942
Published by the Hydrographic Office, Washington, D. C., Sept. 1942

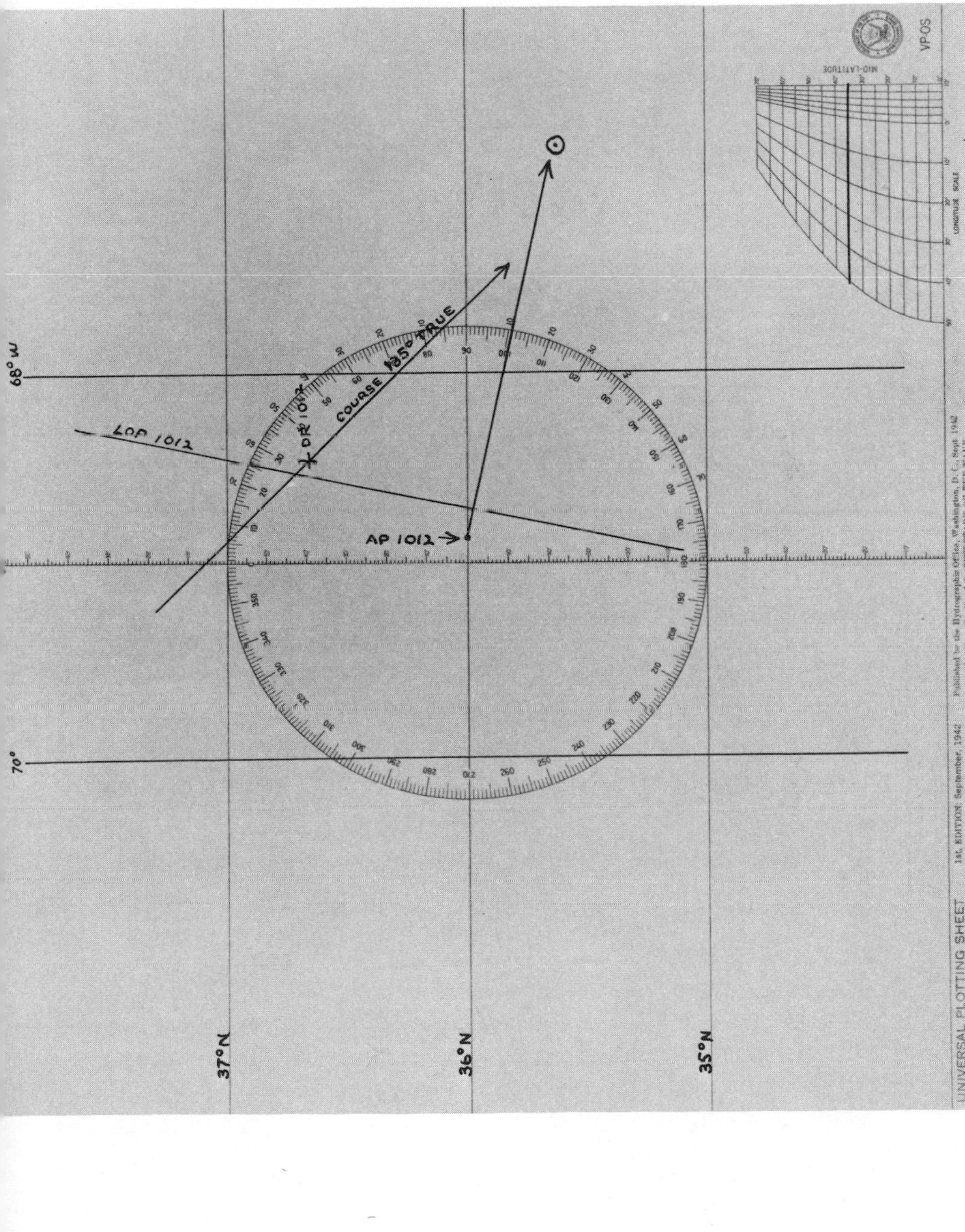
68° W
70°
37°N
36°N
35°N
LOP 1012
DR 1012
COURSE 135° TRUE
AP 1012
UNIVERSAL PLOTTING SHEET
1st EDITION, September, 1942
MID-LATITUDE
LONGITUDE SCALE
VP-OS

Fig. 8.6

Sextant Reading (Hs)	=	66°	26′
Index Correction (IC)	=	+	05′
Dip	=	−	03′
Apparent Altitude (ha)	=	66°	28′
Main Correction	=	+	16′
Observed Altitude (Ho)	=	66°	44′

SUN

GHA 18 Hr.	=	89° 00′	Dec. = N 23°02′	
6 Min. 50 Sec.	=	1° 43′		
GHA	=	90° 43′		
(+360?)	=			
GHA	=	90° 43′		
Assumed Long. W. (−)	=	-67° 43′		
Assumed Long. E. (+)	=		Assumed Lat. =35° N	
LHA	=	23°		
(−360?)	=			
LHA	=	23°		

Hc	=	66°40′	d	=	+34′
Corr.	=	+ 01′	Dec. Inc.	=	2′
Hc	=	66°41′	Corr.	=	1′
Ho	=	66°44′			
		3′ toward sun			360°
					−115°
Z	=	115°	Zn	=	245°

The plot of this is shown in Figure 8.7.

Assume that the log shows that you have run 20 miles since your 10 A.M. observation. You have steered 135° true (see course line plotted from DR position), so you mark off 20 miles from the point where your 1012 LOP intersects your course, and your fix is where your advanced 1012 LOP crosses your 1407 LOP. You start your new DR from this point, so start a new course line from there.

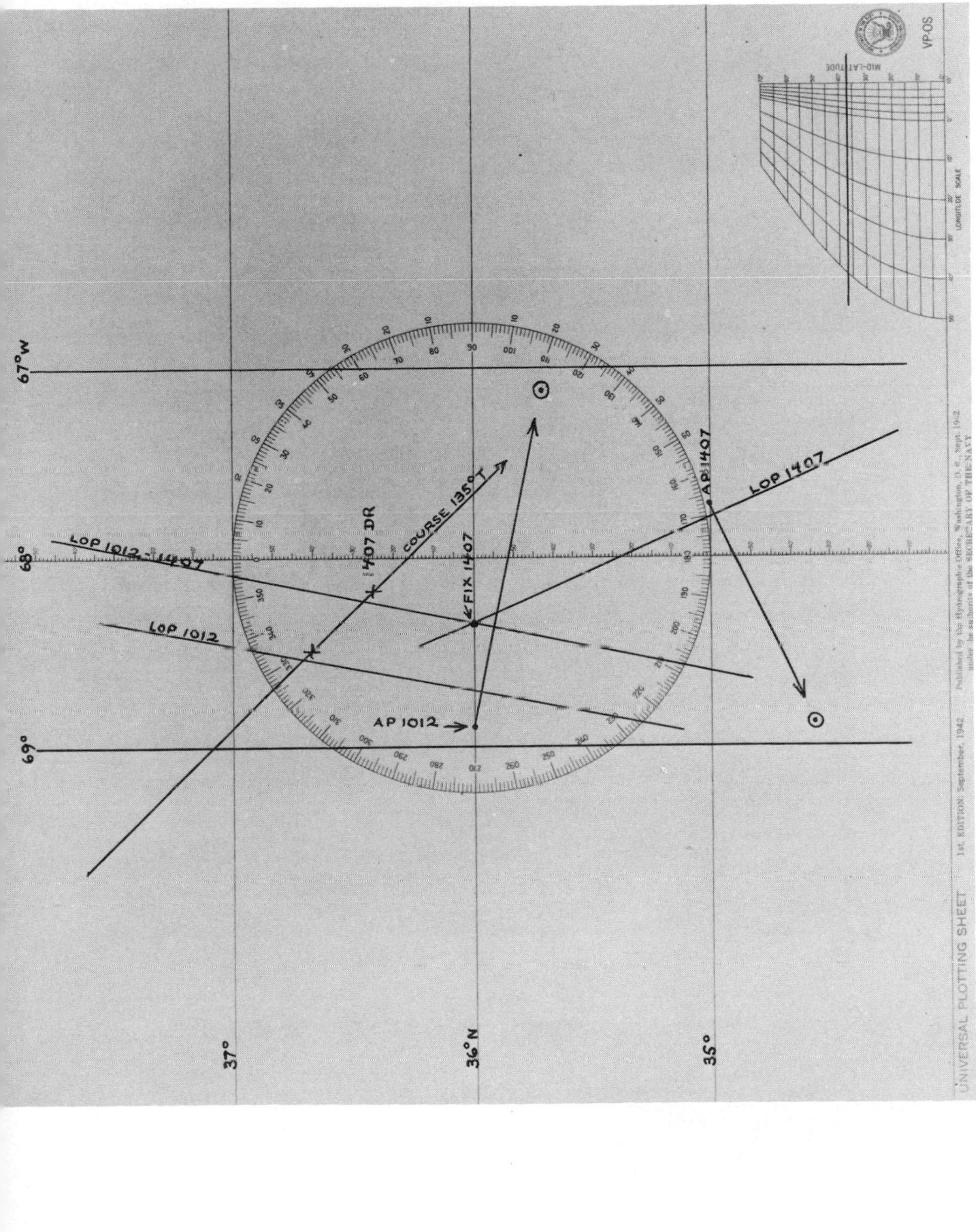
67° W
68°
69°
37°
36° N
35°
LOP 1012
LOP 1012–1407
COURSE 135° T
1407 DR
FIX 1407
AP 1012
AP 1407
LOP 1407
UNIVERSAL PLOTTING SHEET
1st. EDITION: September, 1942
VP-OS
MID-LATITUDE
LONGITUDE SCALE

Fig. 8.7

Many owners of small sailboats (twenty-six to forty feet) go no further than this with celestial navigation, nor do they need to. The process of fixing position by advancing sun LOPs has the advantage of being quick and usable whenever the sun is visible. It is not dependent on a special condition, as is the noon sight, which can be obviated by sudden clouds, squalls, emergencies aboard, and so on.

The weakness of this method is that it depends on a good knowledge of distance run between sights. In a sailboat of moderate length, however, this is relatively easy to determine within acceptable limits. In fact, I have read that many offshore sailors are able to estimate their distance run very accurately without a log or other mechanical or electronic device.

The thing to remember in using this method is that you need to let several hours pass between sights so that the bearing of the sun will change enough for the LOPs to intersect at a broad rather than a narrow angle. If they intersect at a small angle, it is difficult to determine the exact point at which they cross.

The sight reduction tables used in this example, published in the United States as H.O. 249, are also published in England as A.P. 3270. The sources for this book are listed in the last chapter. There is another table available, known in the United States as H.O. 229 and in England as N.P. 401. Since this table is the latest thing and has some advantages over H.O. 249, let's use it to reduce the 1407 sight.

Table 14 shows the relevant page from N.P. 401, the same as H.O. 229. As you see, there is a slight difference in format.

LHA	= 23°		
Assumed Latitude	= 35° N		
Declination	= 23°02′N		
From H.O. 249, Hc	= 66°40′	d = 34′	Z = 115°
From H.O. 229, Hc	= 66°39.9′	d = +34.5′	Z = 114.8°

Right here, you can see a difference between the two tables. H.O. 229 gives you answers to one-tenth of a minute of a degree! The reason for this is that these tables were designed for use by the navy, and the navy likes to know where its ships are to the nearest tenth of a mile.

3°, 337° L.H.A. LATITUDE SAME NAME AS DECLINATION

N. Lat. { L.H.A. greater than 180°......Zn=Z; L.H.A. less than 180°............Zn=360°−Z }

ec.	30°			31°			32°			33°		
	Hc	d	Z	Hc	d	Z	Hc	d	Z	Hc	d	Z
°	° ′	′	°	° ′	′	°	° ′	′	°	° ′	′	°
0	52 51.7	+49.5	139.7	52 05.7	+50.1	140.5	51 19.1	+50.7	141.3	50 32.0	+51.3	142.1
1	53 41.2	49.0	138.7	52 55.8	49.6	139.6	52 09.8	50.3	140.4	51 23.3	50.8	141.2
2	54 30.2	48.5	137.7	53 45.4	49.3	138.7	53 00.1	49.9	139.5	52 14.1	50.5	140.4
3	55 18.7	47.9	136.7	54 34.7	48.7	137.7	53 50.0	49.4	138.6	53 04.6	50.1	139.5
4	56 06.6	47.4	135.6	55 23.4	48.2	136.7	54 39.4	48.9	137.6	53 54.7	49.6	138.6
5	56 54.0	+46.8	134.5	56 11.6	+47.6	135.6	55 28.3	+48.4	136.6	54 44.3	+49.2	137.6
6	57 40.8	46.1	133.4	56 59.2	47.0	134.5	56 16.7	47.9	135.6	55 33.5	48.6	136.6
7	58 26.9	45.4	132.2	57 46.2	46.3	133.3	57 04.6	47.2	134.5	56 22.1	48.1	135.6
8	59 12.3	44.7	130.9	58 32.5	45.7	132.1	57 51.8	46.6	133.3	57 10.2	47.5	134.5
9	59 57.0	43.8	129.6	59 18.2	44.9	130.9	58 38.4	45.9	132.1	57 57.7	46.9	133.3
10	60 40.8	+42.9	128.2	60 03.1	+44.1	129.6	59 24.3	+45.2	130.9	58 44.6	+46.1	132.1
11	61 23.7	41.9	126.8	60 47.2	43.2	128.2	60 09.5	44.3	129.6	59 30.7	45.5	130.9
12	62 05.6	41.0	125.3	61 30.4	42.2	126.8	60 53.8	43.5	128.2	60 16.2	44.6	129.6
13	62 46.6	39.8	123.7	62 12.6	41.2	125.3	61 37.3	42.5	126.8	61 00.8	43.7	128.2
14	63 26.4	38.6	122.0	62 53.8	40.1	123.7	62 19.8	41.6	125.3	61 44.5	42.9	126.8
15	64 05.0	+37.3	120.3	63 33.9	+38.9	122.0	63 01.4	+40.4	123.7	62 27.4	+41.8	125.3
16	64 42.3	35.9	118.5	64 12.8	37.7	120.3	63 41.8	39.2	122.0	63 09.2	40.7	123.7
17	65 18.2	34.4	116.6	64 50.5	36.2	118.5	64 21.0	38.0	120.3	63 49.9	39.6	122.1
18	65 52.6	32.9	114.6	65 26.7	34.8	116.6	64 59.0	36.5	118.5	64 29.5	38.2	120.4
19	66 25.5	31.1	112.5	66 01.5	33.1	114.6	65 35.5	35.1	116.6	65 07.7	37.0	118.5
20	66 56.6	+29.2	110.4	66 34.6	+31.5	112.5	66 10.6	+33.5	114.6	65 44.7	+35.4	116.7
21	67 25.8	27.4·	108.1	67 06.1	29.6	110.4	66 44.1	31.8	112.6	66 20.1	33.8	114.7
22	67 53.2	25.2·	105.8	67 35.7	27.6·	108.1	67 15.9	29.9	110.4	66 53.9	32.1	112.6
23	68 18.4	23.0·	103.3	68 03.3	25.6·	105.8	67 45.8	28.0·	108.1	67 26.0	30.3	110.4
24	68 41.4	20.7·	100.8	68 28.9	23.3·	103.3	68 13.8	25.9·	105.8	67 56.3	28.3·	108.1
25	69 02.1	+18.3·	98.2	68 52.2	+21.0·	100.8	68 39.7	+23.6·	103.3	68 24.6	+26.3·	105.8
26	69 20.4	15.7·	95.5	69 13.2	18.5·	98.2	69 03.3	21.4·	100.7	68 50.9	23.9·	103.3
27	69 36.1	13.0·	92.8	69 31.7	16.0·	95.5	69 24.7	18.8·	98.1	69 14.8	21.7·	100.7
28	69 49.1	10.3·	90.0	69 47.7	13.3·	92.7	69 43.5	16.2·	95.4	69 36.5	19.1·	98.1
29	69 59.4	7.5·	87.1	70 01.0	10.5·	89.8	69 59.7	13.6·	92.6	69 55.6	16.5·	95.3
30	70 06.9	+4.6·	84.2	70 11.5	+7.7·	87.0	70 13.3	+10.7·	89.7	70 12.1	+13.8·	92.5
31	70 11.5	+1.8·	81.3	70 19.2	4.8·	84.0	70 24.0	7.9·	86.8	70 25.9	11.0·	89.6
32	70 13.3	−1.2·	78.3	70 24.0	+1.9·	81.1	70 31.9	5.0·	83.8	70 36.9	8.1·	86.7
33	70 12.1	4.1·	75.3	70 25.9	−1.1·	78.1	70 36.9	+2.0·	80.8	70 45.0	5.1·	83.7
34	70 08.0	6.9·	72.4	70 24.0	4.0·	75.1	70 38.9	−1.0·	77.8	70 50.1	+2.1·	80.6
35	70 01.1	−9.8·	69.5	70 20.8	−6.9·	72.1	70 37.9	−4.0·	74.8	70 52.2	−0.9·	77.6
36	69 51.3	12.5·	66.6	70 12.9	9.8·	69.2	70 33.0	6.9·	71.8	70 51.0	3.9·	74.6
37	69 38.8	15.2·	63.8	70 04.1	12.5·	66.3	70 27.0	9.7·	68.8	70 47.4	6.9·	71.5
38	69 23.6	17.8·	61.0	69 51.6	15.3·	63.4	70 17.3	12.7·	65.9	70 40.5	9.8·	68.5
39	69 05.8	20.2·	58.3	69 36.3	17.9·	60.6	70 04.6	15.3·	63.0	70 30.7	12.7·	65.5
40	68 45.6	−22.6·	55.7	69 18.4	−20.3·	57.9	69 49.3	−18.0·	60.2	70 18.0	−15.4·	62.6
41	68 23.0	24.8·	53.2	68 58.1	22.7·	55.3	69 31.3	20.4·	57.4	70 02.6	18.1·	59.8
42	67 58.2	26.9·	50.7	68 35.4	25.0·	52.7	69 10.9	22.9·	54.8	69 44.5	20.6·	57.0
43	67 31.3	28.9	48.4	68 10.4	27.1·	50.2	68 48.0	25.2·	52.2	69 23.9	23.1·	54.3
44	67 02.4	30.8	46.1	67 43.3	29.1	47.9	68 22.8	27.2·	49.7	69 00.8	25.3·	51.7
45	66 31.6	−32.5	43.9	67 14.2	−30.9	45.6	67 55.6	−29.3·	47.3	68 35.5	−27.5·	49.2
46	65 59.1	34.1	41.8	66 43.3	32.7	43.4	67 26.3	31.2	45.0	68 08.0	29.5	46.8
47	65 25.0	35.7	39.8	66 10.6	34.4	41.3	66 55.1	32.9	42.8	67 38.5	31.4	44.5
48	64 49.3	37.0	37.9	65 36.2	35.8	39.3	66 22.2	34.6	40.7	67 07.1	33.1	42.3
49	64 12.3	38.4	36.1	65 00.4	37.3	37.4	65 47.6	36.1	38.7	66 34.0	34.8	40.1
50	63 33.9	−39.6	34.3	64 23.1	−38.6	35.5	65 11.5	−37.4	36.8	65 59.2	−36.3	38.1
51	62 54.3	40.7	32.7	63 44.5	39.8	33.8	64 34.1	38.9	34.9	65 22.9	37.8	36.2
52	62 13.6	41.8	31.1	63 04.7	40.9	32.1	63 55.2	40.0	33.2	64 45.1	39.0	34.3
53	61 31.8	42.7	29.6	62 23.8	42.0	30.5	63 15.2	41.1	31.5	64 06.1	40.3	32.6
54	60 49.1	43.6	28.1	61 41.8	42.9	29.0	62 34.1	42.2	29.9	63 25.8	41.4	30.9
55	60 05.5	−44.5	26.7	60 58.9	−43.9	27.5	61 51.9	43.2	28.4	62 44.4	−42.4	29.3
56	59 21.0	45.2	25.4	60 15.0	44.6	26.1	61 08.7	44.0	26.9	62 02.0	43.4	27.8
57	58 35.7	45.9	24.1	59 30.4	45.5	24.8	60 24.7	44.9	25.5	61 18.6	44.3	26.3
58	57 49.8	46.7	22.9	58 44.9	46.2	23.5	59 39.8	45.7	24.2	60 34.3	45.1	24.9
59	57 03.1	47.3	21.7	57 58.7	46.8	22.3	58 54.1	46.4	22.9	59 49.2	45.8	23.6
60	56 15.8	−47.8	20.6	57 11.9	−47.5	21.1	58 07.7	−47.0	21.7	59 03.4	−46.6	22.3
61	55 28.0	48.4	19.5	56 24.4	48.0	20.0	57 20.7	47.6	20.6	58 16.8	47.3	21.1
62	54 39.6	48.9	18.5	55 36.4	48.6	18.9	56 33.1	48.3	19.4	57 29.5	47.8	20.0
63	53 50.7	49.4	17.5	54 47.8	49.1	17.9	55 44.8	48.7	18.4	56 41.7	48.4	18.8
64	53 01.3	49.8	16.5	53 58.7	49.5	16.9	54 56.1	49.3	17.3	55 53.3	49.0	17.8
65	52 11.5	−50.3	15.6	53 09.2	−50.0	16.0	54 06.8	−49.7	16.4	55 04.3	−49.4	16.8
66	51 21.2	50.6	14.7	52 19.2	50.4	15.1	53 17.1	50.2	15.4	54 14.9	49.9	15.8
67	50 30.6	51.0	13.9	51 28.8	50.8	14.2	52 26.9	50.5	14.5	53 25.0	50.3	14.8
68	49 39.6	51.3	13.1	50 38.0	51.1	13.3	51 36.4	50.9	13.6	52 34.7	50.8	13.9
69	48 48.3	51.6	12.3	49 46.9	51.5	12.5	50 45.5	51.3	12.8	51 43.9	51.1	13.1
70	47 56.7	−51.9	11.5	48 55.4	−51.7	11.7	49 54.2	−51.6	12.0	50 52.8	−51.4	12.2
71	47 04.8	52.3	10.8	48 03.7	52.1	11.0	49 02.6	52.0	11.2	50 01.4	51.8	11.4
72	46 12.5	52.4	10.0	47 11.6	52.3	10.2	48 10.6	52.2	10.4	49 09.6	52.0	10.6
73	45 20.1	52.7	9.4	46 19.3	52.6	9.5	47 18.4	52.5	9.7	48 17.6	52.4	9.9
74	44 27.4	53.0	8.7	45 26.7	52.9	8.8	46 25.9	52.7	9.0	47 25.2	52.6	9.2
75	43 34.4	−53.1	8.0	44 33.8	−53.0	8.2	45 33.2	−53.0	8.3	46 32.6	−52.9	8.5
76	42 41.3	53.4	7.4	43 40.8	53.3	7.5	44 40.2	53.2	7.6	45 39.7	53.1	7.8
77	41 47.9	53.6	6.8	42 47.5	53.5	6.9	43 47.0	53.4	7.0	44 46.6	53.3	7.1
78	40 54.3	53.7	6.2	41 54.0	53.7	6.3	42 53.6	53.6	6.4	43 53.3	53.6	6.5
79	40 00.6	53.9	5.6	41 00.3	53.8	5.7	42 00.0	53.8	5.8	42 59.7	53.7	5.9
80	39 06.7	−54.1	5.0	40 06.5	−54.1	5.1	41 06.2	−54.0	5.2	42 06.0	−53.9	5.2
81	38 12.6	54.2	4.5	39 12.4	54.1	4.5	40 12.2	54.1	4.6	41 12.1	54.1	4.7
82	37 18.4	54.4	3.9	38 18.3	54.4	4.0	39 18.1	54.3	4.0	40 18.0	54.3	4.1
83	36 24.0	54.5	3.4	37 23.9	54.5	3.4	38 23.8	54.4	3.5	39 23.7	54.4	3.5
84	35 29.5	54.6	2.9	36 29.4	54.6	2.9	37 29.4	54.6	3.0	38 29.3	54.6	3.0
85	34 34.9	−54.8	2.4	35 34.8	−54.7	2.4	36 34.8	−54.7	2.4	37 34.7	−54.7	2.5
86	33 40.1	54.8	1.9	34 40.1	54.9	1.9	35 40.1	54.9	1.9	36 40.0	54.8	1.9
87	32 45.3	55.0	1.4	33 45.2	54.9	1.4	34 45.2	54.9	1.4	35 45.2	55.0	1.4
88	31 50.3	55.1	0.9	32 50.3	55.1	0.9	33 50.3	55.1	0.9	34 50.2	55.0	1.0
89	30 55.2	55.2	0.5	31 55.2	55.2	0.5	32 55.2	55.2	0.5	33 55.2	55.2	0.5
90	30 00.0	−55.3	0.0	31 00.0	−55.3	0.0	32 00.0	−55.3	0.0	33 00.0	−55.3	0.0
	30°			31°			32°			33°		

34°			35°			36°			37°			Dec.
Hc	d	Z	Hc	d	Z	Hc	d	Z	Hc	d	Z	
° ′	′	°	° ′	′	°	° ′	′	°	° ′	′	°	°
49 44.5	+51.7	142.8	48 56.5	+52.2	143.5	48 08.0	+52.7	144.2	47 19.2	+53.1	144.8	0
50 36.2	51.5	142.0	49 48.7	52.0	142.7	49 00.7	52.5	143.4	48 12.3	52.9	144.1	1
51 27.7	51.0	141.2	50 40.7	51.6	142.0	49 53.2	52.1	142.7	49 05.2	52.6	143.4	2
52 18.7	50.7	140.3	51 32.3	51.2	141.1	50 45.3	51.8	141.9	49 57.8	52.3	142.7	3
53 09.4	50.3	139.5	52 23.5	50.9	140.3	51 37.1	51.5	141.1	50 50.1	52.0	141.9	4
53 59.7	+49.8	138.5	53 14.4	+50.5	139.4	52 28.6	+51.1	140.3	51 42.1	+51.7	141.1	5
54 49.5	49.4	137.6	54 04.9	50.1	138.5	53 19.7	50.7	139.4	52 33.8	51.3	140.3	6
55 38.9	48.9	136.6	54 55.0	49.6	137.6	54 10.4	50.3	138.5	53 25.1	50.9	139.4	7
56 27.8	48.3	135.5	55 44.6	49.1	136.6	55 00.7	49.8	137.6	54 16.0	50.5	138.5	8
57 16.1	47.8	134.5	56 33.7	48.6	135.5	55 50.5	49.3	136.6	55 06.5	50.1	137.6	9
58 03.9	+47.1	133.3	57 22.3	+47.9	134.5	56 39.8	+48.8	135.6	55 56.6	+49.6	136.6	10
58 51.0	46.4	132.1	58 10.2	47.4	133.3	57 28.6	48.3	134.5	56 46.2	49.0	135.6	11
59 37.4	45.7	130.9	58 57.6	46.7	132.2	58 16.9	47.6	133.4	57 35.2	48.5	134.5	12
60 23.1	44.9	129.6	59 44.3	46.0	130.9	59 04.5	47.0	132.2	58 23.7	47.9	133.4	13
61 08.0	44.1	128.3	60 30.3	45.2	129.6	59 51.5	46.2	131.0	59 11.6	47.3	132.2	14
61 52.1	+43.1	126.8	61 15.5	+44.3	128.3	60 37.7	+45.5	129.7	59 58.9	+46.5	131.0	15
62 35.2	42.1	125.3	61 59.8	43.4	126.9	61 23.2	44.6	128.3	60 45.4	45.8	129.8	16
63 17.3	41.0	123.8	62 43.2	42.5	125.4	62 07.8	43.8	126.9	61 31.2	44.9	128.4	17
63 58.3	39.9	122.1	63 25.7	41.3	123.8	62 51.6	42.7	125.5	62 16.1	44.1	127.0	18
64 38.2	38.6	120.4	64 07.0	40.3	122.2	63 34.3	41.7	123.9	63 00.2	43.0	125.5	19
65 16.8	+37.3	118.6	64 47.3	+38.9	120.5	64 16.0	+40.6	122.3	63 43.2	+42.1	124.0	20
65 54.1	35.7	116.7	65 26.2	37.6	118.7	64 56.6	39.3	120.5	64 25.3	40.8	122.3	21
66 29.8	34.2	114.7	66 03.8	36.1	116.8	65 35.9	37.9	118.7	65 06.1	39.7	120.6	22
67 04.0	32.5	112.6	66 39.9	34.5	114.8	66 13.8	36.5	116.8	65 45.8	38.3	118.8	23
67 36.5	30.6	110.4	67 14.4	32.9	112.7	66 50.3	34.9	114.8	66 24.1	36.8	116.9	24
68 07.1	+28.7·	108.2	67 47.3	+31.0	110.5	67 25.2	+33.1	112.7	67 00.9	+35.3	114.9	25
68 35.8	26.5·	105.8	68 18.3	29.0·	108.2	67 58.3	31.4	110.5	67 36.2	33.5	112.8	26
69 02.3	24.4·	103.3	68 47.3	26.9·	105.8	68 29.7	29.4·	108.3	68 09.7	31.7	110.6	27
69 26.7	21.9·	100.7	69 14.2	24.6·	103.3	68 59.1	27.2·	105.8	68 41.4	29.8·	108.3	28
69 48.6	19.4·	98.0	69 38.8	22.2·	100.7	69 26.3	25.0·	103.3	69 11.2	27.6·	105.9	29
70 08.0	+16.8·	95.3	70 01.1	+19.7·	98.0	69 51.3	+22.6·	100.7	69 38.8	+25.3·	103.4	30
70 24.8	14.1·	92.4	70 20.8	17.1·	95.2	70 13.9	20.0·	98.0	70 04.1	22.9·	100.7	31
70 38.9	11.2·	89.5	70 37.9	14.3·	92.4	70 33.9	17.4·	95.2	70 27.0	20.4·	98.0	32
70 50.1	8.3·	86.5	70 52.2	11.4·	89.4	70 51.3	14.6·	92.3	70 47.4	17.7·	95.2	33
70 58.4	5.2·	83.5	71 03.6	8.5·	86.4	71 05.9	11.6·	89.3	71 05.1	14.8·	92.2	34
71 03.6	+2.3·	80.4	71 12.1	+5.4·	83.3	71 17.5	+8.7·	86.3	71 19.9	+11.9·	89.2	35
71 05.9	−0.8·	77.4	71 17.5	+2.4·	80.2	71 26.2	5.6·	83.2	71 31.8	8.8·	86.2	36
71 05.1	3.9·	74.3	71 19.9	−0.8·	77.1	71 31.8	+2.4·	80.0	71 40.6	5.7·	83.0	37
71 01.2	6.9·	71.2	71 19.1	3.8·	74.0	71 34.2	−0.7·	76.9	71 46.3	+2.6·	79.8	38
70 54.3	9.9·	68.2	71 15.3	6.9·	70.9	71 33.5	3.8·	73.7	71 48.9	−0.6·	76.6	39
70 44.4	−12.7·	65.2	71 08.4	−9.9·	67.8	71 29.7	−6.9·	70.6	71 48.3	−3.8·	73.4	40
70 31.7	15.5·	62.2	70 58.5	12.8·	64.8	71 22.8	10.0·	67.5	71 44.5	7.0·	70.3	41
70 16.2	18.2·	59.3	70 45.7	15.7·	61.8	71 12.8	12.9·	64.4	71 37.5	10.0·	67.1	42
69 58.0	20.8·	56.5	70 30.0	18.4·	58.9	70 59.9	15.8·	61.4	71 27.5	13.1·	64.0	43
69 37.2	23.3·	53.8	70 11.6	20.9·	56.0	70 44.1	18.5·	58.4	71 14.4	15.9·	60.9	44
69 13.9	−25.5·	51.2	69 50.7	−23.4·	53.3	70 25.6	−21.2·	55.6	70 58.5	−18.7·	57.9	45
68 48.4	27.7·	48.7	69 27.3	25.8·	50.7	70 04.4	23.6·	52.8	70 39.8	21.4·	55.1	46
68 20.7	29.7·	46.2	69 01.5	27.9·	48.1	69 40.8	26.0·	50.1	70 18.4	23.9·	52.3	47
67 51.0	31.7	43.9	68 33.6	30.0·	45.7	69 14.8	28.2·	47.5	69 54.5	26.2·	49.6	48
67 19.3	33.4	41.7	68 03.6	31.9	43.3	68 46.6	30.2·	45.1	69 28.3	28.5·	47.0	49
66 45.9	−35.0	39.5	67 31.7	−33.7	41.1	68 16.4	−32.2	42.7	68 59.8	−30.5·	44.5	50
66 10.9	36.6	37.5	66 58.0	35.3	38.9	67 44.2	33.9	40.5	68 29.3	32.5	42.1	51
65 34.3	38.0	35.6	66 22.7	36.8	36.9	67 10.3	35.6	38.3	67 56.8	34.2	39.8	52
64 56.3	39.3	33.7	65 45.9	38.3	34.9	66 34.7	37.2	36.3	67 22.6	35.9	37.7	53
64 17.0	40.5	32.0	65 07.6	39.5	33.1	65 57.5	38.5	34.3	66 46.7	37.4	35.6	54
63 36.5	−41.6	30.3	64 28.1	−40.8	31.3	65 19.0	−39.8	32.5	66 09.3	−38.8	33.7	55
62 54.9	42.7	28.7	63 47.3	41.9	29.8	64 39.2	41.1	30.7	65 30.5	40.2	31.8	56
62 12.2	43.6	27.2	63 05.4	42.9	28.0	63 58.1	42.2	29.0	64 50.3	41.3	30.0	57
61 28.6	44.5	25.7	62 22.5	43.9	26.5	63 15.9	43.1	27.4	64 09.0	42.5	28.4	58
60 44.1	45.4	24.3	61 38.6	44.8	25.1	62 32.8	44.2	25.9	63 26.5	43.4	26.8	59
59 58.7	−46.1	23.0	60 53.8	−45.5	23.7	61 48.6	−45.0	24.4	62 43.1	−44.4	25.2	60
59 12.6	46.8	21.7	60 08.3	46.4	22.4	61 03.6	45.8	23.0	61 58.7	45.3	23.8	61
58 25.8	47.4	20.5	59 21.9	47.0	21.1	60 17.8	46.6	21.7	61 13.4	46.1	22.4	62
57 38.4	48.1	19.4	58 34.9	47.7	19.9	59 31.2	47.2	20.5	60 27.3	46.8	21.1	63
56 50.3	48.6	18.2	57 47.2	48.2	18.7	58 44.0	47.9	19.3	59 40.5	47.5	19.8	64
56 01.7	−49.1	17.2	56 59.0	−48.9	17.6	57 56.1	−48.5	18.1	58 53.0	−48.1	18.6	65
55 12.6	49.7	16.2	56 10.1	49.3	16.6	57 07.6	49.1	17.0	58 04.9	48.8	17.5	66
54 22.9	50.1	15.2	55 20.8	49.8	15.6	56 18.5	49.5	16.0	57 16.1	49.2	16.4	67
53 32.8	50.5	14.3	54 31.0	50.3	14.6	55 29.0	50.1	15.0	56 26.9	49.8	15.4	68
52 42.3	50.9	13.4	53 40.7	50.7	13.7	54 38.9	50.4	14.0	55 37.1	50.2	14.4	69
51 51.4	−51.2	12.5	52 50.0	−51.1	12.8	53 48.5	−50.9	13.1	54 46.9	−50.7	13.4	70
51 00.2	51.6	11.7	51 58.9	51.4	11.9	52 57.6	51.3	12.2	53 56.2	51.1	12.5	71
50 08.6	52.0	10.9	51 07.5	51.8	11.1	52 06.3	51.6	11.3	53 05.1	51.4	11.6	72
49 16.6	52.2	10.1	50 15.7	52.1	10.3	51 14.7	51.9	10.5	52 13.7	51.8	10.7	73
48 24.4	52.5	9.3	49 23.6	52.4	9.5	50 22.8	52.3	9.7	51 21.9	52.1	9.9	74
47 31.9	−52.8	8.6	48 31.2	−52.6	8.8	49 30.5	−52.6	9.0	50 29.8	−52.5	9.1	75
46 39.1	53.0	7.9	47 38.6	53.0	8.1	48 37.9	52.8	8.2	49 37.3	52.7	8.4	76
45 46.1	53.2	7.2	46 45.6	53.1	7.4	47 45.1	53.1	7.5	48 44.6	53.0	7.7	77
44 52.9	53.5	6.6	45 52.5	53.4	6.7	46 52.0	53.3	6.8	47 51.6	53.2	7.0	78
43 59.4	53.7	5.9	44 59.1	53.6	6.1	45 58.7	53.5	6.2	46 58.4	53.5	6.3	79
43 05.7	−53.8	5.3	44 05.5	−53.9	5.4	45 05.2	−53.8	5.5	46 04.9	−53.7	5.6	80
42 11.9	54.1	4.7	43 11.6	54.0	4.8	44 11.4	53.9	4.9	45 11.2	53.9	5.0	81
41 17.8	54.2	4.2	42 17.6	54.1	4.2	43 17.5	54.2	4.3	44 17.3	54.1	4.4	82
40 23.6	54.4	3.6	41 23.5	54.4	3.6	42 23.3	54.3	3.7	43 23.2	54.3	3.8	83
39 29.2	54.5	3.0	40 29.1	54.5	3.1	41 29.0	54.5	3.1	42 28.9	54.4	3.2	84
38 34.7	−54.7	2.5	39 34.6	−54.6	2.5	40 34.5	−54.6	2.6	41 34.5	−54.6	2.6	85
37 40.0	54.8	2.0	38 40.0	54.8	2.0	39 39.9	54.8	2.0	40 39.9	54.8	2.1	86
36 45.2	55.0	1.5	37 45.2	55.0	1.5	38 45.1	54.9	1.5	39 45.1	54.9	1.5	87
35 50.2	55.0	1.0	36 50.2	55.0	1.0	37 50.2	55.0	1.0	38 50.2	55.0	1.0	88
34 55.2	55.2	0.5	35 55.2	55.2	0.5	36 55.2	55.2	0.5	37 55.2	55.2	0.5	89
34 00.0	−55.3	0.0	35 00.0	−55.3	0.0	36 00.0	−55.3	0.0	37 00.0	−55.3	0.0	90
34°			35°			36°			37°			

3°, 337° L.H.A. LATITUDE SAME NAME AS DECLINATION

Table 14

Since large ships are very stable platforms from which to take sights, this is reasonable. It is doubtful, though, that anyone could work to this precision from the deck of a small sailboat.

To proceed, on the inside front cover of H.O. 229 is the part of the interpolation table that you want (Table 15). The declination increment is 2′, which appears in the extreme left-hand column. The *d* value is +34.5, so proceeding to the right you pick out 1′ from the column headed "30′" and 0.2′ from the column headed "4′" to the right of the .5′ in the column headed "Decimals." In other words, to use this table you have taken 34.5 as though it were 30′ and .5′ and 4′. Adding together what you have taken from the columns (1′ plus 0.2′) gives you 1.2 minutes to add to the Hc:

Hc		66°39.9′
Corr.		+ 1.2′
Hc	=	66°41.1′

H.O. 249 gave you an answer that was within 0.1′ of this solution. One-tenth minute is 600 feet. You may feel that the 600 additional feet is not worth the much more cumbersome interpolation. I know I feel that way.

To make matters somewhat more tedious, there is even a further correction called the *double second difference correction*. To illustrate again with the example:

The *d* immediately above the value of 34.5 is 36.1′.

The *d* immediately below the value of 34.5 is 32.9′.

$$\begin{array}{r} 36.1 \\ -32.9 \\ \hline 3.2 \end{array} = \text{double second difference (DSD)}$$

Looking in the DSD column in Table 15 to the right of 2′ ("Dec. Inc."), you find that this value is too slight to be significant. If the DSD were between 5′ and 15′ the correction would be 0.1′. To use H.O. 229 you must use the DSD correction whenever you find *d* in italics with a dot adjacent.

To reiterate, it seems to me that these computations have a place on capital ships of navies but are needlessly fussy for small-boat use. After

INTERPOLATION TABLE

Tens 10′	20′	30′	40′	50′	Decimals	Units 0′	1′	2′	3′	4′	5′	6′	7′	8′	9′	Double Second Diff. and Corr.
0.0	0.0	0.0	0.0	0.0	.0	0.0	0.0	0.0	0.0	0.0	0.0	0.0	0.1	0.1	0.1	0.0 (0.0) 48.2
0.0	0.0	0.0	0.0	0.1	.1	0.0	0.0	0.0	0.0	0.0	0.0	0.1	0.1	0.1	0.1	
0.0	0.0	0.1	0.1	0.1	.2	0.0	0.0	0.0	0.0	0.0	0.0	0.1	0.1	0.1	0.1	
0.0	0.1	0.1	0.2	0.2	.3	0.0	0.0	0.0	0.0	0.0	0.0	0.1	0.1	0.1	0.1	
0.1	0.1	0.2	0.3	0.3	.4	0.0	0.0	0.0	0.0	0.0	0.0	0.1	0.1	0.1	0.1	
0.1	0.2	0.3	0.3	0.4	.5	0.0	0.0	0.0	0.0	0.0	0.0	0.1	0.1	0.1	0.1	16.2 (0.1) 48.6
0.1	0.2	0.3	0.4	0.5	.6	0.0	0.0	0.0	0.0	0.0	0.0	0.1	0.1	0.1	0.1	
0.1	0.3	0.4	0.5	0.6	.7	0.0	0.0	0.0	0.0	0.0	0.0	0.1	0.1	0.1	0.1	
0.2	0.3	0.4	0.6	0.7	.8	0.0	0.0	0.0	0.0	0.0	0.0	0.1	0.1	0.1	0.1	
0.2	0.3	0.5	0.6	0.8	.9	0.0	0.0	0.0	0.0	0.0	0.0	0.1	0.1	0.1	0.1	
0.1	0.3	0.5	0.6	0.8	.0	0.0	0.0	0.0	0.1	0.1	0.1	0.1	0.2	0.2	0.2	8.2 (0.1) 24.6 (0.2) 41.0
0.2	0.3	0.5	0.7	0.9	.1	0.0	0.0	0.1	0.1	0.1	0.1	0.2	0.2	0.2	0.2	
0.2	0.4	0.6	0.8	1.0	.2	0.0	0.0	0.1	0.1	0.1	0.1	0.2	0.2	0.2	0.2	
0.2	0.4	0.6	0.9	1.1	.3	0.0	0.0	0.1	0.1	0.1	0.1	0.2	0.2	0.2	0.2	
0.2	0.5	0.7	0.9	1.2	.4	0.0	0.0	0.1	0.1	0.1	0.1	0.2	0.2	0.2	0.2	
0.3	0.5	0.8	1.0	1.3	.5	0.0	0.0	0.1	0.1	0.1	0.1	0.2	0.2	0.2	0.2	
0.3	0.5	0.8	1.1	1.3	.6	0.0	0.0	0.1	0.1	0.1	0.1	0.2	0.2	0.2	0.2	
0.3	0.6	0.9	1.2	1.4	.7	0.0	0.0	0.1	0.1	0.1	0.1	0.2	0.2	0.2	0.2	
0.3	0.6	0.9	1.2	1.5	.8	0.0	0.0	0.1	0.1	0.1	0.1	0.2	0.2	0.2	0.2	
0.4	0.7	1.0	1.3	1.6	.9	0.0	0.0	0.1	0.1	0.1	0.1	0.2	0.2	0.2	0.2	
0.3	0.6	1.0	1.3	1.6	.0	0.0	0.0	0.1	0.1	0.2	0.2	0.2	0.3	0.3	0.4	5.0 (0.1) 15.0 (0.2) 25.0 (0.3) 35.1
0.3	0.7	1.0	1.4	1.7	.1	0.0	0.0	0.1	0.1	0.2	0.2	0.3	0.3	0.3	0.4	
0.3	0.7	1.1	1.4	1.8	.2	0.0	0.0	0.1	0.1	0.2	0.2	0.3	0.3	0.3	0.4	
0.4	0.8	1.1	1.5	1.9	.3	0.0	0.1	0.1	0.1	0.2	0.2	0.3	0.3	0.3	0.4	
0.4	0.8	1.2	1.6	2.0	.4	0.0	0.1	0.1	0.1	0.2	0.2	0.3	0.3	0.3	0.4	
0.4	0.8	1.3	1.7	2.1	.5	0.0	0.1	0.1	0.1	0.2	0.2	0.3	0.3	0.4	0.4	
0.4	0.9	1.3	1.7	2.2	.6	0.0	0.1	0.1	0.1	0.2	0.2	0.3	0.3	0.4	0.4	
0.5	0.9	1.4	1.8	2.3	.7	0.0	0.1	0.1	0.2	0.2	0.2	0.3	0.3	0.4	0.4	
0.5	1.0	1.4	1.9	2.4	.8	0.0	0.1	0.1	0.2	0.2	0.2	0.3	0.3	0.4	0.4	
0.5	1.0	1.5	2.0	2.5	.9	0.0	0.1	0.1	0.2	0.2	0.2	0.3	0.3	0.4	0.4	
0.5	1.0	1.5	2.0	2.5	.0	0.0	0.1	0.1	0.2	0.2	0.3	0.3	0.4	0.5	0.5	3.6 (0.1) 10.9 (0.2) 18.2 (0.3) 25.5 (0.4) 32.8 (0.5) 40.1
0.5	1.0	1.5	2.0	2.6	.1	0.0	0.1	0.1	0.2	0.2	0.3	0.4	0.4	0.5	0.5	
0.5	1.0	1.6	2.1	2.6	.2	0.0	0.1	0.1	0.2	0.2	0.3	0.4	0.4	0.5	0.5	
0.5	1.1	1.6	2.2	2.7	.3	0.0	0.1	0.1	0.2	0.3	0.3	0.4	0.4	0.5	0.5	
0.6	1.1	1.7	2.3	2.8	.4	0.0	0.1	0.1	0.2	0.3	0.3	0.4	0.4	0.5	0.5	
0.6	1.2	1.8	2.3	2.9	.5	0.0	0.1	0.1	0.2	0.3	0.3	0.4	0.4	0.5	0.6	
0.6	1.2	1.8	2.4	3.0	.6	0.0	0.1	0.2	0.2	0.3	0.3	0.4	0.4	0.5	0.6	
0.6	1.3	1.9	2.5	3.1	.7	0.0	0.1	0.2	0.2	0.3	0.3	0.4	0.4	0.5	0.6	
0.7	1.3	1.9	2.6	3.2	.8	0.0	0.1	0.2	0.2	0.3	0.3	0.4	0.5	0.5	0.6	
0.7	1.3	2.0	2.6	3.3	.9	0.1	0.1	0.2	0.2	0.3	0.3	0.4	0.5	0.5	0.6	
0.6	1.3	2.0	2.6	3.3	.0	0.0	0.1	0.1	0.2	0.3	0.4	0.4	0.5	0.6	0.7	2.9 (0.1) 8.6 (0.2) 14.4 (0.3) 20.2 (0.4) 25.9 (0.5) 31.7 (0.6) 37.5
0.7	1.3	2.0	2.7	3.4	.1	0.0	0.1	0.2	0.2	0.3	0.4	0.5	0.5	0.6	0.7	
0.7	1.4	2.1	2.8	3.5	.2	0.0	0.1	0.2	0.2	0.3	0.4	0.5	0.5	0.6	0.7	
0.7	1.4	2.1	2.9	3.6	.3	0.0	0.1	0.2	0.2	0.3	0.4	0.5	0.5	0.6	0.7	
0.7	1.5	2.2	2.9	3.7	.4	0.0	0.1	0.2	0.3	0.3	0.4	0.5	0.6	0.6	0.7	
0.8	1.5	2.3	3.0	3.8	.5	0.0	0.1	0.2	0.3	0.3	0.4	0.5	0.6	0.6	0.7	
0.8	1.5	2.3	3.1	3.8	.6	0.0	0.1	0.2	0.3	0.3	0.4	0.5	0.6	0.6	0.7	
0.8	1.6	2.4	3.2	3.9	.7	0.1	0.1	0.2	0.3	0.4	0.4	0.5	0.6	0.7	0.7	
0.8	1.6	2.4	3.2	4.0	.8	0.1	0.1	0.2	0.3	0.4	0.4	0.5	0.6	0.7	0.7	
0.9	1.7	2.5	3.3	4.1	.9	0.1	0.1	0.2	0.3	0.4	0.4	0.5	0.6	0.7	0.7	
0.8	1.6	2.5	3.3	4.1	.0	0.0	0.1	0.2	0.3	0.4	0.5	0.5	0.6	0.7	0.8	2.4 (0.1) 7.2 (0.2) 12.0 (0.3) 16.8 (0.4) 21.6 (0.5) 26.4 (0.6) 31.2 (0.7) 36.0
0.8	1.7	2.5	3.4	4.2	.1	0.0	0.1	0.2	0.3	0.4	0.5	0.6	0.7	0.7	0.8	
0.8	1.7	2.6	3.4	4.3	.2	0.0	0.1	0.2	0.3	0.4	0.5	0.6	0.7	0.8	0.8	
0.9	1.8	2.6	3.5	4.4	.3	0.0	0.1	0.2	0.3	0.4	0.5	0.6	0.7	0.8	0.9	
0.9	1.8	2.7	3.6	4.5	.4	0.0	0.1	0.2	0.3	0.4	0.5	0.6	0.7	0.8	0.9	
0.9	1.8	2.8	3.7	4.6	.5	0.0	0.1	0.2	0.3	0.4	0.5	0.6	0.7	0.8	0.9	
0.9	1.9	2.8	3.7	4.7	.6	0.1	0.1	0.2	0.3	0.4	0.5	0.6	0.7	0.8	0.9	
1.0	1.9	2.9	3.8	4.8	.7	0.1	0.2	0.2	0.3	0.4	0.5	0.6	0.7	0.8	0.9	
1.0	2.0	2.9	3.9	4.9	.8	0.1	0.2	0.3	0.3	0.4	0.5	0.6	0.7	0.8	0.9	
1.0	2.0	3.0	4.0	5.0	.9	0.1	0.2	0.3	0.4	0.4	0.5	0.6	0.7	0.8	0.9	
1.0	2.0	3.0	4.0	5.0	.0	0.0	0.1	0.2	0.3	0.4	0.5	0.6	0.8	0.9	1.0	2.1 (0.1) 6.2 (0.2) 10.4 (0.3) 14.5 (0.4) 18.6 (0.5) 22.8 (0.6) 26.9 (0.7) 31.1 (0.8) 35.2
1.0	2.0	3.0	4.0	5.1	.1	0.0	0.1	0.2	0.3	0.4	0.6	0.7	0.8	0.9	1.0	
1.0	2.0	3.1	4.1	5.1	.2	0.0	0.1	0.2	0.3	0.5	0.6	0.7	0.8	0.9	1.0	
1.0	2.1	3.1	4.2	5.2	.3	0.0	0.1	0.2	0.4	0.5	0.6	0.7	0.8	0.9	1.0	
1.1	2.1	3.2	4.3	5.3	.4	0.0	0.2	0.3	0.4	0.5	0.6	0.7	0.8	0.9	1.0	
1.1	2.2	3.3	4.3	5.4	.5	0.1	0.2	0.3	0.4	0.5	0.6	0.7	0.8	0.9	1.0	
1.1	2.2	3.3	4.4	5.5	.6	0.1	0.2	0.3	0.4	0.5	0.6	0.7	0.8	0.9	1.0	
1.1	2.3	3.4	4.5	5.6	.7	0.1	0.2	0.3	0.4	0.5	0.6	0.7	0.8	0.9	1.1	
1.2	2.3	3.4	4.6	5.7	.8	0.1	0.2	0.3	0.4	0.5	0.6	0.7	0.8	1.0	1.1	
1.2	2.3	3.5	4.6	5.8	.9	0.1	0.2	0.3	0.4	0.5	0.6	0.7	0.9	1.0	1.1	
1.1	2.3	3.5	4.6	5.8	.0	0.0	0.1	0.2	0.4	0.5	0.6	0.7	0.9	1.0	1.1	1.8 (0.1) 5.5 (0.2) 9.1 (0.3) 12.8 (0.4) 16.5 (0.5) 20.1 (0.6) 23.8 (0.7) 27.4 (0.8) 31.1 (0.9) 34.7
1.2	2.3	3.5	4.7	5.9	.1	0.0	0.1	0.3	0.4	0.5	0.6	0.8	0.9	1.0	1.1	
1.2	2.4	3.6	4.8	6.0	.2	0.0	0.1	0.3	0.4	0.5	0.6	0.8	0.9	1.0	1.1	
1.2	2.4	3.6	4.9	6.1	.3	0.0	0.2	0.3	0.4	0.5	0.7	0.8	0.9	1.0	1.2	
1.2	2.5	3.7	4.9	6.2	.4	0.0	0.2	0.3	0.4	0.5	0.7	0.8	0.9	1.0	1.2	
1.3	2.5	3.8	5.0	6.3	.5	0.1	0.2	0.3	0.4	0.6	0.7	0.8	0.9	1.1	1.2	
1.3	2.5	3.8	5.1	6.3	.6	0.1	0.2	0.3	0.4	0.6	0.7	0.8	0.9	1.1	1.2	
1.3	2.6	3.9	5.2	6.4	.7	0.1	0.2	0.3	0.5	0.6	0.7	0.8	1.0	1.1	1.2	
1.3	2.6	3.9	5.2	6.5	.8	0.1	0.2	0.3	0.5	0.6	0.7	0.8	1.0	1.1	1.2	
1.4	2.7	4.0	5.3	6.6	.9	0.1	0.2	0.4	0.5	0.6	0.7	0.9	1.0	1.1	1.2	
10′	20′	30′	40′	50′		0′	1′	2′	3′	4′	5′	6′	7′	8′	9′	

Dec. Inc.	Tens 10′	20′	30′	40′	50′	Decimals	Units 0′	1′	2′	3′	4′	5′	6′	7′	8′	9′	Double Second Diff. and Corr.
8.0	1.3	2.6	4.0	5.3	6.6	.0	0.0	0.1	0.3	0.4	0.6	0.7	0.8	1.0	1.1	1.3	1.6 (0.1) 4.8 (0.2) 8.0 (0.3) 11.2 (0.4) 14.5 (0.5) 17.7 (0.6) 20.9 (0.7) 24.1 (0.8) 27.3 (0.9) 30.5 (1.0) 33.7 (1.1) 36.9
8.1	1.3	2.7	4.0	5.4	6.7	.1	0.0	0.2	0.3	0.4	0.6	0.7	0.9	1.0	1.1	1.3	
8.2	1.3	2.7	4.1	5.4	6.8	.2	0.0	0.2	0.3	0.5	0.6	0.7	0.9	1.0	1.2	1.3	
8.3	1.4	2.8	4.1	5.5	6.9	.3	0.0	0.2	0.3	0.5	0.6	0.8	0.9	1.0	1.2	1.3	
8.4	1.4	2.8	4.2	5.6	7.0	.4	0.1	0.2	0.3	0.5	0.6	0.8	0.9	1.0	1.2	1.3	
8.5	1.4	2.8	4.3	5.7	7.1	.5	0.1	0.2	0.4	0.5	0.6	0.8	0.9	1.1	1.2	1.3	
8.6	1.4	2.9	4.3	5.7	7.2	.6	0.1	0.2	0.4	0.5	0.7	0.8	0.9	1.1	1.2	1.4	
8.7	1.5	2.9	4.4	5.8	7.3	.7	0.1	0.2	0.4	0.5	0.7	0.8	0.9	1.1	1.2	1.4	
8.8	1.5	3.0	4.4	5.9	7.4	.8	0.1	0.3	0.4	0.5	0.7	0.8	1.0	1.1	1.2	1.4	
8.9	1.5	3.0	4.5	6.0	7.5	.9	0.1	0.3	0.4	0.6	0.7	0.8	1.0	1.1	1.3	1.4	
9.0	1.5	3.0	4.5	6.0	7.5	.0	0.0	0.2	0.3	0.5	0.6	0.8	0.9	1.1	1.3	1.4	
9.1	1.5	3.0	4.5	6.0	7.6	.1	0.0	0.2	0.3	0.5	0.6	0.8	1.0	1.1	1.3	1.4	
9.2	1.5	3.0	4.6	6.1	7.6	.2	0.0	0.2	0.3	0.5	0.7	0.8	1.0	1.1	1.3	1.5	
9.3	1.5	3.1	4.6	6.2	7.7	.3	0.0	0.2	0.4	0.5	0.7	0.8	1.0	1.2	1.3	1.5	
9.4	1.6	3.1	4.7	6.3	7.8	.4	0.1	0.2	0.4	0.5	0.7	0.9	1.0	1.2	1.3	1.5	
9.5	1.6	3.2	4.8	6.3	7.9	.5	0.1	0.2	0.4	0.6	0.7	0.9	1.0	1.2	1.3	1.5	1.4 (0.1) 4.2 (0.2) 7.1 (0.3) 9.9 (0.4) 12.7 (0.5) 15.5 (0.6) 18.4 (0.7) 21.2 (0.8) 24.0 (0.9) 26.8 (1.0) 29.7 (1.1) 32.5 (1.2) 35.3
9.6	1.6	3.2	4.8	6.4	8.0	.6	0.1	0.3	0.4	0.6	0.7	0.9	1.0	1.2	1.4	1.5	
9.7	1.6	3.3	4.9	6.5	8.1	.7	0.1	0.3	0.4	0.6	0.7	0.9	1.1	1.2	1.4	1.5	
9.8	1.7	3.3	4.9	6.6	8.2	.8	0.1	0.3	0.4	0.6	0.8	0.9	1.1	1.2	1.4	1.6	
9.9	1.7	3.3	5.0	6.6	8.3	.9	0.1	0.3	0.5	0.6	0.8	0.9	1.1	1.3	1.4	1.6	
10.0	1.6	3.3	5.0	6.6	8.3	.0	0.0	0.2	0.3	0.5	0.7	0.9	1.0	1.2	1.4	1.6	
10.1	1.7	3.3	5.0	6.7	8.4	.1	0.0	0.2	0.4	0.5	0.7	0.9	1.1	1.2	1.4	1.6	
10.2	1.7	3.4	5.1	6.8	8.5	.2	0.0	0.2	0.4	0.6	0.7	0.9	1.1	1.3	1.4	1.6	
10.3	1.7	3.4	5.1	6.9	8.6	.3	0.1	0.2	0.4	0.6	0.8	0.9	1.1	1.3	1.5	1.6	
10.4	1.7	3.5	5.2	6.9	8.7	.4	0.1	0.2	0.4	0.6	0.8	0.9	1.1	1.3	1.5	1.6	
10.5	1.8	3.5	5.3	7.0	8.8	.5	0.1	0.3	0.4	0.6	0.8	1.0	1.1	1.3	1.5	1.7	
10.6	1.8	3.5	5.3	7.1	8.8	.6	0.1	0.3	0.5	0.6	0.8	1.0	1.2	1.3	1.5	1.7	
10.7	1.8	3.6	5.4	7.2	8.9	.7	0.1	0.3	0.5	0.6	0.8	1.0	1.2	1.3	1.5	1.7	
10.8	1.8	3.6	5.4	7.2	9.0	.8	0.1	0.3	0.5	0.7	0.8	1.0	1.2	1.4	1.5	1.7	
10.9	1.9	3.7	5.5	7.3	9.1	.9	0.2	0.3	0.5	0.7	0.9	1.0	1.2	1.4	1.6	1.7	
11.0	1.8	3.6	5.5	7.3	9.1	.0	0.0	0.2	0.4	0.6	0.8	1.0	1.1	1.3	1.5	1.7	1.3 (0.1) 3.8 (0.2) 6.3 (0.3) 8.9 (0.4) 11.4 (0.5) 14.0 (0.6) 16.5 (0.7) 19.0 (0.8) 21.6 (0.9) 24.1 (1.0) 26.7 (1.1) 29.2 (1.2) 31.7 (1.3) 34.3
11.1	1.8	3.7	5.5	7.4	9.2	.1	0.0	0.2	0.4	0.6	0.8	1.0	1.2	1.4	1.6	1.7	
11.2	1.8	3.7	5.6	7.4	9.3	.2	0.0	0.2	0.4	0.6	0.8	1.0	1.2	1.4	1.6	1.8	
11.3	1.9	3.8	5.6	7.5	9.4	.3	0.1	0.2	0.4	0.6	0.8	1.0	1.2	1.4	1.6	1.8	
11.4	1.9	3.8	5.7	7.6	9.5	.4	0.1	0.3	0.5	0.7	0.8	1.0	1.2	1.4	1.6	1.8	
11.5	1.9	3.8	5.8	7.7	9.6	.5	0.1	0.3	0.5	0.7	0.9	1.1	1.2	1.4	1.6	1.8	
11.6	1.9	3.9	5.8	7.7	9.7	.6	0.1	0.3	0.5	0.7	0.9	1.1	1.3	1.5	1.6	1.8	
11.7	2.0	3.9	5.9	7.8	9.8	.7	0.1	0.3	0.5	0.7	0.9	1.1	1.3	1.5	1.7	1.9	
11.8	2.0	4.0	5.9	7.9	9.9	.8	0.2	0.3	0.5	0.7	0.9	1.1	1.3	1.5	1.7	1.9	
11.9	2.0	4.0	6.0	8.0	10.0	.9	0.2	0.4	0.6	0.7	0.9	1.1	1.3	1.5	1.7	1.9	
12.0	2.0	4.0	6.0	8.0	10.0	.0	0.0	0.2	0.4	0.6	0.8	1.0	1.2	1.5	1.7	1.9	
12.1	2.0	4.0	6.0	8.0	10.1	.1	0.0	0.2	0.4	0.6	0.9	1.1	1.3	1.5	1.7	1.9	
12.2	2.0	4.0	6.1	8.1	10.1	.2	0.0	0.2	0.5	0.7	0.9	1.1	1.3	1.5	1.7	1.9	
12.3	2.0	4.1	6.1	8.2	10.2	.3	0.1	0.3	0.5	0.7	0.9	1.1	1.3	1.5	1.7	1.9	
12.4	2.1	4.1	6.2	8.3	10.3	.4	0.1	0.3	0.5	0.7	0.9	1.1	1.3	1.5	1.7	2.0	
12.5	2.1	4.2	6.3	8.3	10.4	.5	0.1	0.3	0.5	0.7	0.9	1.1	1.4	1.6	1.8	2.0	1.2 (0.1) 3.5 (0.2) 5.8 (0.3) 8.1 (0.4) 10.5 (0.5) 12.8 (0.6) 15.1 (0.7) 17.4 (0.8) 19.8 (0.9) 22.1 (1.0) 24.4 (1.1) 26.7 (1.2) 29.1 (1.3) 31.4 (1.4) 33.7 (1.5) 36.0
12.6	2.1	4.2	6.3	8.4	10.5	.6	0.1	0.3	0.5	0.7	1.0	1.2	1.4	1.6	1.8	2.0	
12.7	2.1	4.3	6.4	8.5	10.6	.7	0.1	0.4	0.6	0.8	1.0	1.2	1.4	1.6	1.8	2.0	
12.8	2.2	4.3	6.4	8.6	10.7	.8	0.2	0.4	0.6	0.8	1.0	1.2	1.4	1.6	1.8	2.0	
12.9	2.2	4.3	6.5	8.6	10.8	.9	0.2	0.4	0.6	0.8	1.0	1.2	1.4	1.6	1.9	2.1	
13.0	2.1	4.3	6.5	8.6	10.8	.0	0.0	0.2	0.4	0.7	0.9	1.1	1.3	1.6	1.8	2.0	
13.1	2.2	4.3	6.5	8.7	10.9	.1	0.0	0.2	0.5	0.7	0.9	1.1	1.4	1.6	1.8	2.0	
13.2	2.2	4.4	6.6	8.8	11.0	.2	0.0	0.3	0.5	0.7	0.9	1.2	1.4	1.6	1.8	2.1	
13.3	2.2	4.4	6.6	8.9	11.1	.3	0.1	0.3	0.5	0.7	1.0	1.2	1.4	1.6	1.9	2.1	
13.4	2.2	4.5	6.7	8.9	11.2	.4	0.1	0.3	0.5	0.8	1.0	1.2	1.4	1.7	1.9	2.1	
13.5	2.3	4.5	6.8	9.0	11.3	.5	0.1	0.3	0.6	0.8	1.0	1.2	1.5	1.7	1.9	2.1	
13.6	2.3	4.5	6.8	9.1	11.3	.6	0.1	0.4	0.6	0.8	1.0	1.3	1.5	1.7	1.9	2.2	
13.7	2.3	4.6	6.9	9.2	11.4	.7	0.2	0.4	0.6	0.8	1.1	1.3	1.5	1.7	2.0	2.2	
13.8	2.3	4.6	6.9	9.2	11.5	.8	0.2	0.4	0.6	0.9	1.1	1.3	1.5	1.8	2.0	2.2	
13.9	2.4	4.7	7.0	9.3	11.6	.9	0.2	0.4	0.7	0.9	1.1	1.3	1.6	1.8	2.0	2.2	
14.0	2.3	4.6	7.0	9.3	11.6	.0	0.0	0.2	0.5	0.7	1.0	1.2	1.4	1.7	1.9	2.2	1.1 (0.1) 3.2 (0.2) 5.3 (0.3) 7.5 (0.4) 9.6 (0.5) 11.7 (0.6) 13.9 (0.7) 16.0 (0.8) 18.1 (0.9) 20.3 (1.0) 22.4 (1.1) 24.5 (1.2) 26.7 (1.3) 28.8 (1.4) 30.9 (1.5) 33.1 (1.6) 35.2
14.1	2.3	4.7	7.0	9.4	11.7	.1	0.0	0.3	0.5	0.7	1.0	1.2	1.5	1.7	2.0	2.2	
14.2	2.3	4.7	7.1	9.4	11.8	.2	0.0	0.3	0.5	0.8	1.0	1.3	1.5	1.7	2.0	2.2	
14.3	2.4	4.8	7.1	9.5	11.9	.3	0.1	0.3	0.6	0.8	1.0	1.3	1.5	1.8	2.0	2.2	
14.4	2.4	4.8	7.2	9.6	12.0	.4	0.1	0.3	0.6	0.8	1.1	1.3	1.5	1.8	2.0	2.3	
14.5	2.4	4.8	7.3	9.7	12.1	.5	0.1	0.4	0.6	0.8	1.1	1.3	1.6	1.8	2.1	2.3	
14.6	2.4	4.9	7.3	9.7	12.2	.6	0.1	0.4	0.6	0.9	1.1	1.4	1.6	1.8	2.1	2.3	
14.7	2.5	4.9	7.4	9.8	12.3	.7	0.2	0.4	0.7	0.9	1.1	1.4	1.6	1.9	2.1	2.3	
14.8	2.5	5.0	7.4	9.9	12.4	.8	0.2	0.4	0.7	0.9	1.2	1.4	1.6	1.9	2.1	2.4	
14.9	2.5	5.0	7.5	10.0	12.5	.9	0.2	0.5	0.7	0.9	1.2	1.4	1.7	1.9	2.2	2.4	
15.0	2.5	5.0	7.5	10.0	12.5	.0	0.0	0.3	0.5	0.8	1.0	1.3	1.5	1.8	2.1	2.3	
15.1	2.5	5.0	7.5	10.0	12.6	.1	0.0	0.3	0.5	0.8	1.1	1.3	1.6	1.8	2.1	2.4	
15.2	2.5	5.0	7.6	10.1	12.6	.2	0.1	0.3	0.6	0.8	1.1	1.3	1.6	1.9	2.1	2.4	
15.3	2.5	5.1	7.6	10.2	12.7	.3	0.1	0.3	0.6	0.9	1.1	1.4	1.6	1.9	2.1	2.4	
15.4	2.6	5.1	7.7	10.3	12.8	.4	0.1	0.4	0.6	0.9	1.1	1.4	1.7	1.9	2.2	2.4	
15.5	2.6	5.2	7.8	10.3	12.9	.5	0.1	0.4	0.6	0.9	1.2	1.4	1.7	1.9	2.2	2.5	
15.6	2.6	5.2	7.8	10.4	13.0	.6	0.2	0.4	0.7	0.9	1.2	1.4	1.7	2.0	2.2	2.5	
15.7	2.6	5.3	7.9	10.5	13.1	.7	0.2	0.4	0.7	1.0	1.2	1.5	1.7	2.0	2.2	2.5	
15.8	2.7	5.3	7.9	10.6	13.2	.8	0.2	0.5	0.7	1.0	1.2	1.5	1.8	2.0	2.3	2.5	
15.9	2.7	5.3	8.0	10.6	13.3	.9	0.2	0.5	0.7	1.0	1.3	1.5	1.8	2.0	2.3	2.6	
	10′	20′	30′	40′	50′		0′	1′	2′	3′	4′	5′	6′	7′	8′	9′	

The Double-Second-Difference correction (Corr.) is always to be added to the tabulated altitude.

Table 15

all, there is no sense reducing sights to greater precision than that with which you can take them. You probably can take sights to an accuracy of 0.1′ from the bridge of a battleship, but I really doubt whether it can be done from the deck of a small sailboat.

9
Sun-Moon Fix

A way of overcoming the disadvantages of navigation solely by advancing sun lines is to take nearly simultaneous observations of the sun and the moon. This results in two lines of position that cross to give a fix. Of course, the moon and the sun need to be sited as to azimuth so that the LOPs will cross at an angle of 45° or better. Surprisingly, perhaps, the sun and moon are in this sort of relationship about ten days or so every month, so it is possible to take daylight sights of the sun and moon for a very accurate fix of your position.

Suppose that it is July 1, 1974, at about 8 P.M. or 2000 hours. It has been cloudy and blowing all day, but right now the sky clears, and you are able to get the following observations before a wave completely drenches you and the sextant.

MOON	
20-01-30	18°41′
20-02-25	18°45′
20-03-25	18°25′
20-04-40	19°33′
SUN	
19-34-40	3°29′

Since it is too rough to work with graph paper, you average your moon shots:

Hs = 18°53′
Watch Time = 20-03-00

You have only one sun sight, so that one will have to do.

Taking the moon first, you proceed to fill out the worksheet:

Date	= 7-1-74
Log	= 5740
Watch Time	= 20-02-60
Fast (−) or Slow (+)	= -40
Conversion to GMT	= +4
GMT	= 24-02-20

This means that it is 2 minutes and 20 seconds past midnight, so the date for entering the almanac will be July 2, 1974.

Hs	=	18°53′
IC	=	+ 05′
Dip	=	− 03′
ha	=	18°55′

Turn to the page for July 2 (Table 10 on page 57), and you will note that the entry for the moon is more extensive than that for the sun. This is due to the fact that the motion of the moon is not as regular as that of the sun. The GHA of the moon varies, and there is a correction in the column next to GHA headed "*v*." Also, you will note in the "Dec." and "*d*" columns that the declination of the moon changes much more rapidly than that of the sun. Finally, for the moon you will see a column labeled "H.P." The initials stand for *horizontal parallax*, which is an effect of the closeness of the moon to the earth. For purposes of celestial navigation, it is assumed that the angle measured from horizon to body at the surface of the earth is the same as the angle would be if measured from the center of the earth. In the case of the sun, planets, and stars, this assumption is valid; they are so far away that rays of light from them strike the surface of the earth at the same angle as they would (theoretically) strike the center of the earth (Figure 9.1). Because the moon is so close, however, the same is not true (Figure 9.2), and a correction must be made in addition to the correction for refraction and semidiameter. The correction is handled in the almanacs in much the same manner as the corrections for the sun.

Meanwhile, take out the information for 00 hours GMT and fill out the worksheet:

Fig. 9.1

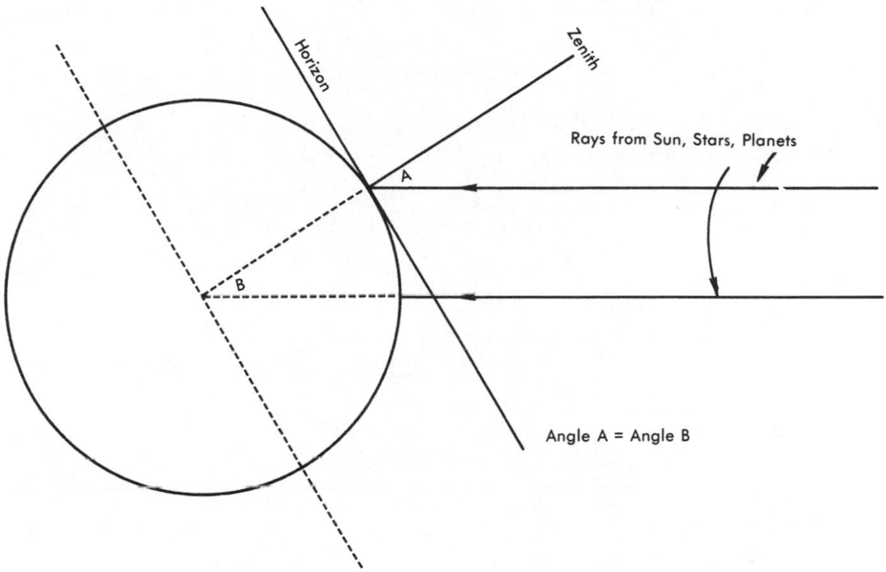
Horizon
Zenith
Rays from Sun, Stars, Planets
A
B
Angle A = Angle B

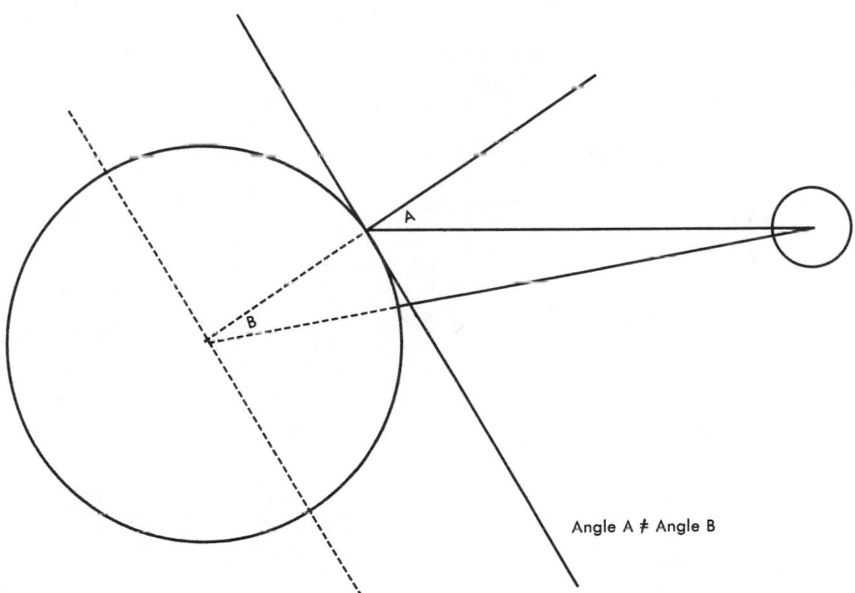
A
B
Angle A ≠ Angle B

Fig. 9.2

MOON			
HP	=	55.1′	
d	=	1.4′	
Dec.	=	S 22°53.4′	Increasing _x_
			Decreasing ___

You note (by looking down the declination column) that the moon is moving farther south as time passes; therefore, the *d* correction is going to be added. To remind yourself, you mark "Increasing."

The *d* correction is tabulated on the same page as the GHA increments, and, because you are dealing with the correction for only 2 minutes of time, does not amount to much.

MOON			
HP	=	55.1′	
d	=	1.4′	
Dec.	=	S 22°53.4′	
d	=	+ 0.1′	Increasing _x_
Dec.	=	S 22°54′	Decreasing ___

Turning to the "Increments and Corrections" page in the back of the almanac (Table 16) for 2 minutes and 20 seconds, you find 33.4′ and proceed to complete this part of the worksheet.

GHA 00 Hr.	=	29°39 ′
2 Min. 20 Sec.	=	34.4′
(v = 10.2) Corr.	=	.4′
GHA	=	29°73′
GHA	=	30°13′
(+ 360?)	=	360°
GHA	=	390°13′
Assumed Long. W. (−)	=	− 69°13′
Assumed Long. E. (+)	=	(+)
LHA	=	321°
(−360?)	=	
Assumed Lat.	=	37° N

2ᵐ INCREMENTS AND CORRECTIONS 3ᵐ

2ᵐ	SUN PLANETS	ARIES	MOON	v or d	Corrⁿ	v or d	Corrⁿ	v or d	Corrⁿ
s	° ′	° ′	° ′	′	′	′	′	′	′
00	0 30·0	0 30·1	0 28·6	0·0	0·0	6·0	0·3	12·0	0·5
01	0 30·3	0 30·3	0 28·9	0·1	0·0	6·1	0·3	12·1	0·5
02	0 30·5	0 30·6	0 29·1	0·2	0·0	6·2	0·3	12·2	0·5
03	0 30·8	0 30·8	0 29·3	0·3	0·0	6·3	0·3	12·3	0·5
04	0 31·0	0 31·1	0 29·6	0·4	0·0	6·4	0·3	12·4	0·5
05	0 31·3	0 31·3	0 29·8	0·5	0·0	6·5	0·3	12·5	0·5
06	0 31·5	0 31·6	0 30·1	0·6	0·0	6·6	0·3	12·6	0·5
07	0 31·8	0 31·8	0 30·3	0·7	0·0	6·7	0·3	12·7	0·5
08	0 32·0	0 32·1	0 30·5	0·8	0·0	6·8	0·3	12·8	0·5
09	0 32·3	0 32·3	0 30·8	0·9	0·0	6·9	0·3	12·9	0·5
10	0 32·5	0 32·6	0 31·0	1·0	0·0	7·0	0·3	13·0	0·5
11	0 32·8	0 32·8	0 31·3	1·1	0·0	7·1	0·3	13·1	0·5
12	0 33·0	0 33·1	0 31·5	1·2	0·1	7·2	0·3	13·2	0·6
13	0 33·3	0 33·3	0 31·7	1·3	0·1	7·3	0·3	13·3	0·6
14	0 33·5	0 33·6	0 32·0	1·4	0·1	7·4	0·3	13·4	0·6
15	0 33·8	0 33·8	0 32·2	1·5	0·1	7·5	0·3	13·5	0·6
16	0 34·0	0 34·1	0 32·5	1·6	0·1	7·6	0·3	13·6	0·6
17	0 34·3	0 34·3	0 32·7	1·7	0·1	7·7	0·3	13·7	0·6
18	0 34·5	0 34·6	0 32·9	1·8	0·1	7·8	0·3	13·8	0·6
19	0 34·8	0 34·8	0 33·2	1·9	0·1	7·9	0·3	13·9	0·6
20	0 35·0	0 35·1	0 33·4	2·0	0·1	8·0	0·3	14·0	0·6
21	0 35·3	0 35·3	0 33·6	2·1	0·1	8·1	0·3	14·1	0·6
22	0 35·5	0 35·6	0 33·9	2·2	0·1	8·2	0·3	14·2	0·6
23	0 35·8	0 35·8	0 34·1	2·3	0·1	8·3	0·3	14·3	0·6
24	0 36·0	0 36·1	0 34·4	2·4	0·1	8·4	0·4	14·4	0·6
25	0 36·3	0 36·3	0 34·6	2·5	0·1	8·5	0·4	14·5	0·6
26	0 36·5	0 36·6	0 34·8	2·6	0·1	8·6	0·4	14·6	0·6
27	0 36·8	0 36·9	0 35·1	2·7	0·1	8·7	0·4	14·7	0·6
28	0 37·0	0 37·1	0 35·3	2·8	0·1	8·8	0·4	14·8	0·6
29	0 37·3	0 37·4	0 35·6	2·9	0·1	8·9	0·4	14·9	0·6
30	0 37·5	0 37·6	0 35·8	3·0	0·1	9·0	0·4	15·0	0·6
31	0 37·8	0 37·9	0 36·0	3·1	0·1	9·1	0·4	15·1	0·6
32	0 38·0	0 38·1	0 36·3	3·2	0·1	9·2	0·4	15·2	0·6
33	0 38·3	0 38·4	0 36·5	3·3	0·1	9·3	0·4	15·3	0·6
34	0 38·5	0 38·6	0 36·7	3·4	0·1	9·4	0·4	15·4	0·6
35	0 38·8	0 38·9	0 37·0	3·5	0·1	9·5	0·4	15·5	0·6
36	0 39·0	0 39·1	0 37·2	3·6	0·2	9·6	0·4	15·6	0·7
37	0 39·3	0 39·4	0 37·5	3·7	0·2	9·7	0·4	15·7	0·7
38	0 39·5	0 39·6	0 37·7	3·8	0·2	9·8	0·4	15·8	0·7
39	0 39·8	0 39·9	0 37·9	3·9	0·2	9·9	0·4	15·9	0·7
40	0 40·0	0 40·1	0 38·2	4·0	0·2	10·0	0·4	16·0	0·7
41	0 40·3	0 40·4	0 38·4	4·1	0·2	10·1	0·4	16·1	0·7
42	0 40·5	0 40·6	0 38·7	4·2	0·2	10·2	0·4	16·2	0·7
43	0 40·8	0 40·9	0 38·9	4·3	0·2	10·3	0·4	16·3	0·7
44	0 41·0	0 41·1	0 39·1	4·4	0·2	10·4	0·4	16·4	0·7
45	0 41·3	0 41·4	0 39·4	4·5	0·2	10·5	0·4	16·5	0·7
46	0 41·5	0 41·6	0 39·6	4·6	0·2	10·6	0·4	16·6	0·7
47	0 41·8	0 41·9	0 39·8	4·7	0·2	10·7	0·4	16·7	0·7
48	0 42·0	0 42·1	0 40·1	4·8	0·2	10·8	0·5	16·8	0·7
49	0 42·3	0 42·4	0 40·3	4·9	0·2	10·9	0·5	16·9	0·7
50	0 42·5	0 42·6	0 40·6	5·0	0·2	11·0	0·5	17·0	0·7
51	0 42·8	0 42·9	0 40·8	5·1	0·2	11·1	0·5	17·1	0·7
52	0 43·0	0 43·1	0 41·0	5·2	0·2	11·2	0·5	17·2	0·7
53	0 43·3	0 43·4	0 41·3	5·3	0·2	11·3	0·5	17·3	0·7
54	0 43·5	0 43·6	0 41·5	5·4	0·2	11·4	0·5	17·4	0·7
55	0 43·8	0 43·9	0 41·8	5·5	0·2	11·5	0·5	17·5	0·7
56	0 44·0	0 44·1	0 42·0	5·6	0·2	11·6	0·5	17·6	0·7
57	0 44·3	0 44·4	0 42·2	5·7	0·2	11·7	0·5	17·7	0·7
58	0 44·5	0 44·6	0 42·5	5·8	0·2	11·8	0·5	17·8	0·7
59	0 44·8	0 44·9	0 42·7	5·9	0·2	11·9	0·5	17·9	0·7
60	0 45·0	0 45·1	0 43·0	6·0	0·3	12·0	0·5	18·0	0·8

3ᵐ	SUN PLANETS	ARIES	MOON	v or d	Corrⁿ	v or d	Corrⁿ	v or d	Corrⁿ
s	° ′	° ′	° ′	′	′	′	′	′	′
00	0 45·0	0 45·1	0 43·0	0·0	0·0	6·0	0·4	12·0	0·7
01	0 45·3	0 45·4	0 43·2	0·1	0·0	6·1	0·4	12·1	0·7
02	0 45·5	0 45·6	0 43·4	0·2	0·0	6·2	0·4	12·2	0·7
03	0 45·8	0 45·9	0 43·7	0·3	0·0	6·3	0·4	12·3	0·7
04	0 46·0	0 46·1	0 43·9	0·4	0·0	6·4	0·4	12·4	0·7
05	0 46·3	0 46·4	0 44·1	0·5	0·0	6·5	0·4	12·5	0·7
06	0 46·5	0 46·6	0 44·4	0·6	0·0	6·6	0·4	12·6	0·7
07	0 46·8	0 46·9	0 44·6	0·7	0·0	6·7	0·4	12·7	0·7
08	0 47·0	0 47·1	0 44·9	0·8	0·0	6·8	0·4	12·8	0·7
09	0 47·3	0 47·4	0 45·1	0·9	0·1	6·9	0·4	12·9	0·8
10	0 47·5	0 47·6	0 45·3	1·0	0·1	7·0	0·4	13·0	0·8
11	0 47·8	0 47·9	0 45·6	1·1	0·1	7·1	0·4	13·1	0·8
12	0 48·0	0 48·1	0 45·8	1·2	0·1	7·2	0·4	13·2	0·8
13	0 48·3	0 48·4	0 46·1	1·3	0·1	7·3	0·4	13·3	0·8
14	0 48·5	0 48·6	0 46·3	1·4	0·1	7·4	0·4	13·4	0·8
15	0 48·8	0 48·9	0 46·5	1·5	0·1	7·5	0·4	13·5	0·8
16	0 49·0	0 49·1	0 46·8	1·6	0·1	7·6	0·4	13·6	0·8
17	0 49·3	0 49·4	0 47·0	1·7	0·1	7·7	0·4	13·7	0·8
18	0 49·5	0 49·6	0 47·2	1·8	0·1	7·8	0·5	13·8	0·8
19	0 49·8	0 49·9	0 47·5	1·9	0·1	7·9	0·5	13·9	0·8
20	0 50·0	0 50·1	0 47·7	2·0	0·1	8·0	0·5	14·0	0·8
21	0 50·3	0 50·4	0 48·0	2·1	0·1	8·1	0·5	14·1	0·8
22	0 50·5	0 50·6	0 48·2	2·2	0·1	8·2	0·5	14·2	0·8
23	0 50·8	0 50·9	0 48·4	2·3	0·1	8·3	0·5	14·3	0·8
24	0 51·0	0 51·1	0 48·7	2·4	0·1	8·4	0·5	14·4	0·8
25	0 51·3	0 51·4	0 48·9	2·5	0·1	8·5	0·5	14·5	0·8
26	0 51·5	0 51·6	0 49·2	2·6	0·2	8·6	0·5	14·6	0·9
27	0 51·8	0 51·9	0 49·4	2·7	0·2	8·7	0·5	14·7	0·9
28	0 52·0	0 52·1	0 49·6	2·8	0·2	8·8	0·5	14·8	0·9
29	0 52·3	0 52·4	0 49·9	2·9	0·2	8·9	0·5	14·9	0·9
30	0 52·5	0 52·6	0 50·1	3·0	0·2	9·0	0·5	15·0	0·9
31	0 52·8	0 52·9	0 50·3	3·1	0·2	9·1	0·5	15·1	0·9
32	0 53·0	0 53·1	0 50·6	3·2	0·2	9·2	0·5	15·2	0·9
33	0 53·3	0 53·4	0 50·8	3·3	0·2	9·3	0·5	15·3	0·9
34	0 53·5	0 53·6	0 51·1	3·4	0·2	9·4	0·5	15·4	0·9
35	0 53·8	0 53·9	0 51·3	3·5	0·2	9·5	0·6	15·5	0·9
36	0 54·0	0 54·1	0 51·5	3·6	0·2	9·6	0·6	15·6	0·9
37	0 54·3	0 54·4	0 51·8	3·7	0·2	9·7	0·6	15·7	0·9
38	0 54·5	0 54·6	0 52·0	3·8	0·2	9·8	0·6	15·8	0·9
39	0 54·8	0 54·9	0 52·3	3·9	0·2	9·9	0·6	15·9	0·9
40	0 55·0	0 55·2	0 52·5	4·0	0·2	10·0	0·6	16·0	0·9
41	0 55·3	0 55·4	0 52·7	4·1	0·2	10·1	0·6	16·1	0·9
42	0 55·5	0 55·7	0 53·0	4·2	0·2	10·2	0·6	16·2	0·9
43	0 55·8	0 55·9	0 53·2	4·3	0·3	10·3	0·6	16·3	1·0
44	0 56·0	0 56·2	0 53·4	4·4	0·3	10·4	0·6	16·4	1·0
45	0 56·3	0 56·4	0 53·7	4·5	0·3	10·5	0·6	16·5	1·0
46	0 56·5	0 56·7	0 53·9	4·6	0·3	10·6	0·6	16·6	1·0
47	0 56·8	0 56·9	0 54·2	4·7	0·3	10·7	0·6	16·7	1·0
48	0 57·0	0 57·2	0 54·4	4·8	0·3	10·8	0·6	16·8	1·0
49	0 57·3	0 57·4	0 54·6	4·9	0·3	10·9	0·6	16·9	1·0
50	0 57·5	0 57·7	0 54·9	5·0	0·3	11·0	0·6	17·0	1·0
51	0 57·8	0 57·9	0 55·1	5·1	0·3	11·1	0·6	17·1	1·0
52	0 58·0	0 58·2	0 55·4	5·2	0·3	11·2	0·7	17·2	1·0
53	0 58·3	0 58·4	0 55·6	5·3	0·3	11·3	0·7	17·3	1·0
54	0 58·5	0 58·7	0 55·8	5·4	0·3	11·4	0·7	17·4	1·0
55	0 58·8	0 58·9	0 56·1	5·5	0·3	11·5	0·7	17·5	1·0
56	0 59·0	0 59·2	0 56·3	5·6	0·3	11·6	0·7	17·6	1·0
57	0 59·3	0 59·4	0 56·6	5·7	0·3	11·7	0·7	17·7	1·0
58	0 59·5	0 59·7	0 56·8	5·8	0·3	11·8	0·7	17·8	1·0
59	0 59·8	0 59·9	0 57·0	5·9	0·3	11·9	0·7	17·9	1·0
60	1 00·0	1 00·2	0 57·3	6·0	0·4	12·0	0·7	18·0	1·1

Table 16

Like the *d* correction, the *v* correction comes from the column to the right of the block for 2 minutes on Table 16, and since only 2 minutes is involved, it is not as large as it could be had the sight been taken toward the end of the hour.

Now turn to the inside back cover of the *Nautical Almanac* (Tables 17 and 18) to complete the correction of the sextant altitude for the effects of refraction, semidiameter, and parallax. The correction table is in two parts, 0-35° apparent altitude and 35°-90°. You want the 0-35° page, since the apparent altitude is 18°55′.

Hs	= 18°53′
IC	= + 05′
Dip	= − 03′
ha	= 18°55′
Main Corr.	= + 62′
Lower Limb	= + 02′
Upper Limb and (−30′)	=
Ho	= 19°59′

The 62′ main correction is taken from the "15°-19°" column opposite 18°50′—62.4′—and rounded off. Then in the column directly below, the lower limb correction of 2′ is taken opposite the horizontal parallax value of 55.2′ (the closest to your value of 55.1). The column gives 1.9′, which again is rounded off. Lower limb ("L") is the lower edge of the moon. Sometimes a half or quarter moon will not have a discernible lower edge, and it is necessary to use the upper edge. Then use the ("U") column and subtract an additional 30′. If you set up your worksheet as shown, you will not forget these factors.

The sun worksheet (using Tables 10, 19, and 20) looks like this:

Date	= 7-1-74
Log	= 5740
Watch Time	= 19-34-40
Fast (−) or Slow (+)	= -40
Conversion to GMT	= +4
GMT	= 23-34-00

ALTITUDE CORRECTION TABLES 0°–35°—MOON

App. Alt.	0°–4° Corrⁿ	5°–9° Corrⁿ	10°–14° Corrⁿ	15°–19° Corrⁿ	20°–24° Corrⁿ	25°–29° Corrⁿ	30°–34° Corrⁿ	App. Alt.
′	° ′	° ′	° ′	° ′	° ′	° ′	° ′	′
00	0 33·8	5 58·2	10 62·1	15 62·8	20 62·2	25 60·8	30 58·9	00
10	35·9	58·5	62·2	62·8	62·1	60·8	58·8	10
20	37·8	58·7	62·2	62·8	62·1	60·7	58·8	20
30	39·6	58·9	62·3	62·8	62·1	60·7	58·7	30
40	41·2	59·1	62·3	62·8	62·0	60·6	58·6	40
50	42·6	59·3	62·4	62·7	62·0	60·6	58·5	50
00	1 44·0	6 59·5	11 62·4	16 62·7	21 62·0	26 60·5	31 58·5	00
10	45·2	59·7	62·4	62·7	61·9	60·4	58·4	10
20	46·3	59·9	62·5	62·7	61·9	60·4	58·3	20
30	47·3	60·0	62·5	62·7	61·9	60·3	58·2	30
40	48·3	60·2	62·5	62·7	61·8	60·3	58·2	40
50	49·2	60·3	62·6	62·7	61·8	60·2	58·1	50
00	2 50·0	7 60·5	12 62·6	17 62·7	22 61·7	27 60·1	32 58·0	00
10	50·8	60·6	62·6	62·6	61·7	60·1	57·9	10
20	51·4	60·7	62·6	62·6	61·6	60·0	57·8	20
30	52·1	60·9	62·7	62·6	61·6	59·9	57·8	30
40	52·7	61·0	62·7	62·6	61·5	59·9	57·7	40
50	53·3	61·1	62·7	62·6	61·5	59·8	57·6	50
00	3 53·8	8 61·2	13 62·7	18 62·5	23 61·5	28 59·7	33 57·5	00
10	54·3	61·3	62·7	62·5	61·4	59·7	57·4	10
20	54·8	61·4	62·7	62·5	61·4	59·6	57·4	20
30	55·2	61·5	62·8	62·5	61·3	59·6	57·3	30
40	55·6	61·6	62·8	62·4	61·3	59·5	57·2	40
50	56·0	61·6	62·8	62·4	61·2	59·4	57·1	50
00	4 56·4	9 61·7	14 62·8	19 62·4	24 61·2	29 59·3	34 57·0	00
10	56·7	61·8	62·8	62·3	61·1	59·3	56·9	10
20	57·1	61·9	62·8	62·3	61·1	59·2	56·9	20
30	57·4	61·9	62·8	62·3	61·0	59·1	56·8	30
40	57·7	62·0	62·8	62·2	60·9	59·1	56·7	40
50	57·9	62·1	62·8	62·2	60·9	59·0	56·6	50

H.P.	L	U	L	U	L	U	L	U	L	U	L	U	L	U	H.P.
′	′	′	′	′	′	′	′	′	′	′	′	′	′	′	′
54·0	0·3	0·9	0·3	0·9	0·4	1·0	0·5	1·1	0·6	1·2	0·7	1·3	0·9	1·5	54·0
54·3	0·7	1·1	0·7	1·2	0·7	1·2	0·8	1·3	0·9	1·4	1·1	1·5	1·2	1·7	54·3
54·6	1·1	1·4	1·1	1·4	1·1	1·4	1·2	1·5	1·3	1·6	1·4	1·7	1·5	1·8	54·6
54·9	1·4	1·6	1·5	1·6	1·5	1·6	1·6	1·7	1·6	1·8	1·8	1·9	1·9	2·0	54·9
55·2	1·8	1·8	1·8	1·8	1·9	1·9	1·9	1·9	2·0	2·0	2·1	2·1	2·2	2·2	55·2
55·5	2·2	2·0	2·2	2·0	2·3	2·1	2·3	2·1	2·4	2·2	2·4	2·3	2·5	2·4	55·5
55·8	2·6	2·2	2·6	2·2	2·6	2·3	2·7	2·3	2·7	2·4	2·8	2·4	2·9	2·5	55·8
56·1	3·0	2·4	3·0	2·5	3·0	2·5	3·0	2·5	3·1	2·6	3·1	2·6	3·2	2·7	56·1
56·4	3·4	2·7	3·4	2·7	3·4	2·7	3·4	2·7	3·4	2·8	3·5	2·8	3·5	2·9	56·4
56·7	3·7	2·9	3·7	2·9	3·8	2·9	3·8	2·9	3·8	3·0	3·8	3·0	3·9	3·0	56·7
57·0	4·1	3·1	4·1	3·1	4·1	3·1	4·1	3·1	4·2	3·1	4·2	3·2	4·2	3·2	57·0
57·3	4·5	3·3	4·5	3·3	4·5	3·3	4·5	3·3	4·5	3·3	4·5	3·4	4·6	3·4	57·3
57·6	4·9	3·5	4·9	3·5	4·9	3·5	4·9	3·5	4·9	3·5	4·9	3·5	4·9	3·6	57·6
57·9	5·3	3·8	5·3	3·8	5·2	3·8	5·2	3·7	5·2	3·7	5·2	3·7	5·2	3·7	57·9
58·2	5·6	4·0	5·6	4·0	5·6	4·0	5·6	4·0	5·6	3·9	5·6	3·9	5·6	3·9	58·2
58·5	6·0	4·2	6·0	4·2	6·0	4·2	6·0	4·2	6·0	4·1	5·9	4·1	5·9	4·1	58·5
58·8	6·4	4·4	6·4	4·4	6·4	4·4	6·3	4·4	6·3	4·3	6·3	4·3	6·2	4·2	58·8
59·1	6·8	4·6	6·8	4·6	6·7	4·6	6·7	4·6	6·7	4·5	6·6	4·5	6·6	4·4	59·1
59·4	7·2	4·8	7·1	4·8	7·1	4·8	7·1	4·8	7·0	4·7	7·0	4·7	6·9	4·6	59·4
59·7	7·5	5·1	7·5	5·0	7·5	5·0	7·5	5·0	7·4	4·9	7·3	4·8	7·2	4·7	59·7
60·0	7·9	5·3	7·9	5·3	7·9	5·2	7·8	5·2	7·8	5·1	7·7	5·0	7·6	4·9	60·0
60·3	8·3	5·5	8·3	5·5	8·2	5·4	8·2	5·4	8·1	5·3	8·0	5·2	7·9	5·1	60·3
60·6	8·7	5·7	8·7	5·7	8·6	5·7	8·6	5·6	8·5	5·5	8·4	5·4	8·2	5·3	60·6
60·9	9·1	5·9	9·0	5·9	9·0	5·9	8·9	5·8	8·8	5·7	8·7	5·6	8·6	5·4	60·9
61·2	9·5	6·2	9·4	6·1	9·4	6·1	9·3	6·0	9·2	5·9	9·1	5·8	8·9	5·6	61·2
61·5	9·8	6·4	9·8	6·3	9·7	6·3	9·7	6·2	9·5	6·1	9·4	5·9	9·2	5·8	61·5

DIP

Ht. of Eye	Corrⁿ	Ht. of Eye	Ht. of Eye	Corrⁿ	Ht. of Eye
m		ft.	m		ft.
2·4	′	8·0	9·5	′	31·5
	−2·8			−5·5	
2·6		8·6	9·9		32·7
	−2·9			−5·6	
2·8		9·2	10·3		33·9
	−3·0			−5·7	
3·0		9·8	10·6		35·1
	−3·1			−5·8	
3·2		10·5	11·0		36·3
	−3·2			−5·9	
3·4		11·2	11·4		37·6
	−3·3			−6·0	
3·6		11·9	11·8		38·9
	−3·4			−6·1	
3·8		12·6	12·2		40·1
	−3·5			−6·2	
4·0		13·3	12·6		41·5
	−3·6			−6·3	
4·3		14·1	13·0		42·8
	−3·7			−6·4	
4·5		14·9	13·4		44·2
	−3·8			−6·5	
4·7		15·7	13·8		45·5
	−3·9			−6·6	
5·0		16·5	14·2		46·9
	−4·0			−6·7	
5·2		17·4	14·7		48·4
	−4·1			−6·8	
5·5		18·3	15·1		49·8
	−4·2			−6·9	
5·8		19·1	15·5		51·3
	−4·3			−7·0	
6·1		20·1	16·0		52·8
	−4·4			−7·1	
6·3		21·0	16·5		54·3
	−4·5			−7·2	
6·6		22·0	16·9		55·8
	−4·6			−7·3	
6·9		22·9	17·4		57·4
	−4·7			−7·4	
7·2		23·9	17·9		58·9
	−4·8			−7·5	
7·5		24·9	18·4		60·5
	−4·9			−7·6	
7·9		26·0	18·8		62·1
	−5·0			−7·7	
8·2		27·1	19·3		63·8
	−5·1			−7·8	
8·5		28·1	19·8		65·4
	−5·2			−7·9	
8·8		29·2	20·4		67·1
	−5·3			−8·0	
9·2		30·4	20·9		68·8
	−5·4			−8·1	
9·5		31·5	21·4		70·5

MOON CORRECTION TABLE

The correction is in two parts; the first correction is taken from the upper part of the table with argument apparent altitude, and the second from the lower part, with argument H.P., in the same column as that from which the first correction was taken. Separate corrections are given in the lower part for lower (L) and upper (U) limbs. All corrections are to be **added** to apparent altitude, *but 30′ is to be subtracted from the altitude of the upper limb.*

For bubble sextant observations ignore dip, take the mean of upper and lower limb corrections and subtract 15′ from the altitude.

App. Alt. = Apparent altitude = Sextant altitude corrected for index error and dip.

Table 17

ALTITUDE CORRECTION TABLES 35°-90°—MOON

App. Alt.	35°–39° Corr^n	40°–44° Corr^n	45°–49° Corr^n	50°–54° Corr^n	55°–59° Corr^n	60°–64° Corr^n	65°–69° Corr^n	70°–74° Corr^n	75°–79° Corr^n	80°–84° Corr^n	85°–89° Corr^n	App. Alt.
00′	35° 56·5′	40° 53·7′	45° 50·5′	50° 46·9′	55° 43·1′	60° 38·9′	65° 34·6′	70° 30·1′	75° 25·3′	80° 20·5′	85° 15·6′	00′
10	56·4	53·6	50·4	46·8	42·9	38·8	34·4	29·9	25·2	20·4	15·5	10
20	56·3	53·5	50·2	46·7	42·8	38·7	34·3	29·7	25·0	20·2	15·3	20
30	56·2	53·4	50·1	46·5	42·7	38·5	34·1	29·6	24·9	20·0	15·1	30
40	56·2	53·3	50·0	46·4	42·5	38·4	34·0	29·4	24·7	19·9	15·0	40
50	56·1	53·2	49·9	46·3	42·4	38·2	33·8	29·3	24·5	19·7	14·8	50
00	36 56·0	41 53·1	46 49·8	51 46·2	56 42·3	61 38·1	66 33·7	71 29·1	76 24·4	81 19·6	86 14·6	00
10	55·9	53·0	49·7	46·0	42·1	37·9	33·5	29·0	24·2	19·4	14·5	10
20	55·8	52·8	49·5	45·9	42·0	37·8	33·4	28·8	24·1	19·2	14·3	20
30	55·7	52·7	49·4	45·8	41·8	37·7	33·2	28·7	23·9	19·1	14·1	30
40	55·6	52·6	49·3	45·7	41·7	37·5	33·1	28·5	23·8	18·9	14·0	40
50	55·5	52·5	49·2	45·5	41·6	37·4	32·9	28·3	23·6	18·7	13·8	50
00	37 55·4	42 52·4	47 49·1	52 45·4	57 41·4	62 37·2	67 32·8	72 28·2	77 23·4	82 18·6	87 13·7	00
10	55·3	52·3	49·0	45·3	41·3	37·1	32·6	28·0	23·3	18·4	13·5	10
20	55·2	52·2	48·8	45·2	41·2	36·9	32·5	27·9	23·1	18·2	13·3	20
30	55·1	52·1	48·7	45·0	41·0	36·8	32·3	27·7	22·9	18·1	13·2	30
40	55·0	52·0	48·6	44·9	40·9	36·6	32·2	27·6	22·8	17·9	13·0	40
50	55·0	51·9	48·5	44·8	40·8	36·5	32·0	27·4	22·6	17·8	12·8	50
00	38 54·9	43 51·8	48 48·4	53 44·6	58 40·6	63 36·4	68 31·9	73 27·2	78 22·5	83 17·6	88 12·7	00
10	54·8	51·7	48·2	44·5	40·5	36·2	31·7	27·1	22·3	17·4	12·5	10
20	54·7	51·6	48·1	44·4	40·3	36·1	31·6	26·9	22·1	17·3	12·3	20
30	54·6	51·5	48·0	44·2	40·2	35·9	31·4	26·8	22·0	17·1	12·2	30
40	54·5	51·4	47·9	44·1	40·1	35·8	31·3	26·6	21·8	16·9	12·0	40
50	54·4	51·2	47·8	44·0	39·9	35·6	31·1	26·5	21·7	16·8	11·8	50
00	39 54·3	44 51·1	49 47·6	54 43·9	59 39·8	64 35·5	69 31·0	74 26·3	79 21·5	84 16·6	89 11·7	00
10	54·2	51·0	47·5	43·7	39·6	35·3	30·8	26·1	21·3	16·5	11·5	10
20	54·1	50·9	47·4	43·6	39·5	35·2	30·7	26·0	21·2	16·3	11·4	20
30	54·0	50·8	47·3	43·5	39·4	35·0	30·5	25·8	21·0	16·1	11·2	30
40	53·9	50·7	47·2	43·3	39·2	34·9	30·4	25·7	20·9	16·0	11·0	40
50	53·8	50·6	47·0	43·2	39·1	34·7	30·2	25·5	20·7	15·8	10·9	50
H.P.	L U	L U	L U	L U	L U	L U	L U	L U	L U	L U	L U	H.P.
54·0′	1·1′ 1·7′	1·3′ 1·9′	1·5′ 2·1′	1·7′ 2·4′	2·0′ 2·6′	2·3′ 2·9′	2·6′ 3·2′	2·9′ 3·5′	3·2′ 3·8′	3·5′ 4·1′	3·8′ 4·5′	54·0′
54·3	1·4 1·8	1·6 2·0	1·8 2·2	2·0 2·5	2·3 2·7	2·5 3·0	2·8 3·2	3·0 3·5	3·3 3·8	3·6 4·1	3·9 4·4	54·3
54·6	1·7 2·0	1·9 2·2	2·1 2·4	2·3 2·6	2·5 2·8	2·7 3·0	3·0 3·3	3·2 3·5	3·5 3·8	3·7 4·1	4·0 4·3	54·6
54·9	2·0 2·2	2·2 2·3	2·3 2·5	2·5 2·7	2·7 2·9	2·9 3·1	3·2 3·3	3·4 3·5	3·6 3·8	3·9 4·0	4·1 4·3	54·9
55·2	2·3 2·3	2·5 2·4	2·6 2·6	2·8 2·8	3·0 2·9	3·2 3·1	3·4 3·3	3·6 3·5	3·8 3·7	4·0 4·0	4·2 4·2	55·2
55·5	2·7 2·5	2·8 2·6	2·9 2·7	3·1 2·9	3·2 3·0	3·4 3·2	3·6 3·4	3·7 3·5	3·9 3·7	4·1 3·9	4·3 4·1	55·5
55·8	3·0 2·6	3·1 2·7	3·2 2·8	3·3 3·0	3·5 3·1	3·6 3·3	3·8 3·4	3·9 3·6	4·1 3·7	4·2 3·9	4·4 4·0	55·8
56·1	3·3 2·8	3·4 2·9	3·5 3·0	3·6 3·1	3·7 3·2	3·8 3·3	4·0 3·4	4·1 3·6	4·2 3·7	4·4 3·8	4·5 4·0	56·1
56·4	3·6 2·9	3·7 3·0	3·8 3·1	3·9 3·2	3·9 3·3	4·0 3·4	4·1 3·5	4·3 3·6	4·4 3·7	4·5 3·8	4·6 3·9	56·4
56·7	3·9 3·1	4·0 3·1	4·1 3·2	4·1 3·3	4·2 3·3	4·3 3·4	4·3 3·5	4·4 3·6	4·5 3·7	4·6 3·8	4·7 3·8	56·7
57·0	4·3 3·2	4·3 3·3	4·3 3·3	4·4 3·4	4·4 3·4	4·5 3·5	4·5 3·5	4·6 3·6	4·7 3·6	4·7 3·7	4·8 3·8	57·0
57·3	4·6 3·4	4·6 3·4	4·6 3·4	4·6 3·5	4·7 3·5	4·7 3·5	4·7 3·6	4·8 3·6	4·8 3·6	4·8 3·7	4·9 3·7	57·3
57·6	4·9 3·6	4·9 3·6	4·9 3·6	4·9 3·6	4·9 3·6	4·9 3·6	4·9 3·6	4·9 3·6	5·0 3·6	5·0 3·6	5·0 3·6	57·6
57·9	5·2 3·7	5·2 3·7	5·2 3·7	5·2 3·7	5·2 3·7	5·1 3·6	5·1 3·6	5·1 3·6	5·1 3·6	5·1 3·6	5·1 3·6	57·9
58·2	5·5 3·9	5·5 3·8	5·5 3·8	5·4 3·8	5·4 3·7	5·4 3·7	5·3 3·7	5·3 3·6	5·2 3·6	5·2 3·5	5·2 3·5	58·2
58·5	5·9 4·0	5·8 4·0	5·8 3·9	5·7 3·9	5·6 3·8	5·6 3·8	5·5 3·7	5·5 3·6	5·4 3·6	5·3 3·5	5·3 3·4	58·5
58·8	6·2 4·2	6·1 4·1	6·0 4·1	6·0 4·0	5·9 3·9	5·8 3·8	5·7 3·7	5·6 3·6	5·5 3·5	5·4 3·5	5·3 3·4	58·8
59·1	6·5 4·3	6·4 4·3	6·3 4·2	6·2 4·1	6·1 4·0	6·0 3·9	5·9 3·8	5·8 3·6	5·7 3·5	5·6 3·4	5·4 3·3	59·1
59·4	6·8 4·5	6·7 4·4	6·6 4·3	6·5 4·2	6·4 4·1	6·2 3·9	6·1 3·8	6·0 3·7	5·8 3·5	5·7 3·4	5·5 3·2	59·4
59·7	7·1 4·6	7·0 4·5	6·9 4·4	6·8 4·3	6·6 4·1	6·5 4·0	6·3 3·8	6·2 3·7	6·0 3·5	5·8 3·3	5·6 3·2	59·7
60·0	7·5 4·8	7·3 4·7	7·2 4·5	7·0 4·4	6·9 4·2	6·7 4·0	6·5 3·9	6·3 3·7	6·1 3·5	5·9 3·3	5·7 3·1	60·0
60·3	7·8 5·0	7·6 4·8	7·5 4·7	7·3 4·5	7·1 4·3	6·9 4·1	6·7 3·9	6·5 3·7	6·3 3·5	6·0 3·2	5·8 3·0	60·3
60·6	8·1 5·1	7·9 5·0	7·7 4·8	7·6 4·6	7·3 4·4	7·1 4·2	6·9 3·9	6·7 3·7	6·4 3·4	6·2 3·2	5·9 2·9	60·6
60·9	8·4 5·3	8·2 5·1	8·0 4·9	7·8 4·7	7·6 4·5	7·3 4·2	7·1 4·0	6·8 3·7	6·6 3·4	6·3 3·2	6·0 2·9	60·9
61·2	8·7 5·4	8·5 5·2	8·3 5·0	8·1 4·8	7·8 4·5	7·6 4·3	7·3 4·0	7·0 3·7	6·7 3·4	6·4 3·1	6·1 2·8	61·2
61·5	9·1 5·6	8·8 5·4	8·6 5·1	8·3 4·9	8·1 4·6	7·8 4·3	7·5 4·0	7·2 3·7	6·9 3·4	6·5 3·1	6·2 2·7	61·5

Table 18

ALTITUDE CORRECTION TABLES 0°–10°—SUN, STARS, PLANETS

App. Alt.	SUN OCT.–MAR. Lower Limb	SUN OCT.–MAR. Upper Limb	SUN APR.–SEPT. Lower Limb	SUN APR.–SEPT. Upper Limb	STARS PLANETS
° ′	′	′	′	′	′
0 00	−18·2	−50·5	−18·4	−50·2	−34·5
03	17·5	49·8	17·8	49·6	33·8
06	16·9	49·2	17·1	48·9	33·2
09	16·3	48·6	16·5	48·3	32·6
12	15·7	48·0	15·9	47·7	32·0
15	15·1	47·4	15·3	47·1	31·4
0 18	−14·5	−46·8	−14·8	−46·6	−30·8
21	14·0	46·3	14·2	46·0	30·3
24	13·5	45·8	13·7	45·5	29·8
27	12·9	45·2	13·2	45·0	29·2
30	12·4	44·7	12·7	44·5	28·7
33	11·9	44·2	12·2	44·0	28·2
0 36	−11·5	−43·8	−11·7	−43·5	−27·8
39	11·0	43·3	11·2	43·0	27·3
42	10·5	42·8	10·8	42·6	26·8
45	10·1	42·4	10·3	42·1	26·4
48	9·6	41·9	9·9	41·7	25·9
51	9·2	41·5	9·5	41·3	25·5
0 54	− 8·8	−41·1	− 9·1	−40·9	−25·1
0 57	8·4	40·7	8·7	40·5	24·7
1 00	8·0	40·3	8·3	40·1	24·3
03	7·7	40·0	7·9	39·7	24·0
06	7·3	39·6	7·5	39·3	23·6
09	6·9	39·2	7·2	39·0	23·2
1 12	− 6·6	−38·9	− 6·8	−38·6	−22·9
15	6·2	38·5	6·5	38·3	22·5
18	5·9	38·2	6·2	38·0	22·2
21	5·6	37·9	5·8	37·6	21·9
24	5·3	37·6	5·5	37·3	21·6
27	4·9	37·2	5·2	37·0	21·2
1 30	− 4·6	−36·9	− 4·9	−36·7	−20·9
35	4·2	36·5	4·4	36·2	20·5
40	3·7	36·0	4·0	35·8	20·0
45	3·2	35·5	3·5	35·3	19·5
50	2·8	35·1	3·1	34·9	19·1
1 55	2·4	34·7	2·6	34·4	18·7
2 00	− 2·0	−34·3	− 2·2	−34·0	−18·3
05	1·6	33·9	1·8	33·6	17·9
10	1·2	33·5	1·5	33·3	17·5
15	0·9	33·2	1·1	32·9	17·2
20	0·5	32·8	0·8	32·6	16·8
25	− 0·2	32·5	0·4	32·2	16·5
2 30	+ 0·2	−32·1	− 0·1	−31·9	−16·1
35	0·5	31·8	+ 0·2	31·6	15·8
40	0·8	31·5	0·5	31·3	15·5
45	1·1	31·2	0·8	31·0	15·2
50	1·4	30·9	1·1	30·7	14·9
2 55	1·6	30·7	1·4	30·4	14·7
3 00	+ 1·9	−30·4	+ 1·7	−30·1	−14·4
05	2·2	30·1	1·9	29·9	14·1
10	2·4	29·9	2·1	29·7	13·9
15	2·6	29·7	2·4	29·4	13·7
20	2·9	29·4	2·6	29·2	13·4
25	3·1	29·2	2·9	28·9	13·2
3 30	+ 3·3	−29·0	+ 3·1	−28·7	−13·0

App. Alt.	SUN OCT.–MAR. Lower Limb	SUN OCT.–MAR. Upper Limb	SUN APR.–SEPT. Lower Limb	SUN APR.–SEPT. Upper Limb	STARS PLANETS
° ′	′	′	′	′	′
3 30	+ 3·3	−29·0	+ 3·1	−28·7	−13·0
35	3·6	28·7	3·3	28·5	12·7
40	3·8	28·5	3·5	28·3	12·5
45	4·0	28·3	3·7	28·1	12·3
50	4·2	28·1	3·9	27·9	12·1
3 55	4·4	27·9	4·1	27·7	11·9
4 00	+ 4·5	−27·8	+ 4·3	−27·5	−11·8
05	4·7	27·6	4·5	27·3	11·6
10	4·9	27·4	4·6	27·2	11·4
15	5·1	27·2	4·8	27·0	11·2
20	5·2	27·1	5·0	26·8	11·1
25	5·4	26·9	5·1	26·7	10·9
4 30	+ 5·6	−26·7	+ 5·3	−26·5	−10·7
35	5·7	26·6	5·5	26·3	10·6
40	5·9	26·4	5·6	26·2	10·4
45	6·0	26·3	5·8	26·0	10·3
50	6·2	26·1	5·9	25·9	10·1
4 55	6·3	26·0	6·0	25·8	10·0
5 00	+ 6·4	−25·9	+ 6·2	−25·6	− 9·9
05	6·6	25·7	6·3	25·5	9·7
10	6·7	25·6	6·4	25·4	9·6
15	6·8	25·5	6·6	25·2	9·5
20	6·9	25·4	6·7	25·1	9·4
25	7·1	25·2	6·8	25·0	9·2
5 30	+ 7·2	−25·1	+ 6·9	−24·9	− 9·1
35	7·3	25·0	7·0	24·8	9·0
40	7·4	24·9	7·2	24·6	8·9
45	7·5	24·8	7·3	24·5	8·8
50	7·6	24·7	7·4	24·4	8·7
5 55	7·7	24·6	7·5	24·3	8·6
6 00	+ 7·8	−24·5	+ 7·6	−24·2	− 8·5
10	8·0	24·3	7·8	24·0	8·3
20	8·2	24·1	8·0	23·8	8·1
30	8·4	23·9	8·1	23·7	7·9
40	8·6	23·7	8·3	23·5	7·7
6 50	8·7	23·6	8·5	23·3	7·6
7 00	+ 8·9	−23·4	+ 8·6	−23·2	− 7·4
10	9·1	23·2	8·8	23·0	7·2
20	9·2	23·1	9·0	22·8	7·1
30	9·3	23·0	9·1	22·7	7·0
40	9·5	22·8	9·2	22·6	6·8
7 50	9·6	22·7	9·4	22·4	6·7
8 00	+ 9·7	−22·6	+ 9·5	−22·3	− 6·6
10	9·9	22·4	9·6	22·2	6·4
20	10·0	22·3	9·7	22·1	6·3
30	10·1	22·2	9·8	22·0	6·2
40	10·2	22·1	10·0	21·8	6·1
8 50	10·3	22·0	10·1	21·7	6·0
9 00	+10·4	−21·9	+10·2	−21·6	− 5·9
10	10·5	21·8	10·3	21·5	5·8
20	10·6	21·7	10·4	21·4	5·7
30	10·7	21·6	10·5	21·3	5·6
40	10·8	21·5	10·6	21·2	5·5
9 50	10·9	21·4	10·6	21·2	5·4
10 00	+11·0	−21·3	+10·7	−21·1	− 5·3

Table 19

34m INCREMENTS AND CORRECTIONS 35m

34m	SUN PLANETS	ARIES	MOON	v or d	Corrn	v or d	Corrn	v or d	Corrn
s	° ′	° ′	° ′	′	′	′	′	′	′
00	8 30·0	8 31·4	8 06·8	0·0	0·0	6·0	3·5	12·0	6·9
01	8 30·3	8 31·6	8 07·0	0·1	0·1	6·1	3·5	12·1	7·0
02	8 30·5	8 31·9	8 07·2	0·2	0·1	6·2	3·6	12·2	7·0
03	8 30·8	8 32·1	8 07·5	0·3	0·2	6·3	3·6	12·3	7·1
04	8 31·0	8 32·4	8 07·7	0·4	0·2	6·4	3·7	12·4	7·1
05	8 31·3	8 32·6	8 08·0	0·5	0·3	6·5	3·7	12·5	7·2
06	8 31·5	8 32·9	8 08·2	0·6	0·3	6·6	3·8	12·6	7·2
07	8 31·8	8 33·2	8 08·4	0·7	0·4	6·7	3·9	12·7	7·3
08	8 32·0	8 33·4	8 08·7	0·8	0·5	6·8	3·9	12·8	7·4
09	8 32·3	8 33·7	8 08·9	0·9	0·5	6·9	4·0	12·9	7·4
10	8 32·5	8 33·9	8 09·2	1·0	0·6	7·0	4·0	13·0	7·5
11	8 32·8	8 34·2	8 09·4	1·1	0·6	7·1	4·1	13·1	7·5
12	8 33·0	8 34·4	8 09·6	1·2	0·7	7·2	4·1	13·2	7·6
13	8 33·3	8 34·7	8 09·9	1·3	0·7	7·3	4·2	13·3	7·6
14	8 33·5	8 34·9	8 10·1	1·4	0·8	7·4	4·3	13·4	7·7
15	8 33·8	8 35·2	8 10·3	1·5	0·9	7·5	4·3	13·5	7·8
16	8 34·0	8 35·4	8 10·6	1·6	0·9	7·6	4·4	13·6	7·8
17	8 34·3	8 35·7	8 10·8	1·7	1·0	7·7	4·4	13·7	7·9
18	8 34·5	8 35·9	8 11·1	1·8	1·0	7·8	4·5	13·8	7·9
19	8 34·8	8 36·2	8 11·3	1·9	1·1	7·9	4·5	13·9	8·0
20	8 35·0	8 36·4	8 11·5	2·0	1·2	8·0	4·6	14·0	8·1
21	8 35·3	8 36·7	8 11·8	2·1	1·2	8·1	4·7	14·1	8·1
22	8 35·5	8 36·9	8 12·0	2·2	1·3	8·2	4·7	14·2	8·2
23	8 35·8	8 37·2	8 12·3	2·3	1·3	8·3	4·8	14·3	8·2
24	8 36·0	8 37·4	8 12·5	2·4	1·4	8·4	4·8	14·4	8·3
25	8 36·3	8 37·7	8 12·7	2·5	1·4	8·5	4·9	14·5	8·3
26	8 36·5	8 37·9	8 13·0	2·6	1·5	8·6	4·9	14·6	8·4
27	8 36·8	8 38·2	8 13·2	2·7	1·6	8·7	5·0	14·7	8·5
28	8 37·0	8 38·4	8 13·4	2·8	1·6	8·8	5·1	14·8	8·5
29	8 37·3	8 38·7	8 13·7	2·9	1·7	8·9	5·1	14·9	8·6
30	8 37·5	8 38·9	8 13·9	3·0	1·7	9·0	5·2	15·0	8·6
31	8 37·8	8 39·2	8 14·2	3·1	1·8	9·1	5·2	15·1	8·7
32	8 38·0	8 39·4	8 14·4	3·2	1·8	9·2	5·3	15·2	8·7
33	8 38·3	8 39·7	8 14·6	3·3	1·9	9·3	5·3	15·3	8·8
34	8 38·5	8 39·9	8 14·9	3·4	2·0	9·4	5·4	15·4	8·9
35	8 38·8	8 40·2	8 15·1	3·5	2·0	9·5	5·5	15·5	8·9
36	8 39·0	8 40·4	8 15·4	3·6	2·1	9·6	5·5	15·6	9·0
37	8 39·3	8 40·7	8 15·6	3·7	2·1	9·7	5·6	15·7	9·0
38	8 39·5	8 40·9	8 15·8	3·8	2·2	9·8	5·6	15·8	9·1
39	8 39·8	8 41·2	8 16·1	3·9	2·2	9·9	5·7	15·9	9·1
40	8 40·0	8 41·4	8 16·3	4·0	2·3	10·0	5·8	16·0	9·2
41	8 40·3	8 41·7	8 16·5	4·1	2·4	10·1	5·8	16·1	9 3
42	8 40·5	8 41·9	8 16·8	4·2	2·4	10·2	5·9	16·2	9·3
43	8 40·8	8 42·2	8 17·0	4·3	2·5	10·3	5·9	16·3	9·4
44	8 41·0	8 42·4	8 17·3	4·4	2·5	10·4	6·0	16·4	9·4
45	8 41·3	8 42·7	8 17·5	4·5	2·6	10·5	6·0	16·5	9·5
46	8 41·5	8 42·9	8 17·7	4·6	2·6	10·6	6·1	16·6	9·5
47	8 41·8	8 43·2	8 18·0	4·7	2·7	10·7	6·2	16·7	9·6
48	8 42·0	8 43·4	8 18·2	4·8	2·8	10·8	6·2	16·8	9·7
49	8 42·3	8 43·7	8 18·5	4·9	2·8	10·9	6·3	16·9	9·7
50	8 42·5	8 43·9	8 18·7	5·0	2·9	11·0	6·3	17·0	9·8
51	8 42·8	8 44·2	8 18·9	5·1	2·9	11·1	6·4	17·1	9·8
52	8 43·0	8 44·4	8 19·2	5·2	3·0	11·2	6·4	17·2	9·9
53	8 43·3	8 44·7	8 19·4	5·3	3·0	11·3	6·5	17·3	9·9
54	8 43·5	8 44·9	8 19·7	5·4	3·1	11·4	6·6	17·4	10·0
55	8 43·8	8 45·2	8 19·9	5·5	3·2	11·5	6·6	17·5	10·1
56	8 44·0	8 45·4	8 20·1	5·6	3·2	11·6	6·7	17·6	10·1
57	8 44·3	8 45·7	8 20·4	5·7	3·3	11·7	6·7	17·7	10·2
58	8 44·5	8 45·9	8 20·6	5·8	3·3	11·8	6·8	17·8	10·2
59	8 44·8	8 46·2	8 20·8	5·9	3·4	11·9	6·8	17·9	10·3
60	8 45·0	8 46·4	8 21·1	6·0	3·5	12·0	6·9	18·0	10·4

35m	SUN PLANETS	ARIES	MOON	v or d	Corrn	v or d	Corrn	v or d	Corrn
s	° ′	° ′	° ′	′	′	′	′	′	′
00	8 45·0	8 46·4	8 21·1	0·0	0·0	6·0	3·6	12·0	7·1
01	8 45·3	8 46·7	8 21·3	0·1	0·1	6·1	3·6	12·1	7·2
02	8 45·5	8 46·9	8 21·6	0·2	0·1	6·2	3·7	12·2	7·2
03	8 45·8	8 47·2	8 21·8	0·3	0·2	6·3	3·7	12·3	7·3
04	8 46·0	8 47·4	8 22·0	0·4	0·2	6·4	3·8	12·4	7·3
05	8 46·3	8 47·7	8 22·3	0·5	0·3	6·5	3·8	12·5	7·4
06	8 46·5	8 47·9	8 22·5	0·6	0·4	6·6	3·9	12·6	7·5
07	8 46·8	8 48·2	8 22·8	0·7	0·4	6·7	4·0	12·7	7·5
08	8 47·0	8 48·4	8 23·0	0·8	0·5	6·8	4·0	12·8	7·6
09	8 47·3	8 48·7	8 23·2	0·9	0·5	6·9	4·1	12·9	7·6
10	8 47·5	8 48·9	8 23·5	1·0	0·6	7·0	4·1	13·0	7·7
11	8 47·8	8 49·2	8 23·7	1·1	0·7	7·1	4·2	13·1	7·8
12	8 48·0	8 49·4	8 23·9	1·2	0·7	7·2	4·3	13·2	7·8
13	8 48·3	8 49·7	8 24·2	1·3	0·8	7·3	4·3	13·3	7·9
14	8 48·5	8 49·9	8 24·4	1·4	0·8	7·4	4·4	13·4	7·9
15	8 48·8	8 50·2	8 24·7	1·5	0·9	7·5	4·4	13·5	8·0
16	8 49·0	8 50·4	8 24·9	1·6	0·9	7·6	4·5	13·6	8·0
17	8 49·3	8 50·7	8 25·1	1·7	1·0	7·7	4·6	13·7	8·1
18	8 49·5	8 50·9	8 25·4	1·8	1·1	7·8	4·6	13·8	8·2
19	8 49·8	8 51·2	8 25·6	1·9	1·1	7·9	4·7	13·9	8·2
20	8 50·0	8 51·5	8 25·9	2·0	1·2	8·0	4·7	14·0	8·3
21	8 50·3	8 51·7	8 26·1	2·1	1·2	8·1	4·8	14·1	8·3
22	8 50·5	8 52·0	8 26·3	2·2	1·3	8·2	4·9	14·2	8·4
23	8 50·8	8 52·2	8 26·6	2·3	1·4	8·3	4·9	14·3	8·5
24	8 51·0	8 52·5	8 26·8	2·4	1·4	8·4	5·0	14·4	8·5
25	8 51·3	8 52·7	8 27·0	2·5	1·5	8·5	5·0	14·5	8·6
26	8 51·5	8 53·0	8 27·3	2·6	1·5	8·6	5·1	14·6	8·6
27	8 51·8	8 53·2	8 27·5	2·7	1·6	8·7	5·1	14·7	8·7
28	8 52·0	8 53·5	8 27·8	2·8	1·7	8·8	5·2	14·8	8·8
29	8 52·3	8 53·7	8 28·0	2·9	1·7	8·9	5·3	14·9	8·8
30	8 52·5	8 54·0	8 28·2	3·0	1·8	9·0	5·3	15·0	8·9
31	8 52·8	8 54·2	8 28·5	3·1	1·8	9·1	5·4	15·1	8·9
32	8 53·0	8 54·5	8 28·7	3·2	1·9	9·2	5·4	15·2	9·0
33	8 53·3	8 54·7	8 29·0	3·3	2·0	9·3	5·5	15·3	9·1
34	8 53·5	8 55·0	8 29·2	3·4	2·0	9·4	5·6	15·4	9·1
35	8 53·8	8 55·2	8 29·4	3·5	2·1	9·5	5·6	15·5	9·2
36	8 54·0	8 55·5	8 29·7	3·6	2·1	9·6	5·7	15·6	9·2
37	8 54·3	8 55·7	8 29·9	3·7	2·2	9·7	5·7	15·7	9·3
38	8 54·5	8 56·0	8 30·2	3·8	2·2	9·8	5·8	15·8	9·3
39	8 54·8	8 56·2	8 30·4	3·9	2·3	9·9	5·9	15·9	9·4
40	8 55·0	8 56·5	8 30·6	4·0	2·4	10·0	5·9	16·0	9·5
41	8 55·3	8 56·7	8 30·9	4·1	2·4	10·1	6·0	16·1	9·5
42	8 55·5	8 57·0	8 31·1	4·2	2·5	10·2	6·0	16·2	9·6
43	8 55·8	8 57·2	8 31·3	4·3	2·5	10·3	6·1	16·3	9·6
44	8 56·0	8 57·5	8 31·6	4·4	2·6	10·4	6·2	16·4	9·7
45	8 56·3	8 57·7	8 31·8	4·5	2·7	10·5	6·2	16·5	9·8
46	8 56·5	8 58·0	8 32·1	4·6	2·7	10·6	6·3	16·6	9·8
47	8 56·8	8 58·2	8 32·3	4·7	2·8	10·7	6·3	16·7	9·9
48	8 57·0	8 58·5	8 32·5	4·8	2·8	10·8	6·4	16·8	9·9
49	8 57·3	8 58·7	8 32·8	4·9	2·9	10·9	6·4	16·9	10·0
50	8 57·5	8 59·0	8 33·0	5·0	3·0	11·0	6·5	17·0	10·1
51	8 57·8	8 59·2	8 33·3	5·1	3·0	11·1	6·6	17·1	10·1
52	8 58·0	8 59·5	8 33·5	5·2	3·1	11·2	6·6	17·2	10·2
53	8 58·3	8 59·7	8 33·7	5·3	3·1	11·3	6·7	17·3	10·2
54	8 58·5	9 00·0	8 34·0	5·4	3·2	11·4	6·7	17·4	10·3
55	8 58·8	9 00·2	8 34·2	5·5	3·3	11·5	6·8	17·5	10·4
56	8 59·0	9 00·5	8 34·4	5·6	3·3	11·6	6·9	17·6	10·4
57	8 59·3	9 00·7	8 34·7	5·7	3·4	11·7	6·9	17·7	10·5
58	8 59·5	9 01·0	8 34·9	5·8	3·4	11·8	7·0	17·8	10·5
59	8 59·8	9 01·2	8 35·2	5·9	3·5	11·9	7·0	17·9	10·6
60	9 00·0	9 01·5	8 35·4	6·0	3·6	12·0	7·1	18·0	10·7

Table 20

Hs	=	3°29′
IC	=	+05′
Dip	=	−03′
ha	=	3°37′
Main Corr.	=	+ 3′
Ho	=	3°40′

SUN

GHA 23 Hr.	=	164° 03′	Dec. = N 23°05′
34 Min. 0 Sec.	=	8° 30′	
GHA	=	172° 33′	
(+360?)	=		
Assumed Long. E.	=	(+)	Assumed Lat. = N 38°
Assumed Long. W.	=	−69° 33′	
LHA	=	103°	
(−360?)	=		

With this information on hand, refer to the sight reduction tables and complete the worksheet, starting with the moon. Since the declination of the moon is south and you are in north latitude, the page in H.O. 249 is one headed "Declination Contrary Name to Latitude" and is reproduced as Table 21. LHA is 321°; declination is 22°54′ S; assumed latitude is 37° N; and Ho is 19°59′.

				d	=	− 50′
Hc	=	20°29′		Dec. Inc.	=	54′
Corr.	=	− 45′		Corr.	=	− 45′
Hc	=	19°44′				
Ho	=	19°59′				
		15′	Toward Moon			

Z = 142° Zn = 142°

Turning to the page from H.O. 249 reproduced as Table 22, do the same thing for the sun:

DECLINATION (15°-29°) CONTRARY NAME TO LATITUDE

LAT 37°

N. Lat. {LHA greater than 180°....... Zn=Z; LHA less than 180°.......... Zn=360−Z}

LHA	15° Hc	15° d	15° Z	16° Hc	16° d	16° Z	17° Hc	17° d	17° Z	18° Hc	18° d	18° Z	19° Hc	19° d	19° Z
	° ′	′	°	° ′	′	°	° ′	′	°	° ′	′	°	° ′	′	°
69	06 56	40	115	06 16	39	116	05 37	40	116	04 57	40	117	04 17	40	118
68	07 39	40	115	06 59	40	116	06 19	40	117	05 39	40	118	04 59	40	118
67	08 23	41	116	07 42	40	117	07 02	40	118	06 22	41	118	05 41	40	119
66	09 06	41	117	08 25	41	117	07 44	40	118	07 04	41	119	06 23	41	120
65	09 48	−41	117	09 07	−40	118	08 27	−41	119	07 46	−41	120	07 05	−41	120
64	10 31	41	118	09 50	42	119	09 08	41	120	08 27	41	120	07 46	42	121
63	11 13	42	119	10 31	41	119	09 50	42	120	09 08	41	121	08 27	42	122
62	11 55	42	119	11 13	42	120	10 31	42	121	09 49	41	122	09 08	43	122
61	12 36	42	120	11 54	42	121	11 12	42	122	10 30	42	122	09 48	42	123
60	13 18	−43	121	12 35	−42	122	11 53	−42	122	11 11	−43	123	10 28	−43	124
59	13 59	43	121	13 16	43	122	12 33	42	123	11 51	43	124	11 08	43	124
58	14 39	42	122	13 57	43	123	13 14	44	124	12 30	43	124	11 47	43	125
57	15 20	43	123	14 37	44	124	13 53	43	124	13 10	44	125	12 26	44	126
56	16 00	44	124	15 16	43	124	14 33	44	125	13 49	44	126	13 05	44	126
55	16 40	−44	124	15 56	−44	125	15 12	−44	126	14 28	−45	126	13 43	−44	127
54	17 19	44	125	16 35	45	126	15 50	44	127	15 06	45	127	14 21	44	128
53	17 58	44	126	17 14	45	127	16 29	45	127	15 44	45	128	14 59	45	129
52	18 37	45	127	17 52	45	127	17 07	45	128	16 22	46	129	15 36	45	129
51	19 15	45	127	18 30	46	128	17 44	45	129	16 59	46	129	16 13	46	130
50	19 53	−46	128	19 07	−45	129	18 22	−46	130	17 36	−46	130	16 50	−47	131
49	20 31	46	129	19 45	47	130	18 58	46	130	18 12	46	131	17 26	47	132
48	21 08	47	130	20 21	46	130	19 35	47	131	18 48	47	132	18 01	47	132
47	21 44	46	131	20 58	47	131	20 11	47	132	19 24	48	133	18 36	47	133
46	22 20	47	131	21 33	47	132	20 46	47	133	19 59	48	133	19 11	48	134
45	22 56	−47	132	22 09	−48	133	21 21	−48	133	20 33	−48	134	19 45	−48	135
44	23 32	48	133	22 44	48	134	21 56	48	134	21 08	49	135	20 19	48	136
43	24 06	48	134	23 18	48	135	22 30	49	135	21 41	48	136	20 53	49	136
42	24 41	49	135	23 52	49	135	23 03	49	136	22 14	49	137	21 25	49	137
41	25 15	49	136	24 26	50	136	23 36	49	137	22 47	49	137	21 58	50	138
40	25 48	−50	136	24 58	−49	137	24 09	−50	138	23 19	−50	138	22 29	−49	139
39	26 21	50	137	25 31	50	138	24 41	50	139	23 51	50	139	23 01	51	140
38	26 53	50	138	26 03	51	139	25 12	50	139	24 22	51	140	23 31	50	141
37	27 25	51	139	26 34	51	140	25 43	50	140	24 53	51	141	24 02	52	142
36	27 56	51	140	27 05	51	141	26 14	52	141	25 22	51	142	24 31	51	142
35	28 26	−51	141	27 35	−52	142	26 43	−51	142	25 52	−52	143	25 00	−52	143
34	28 56	52	142	28 04	51	143	27 13	52	143	26 21	53	144	25 28	52	144
33	29 25	52	143	28 33	52	143	27 41	52	144	26 49	53	145	25 56	52	145
32	29 54	53	144	29 01	52	144	28 09	53	145	27 16	53	146	26 23	53	146
31	30 22	53	145	29 29	53	145	28 36	53	146	27 43	53	146	26 50	53	147
30	30 49	−53	146	29 56	−53	146	29 03	−54	147	28 09	−53	147	27 16	−54	148
29	31 16	54	147	30 22	54	147	29 28	53	148	28 35	54	148	27 41	54	149
28	31 42	54	148	30 48	54	148	29 54	55	149	28 59	54	149	28 05	54	150
27	32 07	54	149	31 13	55	149	30 18	54	150	29 24	55	150	28 29	55	151
26	32 31	54	150	31 37	55	150	30 42	55	151	29 47	55	151	28 52	55	152
25	32 55	−55	151	32 00	−55	151	31 05	−55	152	30 10	−56	152	29 14	−55	153
24	33 18	56	152	32 22	55	152	31 27	55	153	30 32	56	153	29 36	56	154
23	33 40	56	153	32 44	55	154	31 49	56	154	30 53	56	154	29 57	56	155
22	34 01	56	154	33 05	56	155	32 09	56	155	31 13	56	155	30 17	57	156
21	34 22	57	155	33 25	56	156	32 29	56	156	31 33	57	156	30 36	57	157
20	34 41	−56	156	33 45	−57	157	32 48	−57	157	31 51	−57	158	30 54	−56	158
19	35 00	57	157	34 03	57	158	33 06	57	158	32 09	57	159	31 12	57	159
18	35 18	57	159	34 21	57	159	33 24	58	159	32 26	57	160	31 29	57	160
17	35 35	57	160	34 38	58	160	33 40	57	160	32 43	58	161	31 45	58	161
16	35 52	58	161	34 54	58	161	33 56	58	162	32 58	58	162	32 00	58	162
15	36 07	−58	162	35 09	−58	162	34 11	−58	163	33 13	−59	163	32 14	−58	163
14	36 21	58	163	35 23	58	163	34 25	59	164	33 26	58	164	32 28	59	164
13	36 35	59	164	35 36	58	165	34 38	59	165	33 39	59	165	32 40	58	165
12	36 47	59	166	35 48	58	166	34 50	59	166	33 51	59	166	32 52	59	167
11	36 59	59	167	36 00	59	167	35 01	59	167	34 02	59	167	33 03	59	168
10	37 09	−59	168	36 10	−59	168	35 11	−59	168	34 12	−59	169	33 13	−59	169
9	37 19	59	169	36 20	60	169	35 20	59	169	34 21	59	170	33 22	60	170
8	37 27	59	170	36 28	59	170	35 29	60	171	34 29	59	171	33 30	60	171
7	37 35	60	172	36 35	59	172	35 36	60	172	34 36	59	172	33 37	60	172
6	37 42	60	173	36 42	60	173	35 42	59	173	34 43	60	173	33 43	60	173
5	37 47	−60	174	36 47	−59	174	35 48	−60	174	34 48	−60	174	33 48	−60	174
4	37 52	60	175	36 52	60	175	35 52	60	175	34 52	60	175	33 52	59	175
3	37 55	59	176	36 56	60	176	35 56	60	177	34 56	60	177	33 56	60	177
2	37 58	60	178	36 58	60	178	35 58	60	178	34 58	60	178	33 58	60	178
1	38 00	60	179	37 00	60	179	36 00	60	179	35 00	60	179	34 00	60	179
0	38 00	−60	180	37 00	−60	180	36 00	−60	180	35 00	−60	180	34 00	−60	180

LHA	20° Hc	20° d	20° Z	21° Hc	21° d	21° Z	22° Hc	22° d	22° Z	23° Hc	23° d	23° Z	24° Hc	24° d	24° Z
	° ′	′	°	° ′	′	°	° ′	′	°	° ′	′	°	° ′	′	°
69	03 37	40	118	02 57	40	119	02 17	40	120	01 37	40	121	00 57	40	121
68	04 19	40	119	03 39	40	120	02 59	41	121	02 18	40	121	01 38	40	122
67	05 01	41	120	04 20	40	120	03 40	41	121	02 59	40	122	02 19	41	123
66	05 42	40	120	05 02	41	121	04 21	41	122	03 40	41	123	02 59	41	123
65	06 24	−42	121	05 42	−41	122	05 01	−41	123	04 20	−41	123	03 39	−42	124
64	07 04	41	122	06 23	41	122	05 42	42	123	05 00	42	124	04 18	41	125
63	07 45	42	122	07 03	42	123	06 21	41	124	05 40	42	125	04 58	42	125
62	08 25	42	123	07 43	42	124	07 01	42	124	06 19	42	125	05 37	43	126
61	09 06	43	124	08 23	42	124	07 41	43	125	06 58	43	126	06 15	42	127
60	09 45	−43	124	09 02	−42	125	08 20	−43	126	07 37	−43	127	06 54	−43	127
59	10 25	43	125	09 42	44	126	08 58	43	126	08 15	43	127	07 32	44	128
58	11 04	44	126	10 20	43	126	09 37	44	127	08 53	44	128	08 09	43	129
57	11 42	43	126	10 59	44	127	10 15	44	128	09 31	44	129	08 47	45	129
56	12 21	44	127	11 37	45	128	10 52	44	129	10 08	44	129	09 24	45	130
55	12 59	−45	128	12 14	−44	129	11 30	−45	129	10 45	−45	130	10 00	−45	131
54	13 37	45	129	12 52	45	129	12 07	45	130	11 22	46	131	10 36	45	131
53	14 14	45	129	13 29	46	130	12 43	45	131	11 58	46	131	11 12	45	132
52	14 51	46	130	14 05	46	131	13 19	45	131	12 34	46	132	11 48	46	133
51	15 27	46	131	14 41	46	131	13 55	46	132	13 09	46	133	12 23	47	133
50	16 03	−46	132	15 17	−46	132	14 31	−47	133	13 44	−47	134	12 57	−46	134
49	16 39	47	132	15 52	46	133	15 06	47	134	14 19	47	134	13 32	48	135
48	17 14	47	133	16 27	47	134	15 40	47	134	14 53	48	135	14 05	47	136
47	17 49	47	134	17 02	48	134	16 14	48	135	15 26	47	136	14 39	48	136
46	18 23	47	135	17 36	48	135	16 48	48	136	16 00	48	137	15 12	49	137
45	18 57	−48	135	18 09	−48	136	17 21	−49	137	16 32	−48	137	15 44	−49	138
44	19 31	49	136	18 42	48	137	17 54	49	137	17 05	49	138	16 16	49	139
43	20 04	49	137	19 15	49	138	18 26	49	138	17 37	50	139	16 47	49	139
42	20 36	49	138	19 47	50	138	18 57	49	139	18 08	50	140	17 18	50	140
41	21 08	50	139	20 18	49	139	19 29	50	140	18 39	50	140	17 49	51	141
40	21 40	−51	140	20 49	−50	140	19 59	−50	141	19 09	−50	141	18 19	−51	142
39	22 10	50	140	21 20	51	141	20 29	50	142	19 39	51	142	18 48	51	143
38	22 41	51	141	21 50	51	142	20 59	51	142	20 08	51	143	19 17	52	143
37	23 10	51	142	22 19	51	143	21 28	51	143	20 37	52	144	19 45	52	144
36	23 40	52	143	22 48	52	144	21 56	51	144	21 05	52	145	20 13	52	145
35	24 08	−52	144	23 16	−52	144	22 24	−52	145	21 32	−52	145	20 40	−53	146
34	24 36	52	145	23 44	52	145	22 52	53	146	21 59	53	146	21 06	52	147
33	25 04	53	146	24 11	53	146	23 18	53	147	22 25	53	147	21 32	53	148
32	25 30	53	147	24 37	53	147	23 44	53	148	22 51	53	148	21 58	54	149
31	25 57	54	147	25 03	53	148	24 10	54	148	23 16	54	149	22 22	53	149
30	26 22	−54	148	25 28	−54	149	24 34	−54	149	23 40	−54	150	22 46	−54	150
29	26 47	54	149	25 53	55	150	24 58	54	150	24 04	54	151	23 10	55	151
28	27 11	55	150	26 16	54	151	25 22	55	151	24 27	54	152	23 33	55	152
27	27 34	55	151	26 39	54	152	25 45	55	152	24 50	55	153	23 55	56	153
26	27 57	55	152	27 02	55	153	26 07	56	153	25 11	55	154	24 16	55	154
25	28 19	−55	153	27 24	−56	154	26 28	−56	154	25 32	−55	155	24 37	−56	155
24	28 40	56	154	27 44	55	155	26 49	56	155	25 53	56	155	24 57	56	156
23	29 01	56	155	28 05	57	156	27 08	56	156	26 12	56	156	25 16	56	157
22	29 20	56	156	28 24	56	157	27 28	57	157	26 31	56	157	25 35	57	158
21	29 39	56	157	28 43	57	158	27 46	57	158	26 49	57	158	25 52	57	159
20	29 58	−57	158	29 01	−57	159	28 04	−57	159	27 07	−58	159	26 09	−57	160
19	30 15	57	159	29 18	58	160	28 20	57	160	27 23	57	160	26 26	58	161
18	30 32	58	160	29 34	58	161	28 36	57	161	27 39	58	161	26 41	57	162
17	30 47	57	161	29 50	58	162	28 52	58	162	27 54	58	162	26 56	58	163
16	31 02	58	162	30 04	58	163	29 06	58	163	28 08	58	163	27 10	58	164
15	31 16	−58	164	30 18	−58	164	29 20	−59	164	28 21	−58	164	27 23	−58	165
14	31 29	58	165	30 31	59	165	29 32	58	165	28 34	59	165	27 35	58	166
13	31 42	59	166	30 43	59	166	29 44	58	166	28 46	59	166	27 47	59	167
12	31 53	59	167	30 54	58	167	29 56	59	167	28 57	59	167	27 58	59	168
11	32 04	59	168	31 05	59	168	30 06	59	168	29 07	59	168	28 08	60	169
10	32 14	−60	169	31 14	−59	169	30 15	−59	169	29 16	−59	169	28 17	−60	170
9	32 22	59	170	31 23	59	170	30 24	60	170	29 24	59	171	28 25	60	171
8	32 30	59	171	31 31	60	171	30 31	59	171	29 32	60	172	28 32	59	172
7	32 37	59	172	31 38	60	172	30 38	60	173	29 38	59	173	28 39	60	173
6	32 43	59	173	31 44	60	173	30 44	60	174	29 44	60	174	28 44	59	174
5	32 48	−59	174	31 49	−60	175	30 49	−60	175	29 49	−60	175	28 49	−60	175
4	32 53	60	176	31 53	60	176	30 53	60	176	29 53	60	176	28 53	60	176
3	32 56	60	177	31 56	60	177	30 56	60	177	29 56	60	177	28 56	60	177
2	32 58	60	178	31 58	60	178	30 58	60	178	29 58	60	178	28 58	60	178
1	33 00	60	179	32 00	60	179	31 00	60	179	30 00	60	179	29 00	60	179
0	33 00	−60	180	32 00	−60	180	31 00	−60	180	30 00	−60	180	29 00	−60	180

LHA	25° Hc	25° d	25° Z	26° Hc	26° d	26° Z	27° Hc	27° d	27° Z	28° Hc	28° d	28° Z	29° Hc	29° d	29° Z	LHA
	° ′	′	°	° ′	′	°	° ′	′	°	° ′	′	°	° ′	′	°	
69	00 17	40	122	−0 23	40	123	−1 03	40	124	−1 43	40	124	−2 23	39	125	291
68	00 58	41	123	00 17	40	124	−0 23	40	124	−1 03	40	125	−1 43	41	126	292
67	01 38	41	123	00 57	40	124	00 17	41	125	−0 24	41	126	−1 05	40	126	293
66	02 18	41	124	01 37	41	125	00 56	41	126	00 15	41	126	−0 26	41	127	294
65	02 57	−41	125	02 16	−41	125	01 35	−42	126	00 53	−41	127	00 12	−42	128	295
64	03 37	42	125	02 55	42	126	02 13	42	127	01 31	41	127	00 50	42	128	296
63	04 16	43	126	03 33	42	127	02 51	42	127	02 09	42	128	01 27	42	129	297
62	04 54	42	127	04 12	43	127	03 29	42	128	02 47	43	129	02 04	42	129	298
61	05 33	43	127	04 50	43	128	04 07	43	129	03 24	43	129	02 41	43	130	299
60	06 11	−44	128	05 27	−43	129	04 44	−43	129	04 01	−43	130	03 18	−44	131	300
59	06 48	43	129	06 05	44	129	05 21	43	130	04 38	44	131	03 54	44	131	301
58	07 26	44	129	06 42	44	130	05 58	44	131	05 14	44	131	04 30	44	132	302
57	08 02	44	130	07 18	44	131	06 34	44	131	05 50	45	132	05 05	44	133	303
56	08 39	44	131	07 55	45	131	07 10	45	132	06 25	45	133	05 40	44	133	304
55	09 15	−45	131	08 30	−45	132	07 45	−45	133	07 00	−45	133	06 15	−45	134	305
54	09 51	45	132	09 06	46	133	08 20	45	133	07 35	46	134	06 49	45	135	306
53	10 27	46	133	09 41	46	133	08 55	46	134	08 09	46	135	07 23	46	135	307
52	11 02	46	133	10 16	47	134	09 29	46	135	08 43	46	135	07 57	46	136	308
51	11 36	46	134	10 50	47	135	10 03	46	135	09 17	47	136	08 30	47	137	309
50	12 11	−47	135	11 24	−47	135	10 37	−47	136	09 50	−47	137	09 03	−47	137	310
49	12 44	47	136	11 57	47	136	11 10	47	137	10 23	48	137	09 35	47	138	311
48	13 18	48	136	12 30	47	137	11 43	48	138	10 55	48	138	10 07	48	139	312
47	13 51	48	137	13 03	48	138	12 15	48	138	11 27	49	139	10 38	48	139	313
46	14 23	48	138	13 35	49	138	12 46	48	139	11 58	49	140	11 09	48	140	314
45	14 55	−48	139	14 07	−49	139	13 18	−49	140	12 29	−49	140	11 40	−49	141	315
44	15 27	49	139	14 38	50	140	13 48	49	140	12 59	49	141	12 10	50	142	316
43	15 58	50	140	15 08	49	141	14 19	50	141	13 29	50	142	12 39	49	142	317
42	16 28	49	141	15 39	50	141	14 49	50	142	13 59	51	143	13 08	50	143	318
41	16 58	50	142	16 08	50	142	15 18	51	143	14 27	50	143	13 37	51	144	319
40	17 28	−51	142	16 37	−50	143	15 47	−51	144	14 56	−51	144	14 05	−51	145	320
39	17 57	51	143	17 06	51	144	16 15	51	144	15 24	51	145	14 33	52	145	321
38	18 25	51	144	17 34	51	145	16 43	52	145	15 51	52	146	14 59	51	146	322
37	18 53	51	145	18 02	52	145	17 10	52	146	16 18	52	146	15 26	52	147	323
36	19 21	52	146	18 29	53	146	17 36	52	147	16 44	52	147	15 52	53	148	324
35	19 47	−52	147	18 55	−53	147	18 02	−52	148	17 10	−53	148	16 17	−53	149	325
34	20 14	53	147	19 21	53	148	18 28	53	148	17 35	53	149	16 42	53	149	326
33	20 39	53	148	19 46	53	149	18 53	54	149	17 59	53	150	17 06	53	150	327
32	21 04	53	149	20 11	54	150	19 17	54	150	18 23	53	151	17 30	54	151	328
31	21 29	54	150	20 35	54	150	19 41	54	151	18 47	54	151	17 53	54	152	329
30	21 52	−54	151	20 58	−54	151	20 04	−55	152	19 09	−54	152	18 15	−54	153	330
29	22 15	54	152	21 21	55	152	20 26	54	153	19 32	55	153	18 37	55	153	331
28	22 38	55	153	21 43	55	153	20 48	55	153	19 53	55	154	18 58	55	154	332
27	22 59	55	154	22 04	55	154	21 09	55	154	20 14	56	155	19 18	55	155	333
26	23 21	56	154	22 25	55	155	21 30	56	155	20 34	56	156	19 38	56	156	334
25	23 41	−56	155	22 45	−56	156	21 49	−56	156	20 53	−56	157	19 57	−56	157	335
24	24 01	56	156	23 05	57	157	22 08	56	157	21 12	56	157	20 16	56	158	336
23	24 20	57	157	23 23	56	158	22 27	57	158	21 30	56	158	20 34	57	159	337
22	24 38	57	158	23 41	57	158	22 44	56	159	21 48	57	159	20 51	57	160	338
21	24 55	57	159	23 58	57	159	23 01	57	160	22 04	57	160	21 07	57	160	339
20	25 12	−57	160	24 15	−57	160	23 18	−58	161	22 20	−57	161	21 23	−57	161	340
19	25 28	57	161	24 31	58	161	23 33	57	162	22 36	58	162	21 38	58	162	341
18	25 44	58	162	24 46	58	162	23 48	58	163	22 50	58	163	21 52	57	163	342
17	25 58	58	163	25 00	58	163	24 02	58	163	23 04	58	164	22 06	58	164	343
16	26 12	58	164	25 14	59	164	24 15	58	164	23 17	58	165	22 19	58	165	344
15	26 25	−59	165	25 26	−58	165	24 28	−59	165	23 29	−58	166	22 31	−59	166	345
14	26 37	59	166	25 38	58	166	24 40	59	166	23 41	59	167	22 42	58	167	346
13	26 48	59	167	25 49	58	167	24 51	59	167	23 52	59	168	22 53	59	168	347
12	26 59	59	168	26 00	59	168	25 01	59	168	24 02	59	168	23 03	59	169	348
11	27 08	59	169	26 09	59	169	25 10	59	169	24 11	59	169	23 12	59	170	349
10	27 17	−59	170	26 18	−59	170	25 19	−59	170	24 20	−60	170	23 20	−59	171	350
9	27 25	59	171	26 26	59	171	25 27	60	171	24 27	59	171	23 28	60	171	351
8	27 33	60	172	26 33	59	172	25 34	60	172	24 34	59	172	23 35	60	172	352
7	27 39	60	173	26 39	59	173	25 40	60	173	24 40	60	173	23 40	59	173	353
6	27 45	60	174	26 45	60	174	25 45	60	174	24 45	59	174	23 46	60	174	354
5	27 49	−59	175	26 50	−60	175	25 50	−60	175	24 50	−60	175	23 50	−60	175	355
4	27 53	60	176	26 53	60	176	25 53	59	176	24 54	60	176	23 54	60	176	356
3	27 56	60	177	26 56	60	177	25 56	60	177	24 56	60	177	23 56	59	177	357
2	27 58	60	178	26 58	60	178	25 58	60	178	24 58	60	178	23 58	60	178	358
1	28 00	60	179	27 00	60	179	26 00	60	179	25 00	60	179	24 00	60	179	359
0	28 00	−60	180	27 00	−60	180	26 00	−60	180	25 00	−60	180	24 00	−60	180	360

S. Lat. {LHA greater than 180°....... Zn=180−Z; LHA less than 180°.......... Zn=180+Z}

DECLINATION (15°-29°) CONTRARY NAME TO LATITUDE

LAT 37°

LAT 38°

LHA	15° Hc	d	Z	16° Hc	d	Z	17° Hc	d	Z	18° Hc	d	Z	19° Hc	d	Z	20° Hc	d	Z	21° Hc	d	Z	22° Hc	d	Z	23° Hc	d	Z	24° Hc	d	Z	25° Hc	d	Z	26° Hc	d	Z	27° Hc	d	Z	28° Hc	d	Z	29° Hc	d	Z	LHA
	° ′	′	°	° ′	′	°	° ′	′	°	° ′	′	°	° ′	′	°	° ′	′	°	° ′	′	°	° ′	′	°	° ′	′	°	° ′	′	°	° ′	′	°	° ′	′	°	° ′	′	°	° ′	′	°	° ′	′	°	
70	24 49	+34	90	25 23	+35	89	25 58	+33	88	26 31	+34	87	27 05	+33	86	27 38	+33	85	28 11	+3[illegible]	84	28 43	+32	84	29 15	+32	83	29 47	+31	82	30 18	+30	81	30 48	+31	80	31 19	+29	79	31 48	+29	78	32 17	+29	77	290
71	24 02	34	89	24 36	34	89	25 10	34	88	25 44	34	87	26 18	33	86	26 51	33	85	27 24	3[illegible]	84	27 56	32	83	28 28	32	82	29 00	31	81	29 31	31	80	30 02	30	79	30 32	30	78	31 02	30	77	31 32	28	76	289
72	23 14	35	89	23 49	34	88	24 23	34	87	24 57	34	86	25 31	33	85	26 04	33	84	26 37	3[illegible]	83	27 09	33	82	27 42	31	8[illegible]	28 13	32	80	28 45	31	79	29 16	30	79	29 46	30	78	30 16	30	77	30 46	29	76	288
73	22 27	35	88	23 02	34	87	23 36	34	86	24 10	34	86	24 44	33	85	25 17	33	84	25 50	3[illegible]	83	26 23	32	82	26 55	32	8[illegible]	27 27	31	80	27 58	31	79	28 29	31	78	29 00	30	77	29 30	30	76	30 00	29	75	287
74	21 40	34	88	22 14	35	87	22 49	34	86	23 23	34	85	23 57	33	84	24 30	33	83	25 03	3[illegible]	82	25 36	32	81	26 08	32	8[illegible]	26 40	32	79	27 12	31	78	27 43	31	77	28 14	30	76	28 44	30	76	29 14	30	75	286
75	20 53	+34	87	21 27	+35	86	22 02	+34	85	22 36	+34	84	23 10	+33	83	23 43	+33	83	24 16	+3[illegible]	82	24 49	+33	81	25 22	+32	8[illegible]	25 54	−32	79	26 26	+31	78	26 57	+31	77	27 28	+31	76	27 59	+30	75	28 29	+30	74	285
76	20 05	35	86	20 40	35	86	21 15	34	85	21 49	34	84	22 23	33	83	22 56	34	82	23 30	3[illegible]	81	24 02	33	80	24 35	32	7[illegible]	25 07	32	78	25 39	32	77	26 11	31	76	26 42	31	75	27 13	30	74	27 43	31	74	284
77	19 18	35	86	19 53	35	85	20 28	34	84	21 02	34	83	21 36	33	82	22 09	34	81	22 43	3[illegible]	80	23 16	33	80	23 49	32	79	24 21	32	78	24 53	32	77	25 25	32	76	25 57	31	75	26 28	30	74	26 58	30	73	283
78	18 31	35	85	19 06	35	84	19 41	34	83	20 15	34	83	20 49	34	82	21 23	33	81	21 56	3[illegible]	80	22 30	32	79	23 02	33	7[illegible]	23 35	32	77	24 07	32	76	24 39	32	75	25 11	31	74	25 42	31	73	26 13	31	73	282
79	17 44	35	85	18 19	35	84	18 54	34	83	19 28	34	82	20 02	34	81	20 36	34	80	21 10	3[illegible]	79	21 43	33	78	22 16	33	7[illegible]	22 49	32	77	23 21	33	76	23 54	31	75	24 25	32	74	24 57	31	73	25 28	31	72	281
80	16 57	+35	84	17 32	+35	83	18 07	+34	82	18 41	+33	81	19 16	+34	81	19 50	+33	80	20 23	+3[illegible]	79	20 57	+33	78	21 30	+33	7[illegible]	22 03	−33	76	22 36	+32	75	23 08	+32	74	23 40	+32	73	24 12	+31	72	24 43	+31	72	280
81	16 10	35	83	16 45	35	83	17 20	35	82	17 55	34	81	18 29	34	80	19 03	34	79	19 37	3[illegible]	78	20 11	33	77	20 44	33	7[illegible]	21 17	33	76	21 50	33	75	22 23	32	74	22 55	32	73	23 27	31	72	23 58	32	71	279
82	15 23	35	83	15 58	35	82	16 33	35	81	17 08	34	80	17 42	35	79	18 17	34	79	18 51	3[illegible]	78	19 25	33	77	19 58	33	7[illegible]	20 31	34	75	21 05	32	74	21 37	33	73	22 10	32	72	22 42	32	71	23 14	31	71	278
83	14 36	35	82	15 11	35	81	15 46	35	81	16 21	35	80	16 56	34	79	17 30	35	78	18 05	3[illegible]	77	18 39	33	76	19 12	34	75	19 46	33	75	20 19	33	74	20 52	33	73	21 25	32	72	21 57	32	71	22 29	32	70	277
84	13 49	36	82	14 25	35	81	15 00	35	80	15 35	35	79	16 10	34	78	16 44	35	77	17 19	3[illegible]	77	17 53	34	76	18 27	33	75	19 00	34	74	19 34	33	73	20 07	33	72	20 40	33	71	21 13	32	70	21 45	32	70	276
85	13 03	+35	81	13 38	+35	80	14 13	+35	79	14 48	+35	79	15 23	+35	78	15 58	+35	77	16 33	+3[illegible]	76	17 07	+34	75	17 41	+34	74	18 15	−34	73	18 49	+33	73	19 22	+33	72	19 55	+33	71	20 28	+33	70	21 01	+32	69	275
86	12 16	36	80	12 52	35	80	13 27	35	79	14 02	35	78	14 37	35	77	15 12	35	76	15 47	3[illegible]	75	16 21	35	75	16 55	34	74	17 30	34	73	18 04	33	72	18 37	34	71	19 11	33	70	19 44	33	69	20 17	32	69	274
87	11 29	36	80	12 05	36	79	12 41	35	78	13 16	35	77	13 51	35	77	14 26	35	76	15 01	3[illegible]	75	15 36	34	74	16 10	35	73	16 45	34	72	17 19	34	71	17 53	33	71	18 26	34	70	19 00	33	69	19 33	33	68	273
88	10 43	36	79	11 19	35	78	11 54	36	78	12 30	35	77	13 05	36	76	13 41	35	75	14 16	3[illegible]	74	14 50	35	74	15 25	35	73	16 00	34	72	16 34	34	71	17 08	34	70	17 42	34	69	18 16	33	68	18 49	33	67	272
89	09 56	36	79	10 32	36	78	11 08	36	77	11 44	36	76	12 20	35	75	12 55	35	75	13 30	3[illegible]	74	14 05	35	73	14 40	35	72	15 15	34	71	15 49	35	70	16 24	34	70	16 58	34	69	17 32	33	68	18 05	34	67	271
90	09 10	+36	78	09 46	+36	77	10 22	+36	77	10 58	+36	76	11 34	+35	75	12 09	+36	74	12 45	+3[illegible]	73	13 20	+35	72	13 55	+35	72	14 30	+35	71	15 05	+35	70	15 40	+34	69	16 14	+34	68	16 48	+34	67	17 22	+34	66	270
91	08 24	36	78	09 00	36	77	09 36	36	76	10 12	36	75	10 48	36	74	11 24	36	73	12 00	3[illegible]	73	12 35	35	72	13 10	36	7[illegible]	13 46	35	70	14 21	34	69	14 55	35	68	15 30	35	68	16 05	34	67	16 39	34	66	269
92	07 38	36	77	08 14	37	76	08 51	36	75	09 27	36	75	10 03	36	74	10 39	36	73	11 15	3[illegible]	72	11 50	36	71	12 25	35	70	13 01	35	70	13 36	36	69	14 12	34	68	14 46	35	67	15 21	35	66	15 56	34	65	268
93	06 52	36	76	07 28	37	76	08 05	36	75	08 41	36	74	09 17	37	73	09 54	36	72	10 30	3[illegible]	72	11 06	35	71	11 41	36	70	12 17	36	69	12 53	35	68	13 28	35	67	14 03	35	67	14 38	35	66	15 13	34	65	267
94	06 06	37	76	06 43	36	75	07 19	37	74	07 56	36	73	08 32	37	73	09 09	36	72	09 45	3[illegible]	71	10 21	36	70	10 57	36	69	11 33	36	69	12 09	35	68	12 44	36	67	13 20	35	66	13 55	35	65	14 30	35	64	266
95	05 20	+37	75	05 57	+37	74	06 34	+37	74	07 11	+36	73	07 47	+37	72	08 24	+36	71	09 00	+3[illegible]	70	09 37	+36	70	10 13	+36	69	10 49	−36	68	11 25	+36	67	12 01	+36	66	12 37	+35	65	13 12	+36	65	13 48	+35	64	265
96	04 35	37	75	05 12	37	74	05 49	37	73	06 26	36	72	07 02	37	71	07 39	37	71	08 16	3[illegible]	70	08 52	37	69	09 29	36	68	10 05	37	67	10 42	36	67	11 18	36	66	11 54	36	65	12 30	35	64	13 05	36	63	264
97	03 49	37	74	04 26	38	73	05 04	37	72	05 41	37	72	06 18	37	71	06 55	37	70	07 32	3[illegible]	69	08 08	37	68	08 45	37	68	09 22	36	67	09 58	37	66	10 35	36	65	11 11	36	64	11 47	36	64	12 23	36	63	263
98	03 04	37	73	03 41	38	73	04 19	37	72	04 56	37	71	05 33	37	70	06 10	38	69	06 48	3[illegible]	69	07 25	37	68	08 02	36	6[illegible]	08 38	37	66	09 15	37	65	09 52	36	65	10 28	37	64	11 05	36	63	11 41	36	62	262
99	02 18	38	73	02 56	38	72	03 34	37	71	04 11	38	70	04 49	37	70	05 26	38	69	06 04	3[illegible]	68	06 41	37	67	07 18	37	6[illegible]	07 55	37	66	08 32	37	65	09 09	37	64	09 46	37	63	10 23	37	62	11 00	36	62	261
100	01 33	+38	72	02 11	+38	71	02 49	+38	71	03 27	+37	70	04 04	+38	69	04 42	+38	68	05 20	+3[illegible]	67	05 57	+38	67	06 35	+37	6[illegible]	07 12	−38	65	07 50	+37	64	08 27	+37	64	09 04	+37	63	09 41	+37	62	10 18	+37	61	260
101	00 49	38	71	01 27	38	71	02 05	37	70	02 42	38	69	03 20	38	68	03 58	38	68	04 36	3[illegible]	67	05 14	38	66	05 52	38	65	06 30	37	65	07 07	38	64	07 45	37	63	08 22	37	62	08 59	38	61	09 37	37	61	259
102	00 04	38	71	00 42	38	70	01 20	38	69	01 58	39	69	02 37	38	68	03 15	38	67	03 53	3[illegible]	66	04 31	38	65	05 09	38	65	05 47	38	64	06 25	38	63	07 03	37	62	07 40	38	62	08 18	38	61	08 56	37	60	258
103	−0 41	39	70	−0 02	38	69	00 36	38	69	01 14	39	68	01 53	38	67	02 31	39	66	03 10	3[illegible]	66	03 48	38	65	04 25	39	64	05 05	38	63	05 43	38	63	06 21	38	62	06 59	38	61	07 37	38	60	08 15	38	59	257
104	−1 25	38	70	−0 47	39	69	−0 08	39	68	00 31	38	67	01 09	39	67	01 48	39	66	02 27	3[illegible]	65	03 05	39	64	03 44	38	64	04 22	39	63	05 01	38	62	05 39	39	61	06 18	38	60	06 56	38	60	07 34	38	59	256
105	−2 09	+38	69	−1 31	+39	68	−0 52	+39	67	−0 13	+39	67	00 26	+39	66	01 05	+39	65	01 44	+3[illegible]	64	02 23	+39	64	03 02	+38	63	03 40	−39	62	04 19	+39	61	04 58	+39	61	05 37	+38	60	06 15	+39	59	06 54	+38	58	255
106	−2 54	40	68	−2 14	39	68	−1 35	39	67	−0 56	39	66	−0 17	39	65	00 22	39	65	01 01	4[illegible]	64	01 41	39	63	02 20	39	62	02 59	39	62	03 38	39	61	04 17	39	60	04 56	39	59	05 35	39	59	06 14	39	58	254
107	−3 37	39	68	−2 58	39	67	−2 19	40	66	−1 39	39	65	−1 00	40	65	−0 20	39	64	00 19	3[illegible]	63	00 58	40	62	01 38	39	62	02 17	40	61	02 57	39	60	03 36	39	59	04 15	40	59	04 55	39	58	05 34	39	57	253
108	−4 21	40	67	−3 41	39	66	−3 02	40	66	−2 22	40	65	−1 42	39	64	−1 03	40	63	−0 23	4[illegible]	63	00 17	39	62	00 55	40	6[illegible]	01 36	40	60	02 16	39	60	02 55	40	59	03 35	40	58	04 15	39	57	04 54	40	57	252
109	−5 05	40	66	−4 25	40	66	−3 45	40	65	−3 05	40	64	−2 25	40	63	−1 45	40	63	−1 05	4[illegible]	62	−0 25	40	61	00 15	40	6[illegible]	00 55	40	60	01 35	40	59	02 15	40	58	02 55	40	58	03 35	40	57	04 15	40	56	251
110	−5 48	+40	66	−5 08	+40	65	−4 28	+41	64	−3 47	+40	64	−3 07	+40	63	−2 27	+40	62	−1 47	+4[illegible]	61	−1 06	+40	61	−0 25	+40	60	00 14	−41	59	00 55	+40	58	01 35	+40	58	02 15	+41	57	02 56	+40	56	03 36	+40	55	250
111				−5 50	40	64	−5 10	40	64	−4 30	41	63	−3 49	41	62	−3 08	40	61	−2 28	4[illegible]	[illegible]	−1 47	40	60	−1 07	41	59	−0 26	41	59	00 15	40	58	00 55	41	57	01 36	40	56	02 16	41	56	02 57	41	55	249
112							−5 52	40	63	−5 12	41	62	−4 31	41	62	−3 50	41	61	−3 09	4[illegible]	60	−2 28	41	59	−1 47	41	59	−1 06	41	58	−0 25	41	57	00 16	41	56	00 57	41	56	01 38	41	55	02 19	41	54	248
113										−5 53	41	62	−5 12	41	61	−4 31	41	60	−3 50	4[illegible]	59	−3 09	42	59	−2 27	41	58	−1 46	41	57	−1 05	41	57	−0 24	42	56	00 18	41	55	00 59	41	54	01 40	42	54	247
114													−5 53	41	60	−5 12	42	60	−4 30	4[illegible]	59	−3 49	42	58	−3 07	41	5[illegible]	−2 26	42	57	−1 44	41	56	−1 03	42	55	−0 21	42	54	00 21	41	54	01 02	42	53	246
115																−5 53	+42	59	−5 11	+4[illegible]	58	−4 29	+42	57	−3 47	+42	5[illegible]	−3 05	+42	56	−2 23	+42	55	−1 41	+42	55	−0 59	+42	54	−0 17	+42	53	00 25	+42	52	245
116																			−5 51	4[illegible]	58	−5 09	43	57	−4 25	42	5[illegible]	−3 44	42	55	−3 02	42	55	−2 20	43	54	−1 37	42	53	−0 55	42	53	−0 13	43	52	244
117																						−5 48	42	56	−5 05	43	55	−4 23	43	55	−3 40	42	54	−2 58	43	53	−2 15	43	53	−1 32	43	52	−0 49	42	51	243
118																									−5 44	43	55	−5 01	43	54	−4 18	43	53	−3 35	43	53	−2 52	43	52	−2 09	43	51	−1 26	43	51	242
119																												−5 40	44	53	−4 56	43	53	−4 13	44	52	−3 29	43	51	−2 46	43	51	−2 03	44	50	241
120																															−5 34	+44	52	−4 50	+44	51	−4 06	+44	51	−3 22	+43	50	−2 39	+44	49	240
121																															−6 11	44	51	−5 27	44	51	−4 43	45	50	−3 58	44	49	−3 14	44	49	239
122																																		−6 03	44	50	−5 19	45	49	−4 34	44	49	−3 50	45	48	238
123																																					−5 54	44	49	−5 10	45	48	−4 25	45	47	237
124																																								−5 45	46	47	−4 59	45	47	236
125																																											−5 33	+45	46	235
126																																											−6 07	46	45	234

LHA	15° Hc	d	Z	16° Hc	d	Z	17° Hc	d	Z	18° Hc	d	Z	19° Hc	d	Z	20° Hc	d	Z	21° Hc	d	Z	22° Hc	d	Z	23° Hc	d	Z	24° Hc	d	Z	25° Hc	d	Z	26° Hc	d	Z	27° Hc	d	Z	28° Hc	d	Z	29° Hc	d	Z	LHA
86	−6 06	37	104																																											274
85	−5 20	−37	105	−5 57	−37	106																																								275
84	−4 35	37	105	−5 12	37	106	−5 49	37	107	−6 26	36	108																																		276
83	−3 49	37	106	−4 26	38	107	−5 04	37	108	−5 41	37	108	−6 18	37	109																															277
82	−3 04	37	107	−3 41	38	107	−4 19	37	108	−4 56	37	109	−5 33	37	110	−6 10	38	111																												278
81	−2 18	38	107	−2 56	38	108	−3 34	37	109	−4 11	38	110	−4 49	37	110	−5 26	38	111	−6 04	3[illegible]	112																									279
80	−1 33	−38	108	−2 11	−38	109	−2 49	−38	109	−3 27	−37	110	−4 04	−38	111	−4 42	−38	112	−5 20	−3[illegible]	113	−5 57	−38	113																						280
79	−0 49	38	109	−1 27	38	109	−2 05	37	110	−2 42	38	111	−3 20	38	112	−3 58	38	112	−4 36	3[illegible]	113	−5 14	38	114	−5 52	38	115																			281
78	−0 04	38	109	−0 42	38	110	−1 20	38	111	−1 58	39	111	−2 37	38	112	−3 15	38	113	−3 53	3[illegible]	114	−4 31	38	115	−5 09	38	115	−5 47	38	116																282
77	00 41	39	110	00 02	38	111	−0 36	38	111	−1 14	39	112	−1 53	38	113	−2 31	39	114	−3 10	3[illegible]	114	−3 48	38	115	−4 25	39	116	−5 05	38	117	−5 43	38	117	−6 21	38	118										283
76	01 25	38	110	00 47	39	111	00 08	39	112	−0 31	38	113	−1 09	39	113	−1 48	39	114	−2 27	3[illegible]	115	−3 05	39	116	−3 44	38	116	−4 22	39	117	−5 01	38	118	−5 39	39	119	−6 18	38	120							284
75	02 09	−38	111	01 31	−39	112	00 52	−39	113	00 13	−39	113	−0 26	−39	114	−1 05	−39	115	−1 44	−3[illegible]	116	−2 23	−39	116	−3 02	−38	117	−3 40	−39	118	−4 19	−39	119	−4 58	−39	119	−5 37	−38	120	−6 15	−39	121				285
74	02 54	40	112	02 14	39	112	01 35	39	113	00 56	39	114	00 17	39	115	−0 22	39	115	−1 01	4[illegible]	116	−1 41	39	117	−2 20	39	118	−2 59	39	118	−3 38	39	119	−4 17	39	120	−4 56	39	121	−5 35	39	121	−6 14	39	122	286
73	03 37	39	112	02 58	39	113	02 19	40	114	01 39	39	115	01 00	40	115	00 20	39	116	−0 19	3[illegible]	117	−0 58	40	118	−1 38	39	118	−2 17	40	119	−2 57	39	120	−3 36	39	121	−4 15	40	121	−4 55	39	122	−5 34	39	123	287
72	04 21	40	113	03 41	39	114	03 02	40	114	02 22	40	115	01 42	39	116	01 03	40	117	00 23	4[illegible]	117	−0 17	39	118	−0 55	40	119	−1 36	40	120	−2 16	39	120	−2 55	40	121	−3 35	40	122	−4 15	39	123	−4 54	40	123	288
71	05 05	40	114	04 25	40	114	03 45	40	115	03 05	40	116	02 25	40	117	01 45	40	117	01 05	4[illegible]	118	00 25	40	119	−0 15	40	119	−0 55	40	120	−1 35	40	121	−2 15	40	122	−2 55	40	122	−3 35	40	123	−4 15	40	124	289
70	05 48	−40	114	05 08	−40	115	04 28	−41	116	03 47	−40	116	03 07	−40	117	02 27	−40	118	01 47	−4[illegible]	119	01 06	−40	119	00 25	−40	120	−0 14	−41	121	−0 55	−40	122	−1 35	−40	122	−2 15	−41	123	−2 56	−40	124	−3 36	−40	125	290

S. Lat. {LHA greater than 180° .... Zn=180−Z; LHA less than 180° .... Zn=180+Z}

DECLINATION (15°-29°) CONTRARY NAME TO LATITUDE

Table 22

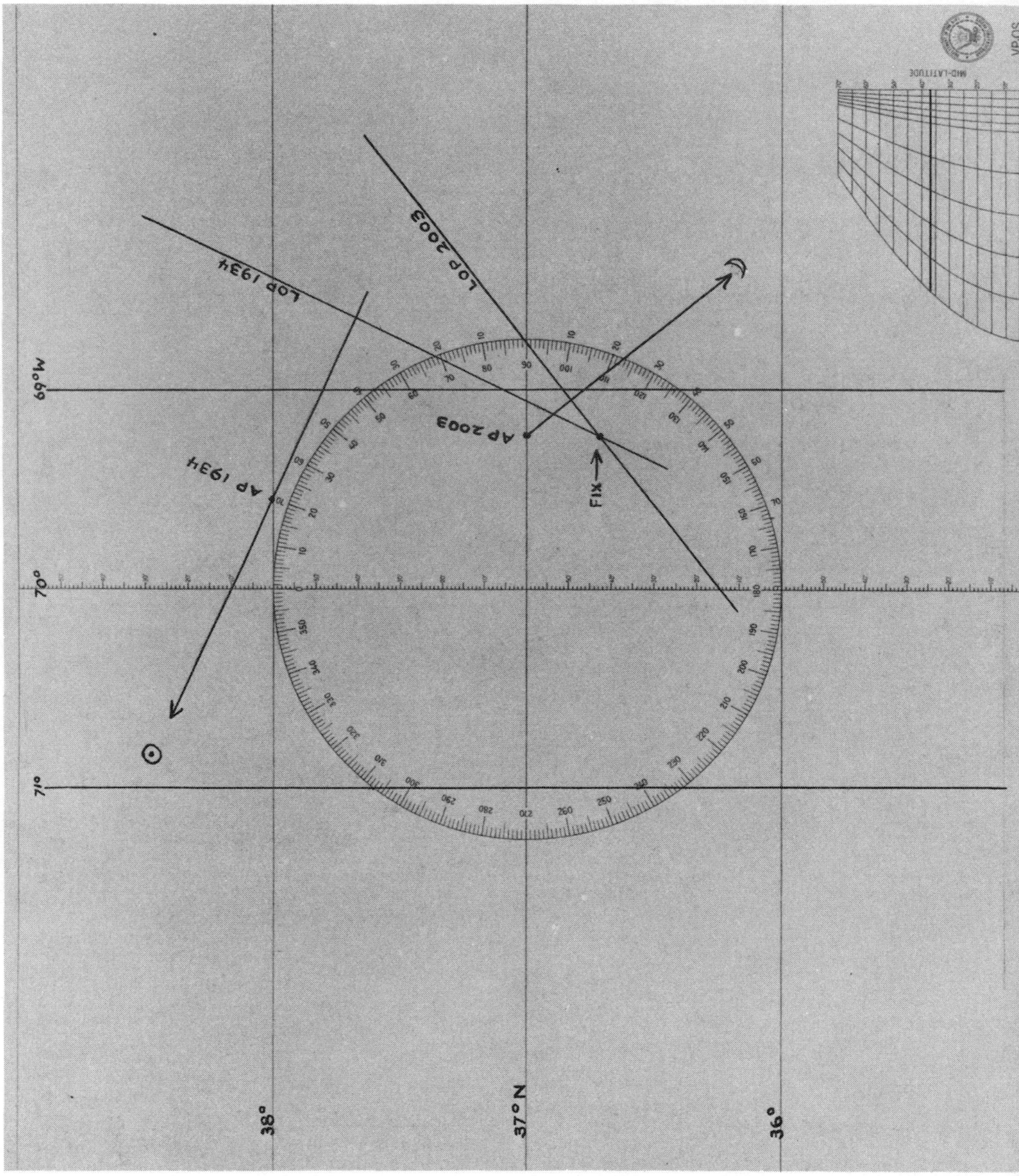
LOP 2003
LOP 1934
AP 2003
AP 1934
FIX
69°W
70°
71°
38°
37°N
36°
VP-OS

Hc	= 4°26′		d	= + 39′
Corr.	= + 03′		Dec. Inc.	= 5′
			Corr.	= + 3′
Hc	= 4°29′			
Ho	= 3°40′			
	49′	Away from Sun		
				360°
				− 64′
Z	= 64°		Zn	= 296°

These sights are plotted in Figure 9.3.

10
Celestial Navigation without Sight Reduction Tables

The noon sight for latitude is the traditional observation at sea. A ship's daily run is still reckoned from noon sight to noon sight, and betting on the distance covered is a common shipboard pastime, as anyone who has taken a voyage by ocean liner will know.

The sight is based on the fact that at noon the sun is on the same line of longitude as you are, and therefore, there is no triangle to solve. Your observation of the altitude of the sun is simply converted to zenith distance, and, with this knowledge and the approximate GMT, you apply declination to determine your latitude. Figures 10.1, 10.2, and 10.3 show the three possible cases of the noon sight. As you can see, all you have to do is remember what zenith distance is and you will have no trouble at all.

Let's take the cases one at a time and illustrate. In all these examples, it is August 15, 1974 (see Table 23).

Declination same name and less than your latitude.

This means that the sun bears south of you in north latitude or north of you in south latitude. You are headed for the Faeroes, and at about 1400 GMT you get a maximum altitude of 45°56′.

$$\begin{aligned}
\text{Lat.} &= \text{ZD} + \text{Declination}\\
\text{ZD} &= 90° - 45°56'\\
\text{ZD} &= 44°04'\\
\text{Lat.} &= 44°04' + 14°04'\\
\text{Lat.} &= 58°08'\,\text{N}
\end{aligned}$$

Declination same name and greater than your latitude.

This means that the sun bears north of you in north latitude or south of you in south latitude. You are near Christmas Island; it is about 2300 GMT, and your Ho is 79°03′.

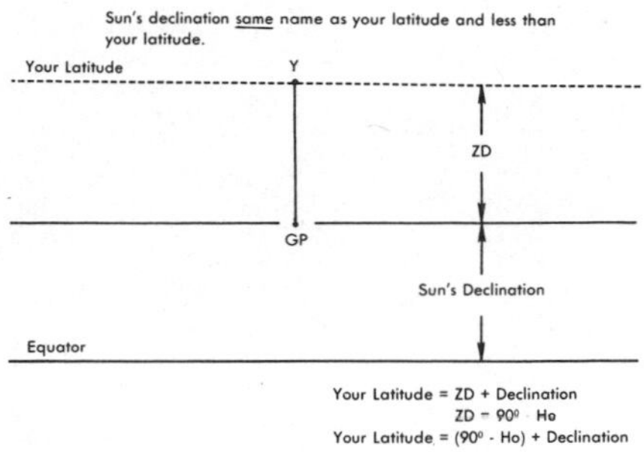
Sun's declination same name as your latitude and less than your latitude.
Your Latitude
Y
ZD
GP
Sun's Declination
Equator
Your Latitude = ZD + Declination
ZD = 90° - Ho
Your Latitude = (90° - Ho) + Declination

Fig. 10.1

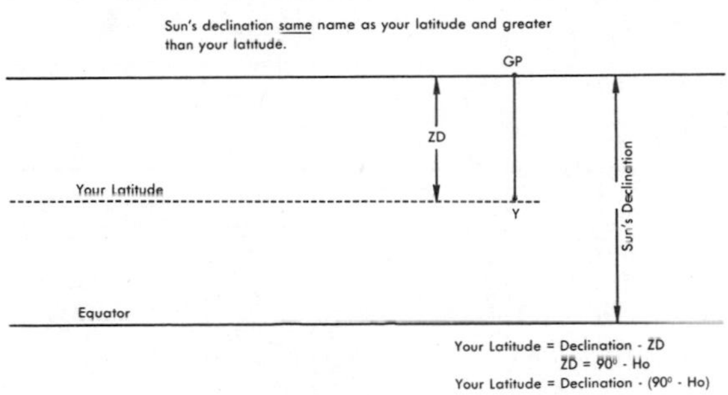
Sun's declination same name as your latitude and greater than your latitude.
GP
ZD
Your Latitude
Y
Sun's Declination
Equator
Your Latitude = Declination - ZD
ZD = 90° - Ho
Your Latitude = Declination - (90° - Ho)

Fig. 10.2

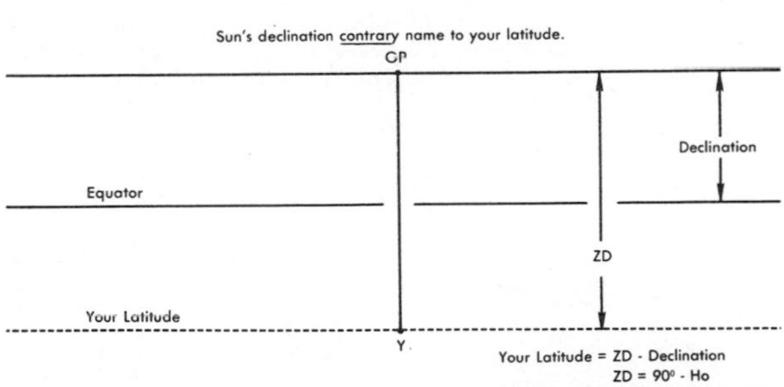
Sun's declination contrary name to your latitude.
GP
Declination
Equator
ZD
Your Latitude
Y
Your Latitude = ZD - Declination
ZD = 90° - Ho
Your Latitude = (90° - Ho) - Declination

Fig. 10.3

1974 AUGUST 14, 15, 16 (WED., THURS., FRI.)

G.M.T. d	h	SUN G.H.A. ° ′	SUN Dec. ° ′	MOON G.H.A. ° ′	v ′	MOON Dec. ° ′	d ′	H.P. ′
14	00	178 48·3	N14 33·0	234 30·5	4·6	N22 30·5	2·6	59·8
	01	193 48·4	32·3	248 54·1	4·5	22 27·9	2·8	59·8
	02	208 48·5	31·5	263 17·6	4·5	22 25·1	2·9	59·8
	03	223 48·7	· · 30·7	277 41·1	4·5	22 22·2	3·1	59·9
	04	238 48·8	30·0	292 04·6	4·5	22 19·1	3·2	59·9
	05	253 48·9	29·2	306 28·1	4·4	22 15·9	3·4	59·9
	06	268 49·0	N14 28·4	320 51·5	4·5	N22 12·5	3·6	60·0
W	07	283 49·1	27·7	335 15·0	4·4	22 08·9	3·7	60·0
E	08	298 49·2	26·9	349 38·4	4·4	22 05·2	3·9	60·0
D	09	313 49·3	· · 26·1	4 01·8	4·4	22 01·3	4·0	60·1
N	10	328 49·5	25·3	18 25·2	4·4	21 57·3	4·2	60·1
E	11	343 49·6	24·6	32 48·6	4·4	21 53·1	4·3	60·1
S	12	358 49·7	N14 23·8	47 12·0	4·4	N21 48·8	4·5	60·2
D	13	13 49·8	23·0	61 35·4	4·4	21 44·3	4·7	60·2
A	14	28 49·9	22·3	75 58·8	4·4	21 39·6	4·8	60·2
Y	15	43 50·0	· · 21·5	90 22·2	4·4	21 34·8	5·0	60·2
	16	58 50·1	20·7	104 45·6	4·4	21 29·8	5·1	60·3
	17	73 50·3	19·9	119 09·0	4·4	21 24·7	5·3	60·3
	18	88 50·4	N14 19·2	133 32·4	4·4	N21 19·4	5·4	60·3
	19	103 50·5	18·4	147 55·8	4·5	21 14·0	5·6	60·4
	20	118 50·6	17·6	162 19·3	4·4	21 08·4	5·8	60·4
	21	133 50·7	· · 16·9	176 42·7	4·4	21 02·6	5·9	60·4
	22	148 50·8	16·1	191 06·1	4·5	20 56·7	6·1	60·5
	23	163 51·0	15·3	205 29·6	4·5	20 50·6	6·2	60·5
15	00	178 51·1	N14 14·5	219 53·1	4·5	N20 44·4	6·3	60·5
	01	193 51·2	13·8	234 16·6	4·5	20 38·1	6·6	60·5
	02	208 51·3	13·0	248 40·1	4·5	20 31·5	6·6	60·6
	03	223 51·4	· · 12·2	263 03·6	4·5	20 24·9	6·8	60·6
	04	238 51·6	11·4	277 27·1	4·6	20 18·1	7·0	60·6
	05	253 51·7	10·7	291 50·7	4·6	20 11·1	7·1	60·6
	06	268 51·8	N14 09·9	306 14·3	4·6	N20 04·0	7·3	60·7
	07	283 51·9	09·1	320 37·9	4·7	19 56·7	7·4	60·7
T	08	298 52·0	08·3	335 01·6	4·6	19 49·3	7·6	60·7
H	09	313 52·2	· · 07·5	349 25·2	4·7	19 41·7	7·7	60·7
U	10	328 52·3	06·8	3 48·9	4·8	19 34·0	7·8	60·8
R	11	343 52·4	06·0	18 12·7	4·7	19 26·2	8·0	60·8
S	12	358 52·5	N14 05·2	32 36·4	4·8	N19 18·2	8·1	60·8
D	13	13 52·6	04·4	47 00·2	4·8	19 10·1	8·3	60·8
A	14	28 52·8	03·6	61 24·0	4·9	19 01·8	8·4	60·9
Y	15	43 52·9	· · 02·9	75 47·9	4·9	18 53·4	8·6	60·9
	16	58 53·0	02·1	90 11·8	4·9	18 44·8	8·6	60·9
	17	73 53·1	01·3	104 35·7	5·0	18 36·2	8·9	60·9
	18	88 53·2	N14 00·5	118 59·7	5·0	N18 27·3	8·9	60·9
	19	103 53·4	13 59·7	133 23·7	5·0	18 18·4	9·1	61·0
	20	118 53·5	59·0	147 47·7	5·1	18 09·3	9·2	61·0
	21	133 53·6	· · 58·2	162 11·8	5·2	18 00·1	9·4	61·0
	22	148 53·7	57·4	176 36·0	5·1	17 50·7	9·5	61·0
	23	163 53·9	56·6	191 00·1	5·3	17 41·2	9·6	61·0
16	00	178 54·0	N13 55·8	205 24·4	5·2	N17 31·6	9·7	61·0
	01	193 54·1	55·0	219 48·6	5·3	17 21·9	9·9	61·1
	02	208 54·2	54·3	234 12·9	5·4	17 12·0	10·0	61·1
	03	223 54·3	· · 53·5	248 37·3	5·4	17 02·0	10·1	61·1
	04	238 54·5	52·7	263 01·7	5·4	16 51·9	10·3	61·1
	05	253 54·6	51·9	277 26·1	5·5	16 41·6	10·3	61·1
	06	268 54·7	N13 51·1	291 50·6	5·6	N16 31·3	10·5	61·1
	07	283 54·8	50·3	306 15·2	5·6	16 20·8	10·6	61·2
	08	298 55·0	49·5	320 39·8	5·6	16 10·2	10·7	61·2
F	09	313 55·1	· · 48·7	335 04·4	5·7	15 59·5	10·9	61·2
R	10	328 55·2	48·0	349 29·1	5·8	15 48·6	10·9	61·2
I	11	343 55·3	47·2	3 53·9	5·8	15 37·7	11·1	61·2
D	12	358 55·5	N13 46·4	18 18·7	5·8	N15 26·6	11·1	61·2
A	13	13 55·6	45·6	32 43·5	5·9	15 15·5	11·3	61·2
Y	14	28 55·7	44·8	47 08·4	6·0	15 04·2	11·4	61·2
	15	43 55·9	· · 44·0	61 33·4	6·0	14 52·8	11·4	61·2
	16	58 56·0	43·2	75 58·4	6·0	14 41·4	11·6	61·3
	17	73 56·1	42·4	90 23·4	6·2	14 29·8	11·7	61·3
	18	88 56·2	N13 41·6	104 48·6	6·1	N14 18·1	11·8	61·3
	19	103 56·4	40·8	119 13·7	6·3	14 06·3	11·9	61·3
	20	118 56·5	40·1	133 39·0	6·2	13 54·4	12·0	61·3
	21	133 56·6	· · 39·3	148 04·2	6·4	13 42·4	12·1	61·3
	22	148 56·7	38·5	162 29·6	6·4	13 30·3	12·1	61·3
	23	163 56·9	37·7	176 55·0	6·4	13 18·2	12·3	61·3
		S.D. 15·8	*d* 0·8	S.D. 16·4		16·6		16·7

Lat. °	Twilight Naut. h m	Twilight Civil h m	Sunrise h m	Moonrise 14 h m	Moonrise 15 h m	Moonrise 16 h m	Moonrise 17 h m
N 72	////	////	02 21	▭	23 14	26 30	02 30
N 70	////	////	02 54	▭	▭	00 24	02 54
68	////	01 39	03 17	22 40	25 01	01 01	03 12
66	////	02 18	03 36	23 28	25 27	01 27	03 26
64	////	02 45	03 50	23 59	25 47	01 47	03 38
62	01 28	03 05	04 02	24 22	00 22	02 03	03 48
60	02 03	03 21	04 13	24 41	00 41	02 16	03 57
N 58	02 28	03 35	04 22	24 56	00 56	02 28	04 04
56	02 47	03 46	04 30	25 09	01 09	02 38	04 11
54	03 02	03 56	04 37	00 04	01 21	02 46	04 16
52	03 15	04 05	04 43	00 16	01 31	02 54	04 22
50	03 27	04 13	04 49	00 26	01 40	03 01	04 26
45	03 49	04 29	05 01	00 47	01 58	03 16	04 37
N 40	04 07	04 42	05 11	01 04	02 14	03 28	04 45
35	04 20	04 53	05 19	01 19	02 27	03 39	04 52
30	04 32	05 02	05 27	01 31	02 38	03 48	04 59
20	04 50	05 17	05 40	01 53	02 57	04 03	05 09
N 10	05 04	05 29	05 51	02 11	03 14	04 17	05 19
0	05 15	05 40	06 01	02 29	03 30	04 30	05 28
S 10	05 25	05 50	06 11	02 46	03 45	04 42	05 37
20	05 33	05 59	06 22	03 05	04 02	04 56	05 46
30	05 41	06 09	06 34	03 26	04 21	05 11	05 57
35	05 45	06 15	06 41	03 39	04 32	05 20	06 03
40	05 49	06 21	06 49	03 53	04 45	05 30	06 10
45	05 53	06 27	06 58	04 10	05 00	05 42	06 18
S 50	05 57	06 35	07 09	04 31	05 18	05 57	06 28
52	05 59	06 39	07 14	04 41	05 27	06 03	06 32
54	06 01	06 42	07 20	04 52	05 37	06 11	06 37
56	06 03	06 47	07 26	05 05	05 47	06 19	06 43
58	06 05	06 51	07 33	05 20	06 00	06 28	06 49
S 60	06 07	06 56	07 41	05 38	06 14	06 39	06 56

Lat. °	Sunset h m	Twilight Civil h m	Twilight Naut. h m	Moonset 14 h m	Moonset 15 h m	Moonset 16 h m	Moonset 17 h m
N 72	21 43	////	////	▭	21 32	20 20	19 50
N 70	21 12	23 51	////	▭	20 21	19 54	19 38
68	20 49	22 23	////	19 57	19 43	19 34	19 27
66	20 31	21 47	////	19 08	19 16	19 18	19 19
64	20 17	21 21	23 52	18 36	18 55	19 05	19 12
62	20 05	21 02	22 35	18 13	18 38	18 54	19 05
60	19 55	20 46	22 02	17 54	18 24	18 44	19 00
N 58	19 46	20 33	21 38	17 38	18 12	18 36	18 55
56	19 38	20 21	21 20	17 25	18 01	18 29	18 50
54	19 31	20 11	21 05	17 13	17 52	18 22	18 46
52	19 25	20 03	20 52	17 03	17 43	18 16	18 43
50	19 19	19 55	20 41	16 53	17 36	18 11	18 40
45	19 07	19 39	20 18	16 34	17 20	17 59	18 32
N 40	18 57	19 26	20 02	16 18	17 06	17 49	18 26
35	18 49	19 16	19 48	16 04	16 55	17 40	18 21
30	18 42	19 07	19 37	15 52	16 45	17 33	18 17
20	18 29	18 52	19 19	15 32	16 28	17 20	18 08
N 10	18 18	18 40	19 05	15 14	16 13	17 08	18 01
0	18 08	18 29	18 54	14 58	15 58	16 58	17 55
S 10	17 58	18 19	18 44	14 41	15 44	16 47	17 48
20	17 47	18 10	18 36	14 23	15 29	16 35	17 40
30	17 35	18 00	18 28	14 02	15 11	16 21	17 32
35	17 28	17 54	18 24	13 50	15 00	16 14	17 27
40	17 21	17 49	18 21	13 36	14 49	16 05	17 21
45	17 12	17 42	18 17	13 19	14 34	15 54	17 15
S 50	17 01	17 34	18 13	12 59	14 17	15 41	17 07
52	16 56	17 31	18 11	12 49	14 09	15 35	17 03
54	16 50	17 27	18 09	12 38	14 00	15 28	16 59
56	16 44	17 23	18 08	12 25	13 49	15 21	16 55
58	16 37	17 19	18 06	12 11	13 37	15 12	16 50
S 60	16 29	17 14	18 04	11 53	13 24	15 03	16 44

Day	SUN Eqn. of Time 00^h m s	SUN Eqn. of Time 12^h m s	SUN Mer. Pass. h m	MOON Mer. Pass. Upper h m	MOON Mer. Pass. Lower h m	MOON Age d	MOON Phase
14	04 47	04 42	12 05	08 43	21 14	26	
15	04 36	04 30	12 05	09 44	22 14	27	
16	04 24	04 18	12 04	10 44	23 13	28	

Table 23

Lat. = Declination − ZD
ZD = 90° − 79°03′
ZD = 10°57′
Lat. = 13°57′ − 10°57′
Lat. = 3° N

Declination <u>*contrary*</u> *name to your latitude.*

You are near Darwin; it is about 0300 GMT; your maximum Ho is 64°17′.

Lat. = ZD − Declination
ZD = 90° − 64°17′
ZD = 25°43′
Lat. = 25°43′ − 14°12′
Lat. = 11°31′ S

In preparing to take the noon sight, you have to determine the approximate GMT of noon. To do this, determine from the almanac the time when the GHA of the sun will be equal to your approximate longitude. In east longitude it occurs when 360° minus GHA is equal to your longitude. The sun moves westward 1° every 4 minutes, so get on deck 15 minutes before you estimate that it will be noon. The sight takes time, so make yourself as comfortable as possible. Sit on the cabin top if you can, and take sights at intervals until you notice that the altitude is not changing much between sights. Carefully rock the sextant and bring the lower edge of the sun tangent to the horizon. Continue to do this, and, as the sun creeps upward, you will notice an appreciable interval of time during which you will not be able to discern any change in altitude. Continue to observe until the sun's disk dips below the horizon. This means that the sun has reached its maximum altitude (noon by definition) and has started descending. Your sextant reading is the maximum altitude and is corrected in the usual way and then used to determine latitude as shown in the three examples.

One of the disadvantages of the noon sight is that you have to sit out in the hot sun for 20 minutes or so. If you are using an EBBCO sextant, the heat will expand the plastic a little and introduce an index error. So when you finish, check your sextant for error. Actually, it is a good practice to check any sextant after an observation or series of observations.

The advantage of the noon sight for latitude is its accuracy. As the sun approaches its maximum altitude, it appears to "hang" at that altitude for an appreciable time (Figure 10.4). Therefore, you have several moments in which to refine the sight and are not bound by a mere instant of tangency as in a normal sun sight. For this reason, once the sun has reached what seems to be its maximum altitude, you can continue to rock the sextant and refine the sight until you notice that the lower edge of the sun's disk has definitely dipped into the horizon.

The noon sight can also be used to determine longitude. As you can see from Figure 10.5, which is a sketch of a graph of the sun's path for an interval before and after noon, if you were to take many sights around and after noon, you could construct such a graph and from it read maximum altitude and time of maximum altitude. Knowing the time of maximum altitude—that is, knowing when the longitude of the sun equaled yours—you could go to the almanac and figure out your longitude. Constructing such a graph, however, is rather tedious, and there is another way to accomplish the same thing. Since you are going to be on deck 15 minutes or so before noon, take a sight in the normal manner and note the time and altitude. Then, after the sun has passed the peak, reset your sextant to the prenoon observation and note the time that the sun reaches that altitude again. Noon must have been halfway between the prenoon and postnoon times, so add the times together, divide by 2, and go to the almanac to determine the longitude. Here is an example:

Prenoon Hs = 45°41′ at 13-31-20 GMT
Postnoon Hs = 45°41′ at 13-51-16 GMT

```
  13-31-20
 +13-51-16
 ---------
  26-82-36
```

26-82-36 ÷ 2 = 13-41-18 = Time of Noon

From the almanac for August 15, 1974 (Table 23) and from Table 24:

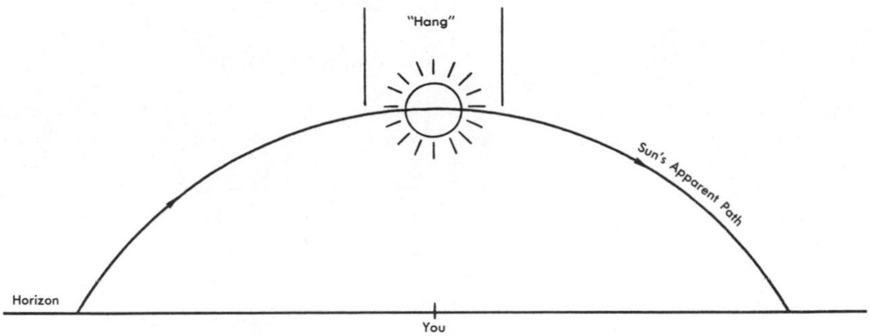

Fig. 10.4

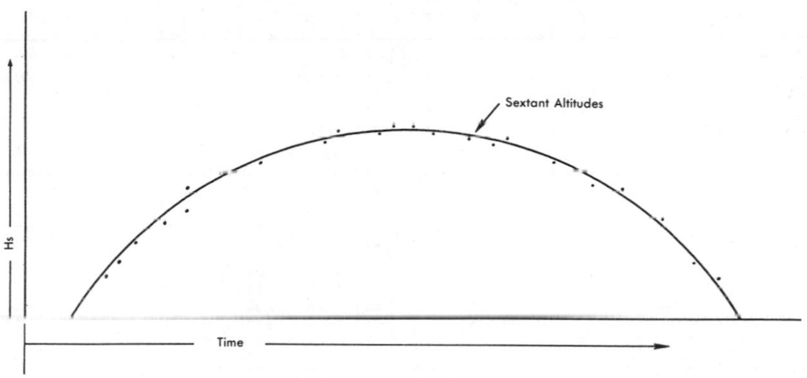

Fig. 10.5

GHA 13 Hr.	=	13°53′
41 Min. 18 Sec.	=	10°20′
GHA	=	23°73′ = 24°13′
Lat.	=	24°13′W

If you want greater accuracy, take more than one prenoon observation and match each one with a postnoon equivalent. Then solve each pair and average the resulting longitudes.

Like any special case, the noon sight has its shortcomings. It takes

40ᵐ INCREMENTS AND CORRECTIONS 41ᵐ

40ᵐ	SUN PLANETS	ARIES	MOON	v or d	Corrⁿ	v or d	Corrⁿ	v or d	Corrⁿ
s	° ′	° ′	° ′	′	′	′	′	′	′
00	10 00·0	10 01·6	9 32·7	0·0	0·0	6·0	4·1	12·0	8·1
01	10 00·3	10 01·9	9 32·9	0·1	0·1	6·1	4·1	12·1	8·2
02	10 00·5	10 02·1	9 33·1	0·2	0·1	6·2	4·2	12·2	8·2
03	10 00·8	10 02·4	9 33·4	0·3	0·2	6·3	4·3	12·3	8·3
04	10 01·0	10 02·6	9 33·6	0·4	0·3	6·4	4·3	12·4	8·4
05	10 01·3	10 02·9	9 33·9	0·5	0·3	6·5	4·4	12·5	8·4
06	10 01·5	10 03·1	9 34·1	0·6	0·4	6·6	4·5	12·6	8·5
07	10 01·8	10 03·4	9 34·3	0·7	0·5	6·7	4·5	12·7	8·6
08	10 02·0	10 03·6	9 34·6	0·8	0·5	6·8	4·6	12·8	8·6
09	10 02·3	10 03·9	9 34·8	0·9	0·6	6·9	4·7	12·9	8·7
10	10 02·5	10 04·1	9 35·1	1·0	0·7	7·0	4·7	13·0	8·8
11	10 02·8	10 04·4	9 35·3	1·1	0·7	7·1	4·8	13·1	8·8
12	10 03·0	10 04·7	9 35·5	1·2	0·8	7·2	4·9	13·2	8·9
13	10 03·3	10 04·9	9 35·8	1·3	0·9	7·3	4·9	13·3	9·0
14	10 03·5	10 05·2	9 36·0	1·4	0·9	7·4	5·0	13·4	9·0
15	10 03·8	10 05·4	9 36·2	1·5	1·0	7·5	5·1	13·5	9·1
16	10 04·0	10 05·7	9 36·5	1·6	1·1	7·6	5·1	13·6	9·2
17	10 04·3	10 05·9	9 36·7	1·7	1·1	7·7	5·2	13·7	9·2
18	10 04·5	10 06·2	9 37·0	1·8	1·2	7·8	5·3	13·8	9·3
19	10 04·8	10 06·4	9 37·2	1·9	1·3	7·9	5·3	13·9	9·4
20	10 05·0	10 06·7	9 37·4	2·0	1·4	8·0	5·4	14·0	9·5
21	10 05·3	10 06·9	9 37·7	2·1	1·4	8·1	5·5	14·1	9·5
22	10 05·5	10 07·2	9 37·9	2·2	1·5	8·2	5·5	14·2	9·6
23	10 05·8	10 07·4	9 38·2	2·3	1·6	8·3	5·6	14·3	9·7
24	10 06·0	10 07·7	9 38·4	2·4	1·6	8·4	5·7	14·4	9·7
25	10 06·3	10 07·9	9 38·6	2·5	1·7	8·5	5·7	14·5	9·8
26	10 06·5	10 08·2	9 38·9	2·6	1·8	8·6	5·8	14·6	9·9
27	10 06·8	10 08·4	9 39·1	2·7	1·8	8·7	5·9	14·7	9·9
28	10 07·0	10 08·7	9 39·3	2·8	1·9	8·8	5·9	14·8	10·0
29	10 07·3	10 08·9	9 39·6	2·9	2·0	8·9	6·0	14·9	10·1
30	10 07·5	10 09·2	9 39·8	3·0	2·0	9·0	6·1	15·0	10·1
31	10 07·8	10 09·4	9 40·1	3·1	2·1	9·1	6·1	15·1	10·2
32	10 08·0	10 09·7	9 40·3	3·2	2·2	9·2	6·2	15·2	10·3
33	10 08·3	10 09·9	9 40·5	3·3	2·2	9·3	6·3	15·3	10·3
34	10 08·5	10 10·2	9 40·8	3·4	2·3	9·4	6·3	15·4	10·4
35	10 08·8	10 10·4	9 41·0	3·5	2·4	9·5	6·4	15·5	10·5
36	10 09·0	10 10·7	9 41·3	3·6	2·4	9·6	6·5	15·6	10·5
37	10 09·3	10 10·9	9 41·5	3·7	2·5	9·7	6·5	15·7	10·6
38	10 09·5	10 11·2	9 41·7	3·8	2·6	9·8	6·6	15·8	10·7
39	10 09·8	10 11·4	9 42·0	3·9	2·6	9·9	6·7	15·9	10·7
40	10 10·0	10 11·7	9 42·2	4·0	2·7	10·0	6·8	16·0	10·8
41	10 10·3	10 11·9	9 42·4	4·1	2·8	10·1	6·8	16·1	10·9
42	10 10·5	10 12·2	9 42·7	4·2	2·8	10·2	6·9	16·2	10·9
43	10 10·8	10 12·4	9 42·9	4·3	2·9	10·3	7·0	16·3	11·0
44	10 11·0	10 12·7	9 43·2	4·4	3·0	10·4	7·0	16·4	11·1
45	10 11·3	10 12·9	9 43·4	4·5	3·0	10·5	7·1	16·5	11·1
46	10 11·5	10 13·2	9 43·6	4·6	3·1	10·6	7·2	16·6	11·2
47	10 11·8	10 13·4	9 43·9	4·7	3·2	10·7	7·2	16·7	11·3
48	10 12·0	10 13·7	9 44·1	4·8	3·2	10·8	7·3	16·8	11·3
49	10 12·3	10 13·9	9 44·4	4·9	3·3	10·9	7·4	16·9	11·4
50	10 12·5	10 14·2	9 44·6	5·0	3·4	11·0	7·4	17·0	11·5
51	10 12·8	10 14·4	9 44·8	5·1	3·4	11·1	7·5	17·1	11·5
52	10 13·0	10 14·7	9 45·1	5·2	3·5	11·2	7·6	17·2	11·6
53	10 13·3	10 14·9	9 45·3	5·3	3·6	11·3	7·6	17·3	11·7
54	10 13·5	10 15·2	9 45·6	5·4	3·6	11·4	7·7	17·4	11·7
55	10 13·8	10 15·4	9 45·8	5·5	3·7	11·5	7·8	17·5	11·8
56	10 14·0	10 15·7	9 46·0	5·6	3·8	11·6	7·8	17·6	11·9
57	10 14·3	10 15·9	9 46·3	5·7	3·8	11·7	7·9	17·7	11·9
58	10 14·5	10 16·2	9 46·5	5·8	3·9	11·8	8·0	17·8	12·0
59	10 14·8	10 16·4	9 46·7	5·9	4·0	11·9	8·0	17·9	12·1
60	10 15·0	10 16·7	9 47·0	6·0	4·1	12·0	8·1	18·0	12·2

41ᵐ	SUN PLANETS	ARIES	MOON	v or d	Corrⁿ	v or d	Corrⁿ	v or d	Corrⁿ
s	° ′	° ′	° ′	′	′	′	′	′	′
00	10 15·0	10 16·7	9 47·0	0·0	0·0	6·0	4·2	12·0	8·3
01	10 15·3	10 16·9	9 47·2	0·1	0·1	6·1	4·2	12·1	8·4
02	10 15·5	10 17·2	9 47·5	0·2	0·1	6·2	4·3	12·2	8·4
03	10 15·8	10 17·4	9 47·7	0·3	0·2	6·3	4·4	12·3	8·5
04	10 16·0	10 17·7	9 47·9	0·4	0·3	6·4	4·4	12·4	8·6
05	10 16·3	10 17·9	9 48·2	0·5	0·3	6·5	4·5	12·5	8·6
06	10 16·5	10 18·2	9 48·4	0·6	0·4	6·6	4·6	12·6	8·7
07	10 16·8	10 18·4	9 48·7	0·7	0·5	6·7	4·6	12·7	8·8
08	10 17·0	10 18·7	9 48·9	0·8	0·6	6·8	4·7	12·8	8·9
09	10 17·3	10 18·9	9 49·1	0·9	0·6	6·9	4·8	12·9	8·9
10	10 17·5	10 19·2	9 49·4	1·0	0·7	7·0	4·8	13·0	9·0
11	10 17·8	10 19·4	9 49·6	1·1	0·8	7·1	4·9	13·1	9·1
12	10 18·0	10 19·7	9 49·8	1·2	0·8	7·2	5·0	13·2	9·1
13	10 18·3	10 19·9	9 50·1	1·3	0·9	7·3	5·0	13·3	9·2
14	10 18·5	10 20·2	9 50·3	1·4	1·0	7·4	5·1	13·4	9·3
15	10 18·8	10 20·4	9 50·6	1·5	1·0	7·5	5·2	13·5	9·3
16	10 19·0	10 20·7	9 50·8	1·6	1·1	7·6	5·3	13·6	9·4
17	10 19·3	10 20·9	9 51·0	1·7	1·2	7·7	5·3	13·7	9·5
18	10 19·5	10 21·2	9 51·3	1·8	1·2	7·8	5·4	13·8	9·5
19	10 19·8	10 21·4	9 51·5	1·9	1·3	7·9	5·5	13·9	9·6
20	10 20·0	10 21·7	9 51·8	2·0	1·4	8·0	5·5	14·0	9·7
21	10 20·3	10 21·9	9 52·0	2·1	1·5	8·1	5·6	14·1	9·8
22	10 20·5	10 22·2	9 52·2	2·2	1·5	8·2	5·7	14·2	9·8
23	10 20·8	10 22·4	9 52·5	2·3	1·6	8·3	5·7	14·3	9·9
24	10 21·0	10 22·7	9 52·7	2·4	1·7	8·4	5·8	14·4	10·0
25	10 21·3	10 23·0	9 52·9	2·5	1·7	8·5	5·9	14·5	10·0
26	10 21·5	10 23·2	9 53·2	2·6	1·8	8·6	5·9	14·6	10·1
27	10 21·8	10 23·5	9 53·4	2·7	1·9	8·7	6·0	14·7	10·2
28	10 22·0	10 23·7	9 53·7	2·8	1·9	8·8	6·1	14·8	10·2
29	10 22·3	10 24·0	9 53·9	2·9	2·0	8·9	6·2	14·9	10·3
30	10 22·5	10 24·2	9 54·1	3·0	2·1	9·0	6·2	15·0	10·4
31	10 22·8	10 24·5	9 54·4	3·1	2·1	9·1	6·3	15·1	10·4
32	10 23·0	10 24·7	9 54·6	3·2	2·2	9·2	6·4	15·2	10·5
33	10 23·3	10 25·0	9 54·9	3·3	2·3	9·3	6·4	15·3	10·6
34	10 23·5	10 25·2	9 55·1	3·4	2·4	9·4	6·5	15·4	10·7
35	10 23·8	10 25·5	9 55·3	3·5	2·4	9·5	6·6	15·5	10·7
36	10 24·0	10 25·7	9 55·6	3·6	2·5	9·6	6·6	15·6	10·8
37	10 24·3	10 26·0	9 55·8	3·7	2·6	9·7	6·7	15·7	10·9
38	10 24·5	10 26·2	9 56·1	3·8	2·6	9·8	6·8	15·8	10·9
39	10 24·8	10 26·5	9 56·3	3·9	2·7	9·9	6·8	15·9	11·0
40	10 25·0	10 26·7	9 56·5	4·0	2·8	10·0	6·9	16·0	11·1
41	10 25·3	10 27·0	9 56·8	4·1	2·8	10·1	7·0	16·1	11·1
42	10 25·5	10 27·2	9 57·0	4·2	2·9	10·2	7·1	16·2	11·2
43	10 25·8	10 27·5	9 57·2	4·3	3·0	10·3	7·1	16·3	11·3
44	10 26·0	10 27·7	9 57·5	4·4	3·0	10·4	7·2	16·4	11·3
45	10 26·3	10 28·0	9 57·7	4·5	3·1	10·5	7·3	16·5	11·4
46	10 26·5	10 28·2	9 58·0	4·6	3·2	10·6	7·3	16·6	11·5
47	10 26·8	10 28·5	9 58·2	4·7	3·3	10·7	7·4	16·7	11·6
48	10 27·0	10 28·7	9 58·4	4·8	3·3	10·8	7·5	16·8	11·6
49	10 27·3	10 29·0	9 58·7	4·9	3·4	10·9	7·5	16·9	11·7
50	10 27·5	10 29·2	9 58·9	5·0	3·5	11·0	7·6	17·0	11·8
51	10 27·8	10 29·5	9 59·2	5·1	3·5	11·1	7·7	17·1	11·8
52	10 28·0	10 29·7	9 59·4	5·2	3·6	11·2	7·7	17·2	11·9
53	10 28·3	10 30·0	9 59·6	5·3	3·7	11·3	7·8	17·3	12·0
54	10 28·5	10 30·2	9 59·9	5·4	3·7	11·4	7·9	17·4	12·0
55	10 28·8	10 30·5	10 00·1	5·5	3·8	11·5	8·0	17·5	12·1
56	10 29·0	10 30·7	10 00·3	5·6	3·9	11·6	8·0	17·6	12·2
57	10 29·3	10 31·0	10 00·6	5·7	3·9	11·7	8·1	17·7	12·2
58	10 29·5	10 31·2	10 00·8	5·8	4·0	11·8	8·2	17·8	12·3
59	10 29·8	10 31·5	10 01·1	5·9	4·1	11·9	8·2	17·9	12·4
60	10 30·0	10 31·7	10 01·3	6·0	4·2	12·0	8·3	18·0	12·5

Table 24

much more time than the methods outlined in previous chapters, for example. Still, the arithmetic involved is simple, and the sight requires only a sextant, an almanac, and a watch—no sight reduction tables. The time available to refine the sight allows great accuracy in determining latitude. Before the advent of chronometers and reliable time, the noon sight *was* celestial navigation. You sailed to the latitude of your destination and then stayed on that latitude (by means of the noon sight) until you arrived. Since an accurate knowledge of time is not necessary, and because the declination of the sun does not change much in a day, this is a good technique to know. If you know what day it is and have a sextant and an almanac, you can navigate.

11
Planets and Stars

At first glance, the possibility of using planets and stars for celestial navigation seems nearly hopeless. You need a horizon to take sights, yet the planets and stars are visible only at night, when the horizon isn't. Actually, what this means is that you have to take your observations of these bodies near dawn and dusk, when you can just see the planet or star and when the sun is still close enough to the horizon line to make the horizon visible. These conditions limit you to periods of about twenty minutes before dawn and twenty minutes after sunset in which to take sights.

Further, because you are working in dim conditions and because the stars and planets do not produce as much light as the sun, you need a sextant with great light-gathering ability. Such a sextant has large superreflective mirrors and a telescope with a large objective lens.

I was unable to get planet or star sights with my EBBCO, and so I bought a Plath. It is a fine instrument, but I soon discovered that the 4-power scope magnified the motion of the boat to such an extent that the sextant was unusable in anything but a fairly smooth sea. I then got a 2.5-power scope, which lets me get the job done in most weathers. The moral of this, I suppose, is that shooting stars and planets is like shooting big game—you need expensive equipment.

The principal advantage of observing the stars is that since there are so many of them it is possible to shoot more than one simultaneously and therefore obtain a highly accurate fix. Planets can be used in conjunction with stars for multibody fixes and have the additional advantage of being very bright so that they can be seen when the sun is lighting the sky enough to define the horizon quite sharply.

The easiest time to start doing star sights is at dawn rather than dusk. At dawn, you can pinpoint your star and be all set when the horizon becomes sharp enough to use. First, of course, you must determine the time at which the horizon will be sharply enough defined. In the top right-hand quarter of the daily pages in the *Nautical Almanac*, you will see a table that contains the necessary information (see Table

25). The *Air Almanac* has this table in an appendix. The column labeled "Civil Twilight" is the one to use. You can see that for latitude 40° N civil twilight (CT) begins at 0403 and ends when the sun rises at 0436. Suppose you were at latitude 37° N on July 4, 1974, and wanted to know what time your watch would say when civil twilight began. From the table, civil twilight at 35° N is at 0421, and at 40° N it is at 0403. Your latitude is approximately halfway between them. The difference is 18 minutes; therefore, add 9 minutes to 0403. CT at your latitude will be at approximately 0412.

Now consider this to be the time of CT at Greenwich, and since the sun moves through 15° of longitude in 1 hour, your longitude divided by 15° gives you the number of hours to add to or subtract from 0412 to arrive at the GMT for civil twilight at your latitude and longitude. Actually, a table that does this for you is reproduced as Table 26. Assume that your DR indicates that you are at approximately 67°30′ west longitude. From the table, 67° is the amount of longitude covered by the sun in 4 hours and 28 minutes. You can figure out that 30′ is the amount of longitude covered in 2 minutes. Since you are in west longitude, you *add* 4 hours and 30 minutes (4 hours and 28 minutes plus 2 minutes) to 0412 to come up with a GMT of 0842 for the beginning of civil twilight at your location. Had you been at 67°30′ east longitude, you would have *subtracted* and the GMT would have been 0412 − 0430 or 2342 hours on July 3, 1974.

Having determined the approximate time of CT, plan to get on deck about 10 minutes early in order to get settled, find your stars, and wait for the proper time to begin shooting. As you can see, like the noon sight, navigating by the planets and stars requires more sitting around than navigating by sun lines.

As you know, the relative positions of the stars do not change much. In the course of a year, for example, the declination of Vega changes about 0.1′, and it moves eastward about 0.5′. For practical day-to-day purposes, you can imagine the stars as imbedded in a crystal sphere that rotates around the earth. There is no need, therefore, to have an hourly tabulation of their declinations and, since they are fixed relative to one another, it is not necessary to calculate the GHA for each star. If you had the calculated GHA for one star, you could figure out the

1974 JULY 3, 4, 5 (WED., THURS., FRI.)

G.M.T. d	h	SUN G.H.A. ° ′	SUN Dec. ° ′	MOON G.H.A. ° ′	v ′	MOON Dec. ° ′	d ′	H.P. ′
3	00	178 59·7	N23 00·6	17 21·6	10·6	S22 53·7	1·4	54·7
	01	193 59·6	00·4	31 51·2	10·5	22 52·3	1·6	54·7
	02	208 59·5	00·2	46 20·7	10·6	22 50·7	1·6	54·7
	03	223 59·4	23 00·1	60 50·3	10·7	22 49·1	1·8	54·6
	04	238 59·3	22 59·9	75 20·0	10·6	22 47·3	1·8	54·6
	05	253 59·2	59·7	89 49·6	10·7	22 45·5	2·0	54·6
	06	268 59·0	N22 59·5	104 19·3	10·7	S22 43·5	2·1	54·6
W	07	283 58·9	59·3	118 49·0	10·8	22 41·4	2·2	54·6
E	08	298 58·8	59·1	133 18·8	10·8	22 39·2	2·3	54·6
D	09	313 58·7	· · 58·9	147 48·6	10·8	22 36·9	2·4	54·6
N	10	328 58·6	58·7	162 18·4	10·8	22 34·5	2·5	54·6
E	11	343 58·5	58·5	176 48·2	10·9	22 32·0	2·6	54·5
S	12	358 58·3	N22 58·3	191 18·1	10·9	S22 29·4	2·8	54·5
D	13	13 58·2	58·1	205 48·0	11·0	22 26·6	2·8	54·5
A	14	28 58·1	57·9	220 18·0	11·0	22 23·8	3·0	54·5
Y	15	43 58·0	· · 57·7	234 48·0	11·0	22 20·8	3·0	54·5
	16	58 57·9	57·5	249 18·0	11·1	22 17·8	3·2	54·5
	17	73 57·8	57·3	263 48·1	11·1	22 14·6	3·3	54·5
	18	88 57·7	N22 57·1	278 18·2	11·1	S22 11·3	3·3	54·5
	19	103 57·6	56·9	292 48·3	11·2	22 08·0	3·5	54·4
	20	118 57·4	56·7	307 18·5	11·2	22 04·5	3·6	54·4
	21	133 57·3	· · 56·5	321 48·7	11·3	22 00·9	3·7	54·4
	22	148 57·2	56·3	336 19·0	11·3	21 57·2	3·8	54·4
	23	163 57·1	56·0	350 49·3	11·3	21 53·4	3·8	54·4
4	00	178 57·0	N22 55·8	5 19·6	11·4	S21 49·6	4·0	54·4
	01	193 56·9	55·6	19 50·0	11·4	21 45·6	4·1	54·4
	02	208 56·8	55·4	34 20·4	11·5	21 41·5	4·2	54·4
	03	223 56·6	· · 55·2	48 50·9	11·5	21 37·3	4·3	54·4
	04	238 56·5	55·0	63 21·4	11·5	21 33·0	4·4	54·4
	05	253 56·4	54·8	77 51·9	11·6	21 28·6	4·5	54·3
	06	268 56·3	N22 54·6	92 22·5	11·7	S21 24·1	4·6	54·3
T	07	283 56·2	54·4	106 53·2	11·6	21 19·5	4·7	54·3
H	08	298 56·1	54·2	121 23·8	11·8	21 14·8	4·8	54·3
U	09	313 56·0	· · 53·9	135 54·6	11·7	21 10·0	4·9	54·3
R	10	328 55·9	53·7	150 25·3	11·9	21 05·1	4·9	54·3
S	11	343 55·8	53·5	164 56·2	11·8	21 00·2	5·1	54·3
D	12	358 55·6	N22 53·3	179 27·0	11·9	S20 55·1	5·2	54·3
A	13	13 55·5	53·1	193 57·9	12·0	20 49·9	5·2	54·3
Y	14	28 55·4	52·9	208 28·9	12·0	20 44·7	5·4	54·3
	15	43 55·3	· · 52·6	222 59·9	12·1	20 39·3	5·4	54·2
	16	58 55·2	52·4	237 31·0	12·1	20 33·9	5·6	54·2
	17	73 55·1	52·2	252 02·1	12·1	20 28·3	5·6	54·2
	18	88 55·0	N22 52·0	266 33·2	12·2	S20 22·7	5·7	54·2
	19	103 54·9	51·8	281 04·4	12·3	20 17·0	5·9	54·2
	20	118 54·8	51·5	295 35·7	12·3	20 11·1	5·9	54·2
	21	133 54·6	· · 51·3	310 07·0	12·3	20 05·2	6·0	54·2
	22	148 54·5	51·1	324 38·3	12·4	19 59·2	6·0	54·2
	23	163 54·4	50·9	339 09·7	12·4	19 53·2	6·2	54·2
5	00	178 54·3	N22 50·7	353 41·1	12·5	S19 47·0	6·3	54·2
	01	193 54·2	50·4	8 12·6	12·6	19 40·7	6·3	54·2
	02	208 54·1	50·2	22 44·2	12·6	19 34·4	6·4	54·2
	03	223 54·0	· · 50·0	37 15·8	12·6	19 28·0	6·5	54·2
	04	238 53·9	49·7	51 47·4	12·7	19 21·5	6·6	54·1
	05	253 53·8	49·5	66 19·1	12·7	19 14·9	6·7	54·1
	06	268 53·7	N22 49·3	80 50·8	12·8	S19 08·2	6·8	54·1
	07	283 53·6	49·1	95 22·6	12·8	19 01·4	6·8	54·1
	08	298 53·4	48·8	109 54·4	12·9	18 54·6	7·0	54·1
F	09	313 53·3	· · 48·6	124 26·3	13·0	18 47·6	7·0	54·1
R	10	328 53·2	48·4	138 58·3	13·0	18 40·6	7·1	54·1
I	11	343 53·1	48·1	153 30·3	13·0	18 33·5	7·1	54·1
D	12	358 53·0	N22 47·9	168 02·3	13·1	S18 26·4	7·3	54·1
A	13	13 52·9	47·7	182 34·4	13·1	18 19·1	7·3	54·1
Y	14	28 52·8	47·4	197 06·5	13·2	18 11·8	7·4	54·1
	15	43 52·7	· · 47·2	211 38·7	13·2	18 04·4	7·5	54·1
	16	58 52·6	47·0	226 10·9	13·3	17 56·9	7·5	54·1
	17	73 52·5	46·7	240 43·2	13·4	17 49·4	7·7	54·1
	18	88 52·4	N22 46·5	255 15·6	13·3	S17 41·7	7·7	54·1
	19	103 52·3	46·3	269 47·9	13·5	17 34·0	7·8	54·1
	20	118 52·1	46·0	284 20·4	13·5	17 26·2	7·8	54·1
	21	133 52·0	· · 45·8	298 52·9	13·5	17 18·4	7·9	54·1
	22	148 51·9	45·5	313 25·4	13·6	17 10·5	8·0	54·0
	23	163 51·8	45·3	327 58·0	13·6	17 02·5	8·1	54·0
		S.D. 15·8	d 0·2	S.D. 14·9		14·8		14·7

Lat. °	Twilight Naut. h m	Twilight Civil h m	Sunrise h m	Moonrise 3 h m	Moonrise 4 h m	Moonrise 5 h m	Moonrise 6 h m
N 72	☐	☐	☐	■	■	23 51	23 04
N 70	☐	☐	☐	■	23 59	23 00	22 38
68	☐	☐	☐	23 23	22 40	22 27	22 19
66	////	////	00 38	21 57	22 02	22 03	22 03
64	////	////	01 46	21 19	21 35	21 45	21 50
62	////	////	02 20	20 52	21 15	21 29	21 39
60	////	01 09	02 45	20 31	20 58	21 16	21 30
N 58	////	01 52	03 04	20 14	20 43	21 05	21 21
56	////	02 20	03 20	19 59	20 31	20 55	21 14
54	01 04	02 41	03 34	19 47	20 20	20 47	21 08
52	01 43	02 58	03 46	19 36	20 11	20 39	21 02
50	02 09	03 13	03 57	19 26	20 02	20 32	20 56
45	02 52	03 41	04 18	19 06	19 44	20 17	20 45
N 40	03 22	04 03	04 36	18 49	19 29	20 04	20 35
35	03 45	04 21	04 51	18 35	19 17	19 54	20 27
30	04 03	04 36	05 03	18 22	19 06	19 44	20 20
20	04 31	05 00	05 25	18 01	18 47	19 28	20 07
N 10	04 53	05 20	05 43	17 43	18 30	19 14	19 56
0	05 12	05 38	06 01	17 26	18 15	19 01	19 46
S 10	05 29	05 55	06 18	17 09	17 59	18 48	19 35
20	05 44	06 12	06 36	16 51	17 43	18 34	19 24
30	06 00	06 30	06 56	16 30	17 23	18 18	19 11
35	06 09	06 40	07 08	16 17	17 12	18 08	19 04
40	06 18	06 52	07 22	16 03	16 59	17 57	18 56
45	06 28	07 05	07 39	15 46	16 44	17 44	18 46
S 50	06 39	07 21	07 59	15 25	16 25	17 29	18 34
52	06 44	07 28	08 08	15 15	16 17	17 22	18 28
54	06 50	07 36	08 19	15 04	16 07	17 13	18 22
56	06 56	07 44	08 31	14 51	15 55	17 04	18 15
58	07 02	07 54	08 45	14 36	15 42	16 54	18 07
S 60	07 09	08 06	09 02	14 19	15 27	16 42	17 58

Lat. °	Sunset h m	Twilight Civil h m	Twilight Naut. h m	Moonset 3 h m	Moonset 4 h m	Moonset 5 h m	Moonset 6 h m
N 72	☐	☐	☐	■	■	■	02 48
N 70	☐	☐	☐	■	■	01 04	03 39
68	☐	☐	☐	23 59	26 22	02 22	04 10
66	23 25	////	////	00 01	01 25	02 59	04 33
64	22 21	////	////	00 51	02 03	03 25	04 51
62	21 47	////	////	01 22	02 29	03 46	05 06
60	21 23	22 57	////	01 45	02 50	04 02	05 18
N 58	21 03	22 15	////	02 04	03 06	04 16	05 29
56	20 47	21 48	////	02 20	03 21	04 28	05 38
54	20 34	21 27	23 02	02 33	03 33	04 38	05 46
52	20 22	21 10	22 24	02 45	03 44	04 47	05 53
50	20 11	20 55	21 59	02 55	03 53	04 56	06 00
45	19 50	20 27	21 16	03 17	04 13	05 13	06 14
N 40	19 32	20 05	20 46	03 35	04 30	05 27	06 25
35	19 18	19 47	20 23	03 49	04 43	05 39	06 35
30	19 05	19 32	20 05	04 02	04 55	05 50	06 44
20	18 43	19 08	19 37	04 24	05 16	06 07	06 58
N 10	18 25	18 48	19 15	04 43	05 33	06 23	07 11
0	18 08	18 30	18 56	05 00	05 50	06 37	07 23
S 10	17 51	18 14	18 40	05 18	06 06	06 52	07 35
20	17 33	17 57	18 24	05 37	06 24	07 07	07 47
30	17 12	17 38	18 08	05 58	06 44	07 25	08 02
35	17 00	17 28	18 00	06 11	06 55	07 35	08 10
40	16 46	17 17	17 51	06 25	07 09	07 46	08 19
45	16 30	17 04	17 41	06 42	07 24	08 00	08 30
S 50	16 10	16 48	17 29	07 04	07 44	08 16	08 43
52	16 00	16 41	17 24	07 14	07 53	08 24	08 49
54	15 50	16 33	17 19	07 25	08 03	08 33	08 56
56	15 38	16 24	17 13	07 38	08 15	08 42	09 04
58	15 23	16 14	17 06	07 53	08 28	08 53	09 12
S 60	15 07	16 03	16 59	08 11	08 43	09 06	09 22

Day	SUN Eqn. of Time 00h m s	SUN Eqn. of Time 12h m s	SUN Mer. Pass. h m	MOON Mer. Pass. Upper h m	MOON Mer. Pass. Lower h m	MOON Age d	Phase
3	04 01	04 06	12 04	23 38	11 13	13	○
4	04 12	04 17	12 04	24 26	12 02	14	
5	04 23	04 28	12 04	00 26	12 49	15	

Table 25

CONVERSION OF ARC TO TIME

0°–59°		60°–119°		120°–179°		180°–239°		240°–299°		300°–359°			0′·00	0′·25	0′·50	0′·75
°	h m	°	h m	°	h m	°	h m	°	h m	°	h m	′	m s	m s	m s	m s
0	0 00	60	4 00	120	8 00	180	12 00	240	16 00	300	20 00	0	0 00	0 01	0 02	0 03
1	0 04	61	4 04	121	8 04	181	12 04	241	16 04	301	20 04	1	0 04	0 05	0 06	0 07
2	0 08	62	4 08	122	8 08	182	12 08	242	16 08	302	20 08	2	0 08	0 09	0 10	0 11
3	0 12	63	4 12	123	8 12	183	12 12	243	16 12	303	20 12	3	0 12	0 13	0 14	0 15
4	0 16	64	4 16	124	8 16	184	12 16	244	16 16	304	20 16	4	0 16	0 17	0 18	0 19
5	0 20	65	4 20	125	8 20	185	12 20	245	16 20	305	20 20	5	0 20	0 21	0 22	0 23
6	0 24	66	4 24	126	8 24	186	12 24	246	16 24	306	20 24	6	0 24	0 25	0 26	0 27
7	0 28	67	4 28	127	8 28	187	12 28	247	16 28	307	20 28	7	0 28	0 29	0 30	0 31
8	0 32	68	4 32	128	8 32	188	12 32	248	16 32	308	20 32	8	0 32	0 33	0 34	0 35
9	0 36	69	4 36	129	8 36	189	12 36	249	16 36	309	20 36	9	0 36	0 37	0 38	0 39
10	0 40	70	4 40	130	8 40	190	12 40	250	16 40	310	20 40	10	0 40	0 41	0 42	0 43
11	0 44	71	4 44	131	8 44	191	12 44	251	16 44	311	20 44	11	0 44	0 45	0 46	0 47
12	0 48	72	4 48	132	8 48	192	12 48	252	16 48	312	20 48	12	0 48	0 49	0 50	0 51
13	0 52	73	4 52	133	8 52	193	12 52	253	16 52	313	20 52	13	0 52	0 53	0 54	0 55
14	0 56	74	4 56	134	8 56	194	12 56	254	16 56	314	20 56	14	0 56	0 57	0 58	0 59
15	1 00	75	5 00	135	9 00	195	13 00	255	17 00	315	21 00	15	1 00	1 01	1 02	1 03
16	1 04	76	5 04	136	9 04	196	13 04	256	17 04	316	21 04	16	1 04	1 05	1 06	1 07
17	1 08	77	5 08	137	9 08	197	13 08	257	17 08	317	21 08	17	1 08	1 09	1 10	1 11
18	1 12	78	5 12	138	9 12	198	13 12	258	17 12	318	21 12	18	1 12	1 13	1 14	1 15
19	1 16	79	5 16	139	9 16	199	13 16	259	17 16	319	21 16	19	1 16	1 17	1 18	1 19
20	1 20	80	5 20	140	9 20	200	13 20	260	17 20	320	21 20	20	1 20	1 21	1 22	1 23
21	1 24	81	5 24	141	9 24	201	13 24	261	17 24	321	21 24	21	1 24	1 25	1 26	1 27
22	1 28	82	5 28	142	9 28	202	13 28	262	17 28	322	21 28	22	1 28	1 29	1 30	1 31
23	1 32	83	5 32	143	9 32	203	13 32	263	17 32	323	21 32	23	1 32	1 33	1 34	1 35
24	1 36	84	5 36	144	9 36	204	13 36	264	17 36	324	21 36	24	1 36	1 37	1 38	1 39
25	1 40	85	5 40	145	9 40	205	13 40	265	17 40	325	21 40	25	1 40	1 41	1 42	1 43
26	1 44	86	5 44	146	9 44	206	13 44	266	17 44	326	21 44	26	1 44	1 45	1 46	1 47
27	1 48	87	5 48	147	9 48	207	13 48	267	17 48	327	21 48	27	1 48	1 49	1 50	1 51
28	1 52	88	5 52	148	9 52	208	13 52	268	17 52	328	21 52	28	1 52	1 53	1 54	1 55
29	1 56	89	5 56	149	9 56	209	13 56	269	17 56	329	21 56	29	1 56	1 57	1 58	1 59
30	2 00	90	6 00	150	10 00	210	14 00	270	18 00	330	22 00	30	2 00	2 01	2 02	2 03
31	2 04	91	6 04	151	10 04	211	14 04	271	18 04	331	22 04	31	2 04	2 05	2 06	2 07
32	2 08	92	6 08	152	10 08	212	14 08	272	18 08	332	22 08	32	2 08	2 09	2 10	2 11
33	2 12	93	6 12	153	10 12	213	14 12	273	18 12	333	22 12	33	2 12	2 13	2 14	2 15
34	2 16	94	6 16	154	10 16	214	14 16	274	18 16	334	22 16	34	2 16	2 17	2 18	2 19
35	2 20	95	6 20	155	10 20	215	14 20	275	18 20	335	22 20	35	2 20	2 21	2 22	2 23
36	2 24	96	6 24	156	10 24	216	14 24	276	18 24	336	22 24	36	2 24	2 25	2 26	2 27
37	2 28	97	6 28	157	10 28	217	14 28	277	18 28	337	22 28	37	2 28	2 29	2 30	2 31
38	2 32	98	6 32	158	10 32	218	14 32	278	18 32	338	22 32	38	2 32	2 33	2 34	2 35
39	2 36	99	6 36	159	10 36	219	14 36	279	18 36	339	22 36	39	2 36	2 37	2 38	2 39
40	2 40	100	6 40	160	10 40	220	14 40	280	18 40	340	22 40	40	2 40	2 41	2 42	2 43
41	2 44	101	6 44	161	10 44	221	14 44	281	18 44	341	22 44	41	2 44	2 45	2 46	2 47
42	2 48	102	6 48	162	10 48	222	14 48	282	18 48	342	22 48	42	2 48	2 49	2 50	2 51
43	2 52	103	6 52	163	10 52	223	14 52	283	18 52	343	22 52	43	2 52	2 53	2 54	2 55
44	2 56	104	6 56	164	10 56	224	14 56	284	18 56	344	22 56	44	2 56	2 57	2 58	2 59
45	3 00	105	7 00	165	11 00	225	15 00	285	19 00	345	23 00	45	3 00	3 01	3 02	3 03
46	3 04	106	7 04	166	11 04	226	15 04	286	19 04	346	23 04	46	3 04	3 05	3 06	3 07
47	3 08	107	7 08	167	11 08	227	15 08	287	19 08	347	23 08	47	3 08	3 09	3 10	3 11
48	3 12	108	7 12	168	11 12	228	15 12	288	19 12	348	23 12	48	3 12	3 13	3 14	3 15
49	3 16	109	7 16	169	11 16	229	15 16	289	19 16	349	23 16	49	3 16	3 17	3 18	3 19
50	3 20	110	7 20	170	11 20	230	15 20	290	19 20	350	23 20	50	3 20	3 21	3 22	3 23
51	3 24	111	7 24	171	11 24	231	15 24	291	19 24	351	23 24	51	3 24	3 25	3 26	3 27
52	3 28	112	7 28	172	11 28	232	15 28	292	19 28	352	23 28	52	3 28	3 29	3 30	3 31
53	3 32	113	7 32	173	11 32	233	15 32	293	19 32	353	23 32	53	3 32	3 33	3 34	3 35
54	3 36	114	7 36	174	11 36	234	15 36	294	19 36	354	23 36	54	3 36	3 37	3 38	3 39
55	3 40	115	7 40	175	11 40	235	15 40	295	19 40	355	23 40	55	3 40	3 41	3 42	3 43
56	3 44	116	7 44	176	11 44	236	15 44	296	19 44	356	23 44	56	3 44	3 45	3 46	3 47
57	3 48	117	7 48	177	11 48	237	15 48	297	19 48	357	23 48	57	3 48	3 49	3 50	3 51
58	3 52	118	7 52	178	11 52	238	15 52	298	19 52	358	23 52	58	3 52	3 53	3 54	3 55
59	3 56	119	7 56	179	11 56	239	15 56	299	19 56	359	23 56	59	3 56	3 57	3 58	3 59

The above table is for converting expressions in arc to their equivalent in time ; its main use in this Almanac is for the conversion of longitude for application to L.M.T. (*added* if *west*, *subtracted* if *east*) to give G.M.T. or vice versa, particularly in the case of sunrise, sunset, etc.

Table 26

GHA for any given star, knowing their relative positions. As a matter of fact, the almanac does just this—it tabulates the GHA of Aries (see Table 28) and gives a list of the 57 brightest stars. For each, it gives the declination and what is called the *sidereal hour angle* (SHA). SHA is the meridian of the star measured westward from Aries. Therefore, to find the GHA of a star, you look up the GHA of Aries and add the SHA of the star.

As a matter of fact, you don't have to go through this process to navigate by the stars. Volume I of H.O. 249 does the work for you and presents a list of seven stars best suited for use at your assumed position. Table 27 shows the two pages for latitude 35° north.

One of the best features of H.O. 249 is that it does not matter whether or not you know your stars. For example, look at the line for LHA Aries (♈) = 347°. In the column headed "Vega" you will see "Hc" and "Zn." The Zn is the true bearing to the star, and Hc is its approximate altitude. Convert true bearing to compass bearing by applying the variation for your area, and look out in that direction. You'll see the star, and it will be quite unmistakable. Not only are the navigational stars bright; they are also well spread out from one another. Set your sextant to the Hc given, point it toward the star, and it will be in your field of view and quite near the horizon.

Conversely, if you see a bright star that you can't identify, take a sight and a compass bearing, use the time of the sight and your approximate longitude to figure out LHA Aries, and look at the table to find the star with that altitude and bearing.

What you need to do before going on deck to take a star sight is to figure out approximately what LHA Aries is going to be; look at that line in the book and, based on the true bearings given, figure out which stars will be most convenient for your use—that is, which ones won't be blocked by the sails. The three stars marked by the asterisk (*) are the best to use if you can and will give about a 360° spread in azimuth. Try in any case to get at least a 270° spread for best accuracy.

Here's an example. It is July 9, 1974. You figure that your DR position on the morning of July 10 will be about 35° N, 65° W, and you would like to take a round of star sights. From the page in the almanac (Table 28), civil twilight begins at 0425 GMT. Since you are going to

35°N

Hc Zn	Hc Zn	Hc Zn	Hc Zn	Hc Zn	Hc Zn	Hc Zn
*CAPELLA	ALDEBARAN	Diphda	*FOMALHAUT	ALTAIR	*VEGA	Kochab
31 39 053	26 44 088	35 55 168	23 29 195	27 20 261	27 18 300	22 53 348
32 18 053	27 33 089	36 05 169	23 16 196	26 32 262	26 36 300	22 44 349
32 58 054	28 22 089	36 14 170	23 02 197	25 43 263	25 54 301	22 34 349
33 38 054	29 11 090	36 22 171	22 47 198	24 54 263	25 11 301	22 25 349
34 18 054	30 00 090	36 29 172	22 32 199	24 06 264	24 29 301	22 16 349
34 57 054	30 49 091	36 35 173	22 16 200	23 17 264	23 47 302	22 07 350
35 37 055	31 38 091	36 41 175	21 59 201	22 28 265	23 06 302	21 58 350
36 18 055	32 28 092	36 45 176	21 41 201	21 39 266	22 24 303	21 50 350
36 58 055	33 17 093	36 48 177	21 23 202	20 50 266	21 43 303	21 42 350
37 38 055	34 06 093	36 50 178	21 04 203	20 01 267	21 02 303	21 33 351
38 19 056	34 55 094	36 51 179	20 44 204	19 12 267	20 21 304	21 26 351
38 59 056	35 44 094	36 51 181	20 24 205	18 23 268	19 40 304	21 18 351
39 40 056	36 33 095	36 50 182	20 03 206	17 33 269	18 59 305	21 11 351
40 21 056	37 22 096	36 48 183	19 41 207	16 44 269	18 19 305	21 03 352
41 02 056	38 11 096	36 45 184	19 19 207	15 55 270	17 39 305	20 56 352
*CAPELLA	RIGEL	*Diphda	FOMALHAUT	Enif	*DENEB	Kochab
41 43 057	16 24 113	36 41 185	18 56 208	38 33 253	40 41 303	20 50 352
42 24 057	17 10 114	36 36 186	18 33 209	37 46 254	39 59 303	20 43 353
43 05 057	17 54 114	36 30 188	18 08 210	36 59 254	39 18 303	20 37 353
43 46 057	18 39 115	36 23 189	17 44 211	36 11 255	38 37 303	20 31 353
44 27 057	19 24 116	36 15 190	17 18 211	35 24 256	37 56 304	20 25 353
45 09 057	20 08 116	36 05 191	16 52 212	34 36 257	37 15 304	20 19 354
45 50 057	20 52 117	35 55 192	16 26 213	33 48 257	36 34 304	20 14 354
46 31 058	21 35 118	35 45 193	15 59 214	33 00 258	35 53 304	20 09 354
47 13 058	22 19 118	35 33 195	15 31 215	32 12 259	35 13 304	20 04 354
47 54 058	23 02 119	35 20 196	15 03 215	31 24 259	34 32 305	19 59 355
48 36 058	23 45 120	35 06 197	14 35 216	30 36 260	33 52 305	19 55 355
49 18 058	24 27 121	34 51 198	14 05 217	29 47 261	33 12 305	19 51 355
49 59 058	25 09 121	34 36 199	13 36 218	28 59 261	32 32 305	19 47 356
50 41 058	25 51 122	34 19 200	13 06 218	28 10 262	31 52 306	19 43 356
51 23 058	26 32 123	34 02 201	12 35 219	27 21 263	31 12 306	19 40 356
*CAPELLA	BETELGEUSE	RIGEL	*Diphda	Alpheratz	*DENEB	Kochab
52 04 058	29 59 103	27 13 124	33 43 202	65 20 264	30 32 306	19 36 356
52 46 058	30 47 104	27 54 125	33 24 203	64 31 264	29 53 307	19 33 357
53 28 058	31 34 104	28 34 125	33 04 205	63 42 265	29 13 307	19 31 357
54 09 058	32 22 105	29 14 126	32 43 206	62 53 266	28 34 307	19 28 357
54 51 058	33 09 106	29 53 127	32 22 207	62 04 267	27 55 308	19 26 358
55 33 058	33 57 106	30 32 128	31 59 208	61 15 267	27 16 308	19 24 358
56 14 058	34 44 107	31 11 129	31 36 209	60 26 268	26 37 308	19 22 358
56 56 058	35 31 108	31 49 130	31 12 210	59 37 268	25 59 308	19 21 358
57 37 058	36 17 109	32 26 131	30 47 211	58 48 269	25 20 309	19 19 359
58 19 058	37 04 109	33 03 132	30 22 212	57 58 270	24 42 309	19 18 359
59 00 057	37 50 110	33 40 133	29 56 213	57 09 270	24 04 309	19 18 359
59 42 057	38 36 111	34 16 134	29 29 214	56 20 271	23 26 310	19 17 000
60 23 057	39 22 112	34 51 134	29 01 215	55 31 271	22 49 310	19 17 000
61 04 057	40 07 113	35 26 135	28 33 216	54 42 272	22 11 311	19 17 000
61 45 057	40 53 113	36 00 136	28 04 216	53 53 272	21 34 311	19 17 000
*Dubhe	POLLUX	SIRIUS	*RIGEL	Diphda	*Alpheratz	Schedar
18 05 025	30 28 075	15 57 124	36 34 137	27 35 217	53 04 273	57 58 323
18 26 026	31 15 076	16 37 125	37 06 138	27 05 218	52 15 273	57 29 322
18 48 026	32 03 076	17 17 126	37 39 140	26 34 219	51 26 274	56 59 322
19 09 026	32 51 077	17 57 127	38 10 141	26 03 220	50 37 274	56 28 322
19 31 027	33 39 077	18 36 127	38 41 142	25 31 221	49 48 275	55 57 321
19 53 027	34 26 077	19 15 128	39 11 143	24 58 222	48 59 275	55 27 321
20 16 027	35 14 078	19 54 129	39 40 144	24 25 223	48 10 276	54 55 321
20 38 027	36 03 078	20 32 130	40 09 145	23 52 223	47 21 276	54 24 320
21 01 028	36 51 079	21 09 130	40 37 146	23 17 224	46 32 277	53 52 320
21 24 028	37 39 079	21 47 131	41 04 147	22 43 225	45 43 277	53 21 320
21 47 028	38 27 080	22 24 132	41 30 149	22 08 226	44 54 278	52 49 319
22 11 029	39 16 080	23 00 133	41 55 150	21 32 227	44 06 278	52 17 319
22 34 029	40 04 081	23 36 133	42 19 151	20 56 228	43 17 279	51 44 319
22 58 029	40 53 081	24 11 134	42 43 152	20 20 228	42 29 279	51 12 319
23 22 029	41 41 082	24 46 135	43 05 153	19 43 229	41 40 280	50 40 319
*Dubhe	POLLUX	PROCYON	*SIRIUS	RIGEL	*Hamal	Schedar
23 46 030	42 30 082	31 50 108	25 21 136	43 27 155	62 32 253	50 07 318
24 11 030	43 19 083	32 37 108	25 55 137	43 47 156	61 45 254	49 34 318
24 36 030	44 07 083	33 23 109	26 28 138	44 07 157	60 58 255	49 01 318
25 00 030	44 56 083	34 10 110	27 01 139	44 25 159	60 11 256	48 29 318
25 25 031	45 45 084	34 56 111	27 33 139	44 43 160	59 23 256	47 56 318
25 51 031	46 34 084	35 42 111	28 04 140	44 59 161	58 35 257	47 23 318
26 16 031	47 23 085	36 27 112	28 36 141	45 14 163	57 47 258	46 50 318
26 42 031	48 12 085	37 13 113	29 06 142	45 28 164	56 59 259	46 16 318
27 07 032	49 01 086	37 58 114	29 36 143	45 41 165	56 10 260	45 43 318
27 33 032	49 50 086	38 43 115	30 05 144	45 53 167	55 22 261	45 10 318
27 59 032	50 39 087	39 27 115	30 33 145	46 04 168	54 33 261	44 37 317
28 25 032	51 28 087	40 11 116	31 01 146	46 13 170	53 45 262	44 04 317
28 52 032	52 17 088	40 55 117	31 28 147	46 22 171	52 56 263	43 30 317
29 18 033	53 06 089	41 39 118	31 55 148	46 29 172	52 07 264	42 57 317
29 45 033	53 55 089	42 22 119	32 20 149	46 35 174	51 18 264	42 24 317
*Dubhe	POLLUX	PROCYON	*SIRIUS	RIGEL	*Hamal	Mirfak
30 11 033	54 45 090	43 05 120	32 45 150	46 39 175	50 30 265	66 51 317
30 38 033	55 34 090	43 47 121	33 09 151	46 43 177	49 41 266	66 18 316
31 05 033	56 23 091	44 29 122	33 32 152	46 45 178	48 52 266	65 44 316
31 32 033	57 12 091	45 10 123	33 55 153	46 46 180	48 02 267	65 09 315
31 59 034	58 01 092	45 51 124	34 17 154	46 46 181	47 13 267	64 34 314
32 27 034	58 50 093	46 32 125	34 38 155	46 44 182	46 24 268	63 59 314
32 54 034	59 39 093	47 12 126	34 58 156	46 41 184	45 35 269	63 23 313
33 21 034	60 28 094	47 51 127	35 17 158	46 37 185	44 46 269	62 47 313
33 49 034	61 17 095	48 30 128	35 35 159	46 32 187	43 57 270	62 10 312
34 17 034	62 06 095	49 08 130	35 53 160	46 26 188	43 08 270	61 34 312
34 44 034	62 55 096	49 46 131	36 09 161	46 18 190	42 19 271	60 57 311
35 12 035	63 44 097	50 23 132	36 25 162	46 09 191	41 29 271	60 20 311
35 40 035	64 33 098	50 59 133	36 39 163	45 59 192	40 40 272	59 42 310
36 08 035	65 22 098	51 34 135	36 53 164	45 48 194	39 51 273	59 05 310
36 36 035	66 10 099	52 09 136	37 06 166	45 36 195	39 02 273	58 27 310

LAT 35°N

LHA ♈	Hc Zn	Hc Zn	Hc Zn	Hc Zn	Hc Zn	Hc Zn	Hc Zn
	*Dubhe	REGULUS	PROCYON	*SIRIUS	RIGEL	ALDEBARAN	*Mirfak
90	37 04 035	30 01 096	52 43 137	37 17 167	45 22 197	63 20 231	57 49 310
91	37 32 035	30 49 097	53 16 139	37 28 168	45 07 198	62 41 233	57 11 309
92	38 00 035	31 38 097	53 48 140	37 38 169	44 52 199	62 02 234	56 33 309
93	38 29 035	32 27 098	54 19 141	37 47 170	44 35 201	61 21 236	55 55 309
94	38 57 035	33 16 099	54 49 143	37 54 172	44 17 202	60 40 237	55 17 309
95	39 25 035	34 04 099	55 18 144	38 01 173	43 58 203	59 58 239	54 38 309
96	39 53 035	34 52 100	55 46 146	38 07 174	43 38 205	59 16 240	54 00 308
97	40 21 035	35 41 101	56 13 148	38 11 175	43 17 206	58 33 241	53 21 308
98	40 50 035	36 29 101	56 39 149	38 15 176	42 55 207	57 50 242	52 43 308
99	41 18 035	37 17 102	57 04 151	38 18 178	42 32 208	57 06 244	52 04 308
100	41 46 035	38 05 103	57 27 153	38 19 179	42 08 210	56 22 245	51 26 308
101	42 14 035	38 53 104	57 49 154	38 20 180	41 44 211	55 38 246	50 47 308
102	42 43 035	39 41 104	58 10 156	38 19 181	41 18 212	54 53 247	50 08 308
103	43 11 035	40 28 105	58 29 158	38 17 182	40 52 213	54 07 248	49 29 308
104	43 39 035	41 16 106	58 47 160	38 15 184	40 24 214	53 22 249	48 51 308
	*Alkaid	REGULUS	Alphard	*SIRIUS	RIGEL	ALDEBARAN	*CAPELLA
105	19 12 042	42 03 107	34 28 134	38 11 185	39 56 215	52 36 250	67 17 307
106	19 46 043	42 50 107	35 03 135	38 06 186	39 27 217	51 49 251	66 38 306
107	20 19 043	43 36 108	35 37 136	38 01 187	38 58 218	51 03 252	65 58 306
108	20 53 043	44 23 109	36 10 137	37 54 189	38 27 219	50 16 252	65 18 305
109	21 27 044	45 09 110	36 43 138	37 46 190	37 56 220	49 29 253	64 38 305
110	22 01 044	45 55 111	37 16 139	37 37 191	37 24 221	48 42 254	63 57 304
111	22 35 044	46 41 112	37 47 141	37 27 192	36 52 222	47 55 255	63 17 304
112	23 10 045	47 27 113	38 18 142	37 16 193	36 18 223	47 07 256	62 36 304
113	23 44 045	48 12 114	38 48 143	37 05 195	35 45 224	46 19 257	61 55 304
114	24 19 045	48 57 114	39 18 144	36 52 196	35 10 225	45 32 257	61 14 303
115	24 54 046	49 42 115	39 46 145	36 38 197	34 35 226	44 43 258	60 33 303
116	25 30 046	50 26 116	40 14 146	36 23 198	33 59 227	43 55 259	59 51 303
117	26 05 046	51 10 117	40 41 147	36 08 199	33 23 228	43 07 260	59 10 303
118	26 41 047	51 53 119	41 07 148	35 51 200	32 47 229	42 19 260	58 29 302
119	27 17 047	52 36 120	41 33 150	35 34 201	32 09 230	41 30 261	57 47 302
	*Alkaid	REGULUS	*Alphard	SIRIUS	RIGEL	ALDEBARAN	*CAPELLA
120	27 53 047	53 18 121	41 57 151	35 15 203	31 32 231	40 42 262	57 06 302
121	28 29 047	54 00 122	42 21 152	34 56 204	30 53 231	39 53 262	56 24 302
122	29 05 048	54 42 123	42 43 153	34 36 205	30 15 232	39 04 263	55 42 302
123	29 42 048	55 23 124	43 05 155	34 15 206	29 36 233	38 15 264	55 01 302
124	30 18 048	56 03 126	43 26 156	33 53 207	28 56 234	37 27 264	54 19 302
125	30 55 049	56 43 127	43 45 157	33 30 208	28 16 235	36 38 265	53 37 302
126	31 32 049	57 22 128	44 04 158	33 07 209	27 36 236	35 49 265	52 56 302
127	32 09 049	58 00 130	44 21 160	32 43 210	26 55 237	35 00 266	52 14 302
128	32 46 049	58 37 131	44 38 161	32 18 211	26 14 237	34 11 267	51 32 302
129	33 23 049	59 14 132	44 53 162	31 52 212	25 32 238	33 21 267	50 51 302
130	34 01 050	59 50 134	45 08 164	31 26 213	24 50 239	32 32 268	50 09 302
131	34 38 050	60 25 136	45 21 165	30 59 214	24 08 240	31 43 268	49 27 302
132	35 16 050	60 59 137	45 33 166	30 31 215	23 25 240	30 54 269	48 46 302
133	35 54 050	61 32 139	45 44 168	30 02 216	22 42 241	30 05 270	48 04 302
134	36 32 051	62 03 141	45 54 169	29 33 217	21 59 242	29 16 270	47 23 302
	*Alkaid	ARCTURUS	SPICA	*Alphard	SIRIUS	BETELGEUSE	*CAPELLA
135	37 10 051	20 04 080	12 38 113	46 02 171	29 03 218	39 11 248	46 41 302
136	37 48 051	20 52 080	13 23 114	46 10 172	28 33 219	38 26 249	46 00 303
137	38 26 051	21 41 081	14 08 115	46 16 174	28 02 220	37 40 250	45 18 303
138	39 04 051	22 29 082	14 53 115	46 21 175	27 30 221	36 53 251	44 37 303
139	39 42 051	23 18 082	15 37 116	46 24 176	26 58 221	36 07 252	43 56 303
140	40 21 052	24 07 083	16 21 117	46 27 178	26 25 222	35 20 252	43 15 303
141	40 59 052	24 55 083	17 05 117	46 28 179	25 52 223	34 33 253	42 33 303
142	41 38 052	25 44 084	17 48 118	46 28 181	25 18 224	33 46 254	41 52 303
143	42 17 052	26 33 084	18 32 119	46 27 182	24 43 225	32 59 255	41 11 304
144	42 55 052	27 22 085	19 15 119	46 25 184	24 08 226	32 11 255	40 30 304
145	43 34 052	28 11 085	19 57 120	46 21 185	23 33 227	31 24 256	39 50 304
146	44 13 052	29 00 086	20 40 121	46 16 186	22 57 227	30 36 257	39 09 304
147	44 52 052	29 49 086	21 22 122	46 10 188	22 20 228	29 48 257	38 28 304
148	45 31 052	30 38 087	22 03 122	46 03 189	21 44 229	29 00 258	37 48 305
149	46 10 052	31 27 087	22 45 123	45 54 191	21 06 230	28 12 259	37 07 305
	Alkaid	*ARCTURUS	SPICA	*Alphard	PROCYON	BETELGEUSE	*CAPELLA
150	46 48 052	32 16 088	23 26 124	45 45 192	45 46 236	27 24 259	36 27 305
151	47 27 052	33 05 088	24 06 125	45 34 193	45 05 237	26 35 260	35 47 305
152	48 06 052	33 55 089	24 47 125	45 22 195	44 23 238	25 47 261	35 07 305
153	48 45 053	34 44 090	25 26 126	45 09 196	43 41 239	24 58 261	34 27 306
154	49 24 053	35 33 090	26 06 127	44 54 198	42 59 240	24 10 262	33 47 306
155	50 03 052	36 22 091	26 45 128	44 39 199	42 16 241	23 21 262	33 07 306
156	50 42 052	37 11 091	27 23 129	44 22 200	41 33 242	22 32 263	32 28 306
157	51 21 052	38 00 092	28 02 130	44 05 202	40 49 243	21 43 264	31 48 307
158	52 00 052	38 49 093	28 39 130	43 46 203	40 05 244	20 55 264	31 09 307
159	52 39 052	39 38 093	29 16 131	43 27 204	39 21 245	20 06 265	30 30 307
160	53 18 052	40 28 094	29 53 132	43 06 205	38 37 245	19 17 266	29 51 308
161	53 57 052	41 17 094	30 29 133	42 45 207	37 52 246	18 28 266	29 12 308
162	54 36 052	42 06 095	31 05 134	42 22 208	37 07 247	17 39 267	28 33 308
163	55 14 052	42 54 096	31 40 135	41 59 209	36 21 248	16 49 267	27 55 308
164	55 53 052	43 43 096	32 15 136	41 34 210	35 36 249	16 00 268	27 16 309
	*Dubhe	Alkaid	*ARCTURUS	SPICA	REGULUS	*PROCYON	CAPELLA
165	63 05 000	56 31 051	44 32 097	32 48 137	64 08 211	34 50 250	26 38 309
166	63 05 359	57 10 051	45 21 098	33 22 138	63 42 213	34 03 250	26 00 309
167	63 04 358	57 48 051	46 10 098	33 55 139	63 14 215	33 17 251	25 22 310
168	63 02 357	58 26 051	46 58 099	34 27 140	62 45 217	32 30 252	24 44 310
169	63 00 356	59 04 050	47 47 100	34 58 141	62 15 219	31 44 253	24 07 310
170	62 56 355	59 42 050	48 35 101	35 29 142	61 44 220	30 57 253	23 30 311
171	62 52 354	60 20 050	49 23 101	35 59 143	61 12 222	30 10 254	22 52 311
172	62 46 353	60 57 049	50 11 102	36 28 144	60 38 224	29 22 255	22 16 311
173	62 40 352	61 34 049	50 59 103	36 57 145	60 04 225	28 35 255	21 39 312
174	62 33 351	62 11 049	51 47 104	37 25 146	59 28 227	27 47 256	21 02 312
175	62 25 350	62 48 048	52 35 104	37 52 147	58 52 228	26 59 257	20 26 313
176	62 17 349	63 24 048	53 22 105	38 18 148	58 15 230	26 11 257	19 50 313
177	62 07 348	64 01 047	54 10 106	38 44 149	57 37 231	25 23 258	19 14 313
178	61 57 347	64 36 046	54 57 107	39 08 150	56 58 233	24 35 259	18 38 314
179	61 46 347	65 12 046	55 44 108	39 32 152	56 19 234	23 47 259	18 03 314

Table 27

LAT 35°N

LHA ♈	Hc Zn	Hc Zn	Hc Zn	Hc Zn	Hc Zn	Hc Zn	Hc Zn
	*VEGA	Alphecca	ARCTURUS	*SPICA	REGULUS	*POLLUX	Dubhe
180	15 02 053	44 02 085	56 30 109	39 55 153	55 38 235	35 50 282	61 34 346
181	15 41 053	44 51 085	57 17 110	40 17 154	54 58 236	35 02 282	61 21 345
182	16 21 054	45 40 086	58 03 111	40 38 155	54 17 238	34 14 283	61 08 344
183	17 00 054	46 29 086	58 48 112	40 58 156	53 35 239	33 26 283	60 54 343
184	17 40 055	47 18 087	59 34 113	41 17 158	52 52 240	32 38 284	60 39 342
185	18 21 055	48 07 087	60 19 114	41 36 159	52 10 241	31 51 284	60 24 341
186	19 01 055	48 56 088	61 03 116	41 53 160	51 26 242	31 03 284	60 08 341
187	19 42 056	49 46 089	61 47 117	42 09 161	50 43 243	30 15 285	59 52 340
188	20 22 056	50 35 089	62 31 118	42 24 163	49 59 244	29 28 285	59 34 339
189	21 03 057	51 24 090	63 14 119	42 38 164	49 14 245	28 41 286	59 17 338
190	21 44 057	52 13 090	63 57 121	42 51 165	48 29 246	27 53 286	58 58 338
191	22 26 057	53 02 091	64 39 122	43 03 167	47 44 247	27 06 287	58 39 337
192	23 07 058	53 51 091	65 20 124	43 14 168	46 59 248	26 19 287	58 20 336
193	23 49 058	54 40 092	66 00 125	43 23 169	46 13 249	25 32 288	58 00 336
194	24 31 059	55 29 093	66 40 127	43 32 171	45 27 250	24 46 288	57 39 335
	*Kochab	VEGA	Rasalhague	*ANTARES	SPICA	*REGULUS	Dubhe
195	48 28 011	25 13 059	24 48 092	11 26 134	43 39 172	44 41 251	57 19 335
196	48 38 011	25 55 059	25 37 092	12 01 135	43 46 173	43 55 251	56 57 334
197	48 47 010	26 38 060	26 26 093	12 36 135	43 51 175	43 08 252	56 35 333
198	48 55 010	27 20 060	27 15 094	13 10 136	43 55 176	42 21 253	56 13 333
199	49 03 010	28 03 060	28 04 094	13 44 137	43 58 177	41 34 254	55 51 332
200	49 11 009	28 46 061	28 53 095	14 17 138	43 59 179	40 47 255	55 28 332
201	49 19 009	29 29 061	29 42 095	14 50 138	44 00 180	39 59 255	55 04 331
202	49 27 008	30 12 062	30 31 096	15 23 139	43 59 181	39 11 256	54 41 331
203	49 34 008	30 55 062	31 20 097	15 55 140	43 57 183	38 24 257	54 17 331
204	49 40 008	31 38 062	32 09 097	16 26 141	43 54 184	37 36 258	53 53 330
205	49 47 007	32 22 063	32 57 098	16 57 141	43 50 186	36 48 258	53 28 330
206	49 53 007	33 06 063	33 46 099	17 28 142	43 45 187	35 59 259	53 03 329
207	49 59 007	33 50 063	34 35 099	17 58 143	43 39 188	35 11 260	52 38 329
208	50 04 006	34 33 064	35 23 100	18 27 144	43 30 190	34 23 260	52 13 329
209	50 09 006	35 18 064	36 12 101	18 56 144	43 22 191	33 34 261	51 47 328
	*Kochab	VEGA	Rasalhague	*ANTARES	SPICA	*REGULUS	Dubhe
210	50 14 005	36 02 064	37 00 101	19 24 145	43 12 192	32 46 262	51 21 328
211	50 18 005	36 46 064	37 48 102	19 52 146	43 01 194	31 57 262	50 55 328
212	50 22 005	37 30 065	38 36 103	20 19 147	42 49 195	31 08 263	50 29 328
213	50 26 004	38 15 065	39 24 103	20 46 148	42 35 196	30 20 264	50 02 327
214	50 30 004	39 00 065	40 12 104	21 12 149	42 21 198	29 31 264	49 35 327
215	50 32 003	39 44 066	40 59 105	21 37 149	42 06 199	28 42 265	49 09 327
216	50 35 003	40 29 066	41 47 106	22 02 150	41 49 200	27 53 265	48 42 327
217	50 37 002	41 14 066	42 34 106	22 26 151	41 32 201	27 04 266	48 15 326
218	50 39 002	41 59 067	43 21 107	22 49 152	41 14 203	26 15 267	47 47 326
219	50 41 002	42 44 067	44 08 108	23 12 153	40 54 204	25 26 267	47 20 326
220	50 42 001	43 29 067	44 54 109	23 34 154	40 34 205	24 37 268	46 52 326
221	50 43 001	44 15 067	45 41 110	23 55 155	40 13 206	23 47 268	46 25 326
222	50 43 000	45 00 068	46 27 111	24 16 156	39 51 207	22 58 269	45 57 326
223	50 43 000	45 46 068	47 13 111	24 36 156	39 28 209	22 09 270	45 29 325
224	50 43 359	46 31 068	47 59 112	24 55 157	39 04 210	21 20 270	45 01 325
	DENEB	VEGA	*ALTAIR	ANTARES	*SPICA	Denebola	*Dubhe
225	27 08 052	47 17 068	19 28 093	25 14 158	38 39 211	42 26 257	44 33 325
226	27 47 052	48 03 069	20 17 093	25 32 159	38 13 212	41 38 258	44 05 325
227	28 26 053	48 48 069	21 06 094	25 49 160	37 47 213	40 50 259	43 37 325
228	29 05 053	49 34 069	21 55 095	26 05 161	37 19 214	40 02 259	43 09 325
229	29 45 053	50 20 069	22 44 095	26 20 162	36 51 215	39 14 260	42 41 325
230	30 24 054	51 06 070	23 33 096	26 35 163	36 23 216	38 25 261	42 13 325
231	31 04 054	51 52 070	24 22 096	26 49 164	35 53 217	37 36 262	41 45 325
232	31 44 054	52 39 070	25 11 097	27 02 165	35 23 218	36 48 262	41 16 325
233	32 24 054	53 25 070	26 00 098	27 14 166	34 52 219	35 59 263	40 48 325
234	33 04 055	54 11 070	26 48 098	27 26 167	34 20 220	35 10 263	40 20 325
235	33 44 055	54 57 071	27 37 099	27 37 168	33 48 221	34 21 264	39 52 325
236	34 24 055	55 44 071	28 26 100	27 46 169	33 15 222	33 33 265	39 23 325
237	35 05 056	56 30 071	29 14 100	27 55 170	32 42 223	32 44 265	38 55 325
238	35 45 056	57 17 071	30 02 101	28 04 171	32 08 224	31 55 266	38 27 325
239	36 26 056	58 03 071	30 50 102	28 11 172	31 33 225	31 06 267	37 59 325
	*DENEB	ALTAIR	Nunki	*ANTARES	SPICA	ARCTURUS	*Alkaid
240	37 07 056	31 39 102	16 13 140	28 17 173	30 58 226	61 54 243	61 41 311
241	37 48 056	32 27 103	16 45 141	28 23 174	30 22 227	61 10 244	61 04 311
242	38 29 057	33 14 104	17 15 142	28 28 175	29 46 228	60 25 245	60 26 310
243	39 10 057	34 02 104	17 46 142	28 31 176	29 09 229	59 40 247	59 49 310
244	39 51 057	34 50 105	18 15 143	28 34 177	28 32 230	58 55 248	59 11 310
245	40 32 057	35 37 106	18 44 144	28 36 178	27 54 231	58 09 249	58 33 309
246	41 14 057	36 24 107	19 13 145	28 38 179	27 16 231	57 23 250	57 55 309
247	41 55 058	37 11 107	19 41 146	28 38 180	26 37 232	56 37 251	57 17 309
248	42 37 058	37 58 108	20 08 146	28 37 181	25 58 233	55 50 252	56 38 309
249	43 18 058	38 45 109	20 35 147	28 36 182	25 19 234	55 04 253	56 00 308
250	44 00 058	39 31 110	21 02 148	28 34 183	24 39 235	54 17 254	55 21 308
251	44 42 058	40 17 110	21 27 149	28 31 184	23 58 235	53 29 255	54 43 308
252	45 24 058	41 03 111	21 52 150	28 27 185	23 18 236	52 42 255	54 04 308
253	46 06 059	41 49 112	22 17 151	28 22 186	22 37 237	51 54 256	53 25 308
254	46 48 059	42 34 113	22 40 152	28 16 187	21 55 238	51 06 257	52 46 308
	*DENEB	ALTAIR	Nunki	*ANTARES	ARCTURUS	*Alkaid	Kochab
255	47 30 059	43 19 114	23 03 152	28 09 188	50 18 258	52 07 308	47 42 348
256	48 12 059	44 04 115	23 26 153	28 02 189	49 30 259	51 28 308	47 32 347
257	48 54 059	44 49 116	23 47 154	27 54 190	48 42 259	50 49 308	47 21 347
258	49 36 059	45 33 117	24 08 155	27 44 191	47 54 260	50 10 308	47 10 347
259	50 18 059	46 17 118	24 29 156	27 34 192	47 05 261	49 31 307	46 58 346
260	51 00 059	47 00 119	24 48 157	27 24 193	46 17 262	48 52 307	46 47 346
261	51 43 059	47 43 120	25 07 158	27 12 194	45 28 262	48 13 308	46 35 346
262	52 25 059	48 25 121	25 25 159	26 59 195	44 39 263	47 34 308	46 22 346
263	53 07 059	49 08 122	25 43 160	26 46 196	43 50 264	46 55 308	46 10 345
264	53 49 059	49 49 123	25 59 161	26 32 197	43 02 264	46 17 308	45 58 345
265	54 32 059	50 30 124	26 15 162	26 17 198	42 13 265	45 38 308	45 45 345
266	55 14 059	51 11 125	26 30 163	26 01 199	41 24 266	44 59 308	45 32 345
267	55 56 059	51 51 126	26 45 164	25 45 200	40 35 266	44 20 308	45 19 344
268	56 39 059	52 30 127	26 58 165	25 28 201	39 46 267	43 41 308	45 05 344
269	57 21 059	53 09 129	27 11 166	25 10 202	38 57 267	43 02 308	44 52 344

LAT 3

LHA ♈	Hc Zn	Hc Zn	Hc Zn	Hc Zn	Hc Zn	Hc Zn	Hc
	*DENEB	ALTAIR	*Nunki	ANTARES	ARCTURUS	*Alkaid	Ko
270	58 03 059	53 47 130	27 23 167	24 51 203	38 07 268	42 24 308	44 38
271	58 45 059	54 24 131	27 34 168	24 32 204	37 18 269	41 45 308	44 24
272	59 27 059	55 01 132	27 44 169	24 12 205	36 29 269	41 06 308	44 10
273	60 09 059	55 37 134	27 53 170	23 51 206	35 40 270	40 28 308	43 56
274	60 51 059	56 12 135	28 02 171	23 29 206	34 51 270	39 49 309	43 42
275	61 33 058	56 46 137	28 09 172	23 07 207	34 02 271	39 11 309	43 28
276	62 15 058	57 19 138	28 16 173	22 44 208	33 13 271	38 33 309	43 13
277	62 57 058	57 52 140	28 22 174	22 21 209	32 23 272	37 55 309	42 58
278	63 38 058	58 23 141	28 27 175	21 56 210	31 34 273	37 16 309	42 43
279	64 20 057	58 53 143	28 32 176	21 32 211	30 45 273	36 38 309	42 28
280	65 01 057	59 22 145	28 35 177	21 06 212	29 56 274	36 01 310	42 13
281	65 42 056	59 50 146	28 37 178	20 40 212	29 07 274	35 23 310	41 58
282	66 23 056	60 16 148	28 39 179	20 13 213	28 18 275	34 45 310	41 43
283	67 03 055	60 41 150	28 40 180	19 46 214	27 29 275	34 08 310	41 27
284	67 44 055	61 05 152	28 40 181	19 18 215	26 40 276	33 30 310	41 12
	*Alpheratz	Enif	ALTAIR	*ANTARES	ARCTURUS	*Alkaid	Ko
285	26 15 072	45 11 114	61 28 154	18 50 216	25 51 276	32 53 311	40 56
286	27 02 072	45 55 115	61 49 156	18 21 217	25 03 277	32 16 311	40 41
287	27 48 073	46 40 116	62 08 158	17 51 217	24 14 277	31 38 311	40 25
288	28 35 073	47 23 117	62 26 160	17 21 218	23 25 278	31 02 311	40 09
289	29 23 074	48 07 118	62 42 162	16 51 219	22 36 278	30 25 312	39 53
290	30 10 074	48 50 119	62 57 164	16 19 220	21 48 279	29 48 312	39 37
291	30 57 074	49 33 120	63 09 166	15 48 220	20 59 279	29 12 312	39 21
292	31 44 075	50 15 121	63 20 168	15 16 221	20 11 280	28 35 312	39 05
293	32 32 075	50 57 122	63 29 170	14 43 222	19 23 281	27 59 313	38 49
294	33 19 076	51 38 123	63 37 173	14 10 223	18 34 281	27 23 313	38 33
295	34 07 076	52 19 125	63 42 175	13 37 223	17 46 282	26 47 313	38 17
296	34 55 077	52 59 126	63 46 177	13 03 224	16 58 282	26 12 314	38 01
297	35 43 077	53 38 127	63 47 179	12 28 225	16 10 283	25 36 314	37 45
298	36 31 078	54 17 128	63 47 181	11 53 225	15 22 283	25 01 314	37 29
299	37 19 078	54 56 130	63 45 184	11 18 226	14 34 284	24 26 315	37 12
	*Mirfak	Alpheratz	*FOMALHAUT	ALTAIR	Rasalhague	*Alphecca	Koc
300	14 36 039	38 07 078	13 06 142	63 40 186	50 03 245	33 16 281	36 56
301	15 07 039	38 55 079	13 36 142	63 34 188	49 18 246	32 28 282	36 40
302	15 38 039	39 43 079	14 05 143	63 26 190	48 33 247	31 40 282	36 24
303	16 10 040	40 32 080	14 35 144	63 17 193	47 48 248	30 52 283	36 07
304	16 41 040	41 20 080	15 03 145	63 05 195	47 02 249	30 04 283	35 51
305	17 13 041	42 08 081	15 31 145	62 52 197	46 16 250	29 16 284	35 35
306	17 45 041	42 57 081	15 59 146	62 37 199	45 30 251	28 29 284	35 19
307	18 18 041	43 46 082	16 26 147	62 20 201	44 44 251	27 41 285	35 02
308	18 50 042	44 34 082	16 53 148	62 02 203	43 57 252	26 54 285	34 46
309	19 23 042	45 23 082	17 18 149	61 42 205	43 10 253	26 06 286	34 30
310	19 56 043	46 12 083	17 44 149	61 20 207	42 23 254	25 19 286	34 14
311	20 30 043	47 01 083	18 09 150	60 57 209	41 36 255	24 32 287	33 58
312	21 03 043	47 49 084	18 33 151	60 33 211	40 48 255	23 45 287	33 42
313	21 37 044	48 38 084	18 56 152	60 07 212	40 01 256	22 58 288	33 26
314	22 11 044	49 27 085	19 19 153	59 40 214	39 13 257	22 11 288	33 10
	*Mirfak	Hamal	Diphda	*FOMALHAUT	ALTAIR	*VEGA	Koc
315	22 45 044	23 51 077	15 11 126	19 41 153	59 12 216	61 08 288	32 54
316	23 19 045	24 39 078	15 51 126	20 03 154	58 43 218	60 21 288	32 38
317	23 54 045	25 27 078	16 30 127	20 24 155	58 12 219	59 35 288	32 22
318	24 29 045	26 15 079	17 09 128	20 44 156	57 41 221	58 48 289	32 07
319	25 04 045	27 04 079	17 48 129	21 04 157	57 08 222	58 01 289	31 51
320	25 39 046	27 52 080	18 26 129	21 23 158	56 35 224	57 15 289	31 36
321	26 14 046	28 40 080	19 04 130	21 41 159	56 00 225	56 28 289	31 20
322	26 49 046	29 29 081	19 41 131	21 59 159	55 25 227	55 42 289	31 05
323	27 25 047	30 17 081	20 18 132	22 16 160	54 49 228	54 56 289	30 49
324	28 01 047	31 06 082	20 55 132	22 32 161	54 12 229	54 09 290	30 34
325	28 37 047	31 55 082	21 31 133	22 47 162	53 34 231	53 23 290	30 19
326	29 13 047	32 43 083	22 06 134	23 02 163	52 56 232	52 37 290	30 04
327	29 49 048	33 32 083	22 41 135	23 16 164	52 17 233	51 51 290	29 49
328	30 26 048	34 21 084	23 16 136	23 29 165	51 37 234	51 05 290	29 34
329	31 02 048	35 10 084	23 50 136	23 42 166	50 57 235	50 19 291	29 20
	*CAPELLA	Hamal	Diphda	*FOMALHAUT	ALTAIR	*VEGA	Koc
330	13 20 043	35 59 085	24 24 137	23 53 167	50 16 237	49 33 291	29 05
331	13 53 043	36 48 085	24 57 138	24 04 168	49 35 238	48 47 291	28 51
332	14 27 043	37 37 086	25 29 139	24 14 169	48 53 239	48 01 291	28 36
333	15 01 044	38 26 086	26 01 140	24 24 170	48 11 240	47 15 292	28 22
334	15 35 044	39 15 087	26 33 141	24 32 170	47 28 241	46 30 292	28 08
335	16 10 045	40 04 088	27 03 142	24 40 171	46 45 242	45 44 292	27 54
336	16 44 045	40 53 088	27 33 143	24 47 172	46 02 243	44 58 292	27 40
337	17 19 045	41 42 089	28 03 144	24 53 173	45 18 244	44 13 293	27 26
338	17 54 046	42 31 089	28 32 144	24 58 174	44 34 245	43 28 293	27 13
339	18 30 046	43 21 090	29 00 145	25 03 175	43 49 246	42 42 293	26 59
340	19 05 047	44 10 090	29 28 146	25 07 176	43 04 246	41 57 293	26 46
341	19 41 047	44 59 091	29 55 147	25 09 177	42 19 247	41 12 294	26 33
342	20 17 047	45 48 092	30 21 148	25 11 178	41 33 248	40 27 294	26 20
343	20 54 048	46 37 092	30 46 149	25 13 179	40 48 249	39 43 294	26 07
344	21 30 048	47 26 093	31 11 150	25 13 180	40 02 250	38 58 295	25 54
	*CAPELLA	ALDEBARAN	Diphda	*FOMALHAUT	ALTAIR	*VEGA	Koc
345	22 07 048	14 31 080	31 35 151	25 13 181	39 15 251	38 13 295	25 41
346	22 44 049	15 20 080	31 58 152	25 11 182	38 29 251	37 29 295	25 29
347	23 21 049	16 08 081	32 21 153	25 09 183	37 42 252	36 44 296	25 17
348	23 58 050	16 57 081	32 42 154	25 07 184	36 55 253	36 00 296	25 05
349	24 36 050	17 45 082	33 03 155	25 03 185	36 08 254	35 16 296	24 53
350	25 13 050	18 34 083	33 23 156	24 58 186	35 21 254	34 32 296	24 41
351	25 51 051	19 23 083	33 42 158	24 53 187	34 33 255	33 48 297	24 29
352	26 29 051	20 12 084	34 01 159	24 47 188	33 46 256	33 04 297	24 18
353	27 07 051	21 00 084	34 18 160	24 40 189	32 58 257	32 20 297	24 07
354	27 46 051	21 49 085	34 35 161	24 32 190	32 10 257	31 37 298	23 56
355	28 24 052	22 38 085	34 51 162	24 24 190	31 22 258	30 53 298	23 45
356	29 03 052	23 27 086	35 05 163	24 14 191	30 34 259	30 10 298	23 34
357	29 42 052	24 16 086	35 19 164	24 04 192	29 46 259	29 27 299	23 24
358	30 21 053	25 05 087	35 32 165	23 53 193	28 57 260	28 44 299	23 13
359	31 00 053	25 54 087	35 44 166	23 42 194	28 09 261	28 01 300	23 03

Table 27, continued

be 65° west of Greenwich, you add 4 hours and 20 minutes (see Table 26) and find that GMT of civil twilight for you is 0845. Using Tables 28 and 29, you figure the following:

ARIES		
GHA 8 Hr.	= 47°52′	
45 Min. 0 Sec.	= 11°17′	
GHA	= 58°69′	
GHA	= 59°09′	
(+360?)	= 360°	
	= 419°09′	
Assumed Long. W.	= −65°09′	
Assumed Long. E.	= (+)	
LHA	= 354°	Assumed Lat. = 35° N

Since each line on Table 27 is 4 minutes, back up about 5 lines to give yourself time on deck. Reading across the line on Table 27 for LHA Aries (♈) = 349°, you find that the best stars are Capella (Hc 24°36′, Zn 050°), Fomalhaut (Hc 25°03′, Zn 185°), and Vega (Hc 35°16′, Zn 296°).

At the proper time, you go on deck and get the following round of sights:

Capella	= 08-32-10	Ho = 25°36′
Fomalhaut	= 08-33-15	Ho = 25°01′
Vega	= 08-33-53	Ho = 33°25′

1974 JULY 9, 10, 11 (TUES., WED., THURS.)

G.M.T.		ARIES	VENUS −3·4		MARS +2·0		JUPITER −2·2		SATURN +0·3	
		G.H.A.	G.H.A.	Dec.	G.H.A.	Dec.	G.H.A.	Dec.	G.H.A.	Dec.
d	h	° ′	° ′	° ′	° ′	° ′	° ′	° ′	° ′	° ′
9	00	286 33·1	212 13·4	N21 20·0	145 18·1	N16 27·0	297 13·6	S 5 56·4	186 20·0	N22 32·1
	01	301 35·6	227 12·7	20·4	160 19·0	26·5	312 16·0	56·4	201 22·1	32·1
	02	316 38·1	242 12·0	20·8	175 20·0	26·0	327 18·5	56·4	216 24·2	32·1
	03	331 40·5	257 11·3	· · 21·2	190 20·9	· · 25·5	342 21·0	· · 56·4	231 26·3	· · 32·1
	04	346 43·0	272 10·6	21·6	205 21·8	25·0	357 23·4	56·4	246 28·4	32·0
	05	1 45·4	287 09·9	22·1	220 22·8	24·5	12 25·9	56·4	261 30·6	32·0
	06	16 47·9	302 09·2	N21 22·5	235 23·7	N16 24·0	27 28·4	S 5 56·4	276 32·7	N22 32·0
	07	31 50·4	317 08·5	22·9	250 24·6	23·5	42 30·9	56·5	291 34·8	32·0
T	08	46 52·8	332 07·8	23·3	265 25·6	23·0	57 33·3	56·5	306 36·9	32·0
U	09	61 55·3	347 07·1	· · 23·7	280 26·5	· · 22·5	72 35·8	· · 56·5	321 39·0	· · 31·9
E	10	76 57·8	2 06·4	24·1	295 27·4	22·0	87 38·3	56·5	336 41·1	31·9
S	11	92 00·2	17 05·7	24·5	310 28·4	21·5	102 40·7	56·5	351 43·2	31·9
D	12	107 02·7	32 05·0	N21 24·9	325 29·3	N16 21·0	117 43·2	S 5 56·6	6 45·4	N22 31·9
A	13	122 05·2	47 04·3	25·3	340 30·2	20·5	132 45·7	56·6	21 47·5	31·9
Y	14	137 07·6	62 03·6	25·7	355 31·2	20·0	147 48·2	56·6	36 49·6	31·8
	15	152 10·1	77 02·9	· · 26·1	10 32·1	· · 19·5	162 50·6	· · 56·6	51 51·7	· · 31·8
	16	167 12·5	92 02·1	26·5	25 33·0	19·0	177 53·1	56·6	66 53·8	31·8
	17	182 15·0	107 01·4	26·9	40 34·0	18·5	192 55·6	56·6	81 55·9	31·8
	18	197 17·5	122 00·7	N21 27·3	55 34·9	N16 18·1	207 58·1	S 5 56·7	96 58·0	N22 31·8
	19	212 19·9	137 00·0	27·7	70 35·8	17·6	223 00·5	56·7	112 00·2	31·7
	20	227 22·4	151 59·3	28·1	85 36·8	17·1	238 03·0	56·7	127 02·3	31·7
	21	242 24·9	166 58·6	· · 28·5	100 37·7	· · 16·6	253 05·5	· · 56·7	142 04·4	· · 31·7
	22	257 27·3	181 57·9	28·9	115 38·7	16·1	268 08·0	56·7	157 06·5	31·7
	23	272 29·8	196 57·2	29·3	130 39·6	15·6	283 10·4	56·7	172 08·6	31·7
10	00	287 32·3	211 56·5	N21 29·7	145 40·5	N16 15·1	298 12·9	S 5 56·8	187 10·7	N22 31·6
	01	302 34·7	226 55·8	30·1	160 41·5	14·6	313 15·4	56·8	202 12·8	31·6
	02	317 37·2	241 55·1	30·5	175 42·4	14·1	328 17·9	56·8	217 15·0	31·6
	03	332 39·7	256 54·3	· · 30·9	190 43·3	· · 13·6	343 20·3	· · 56·8	232 17·1	· · 31·6
	04	347 42·1	271 53·6	31·3	205 44·3	13·1	358 22·8	56·8	247 19·2	31·6
	05	2 44·6	286 52·9	31·7	220 45·2	12·6	13 25·3	56·9	262 21·3	31·5
	06	17 47·0	301 52·2	N21 32·1	235 46·1	N16 12·1	28 27·8	S 5 56·9	277 23·4	N22 31·5
W	07	32 49·5	316 51·5	32·5	250 47·1	11·6	43 30·3	56·9	292 25·5	31·5
E	08	47 52·0	331 50·8	32·9	265 48·0	11·1	58 32·7	56·9	307 27·6	31·5
D	09	62 54·4	346 50·1	· · 33·2	280 49·0	· · 10·6	73 35·2	· · 56·9	322 29·8	· · 31·5
N	10	77 56·9	1 49·4	33·6	295 49·9	10·1	88 37·7	57·0	337 31·9	31·4
E	11	92 59·4	16 48·6	34·0	310 50·8	09·6	103 40·2	57·0	352 34·0	31·4
S	12	108 01·8	31 47·9	N21 34·4	325 51·8	N16 09·1	118 42·7	S 5 57·0	7 36·1	N22 31·4
D	13	123 04·3	46 47·2	34·8	340 52·7	08·6	133 45·1	57·0	22 38·2	31·4
A	14	138 06·8	61 46·5	35·2	355 53·6	08·1	148 47·6	57·0	37 40·3	31·4
Y	15	153 09·2	76 45·8	· · 35·5	10 54·6	· · 07·6	163 50·1	· · 57·1	52 42·5	· · 31·3
	16	168 11·7	91 45·1	35·9	25 55·5	07·1	178 52·6	57·1	67 44·6	31·3
	17	183 14·2	106 44·3	36·3	40 56·5	06·6	193 55·1	57·1	82 46·7	31·3
	18	198 16·6	121 43·6	N21 36·7	55 57·4	N16 06·1	208 57·5	S 5 57·1	97 48·8	N22 31·3
	19	213 19·1	136 42·9	37·1	70 58·3	05·6	224 00·0	57·2	112 50·9	31·3
	20	228 21·5	151 42·2	37·4	85 59·3	05·1	239 02·5	57·2	127 53·0	31·2
	21	243 24·0	166 41·5	· · 37·8	101 00·2	· · 04·6	254 05·0	· · 57·2	142 55·1	· · 31·2
	22	258 26·5	181 40·8	38·2	116 01·1	04·1	269 07·5	57·2	157 57·3	31·2
	23	273 28·9	196 40·0	38·6	131 02·1	03·6	284 10·0	57·2	172 59·4	31·2
11	00	288 31·4	211 39·3	N21 38·9	146 03·0	N16 03·1	299 12·4	S 5 57·3	188 01·5	N22 31·2
	01	303 33·9	226 38·6	39·3	161 04·0	02·6	314 14·9	57·3	203 03·6	31·1
	02	318 36·3	241 37·9	39·7	176 04·9	02·1	329 17·4	57·3	218 05·7	31·1
	03	333 38·8	256 37·2	· · 40·0	191 05·8	· · 01·6	344 19·9	· · 57·3	233 07·8	· · 31·1
	04	348 41·3	271 36·4	40·4	206 06·8	01·1	359 22·4	57·3	248 09·9	31·1
	05	3 43·7	286 35·7	40·8	221 07·7	00·6	14 24·9	57·4	263 12·1	31·1
	06	18 46·2	301 35·0	N21 41·1	236 08·7	N16 00·1	29 27·4	S 5 57·4	278 14·2	N22 31·0
	07	33 48·6	316 34·3	41·5	251 09·6	15 59·6	44 29·8	57·4	293 16·3	31·0
T	08	48 51·1	331 33·5	41·9	266 10·5	59·1	59 32·3	57·4	308 18·4	31·0
H	09	63 53·6	346 32·8	· · 42·2	281 11·5	· · 58·6	74 34·8	· · 57·5	323 20·5	· · 31·0
U	10	78 56·0	1 32·1	42·6	296 12·4	58·1	89 37·3	57·5	338 22·6	31·0
R	11	93 58·5	16 31·4	43·0	311 13·4	57·5	104 39·8	57·5	353 24·8	30·9
S	12	109 01·0	31 30·7	N21 43·3	326 14·3	N15 57·0	119 42·3	S 5 57·5	8 26·9	N22 30·9
D	13	124 03·4	46 29·9	43·7	341 15·2	56·5	134 44·8	57·6	23 29·0	30·9
A	14	139 05·9	61 29·2	44·0	356 16·2	56·0	149 47·3	57·6	38 31·1	30·9
Y	15	154 08·4	76 28·5	· · 44·4	11 17·1	· · 55·5	164 49·7	· · 57·6	53 33·2	· · 30·9
	16	169 10·8	91 27·8	44·7	26 18·1	55·0	179 52·2	57·6	68 35·3	30·8
	17	184 13·3	106 27·0	45·1	41 19·0	54·5	194 54·7	57·7	83 37·4	30·8
	18	199 15·8	121 26·3	N21 45·4	56 19·9	N15 54·0	209 57·2	S 5 57·7	98 39·6	N22 30·8
	19	214 18·2	136 25·6	45·8	71 20·9	53·5	224 59·7	57·7	113 41·7	30·8
	20	229 20·7	151 24·9	46·2	86 21·8	53·0	240 02·2	57·7	128 43·8	30·8
	21	244 23·1	166 24·1	· · 46·5	101 22·8	· · 52·5	255 04·7	· · 57·8	143 45·9	· · 30·7
	22	259 25·6	181 23·4	46·9	116 23·7	52·0	270 07·2	57·8	158 48·0	30·7
	23	274 28·1	196 22·7	47·2	131 24·6	51·5	285 09·7	57·8	173 50·1	30·7
Mer. Pass.		h m 4 49·1	v −0·7	d 0·4	v 0·9	d 0·5	v 2·5	d 0·0	v 2·1	d 0·0

STARS		
Name	S.H.A.	Dec.
	° ′	° ′
Acamar	315 40·6	S 40 24·1
Achernar	335 48·4	S 57 21·6
Acrux	173 42·2	S 62 57·9
Adhara	255 35·8	S 28 56·2
Aldebaran	291 23·1	N 16 27·5
Alioth	166 46·2	N 56 06·1
Alkaid	153 21·7	N 49 26·6
Al Na'ir	28 19·8	S 47 04·8
Alnilam	276 16·3	S 1 13·0
Alphard	218 25·0	S 8 33·0
Alphecca	126 35·4	N 26 48·1
Alpheratz	358 13·6	N 28 57·0
Altair	62 36·3	N 8 48·1
Ankaa	353 44·3	S 42 26·3
Antares	113 01·8	S 26 22·7
Arcturus	146 22·2	N 19 18·9
Atria	108 29·3	S 68 59·2
Avior	234 30·6	S 59 25·8
Bellatrix	279 03·6	N 6 19·7
Betelgeuse	271 33·2	N 7 24·2
Canopus	264 09·6	S 52 40·8
Capella	281 18·0	N 45 58·3
Deneb	49 50·9	N 45 11·4
Denebola	183 03·5	N 14 42·9
Diphda	349 25·1	S 18 07·3
Dubhe	194 27·6	N 61 53·5
Elnath	278 49·8	N 28 35·2
Eltanin	90 59·2	N 51 29·6
Enif	34 15·5	N 9 45·6
Fomalhaut	15 55·8	S 29 45·2
Gacrux	172 33·6	S 56 58·6
Gienah	176 22·4	S 17 24·2
Hadar	149 29·2	S 60 15·4
Hamal	328 33·8	N 23 20·5
Kaus Aust.	84 22·0	S 34 23·8
Kochab	137 18·1	N 74 15·8
Markab	14 07·2	N 15 04·2
Menkar	314 45·7	N 3 59·5
Menkent	148 41·9	S 36 15·0
Miaplacidus	221 46·8	S 69 37·0
Mirfak	309 22·5	N 49 46·2
Nunki	76 34·0	S 26 19·7
Peacock	54 04·4	S 56 48·9
Pollux	244 03·7	N 28 05·3
Procyon	245 30·5	N 5 17·4
Rasalhague	96 33·2	N 12 34·7
Regulus	208 14·7	N 12 05·5
Rigel	281 40·4	S 8 13·7
Rigil Kent.	140 31·3	S 60 44·1
Sabik	102 45·7	S 15 41·7
Schedar	350 13·8	N 56 23·7
Shaula	97 01·1	S 37 05·2
Sirius	258 59·8	S 16 40·8
Spica	159 02·0	S 11 01·8
Suhail	223 14·3	S 43 19·9
Vega	80 58·3	N 38 45·7
Zuben'ubi	137 37·6	S 15 56·3
	S.H.A.	Mer. Pass.
	° ′	h m
Venus	284 24·2	9 53
Mars	218 08·3	14 16
Jupiter	10 40·7	4 06
Saturn	259 38·5	11 30

Table 28

1974 JULY 9, 10, 11 (TUES., WED., THURS.)

G.M.T.		SUN G.H.A.	SUN Dec.	MOON G.H.A.	*v*	Dec.	*d*	H.P.
d	h	° ′	° ′	° ′	′	° ′	′	′
9	00	178 44·5	N22 25·9	311 20·2	15·9	S 4 52·0	11·4	54·3
	01	193 44·4	25·7	325 55·1	15·9	4 40·6	11·4	54·3
	02	208 44·3	25·4	340 30·0	15·9	4 29·2	11·4	54·3
	03	223 44·2	· · 25·1	355 04·9	16·0	4 17·8	11·5	54·3
	04	238 44·1	24·8	9 39·9	15·9	4 06·3	11·5	54·3
	05	253 44·0	24·5	24 14·8	15·9	3 54·8	11·4	54·3
	06	268 43·9	N22 24·2	38 49·7	15·9	S 3 43·4	11·6	54·3
	07	283 43·8	23·9	53 24·6	15·9	3 31·8	11·5	54·3
T	08	298 43·7	23·6	67 59·5	16·0	3 20·3	11·5	54·4
U	09	313 43·6	· · 23·3	82 34·5	15·9	3 08·8	11·6	54·4
E	10	328 43·5	23·0	97 09·4	15·9	2 57·2	11·6	54·4
S	11	343 43·4	22·7	111 44·3	15·9	2 45·6	11·6	54·4
D	12	358 43·4	N22 22·4	126 19·2	15·9	S 2 34·0	11·6	54·4
A	13	13 43·3	22·1	140 54·1	15·9	2 22·4	11·6	54·4
Y	14	28 43·2	21·8	155 29·0	15·9	2 10·8	11·7	54·4
	15	43 43·1	· · 21·5	170 03·9	15·9	1 59·1	11·6	54·4
	16	58 43·0	21·2	184 38·8	15·9	1 47·5	11·7	54·5
	17	73 42·9	20·9	199 13·7	15·9	1 35·8	11·7	54·5
	18	88 42·8	N22 20·6	213 48·6	15·9	S 1 24·1	11·6	54·5
	19	103 42·7	20·3	228 23·5	15·9	1 12·5	11·7	54·5
	20	118 42·6	20·0	242 58·4	15·8	1 00·8	11·7	54·5
	21	133 42·5	· · 19·7	257 33·2	15·9	0 49·1	11·8	54·5
	22	148 42·4	19·4	272 08·1	15·8	0 37·3	11·7	54·6
	23	163 42·3	19·1	286 42·9	15·8	0 25·6	11·7	54·6
10	00	178 42·3	N22 18·8	301 17·7	15·8	S 0 13·9	11·7	54·6
	01	193 42·2	18·5	315 52·5	15·8	S 0 02·2	11·8	54·6
	02	208 42·1	18·2	330 27·3	15·8	N 0 09·6	11·7	54·6
	03	223 42·0	· · 17·9	345 02·1	15·7	0 21·3	11·8	54·6
	04	238 41·9	17·6	359 36·8	15·8	0 33·1	11·8	54·7
	05	253 41·8	17·3	14 11·6	15·7	0 44·9	11·7	54·7
	06	268 41·7	N22 16·9	28 46·3	15·7	N 0 56·6	11·8	54·7
W	07	283 41·6	16·6	43 21·0	15·7	1 08·4	11·7	54·7
E	08	298 41·5	16·3	57 55·7	15·6	1 20·1	11·8	54·7
D	09	313 41·4	· · 16·0	72 30·3	15·7	1 31·9	11·8	54·7
N	10	328 41·4	15·7	87 05·0	15·6	1 43·7	11·8	54·8
E	11	343 41·3	15·4	101 39·6	15·6	1 55·5	11·7	54·8
S	12	358 41·2	N22 15·1	116 14·2	15·6	N 2 07·2	11·8	54·8
D	13	13 41·1	14·8	130 48·8	15·5	2 19·0	11·8	54·8
A	14	28 41·0	14·4	145 23·3	15·5	2 30·8	11·7	54·8
Y	15	43 40·9	· · 14·1	159 57·8	15·5	2 42·5	11·8	54·9
	16	58 40·8	13·8	174 32·3	15·4	2 54·3	11·8	54·9
	17	73 40·7	13·5	189 06·7	15·5	3 06·1	11·7	54·9
	18	88 40·7	N22 13·2	203 41·2	15·4	N 3 17·8	11·8	54·9
	19	103 40·6	12·9	218 15·6	15·3	3 29·6	11·7	54·9
	20	118 40·5	12·5	232 49·9	15·3	3 41·3	11·8	55·0
	21	133 40·4	· · 12·2	247 24·2	15·3	3 53·1	11·7	55·0
	22	148 40·3	11·9	261 58·5	15·3	4 04·8	11·7	55·0
	23	163 40·2	11·6	276 32·8	15·2	4 16·5	11·8	55·0
11	00	178 40·1	N22 11·3	291 07·0	15·2	N 4 28·3	11·7	55·1
	01	193 40·0	10·9	305 41·2	15·2	4 40·0	11·7	55·1
	02	208 40·0	10·6	320 15·4	15·1	4 51·7	11·6	55·1
	03	223 39·9	· · 10·3	334 49·5	15·0	5 03·3	11·7	55·1
	04	238 39·8	10·0	349 23·5	15·1	5 15·0	11·7	55·1
	05	253 39·7	09·6	3 57·6	14·9	5 26·7	11·6	55·2
	06	268 39·6	N22 09·3	18 31·5	15·0	N 5 38·3	11·7	55·2
	07	283 39·5	09·0	33 05·5	14·9	5 50·0	11·6	55·2
T	08	298 39·4	08·7	47 39·4	14·8	6 01·6	11·6	55·2
H	09	313 39·4	· · 08·3	62 13·2	14·9	6 13·2	11·6	55·3
U	10	328 39·3	08·0	76 47·1	14·7	6 24·8	11·6	55·3
R	11	343 39·2	07·7	91 20·8	14·7	6 36·4	11·6	55·3
S	12	358 39·1	N22 07·3	105 54·5	14·7	N 6 48·0	11·5	55·3
D	13	13 39·0	07·0	120 28·2	14·6	6 59·5	11·5	55·4
A	14	28 38·9	06·7	135 01·8	14·6	7 11·0	11·5	55·4
Y	15	43 38·9	· · 06·4	149 35·4	14·5	7 22·5	11·5	55·4
	16	58 38·8	06·0	164 08·9	14·5	7 34·0	11·5	55·5
	17	73 38·7	05·7	178 42·4	14·4	7 45·5	11·4	55·5
	18	88 38·6	N22 05·4	193 15·8	14·3	N 7 56·9	11·4	55·5
	19	103 38·5	05·0	207 49·1	14·3	8 08·3	11·4	55·5
	20	118 38·4	04·7	222 22·4	14·2	8 19·7	11·4	55·6
	21	133 38·4	· · 04·4	236 55·6	14·2	8 31·1	11·3	55·6
	22	148 38·3	04·0	251 28·8	14·2	8 42·4	11·3	55·6
	23	163 38·2	03·7	266 02·0	14·0	8 53·7	11·3	55·6
		S.D. 15·8	*d* 0·3	S.D. 14·8		14·9		15·1

Lat.	Twilight Naut.	Twilight Civil	Sunrise	Moonrise 9	Moonrise 10	Moonrise 11	Moonrise 12
°	h m	h m	h m	h m	h m	h m	h m
N 72	▭	▭	▭	22 00	21 43	21 24	20 59
N 70	▭	▭	▭	21 59	21 48	21 36	21 22
68	▭	▭	▭	21 59	21 53	21 46	21 40
66	////	////	01 10	21 58	21 56	21 55	21 54
64	////	////	02 00	21 57	21 59	22 02	22 06
62	////	////	02 31	21 57	22 02	22 08	22 16
60	////	01 27	02 54	21 57	22 05	22 14	22 25
N 58	////	02 03	03 12	21 56	22 07	22 18	22 32
56	////	02 29	03 27	21 56	22 09	22 23	22 39
54	01 20	02 48	03 40	21 55	22 10	22 26	22 45
52	01 54	03 05	03 52	21 55	22 12	22 30	22 51
50	02 17	03 19	04 02	21 55	22 13	22 33	22 56
45	02 58	03 46	04 23	21 54	22 17	22 40	23 07
N 40	03 27	04 07	04 40	21 54	22 19	22 46	23 16
35	03 48	04 25	04 54	21 54	22 22	22 51	23 23
30	04 06	04 39	05 06	21 53	22 24	22 56	23 30
20	04 34	05 03	05 27	21 53	22 27	23 03	23 42
N 10	04 55	05 22	05 45	21 52	22 30	23 10	23 53
0	05 13	05 39	06 02	21 52	22 33	23 17	24 03
S 10	05 29	05 55	06 18	21 51	22 37	23 23	24 13
20	05 45	06 12	06 36	21 51	22 40	23 30	24 23
30	06 00	06 30	06 56	21 50	22 44	23 39	24 36
35	06 08	06 39	07 07	21 50	22 46	23 43	24 43
40	06 17	06 50	07 21	21 49	22 48	23 49	24 51
45	06 26	07 03	07 36	21 49	22 51	23 55	25 01
S 50	06 37	07 18	07 56	21 49	22 55	24 02	00 02
52	06 42	07 25	08 05	21 48	22 56	24 06	00 06
54	06 47	07 32	08 15	21 48	22 58	24 10	00 10
56	06 53	07 41	08 27	21 48	23 00	24 14	00 14
58	06 59	07 50	08 40	21 48	23 02	24 19	00 19
S 60	07 06	08 01	08 56	21 47	23 05	24 24	00 24

Lat.	Sunset	Twilight Civil	Twilight Naut.	Moonset 9	Moonset 10	Moonset 11	Moonset 12
°	h m	h m	h m	h m	h m	h m	h m
N 72	▭	▭	▭	08 49	10 34	12 23	14 21
N 70	▭	▭	▭	08 53	10 32	12 12	14 00
68	▭	▭	▭	08 57	10 30	12 04	13 44
66	22 56	////	////	09 00	10 28	11 57	13 31
64	22 08	////	////	09 03	10 26	11 51	13 20
62	21 30	////	////	09 05	10 25	11 47	13 11
60	21 16	22 41	////	09 07	10 24	11 42	13 03
N 58	20 58	22 05	////	09 09	10 23	11 38	12 56
56	20 43	21 41	////	09 11	10 22	11 35	12 50
54	20 30	21 21	22 48	09 12	10 22	11 32	12 45
52	20 18	21 05	22 15	09 14	10 21	11 29	12 40
50	20 08	20 51	21 52	09 15	10 20	11 27	12 36
45	19 47	20 24	21 11	09 17	10 19	11 22	12 26
N 40	19 30	20 03	20 43	09 19	10 18	11 17	12 18
35	19 16	19 46	20 21	09 21	10 17	11 13	12 12
30	19 04	19 31	20 04	09 23	10 16	11 10	12 06
20	18 43	19 08	19 36	09 26	10 14	11 04	11 56
N 10	18 26	18 48	19 15	09 28	10 13	10 59	11 47
0	18 09	18 31	18 57	09 30	10 12	10 54	11 38
S 10	17 52	18 15	18 41	09 32	10 10	10 49	11 30
20	17 35	17 59	18 26	09 35	10 09	10 44	11 21
30	17 15	17 41	18 11	09 37	10 07	10 38	11 11
35	17 03	17 31	18 03	09 39	10 06	10 35	11 06
40	16 50	17 20	17 54	09 40	10 05	10 31	10 59
45	16 34	17 08	17 44	09 42	10 04	10 27	10 52
S 50	16 15	16 53	17 34	09 45	10 03	10 22	10 43
52	16 06	16 46	17 29	09 46	10 02	10 19	10 38
54	15 56	16 38	17 24	09 47	10 01	10 17	10 34
56	15 44	16 30	17 18	09 48	10 01	10 14	10 29
58	15 31	16 21	17 12	09 50	10 00	10 11	10 23
S 60	15 15	16 10	17 05	09 51	09 59	10 07	10 17

Day	SUN Eqn. of Time 00ʰ	SUN Eqn. of Time 12ʰ	SUN Mer. Pass.	MOON Mer. Pass. Upper	MOON Mer. Pass. Lower	Age	Phase
	m s	m s	h m	h m	h m	d	
9	05 02	05 06	12 05	03 20	15 41	19	◐
10	05 11	05 15	12 05	04 02	16 22	20	
11	05 19	05 23	12 05	04 44	17 05	21	

Table 28, continued

44^m INCREMENTS AND CORRECTIONS 45^m

44^m	SUN PLANETS	ARIES	MOON	v or d	Corr^n	v or d	Corr^n	v or d	Corr^n
s	° ′	° ′	° ′	′	′	′	′	′	′
00	11 00·0	11 01·8	10 29·9	0·0	0·0	6·0	4·5	12·0	8·9
01	11 00·3	11 02·1	10 30·2	0·1	0·1	6·1	4·5	12·1	9·0
02	11 00·5	11 02·3	10 30·4	0·2	0·1	6·2	4·6	12·2	9·0
03	11 00·8	11 02·6	10 30·6	0·3	0·2	6·3	4·7	12·3	9·1
04	11 01·0	11 02·8	10 30·9	0·4	0·3	6·4	4·7	12·4	9·2
05	11 01·3	11 03·1	10 31·1	0·5	0·4	6·5	4·8	12·5	9·3
06	11 01·5	11 03·3	10 31·4	0·6	0·4	6·6	4·9	12·6	9·3
07	11 01·8	11 03·6	10 31·6	0·7	0·5	6·7	5·0	12·7	9·4
08	11 02·0	11 03·8	10 31·8	0·8	0·6	6·8	5·0	12·8	9·5
09	11 02·3	11 04·1	10 32·1	0·9	0·7	6·9	5·1	12·9	9·6
10	11 02·5	11 04·3	10 32·3	1·0	0·7	7·0	5·2	13·0	9·6
11	11 02·8	11 04·6	10 32·6	1·1	0·8	7·1	5·3	13·1	9·7
12	11 03·0	11 04·8	10 32·8	1·2	0·9	7·2	5·3	13·2	9·8
13	11 03·3	11 05·1	10 33·0	1·3	1·0	7·3	5·4	13·3	9·9
14	11 03·5	11 05·3	10 33·3	1·4	1·0	7·4	5·5	13·4	9·9
15	11 03·8	11 05·6	10 33·5	1·5	1·1	7·5	5·6	13·5	10·0
16	11 04·0	11 05·8	10 33·8	1·6	1·2	7·6	5·6	13·6	10·1
17	11 04·3	11 06·1	10 34·0	1·7	1·3	7·7	5·7	13·7	10·2
18	11 04·5	11 06·3	10 34·2	1·8	1·3	7·8	5·8	13·8	10·2
19	11 04·8	11 06·6	10 34·5	1·9	1·4	7·9	5·9	13·9	10·3
20	11 05·0	11 06·8	10 34·7	2·0	1·5	8·0	5·9	14·0	10·4
21	11 05·3	11 07·1	10 34·9	2·1	1·6	8·1	6·0	14·1	10·5
22	11 05·5	11 07·3	10 35·2	2·2	1·6	8·2	6·1	14·2	10·5
23	11 05·8	11 07·6	10 35·4	2·3	1·7	8·3	6·2	14·3	10·6
24	11 06·0	11 07·8	10 35·7	2·4	1·8	8·4	6·2	14·4	10·7
25	11 06·3	11 08·1	10 35·9	2·5	1·9	8·5	6·3	14·5	10·8
26	11 06·5	11 08·3	10 36·1	2·6	1·9	8·6	6·4	14·6	10·8
27	11 06·8	11 08·6	10 36·4	2·7	2·0	8·7	6·5	14·7	10·9
28	11 07·0	11 08·8	10 36·6	2·8	2·1	8·8	6·5	14·8	11·0
29	11 07·3	11 09·1	10 36·9	2·9	2·2	8·9	6·6	14·9	11·1
30	11 07·5	11 09·3	10 37·1	3·0	2·2	9·0	6·7	15·0	11·1
31	11 07·8	11 09·6	10 37·3	3·1	2·3	9·1	6·7	15·1	11·2
32	11 08·0	11 09·8	10 37·6	3·2	2·4	9·2	6·8	15·2	11·3
33	11 08·3	11 10·1	10 37·8	3·3	2·4	9·3	6·9	15·3	11·3
34	11 08·5	11 10·3	10 38·0	3·4	2·5	9·4	7·0	15·4	11·4
35	11 08·8	11 10·6	10 38·3	3·5	2·6	9·5	7·0	15·5	11·5
36	11 09·0	11 10·8	10 38·5	3·6	2·7	9·6	7·1	15·6	11·6
37	11 09·3	11 11·1	10 38·8	3·7	2·7	9·7	7·2	15·7	11·6
38	11 09·5	11 11·3	10 39·0	3·8	2·8	9·8	7·3	15·8	11·7
39	11 09·8	11 11·6	10 39·2	3·9	2·9	9·9	7·3	15·9	11·8
40	11 10·0	11 11·8	10 39·5	4·0	3·0	10·0	7·4	16·0	11·9
41	11 10·3	11 12·1	10 39·7	4·1	3·0	10·1	7·5	16·1	11·9
42	11 10·5	11 12·3	10 40·0	4·2	3·1	10·2	7·6	16·2	12·0
43	11 10·8	11 12·6	10 40·2	4·3	3·2	10·3	7·6	16·3	12·1
44	11 11·0	11 12·8	10 40·4	4·4	3·3	10·4	7·7	16·4	12·2
45	11 11·3	11 13·1	10 40·7	4·5	3·3	10·5	7·8	16·5	12·2
46	11 11·5	11 13·3	10 40·9	4·6	3·4	10·6	7·9	16·6	12·3
47	11 11·8	11 13·6	10 41·1	4·7	3·5	10·7	7·9	16·7	12·4
48	11 12·0	11 13·8	10 41·4	4·8	3·6	10·8	8·0	16·8	12·5
49	11 12·3	11 14·1	10 41·6	4·9	3·6	10·9	8·1	16·9	12·5
50	11 12·5	11 14·3	10 41·9	5·0	3·7	11·0	8·2	17·0	12·6
51	11 12·8	11 14·6	10 42·1	5·1	3·8	11·1	8·2	17·1	12·7
52	11 13·0	11 14·8	10 42·3	5·2	3·9	11·2	8·3	17·2	12·8
53	11 13·3	11 15·1	10 42·6	5·3	3·9	11·3	8·4	17·3	12·8
54	11 13·5	11 15·3	10 42·8	5·4	4·0	11·4	8·5	17·4	12·9
55	11 13·8	11 15·6	10 43·1	5·5	4·1	11·5	8·5	17·5	13·0
56	11 14·0	11 15·8	10 43·3	5·6	4·2	11·6	8·6	17·6	13·1
57	11 14·3	11 16·1	10 43·5	5·7	4·2	11·7	8·7	17·7	13·1
58	11 14·5	11 16·3	10 43·8	5·8	4·3	11·8	8·8	17·8	13·2
59	11 14·8	11 16·6	10 44·0	5·9	4·4	11·9	8·8	17·9	13·3
60	11 15·0	11 16·8	10 44·3	6·0	4·5	12·0	8·9	18·0	13·4

45^m	SUN PLANETS	ARIES	MOON	v or d	Corr^n	v or d	Corr^n	v or d	Corr^n
s	° ′	° ′	° ′	′	′	′	′	′	′
00	11 15·0	11 16·8	10 44·3	0·0	0·0	6·0	4·6	12·0	9·1
01	11 15·3	11 17·1	10 44·5	0·1	0·1	6·1	4·6	12·1	9·2
02	11 15·5	11 17·3	10 44·7	0·2	0·2	6·2	4·7	12·2	9·3
03	11 15·8	11 17·6	10 45·0	0·3	0·2	6·3	4·8	12·3	9·3
04	11 16·0	11 17·9	10 45·2	0·4	0·3	6·4	4·9	12·4	9·4
05	11 16·3	11 18·1	10 45·4	0·5	0·4	6·5	4·9	12·5	9·5
06	11 16·5	11 18·4	10 45·7	0·6	0·5	6·6	5·0	12·6	9·6
07	11 16·8	11 18·6	10 45·9	0·7	0·5	6·7	5·1	12·7	9·6
08	11 17·0	11 18·9	10 46·2	0·8	0·6	6·8	5·2	12·8	9·7
09	11 17·3	11 19·1	10 46·4	0·9	0·7	6·9	5·2	12·9	9·8
10	11 17·5	11 19·4	10 46·6	1·0	0·8	7·0	5·3	13·0	9·9
11	11 17·8	11 19·6	10 46·9	1·1	0·8	7·1	5·4	13·1	9·9
12	11 18·0	11 19·9	10 47·1	1·2	0·9	7·2	5·5	13·2	10·0
13	11 18·3	11 20·1	10 47·4	1·3	1·0	7·3	5·5	13·3	10·1
14	11 18·5	11 20·4	10 47·6	1·4	1·1	7·4	5·6	13·4	10·2
15	11 18·8	11 20·6	10 47·8	1·5	1·1	7·5	5·7	13·5	10·2
16	11 19·0	11 20·9	10 48·1	1·6	1·2	7·6	5·8	13·6	10·3
17	11 19·3	11 21·1	10 48·3	1·7	1·3	7·7	5·8	13·7	10·4
18	11 19·5	11 21·4	10 48·5	1·8	1·4	7·8	5·9	13·8	10·5
19	11 19·8	11 21·6	10 48·8	1·9	1·4	7·9	6·0	13·9	10·5
20	11 20·0	11 21·9	10 49·0	2·0	1·5	8·0	6·1	14·0	10·6
21	11 20·3	11 22·1	10 49·3	2·1	1·6	8·1	6·1	14·1	10·7
22	11 20·5	11 22·4	10 49·5	2·2	1·7	8·2	6·2	14·2	10·8
23	11 20·8	11 22·6	10 49·7	2·3	1·7	8·3	6·3	14·3	10·8
24	11 21·0	11 22·9	10 50·0	2·4	1·8	8·4	6·4	14·4	10·9
25	11 21·3	11 23·1	10 50·2	2·5	1·9	8·5	6·4	14·5	11·0
26	11 21·5	11 23·4	10 50·5	2·6	2·0	8·6	6·5	14·6	11·1
27	11 21·8	11 23·6	10 50·7	2·7	2·0	8·7	6·6	14·7	11·1
28	11 22·0	11 23·9	10 50·9	2·8	2·1	8·8	6·7	14·8	11·2
29	11 22·3	11 24·1	10 51·2	2·9	2·2	8·9	6·7	14·9	11·3
30	11 22·5	11 24·4	10 51·4	3·0	2·3	9·0	6·8	15·0	11·4
31	11 22·8	11 24·6	10 51·6	3·1	2·4	9·1	6·9	15·1	11·5
32	11 23·0	11 24·9	10 51·9	3·2	2·4	9·2	7·0	15·2	11·5
33	11 23·3	11 25·1	10 52·1	3·3	2·5	9·3	7·1	15·3	11·6
34	11 23·5	11 25·4	10 52·4	3·4	2·6	9·4	7·1	15·4	11·7
35	11 23·8	11 25·6	10 52·6	3·5	2·7	9·5	7·2	15·5	11·8
36	11 24·0	11 25·9	10 52·8	3·6	2·7	9·6	7·3	15·6	11·8
37	11 24·3	11 26·1	10 53·1	3·7	2·8	9·7	7·4	15·7	11·9
38	11 24·5	11 26·4	10 53·3	3·8	2·9	9·8	7·4	15·8	12·0
39	11 24·8	11 26·6	10 53·6	3·9	3·0	9·9	7·5	15·9	12·1
40	11 25·0	11 26·9	10 53·8	4·0	3·0	10·0	7·6	16·0	12·1
41	11 25·3	11 27·1	10 54·0	4·1	3·1	10·1	7·7	16·1	12·2
42	11 25·5	11 27·4	10 54·3	4·2	3·2	10·2	7·7	16·2	12·3
43	11 25·8	11 27·6	10 54·5	4·3	3·3	10·3	7·8	16·3	12·4
44	11 26·0	11 27·9	10 54·7	4·4	3·3	10·4	7·9	16·4	12·4
45	11 26·3	11 28·1	10 55·0	4·5	3·4	10·5	8·0	16·5	12·5
46	11 26·5	11 28·4	10 55·2	4·6	3·5	10·6	8·0	16·6	12·6
47	11 26·8	11 28·6	10 55·5	4·7	3·6	10·7	8·1	16·7	12·7
48	11 27·0	11 28·9	10 55·7	4·8	3·6	10·8	8·2	16·8	12·7
49	11 27·3	11 29·1	10 55·9	4·9	3·7	10·9	8·3	16·9	12·8
50	11 27·5	11 29·4	10 56·2	5·0	3·8	11·0	8·3	17·0	12·9
51	11 27·8	11 29·6	10 56·4	5·1	3·9	11·1	8·4	17·1	13·0
52	11 28·0	11 29·9	10 56·7	5·2	3·9	11·2	8·5	17·2	13·0
53	11 28·3	11 30·1	10 56·9	5·3	4·0	11·3	8·6	17·3	13·1
54	11 28·5	11 30·4	10 57·1	5·4	4·1	11·4	8·6	17·4	13·2
55	11 28·8	11 30·6	10 57·4	5·5	4·2	11·5	8·7	17·5	13·3
56	11 29·0	11 30·9	10 57·6	5·6	4·2	11·6	8·8	17·6	13·3
57	11 29·3	11 31·1	10 57·9	5·7	4·3	11·7	8·9	17·7	13·4
58	11 29·5	11 31·4	10 58·1	5·8	4·4	11·8	8·9	17·8	13·5
59	11 29·8	11 31·6	10 58·3	5·9	4·5	11·9	9·0	17·9	13·6
60	11 30·0	11 31·9	10 58·6	6·0	4·6	12·0	9·1	18·0	13·7

Table 29

32^m INCREMENTS AND CORRECTIONS 33^m

32^m	SUN PLANETS	ARIES	MOON	v or d	Corr^n	v or d	Corr^n	v or d	Corr^n
s	° ′	° ′	° ′	′	′	′	′	′	′
00	8 00·0	8 01·3	7 38·1	0·0	0·0	6·0	3·3	12·0	6·5
01	8 00·3	8 01·6	7 38·4	0·1	0·1	6·1	3·3	12·1	6·6
02	8 00·5	8 01·8	7 38·6	0·2	0·1	6·2	3·4	12·2	6·6
03	8 00·8	8 02·1	7 38·8	0·3	0·2	6·3	3·4	12·3	6·7
04	8 01·0	8 02·3	7 39·1	0·4	0·2	6·4	3·5	12·4	6·7
05	8 01·3	8 02·6	7 39·3	0·5	0·3	6·5	3·5	12·5	6·8
06	8 01·5	8 02·8	7 39·6	0·6	0·3	6·6	3·6	12·6	6·8
07	8 01·8	8 03·1	7 39·8	0·7	0·4	6·7	3·6	12·7	6·9
08	8 02·0	8 03·3	7 40·0	0·8	0·4	6·8	3·7	12·8	6·9
09	8 02·3	8 03·6	7 40·3	0·9	0·5	6·9	3·7	12·9	7·0
10	8 02·5	8 03·8	7 40·5	1·0	0·5	7·0	3·8	13·0	7·0
11	8 02·8	8 04·1	7 40·8	1·1	0·6	7·1	3·8	13·1	7·1
12	8 03·0	8 04·3	7 41·0	1·2	0·7	7·2	3·9	13·2	7·2
13	8 03·3	8 04·6	7 41·2	1·3	0·7	7·3	4·0	13·3	7·2
14	8 03·5	8 04·8	7 41·5	1·4	0·8	7·4	4·0	13·4	7·3
15	8 03·8	8 05·1	7 41·7	1·5	0·8	7·5	4·1	13·5	7·3
16	8 04·0	8 05·3	7 42·0	1·6	0·9	7·6	4·1	13·6	7·4
17	8 04·3	8 05·6	7 42·2	1·7	0·9	7·7	4·2	13·7	7·4
18	8 04·5	8 05·8	7 42·4	1·8	1·0	7·8	4·2	13·8	7·5
19	8 04·8	8 06·1	7 42·7	1·9	1·0	7·9	4·3	13·9	7·5
20	8 05·0	8 06·3	7 42·9	2·0	1·1	8·0	4·3	14·0	7·6
21	8 05·3	8 06·6	7 43·1	2·1	1·1	8·1	4·4	14·1	7·6
22	8 05·5	8 06·8	7 43·4	2·2	1·2	8·2	4·4	14·2	7·7
23	8 05·8	8 07·1	7 43·6	2·3	1·2	8·3	4·5	14·3	7·7
24	8 06·0	8 07·3	7 43·9	2·4	1·3	8·4	4·6	14·4	7·8
25	8 06·3	8 07·6	7 44·1	2·5	1·4	8·5	4·6	14·5	7·9
26	8 06·5	8 07·8	7 44·3	2·6	1·4	8·6	4·7	14·6	7·9
27	8 06·8	8 08·1	7 44·6	2·7	1·5	8·7	4·7	14·7	8·0
28	8 07·0	8 08·3	7 44·8	2·8	1·5	8·8	4·8	14·8	8·0
29	8 07·3	8 08·6	7 45·1	2·9	1·6	8·9	4·8	14·9	8·1
30	8 07·5	8 08·8	7 45·3	3·0	1·6	9·0	4·9	15·0	8·1
31	8 07·8	8 09·1	7 45·5	3·1	1·7	9·1	4·9	15·1	8·2
32	8 08·0	8 09·3	7 45·8	3·2	1·7	9·2	5·0	15·2	8·2
33	8 08·3	8 09·6	7 46·0	3·3	1·8	9·3	5·0	15·3	8·3
34	8 08·5	8 09·8	7 46·2	3·4	1·8	9·4	5·1	15·4	8·3
35	8 08·8	8 10·1	7 46·5	3·5	1·9	9·5	5·1	15·5	8·4
36	8 09·0	8 10·3	7 46·7	3·6	2·0	9·6	5·2	15·6	8·5
37	8 09·3	8 10·6	7 47·0	3·7	2·0	9·7	5·3	15·7	8·5
38	8 09·5	8 10·8	7 47·2	3·8	2·1	9·8	5·3	15·8	8·6
39	8 09·8	8 11·1	7 47·4	3·9	2·1	9·9	5·4	15·9	8·6
40	8 10·0	8 11·3	7 47·7	4·0	2·2	10·0	5·4	16·0	8·7
41	8 10·3	8 11·6	7 47·9	4·1	2·2	10·1	5·5	16·1	8·7
42	8 10·5	8 11·8	7 48·2	4·2	2·3	10·2	5·5	16·2	8·8
43	8 10·8	8 12·1	7 48·4	4·3	2·3	10·3	5·6	16·3	8·8
44	8 11·0	8 12·3	7 48·6	4·4	2·4	10·4	5·6	16·4	8·9
45	8 11·3	8 12·6	7 48·9	4·5	2·4	10·5	5·7	16·5	8·9
46	8 11·5	8 12·8	7 49·1	4·6	2·5	10·6	5·7	16·6	9·0
47	8 11·8	8 13·1	7 49·3	4·7	2·5	10·7	5·8	16·7	9·0
48	8 12·0	8 13·3	7 49·6	4·8	2·6	10·8	5·9	16·8	9·1
49	8 12·3	8 13·6	7 49·8	4·9	2·7	10·9	5·9	16·9	9·2
50	8 12·5	8 13·8	7 50·1	5·0	2·7	11·0	6·0	17·0	9·2
51	8 12·8	8 14·1	7 50·3	5·1	2·8	11·1	6·0	17·1	9·3
52	8 13·0	8 14·3	7 50·5	5·2	2·8	11·2	6·1	17·2	9·3
53	8 13·3	8 14·6	7 50·8	5·3	2·9	11·3	6·1	17·3	9·4
54	8 13·5	8 14·9	7 51·0	5·4	2·9	11·4	6·2	17·4	9·4
55	8 13·8	8 15·1	7 51·3	5·5	3·0	11·5	6·2	17·5	9·5
56	8 14·0	8 15·4	7 51·5	5·6	3·0	11·6	6·3	17·6	9·5
57	8 14·3	8 15·6	7 51·7	5·7	3·1	11·7	6·3	17·7	9·6
58	8 14·5	8 15·9	7 52·0	5·8	3·1	11·8	6·4	17·8	9·6
59	8 14·8	8 16·1	7 52·2	5·9	3·2	11·9	6·4	17·9	9·7
60	8 15·0	8 16·4	7 52·5	6·0	3·3	12·0	6·5	18·0	9·8

33^m	SUN PLANETS	ARIES	MOON	v or d	Corr^n	v or d	Corr^n	v or d	Corr^n
s	° ′	° ′	° ′	′	′	′	′	′	′
00	8 15·0	8 16·4	7 52·5	0·0	0·0	6·0	3·4	12·0	6·7
01	8 15·3	8 16·6	7 52·7	0·1	0·1	6·1	3·4	12·1	6·8
02	8 15·5	8 16·9	7 52·9	0·2	0·1	6·2	3·5	12·2	6·8
03	8 15·8	8 17·1	7 53·2	0·3	0·2	6·3	3·5	12·3	6·9
04	8 16·0	8 17·4	7 53·4	0·4	0·2	6·4	3·6	12·4	6·9
05	8 16·3	8 17·6	7 53·6	0·5	0·3	6·5	3·6	12·5	7·0
06	8 16·5	8 17·9	7 53·9	0·6	0·3	6·6	3·7	12·6	7·0
07	8 16·8	8 18·1	7 54·1	0·7	0·4	6·7	3·7	12·7	7·1
08	8 17·0	8 18·4	7 54·4	0·8	0·4	6·8	3·8	12·8	7·1
09	8 17·3	8 18·6	7 54·6	0·9	0·5	6·9	3·9	12·9	7·2
10	8 17·5	8 18·9	7 54·8	1·0	0·6	7·0	3·9	13·0	7·3
11	8 17·8	8 19·1	7 55·1	1·1	0·6	7·1	4·0	13·1	7·3
12	8 18·0	8 19·4	7 55·3	1·2	0·7	7·2	4·0	13·2	7·4
13	8 18·3	8 19·6	7 55·6	1·3	0·7	7·3	4·1	13·3	7·4
14	8 18·5	8 19·9	7 55·8	1·4	0·8	7·4	4·1	13·4	7·5
15	8 18·8	8 20·1	7 56·0	1·5	0·8	7·5	4·2	13·5	7·5
16	8 19·0	8 20·4	7 56·3	1·6	0·9	7·6	4·2	13·6	7·6
17	8 19·3	8 20·6	7 56·5	1·7	0·9	7·7	4·3	13·7	7·6
18	8 19·5	8 20·9	7 56·7	1·8	1·0	7·8	4·4	13·8	7·7
19	8 19·8	8 21·1	7 57·0	1·9	1·1	7·9	4·4	13·9	7·8
20	8 20·0	8 21·4	7 57·2	2·0	1·1	8·0	4·5	14·0	7·8
21	8 20·3	8 21·6	7 57·5	2·1	1·2	8·1	4·5	14·1	7·9
22	8 20·5	8 21·9	7 57·7	2·2	1·2	8·2	4·6	14·2	7·9
23	8 20·8	8 22·1	7 57·9	2·3	1·3	8·3	4·6	14·3	8·0
24	8 21·0	8 22·4	7 58·2	2·4	1·3	8·4	4·7	14·4	8·0
25	8 21·3	8 22·6	7 58·4	2·5	1·4	8·5	4·7	14·5	8·1
26	8 21·5	8 22·9	7 58·7	2·6	1·5	8·6	4·8	14·6	8·2
27	8 21·8	8 23·1	7 58·9	2·7	1·5	8·7	4·9	14·7	8·2
28	8 22·0	8 23·4	7 59·1	2·8	1·6	8·8	4·9	14·8	8·3
29	8 22·3	8 23·6	7 59·4	2·9	1·6	8·9	5·0	14·9	8·3
30	8 22·5	8 23·9	7 59·6	3·0	1·7	9·0	5·0	15·0	8·4
31	8 22·8	8 24·1	7 59·8	3·1	1·7	9·1	5·1	15·1	8·4
32	8 23·0	8 24·4	8 00·1	3·2	1·8	9·2	5·1	15·2	8·5
33	8 23·3	8 24·6	8 00·3	3·3	1·8	9·3	5·2	15·3	8·5
34	8 23·5	8 24·9	8 00·6	3·4	1·9	9·4	5·2	15·4	8·6
35	8 23·8	8 25·1	8 00·8	3·5	2·0	9·5	5·3	15·5	8·7
36	8 24·0	8 25·4	8 01·0	3·6	2·0	9·6	5·4	15·6	8·7
37	8 24·3	8 25·6	8 01·3	3·7	2·1	9·7	5·4	15·7	8·8
38	8 24·5	8 25·9	8 01·5	3·8	2·1	9·8	5·5	15·8	8·8
39	8 24·8	8 26·1	8 01·8	3·9	2·2	9·9	5·5	15·9	8·9
40	8 25·0	8 26·4	8 02·0	4·0	2·2	10·0	5·6	16·0	8·9
41	8 25·3	8 26·6	8 02·2	4·1	2·3	10·1	5·6	16·1	9·0
42	8 25·5	8 26·9	8 02·5	4·2	2·3	10·2	5·7	16·2	9·0
43	8 25·8	8 27·1	8 02·7	4·3	2·4	10·3	5·8	16·3	9·1
44	8 26·0	8 27·4	8 02·9	4·4	2·5	10·4	5·8	16·4	9·2
45	8 26·3	8 27·6	8 03·2	4·5	2·5	10·5	5·9	16·5	9·2
46	8 26·5	8 27·9	8 03·4	4·6	2·6	10·6	5·9	16·6	9·3
47	8 26·8	8 28·1	8 03·7	4·7	2·6	10·7	6·0	16·7	9·3
48	8 27·0	8 28·4	8 03·9	4·8	2·7	10·8	6·0	16·8	9·4
49	8 27·3	8 28·6	8 04·1	4·9	2·7	10·9	6·1	16·9	9·4
50	8 27·5	8 28·9	8 04·4	5·0	2·8	11·0	6·1	17·0	9·5
51	8 27·8	8 29·1	8 04·6	5·1	2·8	11·1	6·2	17·1	9·5
52	8 28·0	8 29·4	8 04·9	5·2	2·9	11·2	6·3	17·2	9·6
53	8 28·3	8 29·6	8 05·1	5·3	3·0	11·3	6·3	17·3	9·7
54	8 28·5	8 29·9	8 05·3	5·4	3·0	11·4	6·4	17·4	9·7
55	8 28·8	8 30·1	8 05·6	5·5	3·1	11·5	6·4	17·5	9·8
56	8 29·0	8 30·4	8 05·8	5·6	3·1	11·6	6·5	17·6	9·8
57	8 29·3	8 30·6	8 06·1	5·7	3·2	11·7	6·5	17·7	9·9
58	8 29·5	8 30·9	8 06·3	5·8	3·2	11·8	6·6	17·8	9·9
59	8 29·8	8 31·1	8 06·5	5·9	3·3	11·9	6·6	17·9	10·0
60	8 30·0	8 31·4	8 06·8	6·0	3·4	12·0	6·7	18·0	10·1

Table 30

CAPELLA

GHA ♈ 8 Hr.	=	47°52′
32 Min. 10 Sec.	=	8°04′
GHA	=	55°56′
	+	360°
GHA	=	415°56′
Assumed Long. W.	=	65°56′
LHA ♈	=	350°

Hc = 25°13′
Ho = 25°36′ *Zn = 50°
23′ Toward

FOMALHAUT

GHA ♈ 8 Hr.	=	47°52′
33 Min. 15 Sec.	=	8°20′
GHA	=	55°72′
	+	360°
GHA	=	416°12′
Assumed Long. W.	=	66°12′
LHA ♈	=	350°

Hc = 24°58′
Ho = 25°01′ *Zn = 186°
3′ Toward

VEGA

GHA ♈ 8 Hr.	=	47°52′
33 Min. 53 Sec.	=	8°30′
GHA	=	55°82′
	+	360°
GHA ♈	=	416°22′
Assumed Long. W.	=	66°22′
LHA	=	350°

Hc = 34°32′
Ho = 33°25′ *Zn = 296°
1°07′ Away

* Note that bearing (Zn) is given. There is no need to figure Zn from Z in using Volume I of H.O. 249.

Fig. 11.1

In order to make most efficient use of the almanac, you set up the worksheet as shown in Figure 11.1, using Tables 27, 28, and 30. This round of sights is plotted in Figure 11.2 and results in a triangle about 8 miles on a side. Most probably you are in the middle of that triangle.

Now, one final thing. Volume I of H.O. 249 is made up in five-year intervals (called epochs), and since the earth wobbles on its axis (through precession and nutation), a correction is necessary. The table for precession and nutation is reproduced as Table 31. Under each year, it gives a distance and a direction in which the fix (or single LOP) is to be moved. The entry information is LHA Aries and your latitude. Thus, your fix in 1974 at 40° N must be moved 3 miles in direction 060°—roughly northeast.

In taking evening twilight sights of stars, the procedure is the same: Figure out GMT of civil twilight and use that time to look up appropriate stars in H.O. 249, Volume I. Set your sextant to Hc and point in the direction of the star. The tricky part here is that you need the light-gathering power of the sextant to pick up the star and will probably not be able to pick it up with your unaided eye.

The planets lie somewhere between the moon and the stars as far as their navigational characteristics are concerned. Like the most useful stars, they are small and bright; like the moon, some of them have a relatively large *v* correction. Through even a 4-power scope, the planets appear as small disks about the size of pinheads.

Table 32 is a reproduction of the planet diagram from the *Nautical Almanac*. You read it from right to left for any given day, and it shows the relation of the planets and the sun. For example, on April 30, Jupiter is 4 hours ahead of the sun, then comes Venus; Mercury is one-half hour ahead of the sun and too close to be seen in the sun's glare (the gray area), Saturn is 3½ hours behind the sun, and Mars is one-half hour after Saturn. Therefore, if you wanted to use planets on April 30, 1974, your morning planets would be Jupiter and Venus, and your evening planets would be Mars and Saturn. The GHAs of these pairs would be too close to allow them to be used in pairs for a single fix, but in July you could get a two-planet fix using Jupiter and Venus. From the chart, they are about 5 to 7 hours apart during July, and since 1 hour of time is 15° of GHA, they are 75° to 105° of

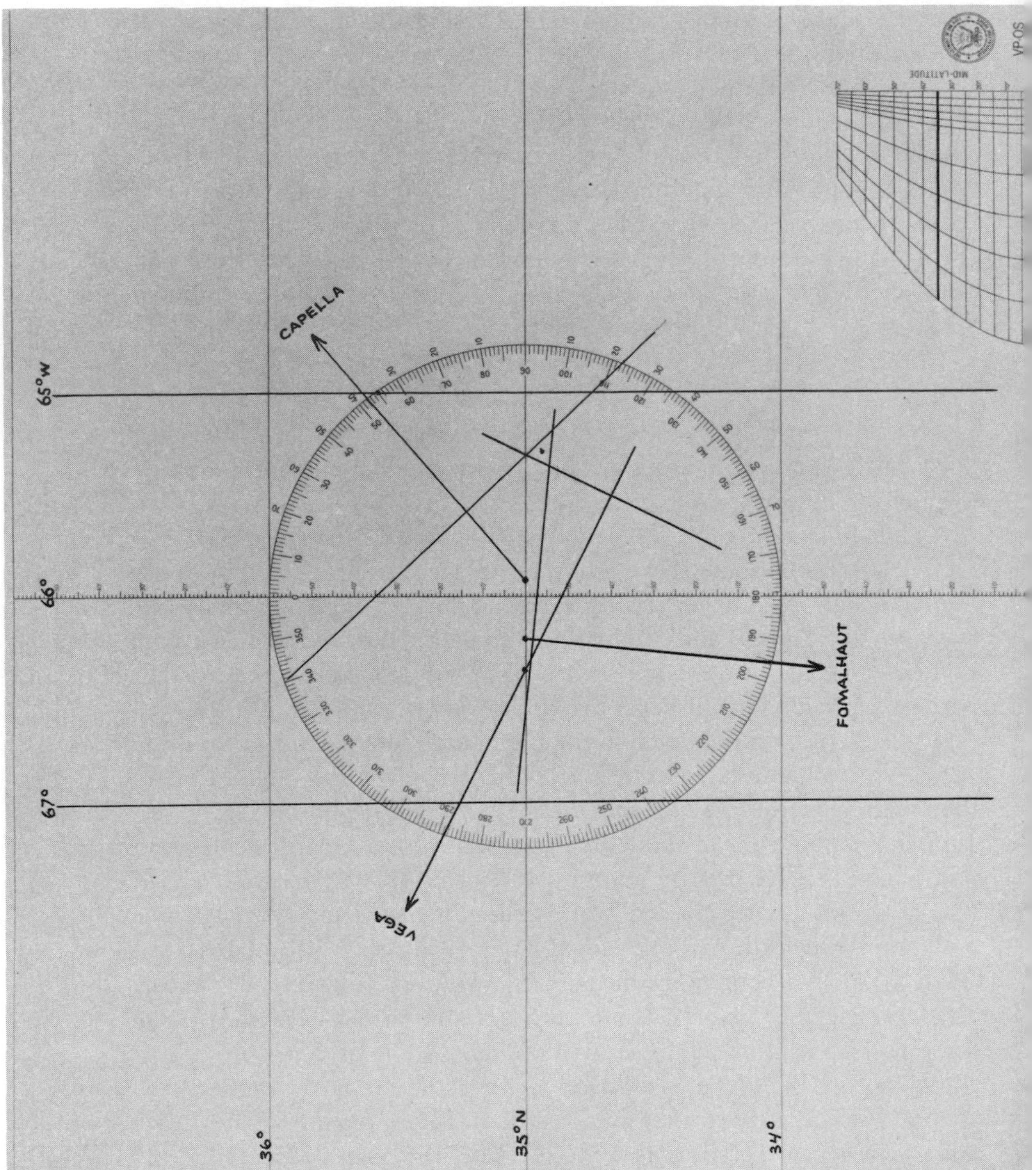

Fig. 11.2

Precession and Nutation Correction

Latitude															
N89°	N80°	N70°	N60°	N50°	N40°	N20°	0°	S20°	S40°	S50°	S60°	S70°	S80°	S89°	LHA ♈
mi. °	mi. °	mi. °	mi. °	mi. °	mi. °	mi. °	mi. °	mi. °	mi. °	mi. °	mi. °	mi. °	mi. °	mi. °	°
1972															
1 000	1 020	1 040	1 050	2 050	2 060	2 070	2 070	2 070	2 060	2 060	1 050	1 040	1 030	1 010	0
1 030	1 040	1 050	2 060	2 060	2 070	2 070	2 070	2 070	2 060	1 050	1 040	1 020	1 000	1 340	30
1 060	1 070	2 070	2 070	2 080	2 080	2 080	2 080	2 070	1 070	1 060	1 040	1 000	1 320	1 310	60
1 090	1 090	2 090	2 090	2 090	2 090	2 090	2 090	2 090	1 090	1 080	0 —	0 —	1 280	1 270	90
1 120	1 110	2 110	2 100	2 100	2 100	2 100	2 100	2 100	1 110	1 120	1 140	0 —	1 230	1 240	120
1 140	1 130	1 120	2 120	2 110	2 110	2 110	2 110	2 110	2 120	1 130	1 140	1 160	1 190	1 210	150
1 170	1 150	1 140	1 130	2 120	2 120	2 110	2 110	2 110	2 120	1 130	1 130	1 140	1 160	1 180	180
1 200	1 180	1 160	1 140	1 130	2 120	2 110	2 110	2 110	2 110	2 120	2 120	1 130	1 140	1 150	210
1 230	1 220	1 180	1 140	1 120	1 110	2 110	2 100	2 100	2 100	2 100	2 110	2 110	1 110	1 120	240
1 270	1 260	0 —	0 —	1 100	1 090	2 090	2 090	2 090	2 090	2 090	2 090	2 090	1 090	1 090	270
1 300	1 310	0 —	1 040	1 060	1 070	2 080	2 080	2 080	2 080	2 080	2 080	2 070	1 070	1 060	300
1 330	1 350	1 020	1 040	1 050	2 060	2 070	2 070	2 070	2 070	2 070	2 060	1 060	1 050	1 040	330
1 000	1 020	1 040	1 050	2 050	2 060	2 070	2 070	2 070	2 060	2 060	1 050	1 040	1 030	1 010	360
1973															
1 000	1 020	2 040	2 050	2 060	3 060	3 070	3 070	3 070	3 060	2 060	2 050	2 040	1 020	1 000	0
1 030	2 050	2 060	2 060	3 070	3 070	3 070	3 070	3 070	2 060	2 050	1 040	1 020	1 350	1 330	30
1 060	2 070	2 070	3 080	3 080	3 080	3 080	3 080	2 080	2 070	1 060	1 040	1 000	1 320	1 300	60
1 090	2 090	2 090	3 090	3 090	3 090	3 090	3 090	2 090	1 090	1 090	0 —	0 —	1 270	1 270	90
1 120	2 110	2 110	3 100	3 100	3 100	3 100	3 100	2 100	2 110	1 120	1 140	1 180	1 220	1 240	120
1 150	2 130	2 120	2 120	3 110	3 110	3 110	3 110	3 110	2 120	2 130	1 140	1 160	1 190	1 210	150
1 180	1 160	2 140	2 130	2 120	3 120	3 110	3 110	3 110	3 120	2 120	2 130	2 140	1 160	1 180	180
1 210	1 190	1 160	1 140	2 130	2 120	3 110	3 110	3 110	3 110	3 110	2 120	2 120	2 130	1 150	210
1 240	1 220	1 180	1 140	1 120	2 110	2 100	3 100	3 100	3 100	3 100	3 100	2 110	2 110	1 120	240
1 270	1 270	0 —	0 —	1 090	1 090	2 090	3 090	3 090	3 090	3 090	3 090	2 090	2 090	1 090	270
1 300	1 320	1 000	1 040	1 060	2 070	2 080	3 080	3 080	3 080	3 080	3 080	2 070	2 070	1 060	300
1 330	1 350	1 020	1 040	2 050	2 060	3 070	3 070	3 070	3 070	3 070	2 060	2 060	2 050	1 030	330
1 000	1 020	2 040	2 050	2 060	3 060	3 070	3 070	3 070	3 060	2 060	2 050	2 040	1 020	1 000	360
1974															
2 000	2 020	2 040	2 050	3 060	3 060	4 070	4 070	4 070	3 060	3 060	2 050	2 040	2 020	2 000	0
2 030	2 050	2 060	3 060	3 070	4 070	4 070	4 070	3 070	3 060	2 050	2 040	1 020	1 350	2 330	30
2 060	2 070	3 070	3 080	4 080	4 080	4 080	4 080	3 080	2 070	1 060	1 040	1 000	1 320	2 300	60
2 090	2 090	3 090	3 090	4 090	4 090	4 090	4 090	3 090	2 090	1 090	0 —	0 —	1 270	2 270	90
2 120	2 110	3 110	3 110	4 100	4 100	4 100	4 100	3 100	2 110	2 120	1 140	1 180	1 220	2 240	120
2 150	2 130	2 130	3 120	3 120	4 110	4 110	4 110	4 110	3 120	2 130	2 140	2 160	1 180	2 210	150
2 180	2 160	2 140	2 130	3 120	3 120	4 110	4 110	4 110	3 120	3 120	2 130	2 140	2 160	2 180	180
2 210	1 190	1 160	2 140	2 130	3 120	3 110	4 110	4 110	4 110	3 110	3 120	2 120	2 130	2 150	210
2 240	1 220	1 180	1 140	1 120	2 110	3 100	4 100	4 100	4 100	4 100	3 100	3 110	2 110	2 120	240
2 270	1 270	0 —	0 —	1 090	2 090	3 090	4 090	4 090	4 090	4 090	3 090	3 090	2 090	2 090	270
2 300	1 320	1 000	1 040	2 060	2 070	3 070	4 080	4 080	4 080	4 080	3 070	3 070	2 070	2 060	300
2 330	1 000	2 020	2 040	2 050	3 060	4 070	4 070	4 070	4 070	3 060	3 060	2 050	2 050	2 030	330
2 000	2 020	2 040	2 050	3 060	3 060	4 070	4 070	4 070	3 060	3 060	2 050	2 040	2 020	2 000	360
1975															
2 000	2 020	2 040	3 050	3 060	4 060	5 070	5 070	5 070	4 060	3 060	3 050	2 040	2 020	2 000	0
2 030	2 050	3 060	4 060	4 070	4 070	5 070	5 070	4 070	3 060	3 050	2 040	2 020	2 350	2 330	30
2 060	3 070	3 070	4 080	4 080	5 080	5 080	5 080	4 080	2 070	2 060	1 040	1 350	1 310	2 300	60
2 090	3 090	3 090	4 090	4 090	5 090	5 090	4 090	4 090	2 090	1 090	1 100	0 —	1 270	2 270	90
2 120	3 110	3 110	4 110	4 100	5 100	5 100	5 100	4 110	3 110	2 120	1 140	1 180	1 220	2 240	120
2 150	2 140	3 130	3 120	4 120	4 110	5 110	5 110	4 110	3 120	3 130	2 140	2 160	2 180	2 210	150
2 180	2 160	2 140	3 130	3 120	4 120	5 110	5 110	5 110	4 120	3 120	3 130	2 140	2 160	2 180	180
2 210	2 190	2 160	2 140	3 130	3 120	4 110	5 110	5 110	4 110	4 110	4 120	3 120	2 130	2 150	210
2 240	1 230	1 190	1 140	2 120	2 110	4 100	5 100	5 100	5 100	4 100	4 100	3 110	3 110	2 120	240
2 270	1 270	0 —	1 080	1 090	2 090	4 090	4 090	5 090	5 090	4 090	4 090	3 090	3 090	2 090	270
2 300	1 320	1 000	1 040	2 060	3 070	4 070	5 080	5 080	5 080	4 080	4 070	3 070	3 070	2 060	300
2 330	2 000	2 020	2 040	3 050	3 060	4 070	5 070	5 070	4 070	4 060	3 060	3 050	2 040	2 030	330
2 000	2 020	2 040	3 050	3 060	4 060	5 070	5 070	5 070	4 060	3 060	3 050	2 040	2 020	2 000	360

Table 31

PLANETS, 1974

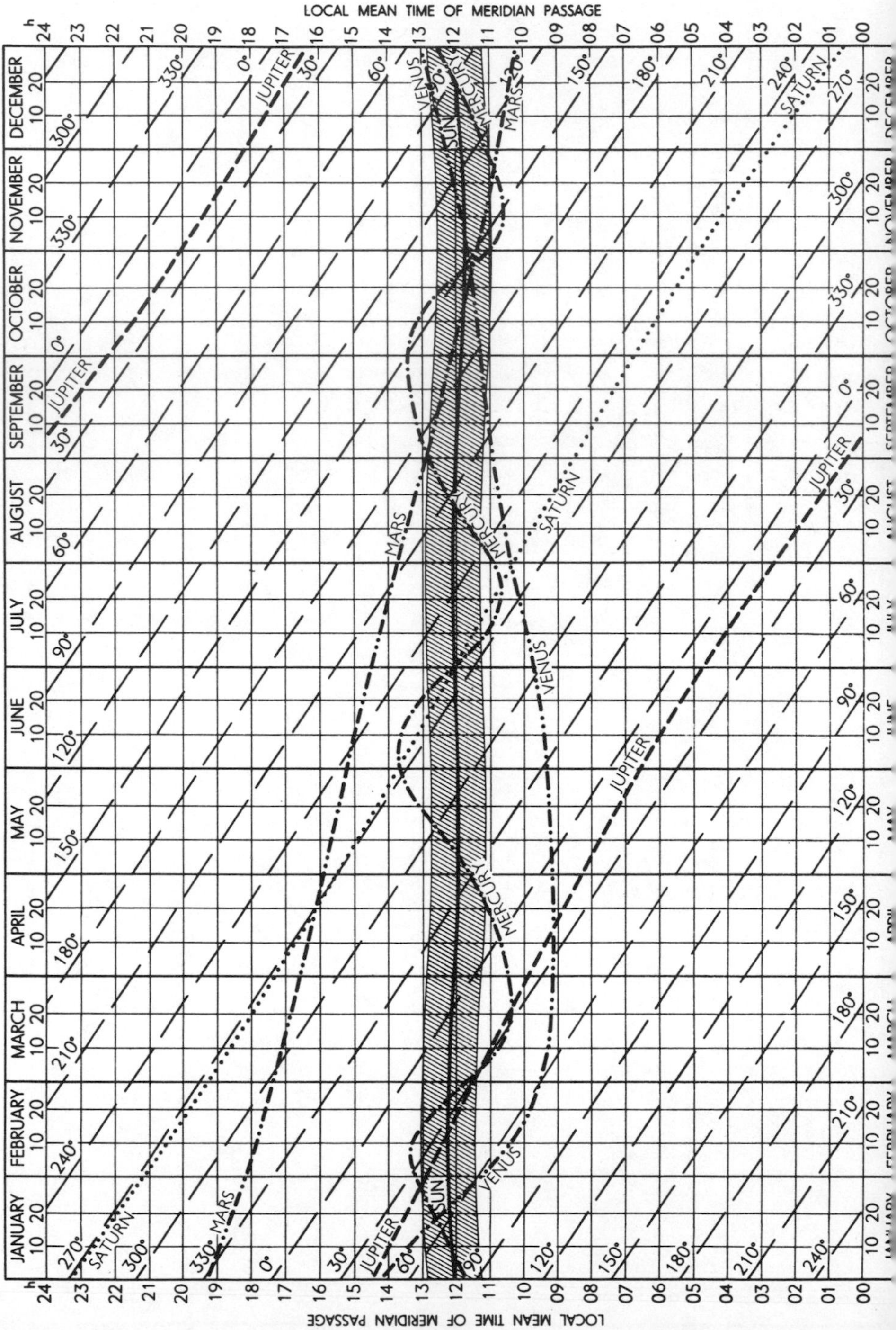

Table 32

longitude apart, and their relative bearings would be well spread.

These two planets are very bright (their magnitudes are given alongside their names in the daily pages of the almanac, and you can compare them with the stars in the Index to Selected Stars in the back pages of the almanac (Table 33). When they are morning planets, there is no doubt as to what you are looking at, because you can compare them to the background of the stars. In the evening, however, and in the case of Mars, which is only of second magnitude, you will need to compute a bearing and altitude.

Here is how you do that. It is April 29, 1974, you are out in the Coral Sea at latitude 20° S, longitude 165° E. From the almanac, civil twilight at Greenwich is at approximately 1800 hours. Your longitude, 165°, is 11 hours of time. Since you are east, you subtract and find that civil twilight for your approximate longitude is 0700 hours April 29, 1974, GMT.

From the daily page for Mars (Table 34), you figure the following:

MARS			
GHA 0700 Hr.	=	225°58′	– Dec. = 24°51′N
Assumed Long. E. (+)	=	+165°	
LHA	=	390°58′	
		−360°	
LHA	=	30°58′	

LHA is about 31°. Since you are in south latitude and the declination of Mars is north, you want the page for Latitude 20°, "Declination (15°-29°) Contrary Name to Latitude" (Table 36). Looking at it, you find that for a body with LHA of 31° and declination of 25° (close enough to the actual value of 24°51′ since you are looking for an approximation) Hc = 35°50′ and Z = 145°. Zn is, therefore, 180° + 145° or 325°. Convert 325° to magnetic, look out over your compass on that bearing, and you will see Mars about four "fists" above the horizon.

INDEX TO SELECTED STARS

Name	No.	Mag.	S.H.A.	Dec.
			°	°
Acamar	7	3·1	316	S. 40
Achernar	5	0·6	336	S. 57
Acrux	30	1·1	174	S. 63
Adhara	19	1·6	256	S. 29
Aldebaran	10	1·1	292	N. 16
Alioth	32	1·7	167	N. 56
Alkaid	34	1·9	153	N. 49
Al Na'ir	55	2·2	28	S. 47
Alnilam	15	1·8	276	S. 1
Alphard	25	2·2	219	S. 9
Alphecca	41	2·3	127	N. 27
Alpheratz	1	2·2	358	N. 29
Altair	51	0·9	63	N. 9
Ankaa	2	2·4	354	S. 42
Antares	42	1·2	113	S. 26
Arcturus	37	0·2	146	N. 19
Atria	43	1·9	109	S. 69
Avior	22	1·7	235	S. 59
Bellatrix	13	1·7	279	N. 6
Betelgeuse	16	Var.*	272	N. 7
Canopus	17	−0·9	264	S. 53
Capella	12	0·2	281	N. 46
Deneb	53	1·3	50	N. 45
Denebola	28	2·2	183	N. 15
Diphda	4	2·2	350	S. 18
Dubhe	27	2·0	195	N. 62
Elnath	14	1·8	279	N. 29
Eltanin	47	2·4	91	N. 51
Enif	54	2·5	34	N. 10
Fomalhaut	56	1·3	16	S. 30
Gacrux	31	1·6	173	S. 57
Gienah	29	2·8	176	S. 17
Hadar	35	0·9	150	S. 60
Hamal	6	2·2	329	N. 23
Kaus Australis	48	2·0	85	S. 34
Kochab	40	2·2	137	N. 74
Markab	57	2·6	14	N. 15
Menkar	8	2·8	315	N. 4
Menkent	36	2·3	149	S. 36
Miaplacidus	24	1·8	222	S. 70
Mirfak	9	1·9	310	N. 50
Nunki	50	2·1	77	S. 26
Peacock	52	2·1	54	S. 57
Pollux	21	1·2	244	N. 28
Procyon	20	0·5	246	N. 5
Rasalhague	46	2·1	97	N. 13
Regulus	26	1·3	208	N. 12
Rigel	11	0·3	282	S. 8
Rigil Kentaurus	38	0·1	141	S. 61
Sabik	44	2·6	103	S. 16
Schedar	3	2·5	350	N. 56
Shaula	45	1·7	97	S. 37
Sirius	18	−1·6	259	S. 17
Spica	33	1·2	159	S. 11
Suhail	23	2·2	223	S. 43
Vega	49	0·1	81	N. 39
Zubenelgenubi	39	2·9	138	S. 16

No.	Name	Mag.	S.H.A.	Dec.
			°	°
1	*Alpheratz*	2·2	358	N. 29
2	*Ankaa*	2·4	354	S. 42
3	*Schedar*	2·5	350	N. 56
4	*Diphda*	2·2	350	S. 18
5	*Achernar*	0·6	336	S. 57
6	*Hamal*	2·2	329	N. 23
7	*Acamar*	3·1	316	S. 40
8	*Menkar*	2·8	315	N. 4
9	*Mirfak*	1·9	310	N. 50
10	*Aldebaran*	1·1	292	N. 16
11	*Rigel*	0·3	282	S. 8
12	*Capella*	0·2	281	N. 46
13	*Bellatrix*	1·7	279	N. 6
14	*Elnath*	1·8	279	N. 29
15	*Alnilam*	1·8	276	S. 1
16	*Betelgeuse*	Var.*	272	N. 7
17	*Canopus*	−0·9	264	S. 53
18	*Sirius*	−1·6	259	S. 17
19	*Adhara*	1·6	256	S. 29
20	*Procyon*	0·5	246	N. 5
21	*Pollux*	1·2	244	N. 28
22	*Avior*	1·7	235	S. 59
23	*Suhail*	2·2	223	S. 43
24	*Miaplacidus*	1·8	222	S. 70
25	*Alphard*	2·2	219	S. 9
26	*Regulus*	1·3	208	N. 12
27	*Dubhe*	2·0	195	N. 62
28	*Denebola*	2·2	183	N. 15
29	*Gienah*	2·8	176	S. 17
30	*Acrux*	1·1	174	S. 63
31	*Gacrux*	1·6	173	S. 57
32	*Alioth*	1·7	167	N. 56
33	*Spica*	1·2	159	S. 11
34	*Alkaid*	1·9	153	N. 49
35	*Hadar*	0·9	150	S. 60
36	*Menkent*	2·3	149	S. 36
37	*Arcturus*	0·2	146	N. 19
38	*Rigil Kentaurus*	0·1	141	S. 61
39	*Zubenelgenubi*	2·9	138	S. 16
40	*Kochab*	2·2	137	N. 74
41	*Alphecca*	2·3	127	N. 27
42	*Antares*	1·2	113	S. 26
43	*Atria*	1·9	109	S. 69
44	*Sabik*	2·6	103	S. 16
45	*Shaula*	1·7	97	S. 37
46	*Rasalhague*	2·1	97	N. 13
47	*Eltanin*	2·4	91	N. 51
48	*Kaus Australis*	2·0	85	S. 34
49	*Vega*	0·1	81	N. 39
50	*Nunki*	2·1	77	S. 26
51	*Altair*	0·9	63	N. 9
52	*Peacock*	2·1	54	S. 57
53	*Deneb*	1·3	50	N. 45
54	*Enif*	2·5	34	N. 10
55	*Al Na'ir*	2·2	28	S. 47
56	*Fomalhaut*	1·3	16	S. 30
57	*Markab*	2·6	14	N. 15

* 0·1 – 1·2

Table 33

1974 APRIL 28, 29, 30 (SUN., MON., TUES.)

G.M.T.		ARIES	VENUS −3·8		MARS +1·7		JUPITER −1·8		SATURN +0·3		STARS		
		G.H.A.	G.H.A.	Dec.	G.H.A.	Dec.	G.H.A.	Dec.	G.H.A.	Dec.	Name	S.H.A.	Dec.
d	h	° ′	° ′	° ′	° ′	° ′	° ′	° ′	° ′	° ′		° ′	° ′
28	00	215 35·1	222 08·5	S 3 48·7	120 33·0	N24 52·5	232 59·6	S 8 23·4	124 42·6	N22 44·1	Acamar	315 40·9	S 40 24·4
	01	230 37·6	237 08·4	47·8	135 33·8	52·4	248 01·6	23·3	139 44·8	44·1	Achernar	335 49·0	S 57 21·9
	02	245 40·0	252 08·3	46·9	150 34·6	52·3	263 03·7	23·1	154 47·0	44·1	Acrux	173 41·7	S 62 57·7
	03	260 42·5	267 08·2	. . 45·9	165 35·4	. . 52·3	278 05·7	. . 22·9	169 49·2	. . 44·1	Adhara	255 35·7	S 28 56·5
	04	275 45·0	282 08·1	45·0	180 36·3	52·2	293 07·7	22·8	184 51·4	44·1	Aldebaran	291 23·3	N 16 27·5
	05	290 47·4	297 08·0	44·1	195 37·1	52·1	308 09·8	22·6	199 53·6	44·1			
	06	305 49·9	312 07·9	S 3 43·2	210 37·9	N24 52·1	323 11·8	S 8 22·5	214 55·8	N22 44·1	Alioth	166 45·8	N 56 05·9
	07	320 52·4	327 07·8	42·2	225 38·7	52·0	338 13·8	22·3	229 58·0	44·1	Alkaid	153 21·4	N 49 26·4
	08	335 54·8	342 07·7	41·3	240 39·5	52·0	353 15·9	22·1	245 00·2	44·1	Al Na'ir	28 20·6	S 47 04·9
S	09	350 57·3	357 07·5	. . 40·4	255 40·3	. . 51·9	8 17·9	. . 22·0	260 02·4	. . 44·1	Alnilam	276 16·4	S 1 13·1
U	10	5 59·7	12 07·4	39·5	270 41·2	51·8	23 20·0	21·8	275 04·6	44·1	Alphard	218 24·8	S 8 33·0
N	11	21 02·2	27 07·3	38·5	285 42·0	51·8	38 22·0	21·6	290 06·8	44·1			
D	12	36 04·7	42 07·2	S 3 37·6	300 42·8	N24 51·7	53 24·0	S 8 21·5	305 09·0	N22 44·1	Alphecca	126 35·5	N 26 47·8
A	13	51 07·1	57 07·1	36·7	315 43·6	51·6	68 26·1	21·3	320 11·2	44·2	Alpheratz	358 14·2	N 28 56·8
Y	14	66 09·6	72 07·0	35·7	330 44·4	51·6	83 28·1	21·1	335 13·4	44·2	Altair	62 36·7	N 8 47·9
	15	81 12·1	87 06·9	. . 34·8	345 45·2	. . 51·5	98 30·2	. . 21·0	350 15·7	. . 44·2	Ankaa	353 44·9	S 42 26·6
	16	96 14·5	102 06·8	33·9	0 46·1	51·5	113 32·2	20·8	5 17·9	44·2	Antares	113 01·9	S 26 22·6
	17	111 17·0	117 06·7	32·9	15 46·9	51·4	128 34·2	20·7	20 20·1	44·2			
	18	126 19·5	132 06·5	S 3 32·0	30 47·7	N24 51·3	143 36·3	S 8 20·5	35 22·3	N22 44·2	Arcturus	146 22·1	N 19 18·8
	19	141 21·9	147 06·4	31·1	45 48·5	51·3	158 38·3	20·3	50 24·5	44·2	Atria	108 29·7	S 68 58·9
	20	156 24·4	162 06·3	30·2	60 49·3	51·2	173 40·4	20·2	65 26·7	44·2	Avior	234 30·1	S 59 26·0
	21	171 26·9	177 06·2	. . 29·2	75 50·1	. . 51·1	188 42·4	. . 20·0	80 28·9	. . 44·2	Bellatrix	279 03·7	N 6 19·6
	22	186 29·3	192 06·1	28·3	90 50·9	51·1	203 44·4	19·8	95 31·1	44·2	Betelgeuse	271 33·3	N 7 24·1
	23	201 31·8	207 06·0	27·4	105 51·8	51·0	218 46·5	19·7	110 33·3	44·2			
29	00	216 34·2	222 05·9	S 3 26·4	120 52·6	N24 50·9	233 48·5	S 8 19·5	125 35·5	N22 44·2	Canopus	264 09·5	S 52 41·2
	01	231 36·7	237 05·0	25·5	135 53·4	50·9	248 50·6	19·4	140 37·7	44·2	Capella	281 18·2	N 45 58·5
	02	246 39·2	252 05·6	24·6	150 54·2	50·8	263 52·6	19·2	155 39·9	44·2	Deneb	49 51·5	N 45 11·1
	03	261 41·6	267 05·5	. . 23·6	165 55·0	. . 50·7	278 54·7	. . 19·0	170 42·1	. . 44·2	Denebola	183 03·3	N 14 42·8
	04	276 44·1	282 05·4	22·7	180 55·8	50·7	293 56·7	18·9	185 44·3	44·2	Diphda	349 25·5	S 10 07·6
	05	291 46·6	297 05·3	21·8	195 56·7	50·6	308 58·7	18·7	200 46·5	44·2			
	06	306 49·0	312 05·2	S 3 20·8	210 57·5	N24 50·5	324 00·8	S 8 18·5	215 48·7	N22 44·2	Dubhe	194 27·0	N 61 53·5
	07	321 51·5	327 05·1	19·9	225 58·3	50·5	339 02·8	18·4	230 50·9	44·2	Elnath	278 50·0	N 28 35·2
M	08	336 54·0	342 05·0	18·9	240 59·1	50·4	354 04·9	18·2	245 53·1	44·2	Eltanin	90 59·4	N 51 29·2
	09	351 56·4	357 04·9	. . 18·0	255 59·9	. . 50·3	9 06·9	. . 18·1	260 55·3	. . 44·2	Enif	34 16·0	N 9 45·3
O	10	6 58·9	12 04·7	17·1	271 00·7	50·2	24 08·9	17·9	275 57·5	44·2	Fomalhaut	15 56·4	S 29 45·4
N	11	22 01·3	27 04·6	16·1	286 01·6	50·2	39 11·0	17·7	290 59·7	44·2			
D	12	37 03·8	42 04·5	S 3 15·2	301 02·4	N24 50·1	54 13·0	S 8 17·6	306 01·9	N22 44·3	Gacrux	172 33·2	S 56 58·5
A	13	52 06·3	57 04·4	14·3	316 03·2	50·0	69 15·1	17·4	321 04·2	44·3	Gienah	176 22·2	S 17 24·2
Y	14	67 08·7	72 04·3	13·3	331 04·0	50·0	84 17·1	17·2	336 06·4	44·3	Hadar	149 29·0	S 60 15·2
	15	82 11·2	87 04·2	. . 12·4	346 04·8	. . 49·9	99 19·2	. . 17·1	351 08·6	. . 44·3	Hamal	328 34·3	N 23 20·5
	16	97 13·7	102 04·1	11·4	1 05·6	49·8	114 21·2	16·9	6 10·8	44·3	Kaus Aust.	84 22·4	S 34 23·8
	17	112 16·1	117 03·9	10·5	16 06·5	49·8	129 23·3	16·8	21 13·0	44·3			
	18	127 18·6	132 03·8	S 3 09·6	31 07·3	N24 49·7	144 25·3	S 8 16·6	36 15·2	N22 44·3	Kochab	137 17·4	N 74 15·5
	19	142 21·1	147 03·7	08·6	46 08·1	49·6	159 27·3	16·4	51 17·4	44·3	Markab	14 07·7	N 15 03·9
	20	157 23·5	162 03·6	07·7	61 08·9	49·5	174 29·4	16·3	66 19·6	44·3	Menkar	314 46·1	N 3 59·4
	21	172 26·0	177 03·5	. . 06·7	76 09·7	. . 49·5	189 31·4	. . 16·1	81 21·8	. . 44·3	Menkent	148 41·8	S 36 14·9
	22	187 28·5	192 03·4	05·8	91 10·5	49·4	204 33·5	16·0	96 24·0	44·3	Miaplacidus	221 45·9	S 69 37·1
	23	202 30·9	207 03·2	04·9	106 11·4	49·3	219 35·5	15·8	111 26·2	44·3			
30	00	217 33·4	222 03·1	S 3 03·9	121 12·2	N24 49·2	234 37·6	S 8 15·6	126 28·4	N22 44·3	Mirfak	309 22·9	N 49 46·3
	01	232 35·8	237 03·0	03·0	136 13·0	49·2	249 39·6	15·5	141 30·6	44·3	Nunki	76 34·4	S 26 19·7
	02	247 38·3	252 02·9	02·0	151 13·8	49·1	264 41·7	15·3	156 32·8	44·3	Peacock	54 05·2	S 56 48·8
	03	262 40·8	267 02·8	. . 01·1	166 14·6	. . 49·0	279 43·7	. . 15·1	171 35·0	. . 44·3	Pollux	244 03·6	N 28 05·4
	04	277 43·2	282 02·7	3 00·1	181 15·4	48·9	294 45·8	15·0	186 37·2	44·3	Procyon	245 30·5	N 5 17·4
	05	292 45·7	297 02·6	2 59·2	196 16·3	48·9	309 47·8	14·8	201 39·4	44·3			
	06	307 48·2	312 02·4	S 2 58·3	211 17·1	N24 48·8	324 49·8	S 8 14·7	216 41·6	N22 44·3	Rasalhague	96 33·4	N 12 34·5
	07	322 50·6	327 02·3	57·3	226 17·9	48·7	339 51·9	14·5	231 43·8	44·3	Regulus	208 14·5	N 12 05·5
T	08	337 53·1	342 02·2	56·4	241 18·7	48·6	354 53·9	14·3	246 46·0	44·3	Rigel	281 40·5	S 8 13·9
U	09	352 55·6	357 02·1	. . 55·4	256 19·5	. . 48·6	9 56·0	. . 14·2	261 48·2	. . 44·3	Rigil Kent.	140 31·2	S 60 43·9
E	10	7 58·0	12 02·0	54·5	271 20·3	48·5	24 58·0	14·0	276 50·4	44·3	Sabik	102 45·9	S 15 41·7
S	11	23 00·5	27 01·9	53·5	286 21·2	48·4	40 00·1	13·9	291 52·6	44·4			
D	12	38 03·0	42 01·7	S 2 52·6	301 22·0	N24 48·3	55 02·1	S 8 13·7	306 54·8	N22 44·4	Schedar	350 14·6	N 56 23·7
A	13	53 05·4	57 01·6	51·6	316 22·8	48·3	70 04·2	13·5	321 57·0	44·4	Shaula	97 01·4	S 37 05·1
Y	14	68 07·9	72 01·5	50·7	331 23·6	48·2	85 06·2	13·4	336 59·2	44·4	Sirius	258 59·8	S 16 41·0
	15	83 10·3	87 01·4	. . 49·7	346 24·4	. . 48·1	100 08·3	. . 13·2	352 01·4	. . 44·4	Spica	159 01·9	S 11 01·9
	16	98 12·8	102 01·3	48·8	1 25·2	48·0	115 10·3	13·1	7 03·6	44·4	Suhail	223 14·0	S 43 20·1
	17	113 15·3	117 01·2	47·9	16 26·1	47·9	130 12·4	12·9	22 05·8	44·4			
	18	128 17·7	132 01·0	S 2 46·9	31 26·9	N24 47·9	145 14·4	S 8 12·7	37 08·0	N22 44·4	Vega	80 58·6	N 38 45·3
	19	143 20·2	147 00·9	46·0	46 27·7	47·8	160 16·5	12·6	52 10·2	44·4	Zuben'ubi	137 37·6	S 15 56·3
	20	158 22·7	162 00·8	45·0	61 28·5	47·7	175 18·5	12·4	67 12·4	44·4			
	21	173 25·1	177 00·7	. . 44·1	76 29·3	. . 47·6	190 20·6	. . 12·3	82 14·6	. . 44·4			
	22	188 27·6	192 00·6	43·1	91 30·1	47·6	205 22·6	12·1	97 16·8	44·4			
	23	203 30·1	207 00·4	42·2	106 31·0	47·5	220 24·7	11·9	112 19·0	44·4			
Mer. Pass.		9h 32·2m	v −0·1	d 0·9	v 0·8	d 0·1	v 2·0	d 0·2	v 2·2	d 0·0			

	S.H.A. (° ′)	Mer. Pass. (h m)
Venus	5 31·6	9 12
Mars	264 18·3	15 56
Jupiter	17 14·3	8 24
Saturn	269 01·3	15 35

Table 34

1974 APRIL 28, 29, 30 (SUN., MON., TUES.)

G.M.T.	SUN		MOON				
d h	G.H.A. ° ′	Dec. ° ′	G.H.A. ° ′	v ′	Dec. ° ′	d ′	H.P. ′
28 00	180 36·0	N13 57·7	104 25·8	6·7	N19 25·2	8·0	59·4
01	195 36·1	58·5	118 51·5	6·8	19 17·2	8·0	59·4
02	210 36·1	13 59·3	133 17·3	6·9	19 09·2	8·1	59·4
03	225 36·2	14 00·1	147 43·2	6·9	19 01·1	8·3	59·4
04	240 36·3	00·9	162 09·1	7·0	18 52·8	8·4	59·4
05	255 36·4	01·7	176 35·1	7·0	18 44·4	8·5	59·4
06	270 36·5	N14 02·5	191 01·1	7·2	N18 35·9	8·6	59·4
07	285 36·6	03·3	205 27·3	7·1	18 27·3	8·7	59·4
08	300 36·7	04·1	219 53·4	7·3	18 18·6	8·9	59·4
S 09	315 36·8	· · 04·8	234 19·7	7·3	18 09·7	8·9	59·3
U 10	330 36·9	05·6	248 46·0	7·4	18 00·8	9·1	59·3
N 11	345 37·0	06·4	263 12·4	7·5	17 51·7	9·2	59·3
D 12	0 37·1	N14 07·2	277 38·9	7·5	N17 42·5	9·3	59·3
A 13	15 37·2	08·0	292 05·4	7·6	17 33·2	9·4	59·3
Y 14	30 37·3	08·8	306 32·0	7·6	17 23·8	9·5	59·3
15	45 37·4	· · 09·6	320 58·6	7·7	17 14·3	9·6	59·3
16	60 37·5	10·4	335 25·3	7·8	17 04·7	9·7	59·3
17	75 37·6	11·1	349 52·1	7·9	16 55·0	9·8	59·3
18	90 37·6	N14 11·9	4 19·0	7·9	N16 45·2	9·9	59·3
19	105 37·7	12·7	18 45·9	8·0	16 35·3	10·0	59·3
20	120 37·8	13·5	33 12·9	8·0	16 25·3	10·1	59·3
21	135 37·9	· · 14·3	47 39·9	8·1	16 15·2	10·2	59·3
22	150 38·0	15·1	62 07·0	8·2	16 05·0	10·3	59·3
23	165 38·1	15·8	76 34·2	8·3	15 54·7	10·3	59·3
29 00	180 38·2	N14 16·6	91 01·5	8·3	N15 44·4	10·5	59·3
01	195 38·3	17·4	105 28·8	8·4	15 33·9	10·6	59·3
02	210 38·4	18·2	119 56·2	8·4	15 23·3	10·6	59·3
03	225 38·5	· · 19·0	134 23·6	8·6	15 12·7	10·8	59·3
04	240 38·6	19·8	148 51·2	8·5	15 01·9	10·8	59·3
05	255 38·6	20·5	163 18·7	8·7	14 51·1	10·9	59·3
06	270 38·7	N14 21·3	177 46·4	8·7	N14 40·2	11·0	59·3
07	285 38·8	22·1	192 14·1	8·8	14 29·2	11·1	59·3
08	300 38·9	22·9	206 41·9	8·8	14 18·1	11·2	59·2
M 09	315 39·0	· · 23·7	221 09·7	8·9	14 06·9	11·2	59·2
O 10	330 39·1	24·4	235 37·6	9·0	13 55·7	11·4	59·2
N 11	345 39·2	25·2	250 05·6	9·0	13 44·3	11·4	59·2
D 12	0 39·3	N14 26·0	264 33·6	9·1	N13 32·9	11·5	59·2
A 13	15 39·4	26·8	279 01·7	9·2	13 21·4	11·5	59·2
Y 14	30 39·4	27·5	293 29·9	9·2	13 09·9	11·6	59·2
15	45 39·5	· · 28·3	307 58·1	9·3	12 58·3	11·7	59·2
16	60 39·6	29·1	322 26·4	9·3	12 46·6	11·8	59·2
17	75 39·7	29·9	336 54·7	9·4	12 34·8	11·8	59·2
18	90 39·8	N14 30·6	351 23·1	9·5	N12 23·0	11·9	59·2
19	105 39·9	31·4	5 51·6	9·5	12 11·1	12·0	59·2
20	120 40·0	32·2	20 20·1	9·6	11 59·1	12·0	59·2
21	135 40·0	· · 33·0	34 48·7	9·6	11 47·1	12·1	59·2
22	150 40·1	33·7	49 17·3	9·7	11 35·0	12·2	59·1
23	165 40·2	34·5	63 46·0	9·8	11 22·8	12·2	59·1
30 00	180 40·3	N14 35·3	78 14·8	9·8	N11 10·6	12·3	59·1
01	195 40·4	36·1	92 43·6	9·8	10 58·3	12·4	59·1
02	210 40·5	36·8	107 12·4	9·9	10 45·9	12·4	59·1
03	225 40·6	· · 37·6	121 41·3	10·0	10 33·5	12·4	59·1
04	240 40·6	38·4	136 10·3	10·0	10 21·1	12·5	59·1
05	255 40·7	39·1	150 39·3	10·1	10 08·6	12·6	59·1
06	270 40·8	N14 39·9	165 08·4	10·1	N 9 56·0	12·6	59·1
07	285 40·9	40·7	179 37·5	10·2	9 43·4	12·7	59·1
T 08	300 41·0	41·5	194 06·7	10·2	9 30·7	12·7	59·1
U 09	315 41·1	· · 42·2	208 35·9	10·3	9 18·0	12·7	59·1
E 10	330 41·1	43·0	223 05·2	10·3	9 05·3	12·8	59·0
S 11	345 41·2	43·8	237 34·5	10·4	8 52·5	12·9	59·0
D 12	0 41·3	N14 44·5	252 03·9	10·4	N 8 39·6	12·9	59·0
A 13	15 41·4	45·3	266 33·3	10·5	8 26·7	12·9	59·0
Y 14	30 41·5	46·1	281 02·8	10·5	8 13·8	13·0	59·0
15	45 41·6	· · 46·8	295 32·3	10·6	8 00·8	13·0	59·0
16	60 41·6	47·6	310 01·9	10·6	7 47·8	13·1	59·0
17	75 41·7	48·4	324 31·5	10·6	7 34·7	13·1	59·0
18	90 41·8	N14 49·1	339 01·1	10·7	N 7 21·6	13·1	59·0
19	105 41·9	49·9	353 30·8	10·7	7 08·5	13·1	59·0
20	120 42·0	50·7	8 00·5	10·8	6 55·4	13·2	58·9
21	135 42·1	· · 51·4	22 30·3	10·8	6 42·2	13·3	58·9
22	150 42·1	52·2	37 00·1	10·9	6 28·9	13·2	58·9
23	165 42·2	52·9	51 30·0	10·8	6 15·7	13·3	58·9
	S.D. 15·9	d 0·8	S.D. 16·2		16·1		16·1

Lat.	Twilight Naut.	Twilight Civil	Sun-rise	Moonrise 28	29	30	1
°	h m	h m	h m	h m	h m	h m	h m
N 72	////	////	02 12		08 55	11 19	13 27
N 70	////	////	02 45	06 57	09 25	11 33	13 31
68	////	01 31	03 09	07 42	09 47	11 44	13 35
66	////	02 10	03 27	08 12	10 04	11 53	13 38
64	////	02 37	03 42	08 34	10 18	12 00	13 40
62	01 20	02 57	03 54	08 51	10 30	12 07	13 42
60	01 56	03 13	04 05	09 06	10 40	12 13	13 44
N 58	02 20	03 27	04 14	09 19	10 48	12 17	13 46
56	02 39	03 38	04 22	09 29	10 56	12 22	13 47
54	02 55	03 48	04 29	09 39	11 02	12 26	13 48
52	03 08	03 57	04 35	09 47	11 08	12 29	13 49
50	03 19	04 05	04 41	09 55	11 14	12 33	13 50
45	03 42	04 21	04 53	10 11	11 25	12 39	13 53
N 40	03 59	04 34	05 03	10 24	11 35	12 45	13 55
35	04 13	04 45	05 12	10 35	11 43	12 50	13 56
30	04 24	04 54	05 19	10 44	11 50	12 55	13 58
20	04 43	05 09	05 32	11 01	12 02	13 02	14 00
N 10	04 56	05 22	05 44	11 15	12 13	13 09	14 03
0	05 08	05 33	05 54	11 29	12 23	13 15	14 05
S 10	05 18	05 43	06 04	11 42	12 33	13 21	14 07
20	05 26	05 52	06 15	11 57	12 44	13 28	14 09
30	05 34	06 03	06 27	12 13	12 56	13 35	14 12
35	05 38	06 08	06 34	12 23	13 03	13 39	14 13
40	05 42	06 14	06 42	12 33	13 11	13 44	14 15
45	05 46	06 21	06 52	12 46	13 20	13 50	14 17
S 50	05 51	06 29	07 03	13 01	13 31	13 56	14 19
52	05 53	06 32	07 08	13 08	13 36	13 59	14 20
54	05 54	06 36	07 14	13 16	13 42	14 03	14 21
56	05 56	06 40	07 20	13 25	13 48	14 07	14 22
58	05 58	06 45	07 27	13 35	13 55	14 11	14 24
S 60	06 01	06 50	07 35	13 47	14 03	14 15	14 25

Lat.	Sun-set	Twilight Civil	Twilight Naut.	Moonset 28	29	30	1
°	h m	h m	h m	h m	h m	h m	h m
N 72	21 49	////	////		03 44	03 11	02 47
N 70	21 14	////	////	03 47	03 13	02 54	02 40
68	20 49	22 32	////	03 00	02 49	02 41	02 34
66	20 30	21 50	////	02 29	02 31	02 31	02 29
64	20 15	21 22	////	02 06	02 16	02 21	02 25
62	20 03	21 01	22 42	01 48	02 03	02 14	02 21
60	19 52	20 44	22 03	01 33	01 53	02 07	02 18
N 58	19 42	20 30	21 38	01 20	01 43	02 01	02 15
56	19 34	20 18	21 18	01 08	01 35	01 55	02 12
54	19 27	20 08	21 02	00 58	01 28	01 51	02 10
52	19 21	19 59	20 49	00 50	01 21	01 46	02 08
50	19 15	19 51	20 37	00 42	01 15	01 42	02 06
45	19 02	19 35	20 14	00 25	01 02	01 34	02 02
N 40	18 52	19 21	19 57	00 11	00 51	01 26	01 58
35	18 43	19 10	19 43	24 42	00 42	01 20	01 55
30	18 36	19 01	19 31	24 34	00 34	01 14	01 52
20	18 23	18 46	19 13	24 19	00 19	01 05	01 47
N 10	18 11	18 33	18 58	24 07	00 07	00 56	01 43
0	18 01	18 22	18 47	23 55	24 48	00 48	01 39
S 10	17 50	18 12	18 37	23 43	24 40	00 40	01 35
20	17 40	18 02	18 28	23 30	24 31	00 31	01 30
30	17 27	17 52	18 20	23 16	24 21	00 21	01 25
35	17 20	17 46	18 16	23 07	24 15	00 15	01 22
40	17 12	17 40	18 12	22 57	24 08	00 08	01 18
45	17 02	17 33	18 08	22 46	24 01	00 01	01 14
S 50	16 51	17 25	18 03	22 32	23 51	25 10	01 10
52	16 46	17 22	18 01	22 25	23 47	25 07	01 07
54	16 40	17 18	18 00	22 18	23 42	25 05	01 05
56	16 34	17 14	17 58	22 10	23 36	25 02	01 02
58	16 27	17 09	17 56	22 00	23 30	24 59	00 59
S 60	16 19	17 04	17 53	21 50	23 23	24 56	00 56

Day	SUN Eqn. of Time 00h	12h	Mer. Pass.	MOON Mer. Pass. Upper	Lower	Age	Phase
	m s	m s	h m	h m	h m	d	
28	02 24	02 28	11 58	17 42	05 14	06	
29	02 33	02 37	11 57	18 36	06 09	07	
30	02 41	02 45	11 57	19 27	07 02	08	

Table 35

LAT 20°

LHA	15° Hc	d	Z	16° Hc	d	Z	17° Hc	d	Z	18° Hc	d	Z	19° Hc	d	Z
	° ′	′	°	° ′	′	°	° ′	′	°	° ′	′	°	° ′	′	°
69	13 42	26	112	13 16	26	113	12 50	26	114	12 24	27	115	11 57	27	116
68	14 34	26	112	14 08	27	113	13 41	26	114	13 15	27	115	12 48	27	116
67	15 26	26	113	15 00	27	114	14 33	27	115	14 06	28	116	13 38	27	116
66	16 18	27	113	15 51	27	114	15 24	28	115	14 56	27	116	14 29	28	117
65	17 10	-28	114	16 42	-27	115	16 15	-28	116	15 47	-28	116	15 19	-28	117
64	18 01	27	114	17 34	28	115	17 06	29	116	16 37	28	117	16 09	29	118
63	18 53	28	115	18 25	29	116	17 56	28	116	17 28	29	117	16 59	29	118
62	19 44	29	115	19 15	28	116	18 47	29	117	18 18	30	118	17 48	29	119
61	20 35	29	116	20 06	29	117	19 37	30	117	19 07	29	118	18 38	30	119
60	21 26	-30	116	20 56	-29	117	20 27	-30	118	19 57	-30	119	19 27	-31	120
59	22 16	29	117	21 47	30	118	21 17	31	118	20 46	30	119	20 16	31	120
58	23 07	31	117	22 36	30	118	22 06	31	119	21 35	31	120	21 04	31	121
57	23 57	31	118	23 26	31	119	22 55	31	120	22 24	32	120	21 52	31	121
56	24 47	31	118	24 16	32	119	23 44	32	120	23 12	31	121	22 41	33	122
55	25 36	-31	119	25 05	-32	120	24 33	-32	121	24 01	-33	122	23 28	-33	122
54	26 25	31	119	25 54	33	120	25 21	32	121	24 49	33	122	24 16	34	123
53	27 14	32	120	26 42	33	121	26 09	33	122	25 36	33	123	25 03	34	124
52	28 03	33	120	27 30	33	121	26 57	33	122	26 24	34	123	25 50	35	124
51	28 52	34	121	28 18	33	122	27 45	34	123	27 11	35	124	26 36	35	125
50	29 40	-34	122	29 06	-34	123	28 32	-35	124	27 57	-35	124	27 22	-35	125
49	30 28	35	122	29 53	34	123	29 19	35	124	28 44	35	125	28 08	36	126
48	31 15	35	123	30 40	35	124	30 05	35	125	29 30	35	126	28 54	37	127
47	32 02	35	124	31 27	36	125	30 51	36	125	30 15	35	126	29 39	37	127
46	32 49	36	124	32 13	36	125	31 37	37	126	31 00	37	127	30 23	37	128
45	33 36	-37	125	32 59	-37	126	32 22	-37	127	31 45	-37	128	31 08	-38	129
44	34 22	37	126	33 45	38	127	33 07	37	128	32 30	39	128	31 51	38	129
43	35 07	37	126	34 30	38	127	33 52	38	128	33 14	39	129	32 35	39	130
42	35 53	39	127	35 14	38	128	34 36	39	129	33 57	39	130	33 18	40	131
41	36 37	38	128	35 59	39	129	35 20	40	130	34 40	40	131	34 00	40	132
40	37 22	-40	129	36 42	-39	130	36 03	-40	131	35 23	-41	131	34 42	-41	132
39	38 05	39	129	37 26	41	130	36 45	40	131	36 05	42	132	35 23	41	133
38	38 49	41	130	38 08	41	131	37 27	41	132	36 46	42	133	36 04	42	134
37	39 31	41	131	38 50	41	132	38 09	42	133	37 27	42	134	36 45	43	135
36	40 14	42	132	39 32	42	133	38 50	43	134	38 07	43	135	37 24	43	136
35	40 55	-42	133	40 13	-43	134	39 30	-43	135	38 47	-43	136	38 04	-44	137
34	41 36	43	134	40 53	43	135	40 10	44	136	39 26	44	137	38 42	44	137
33	42 17	44	135	41 33	44	136	40 49	44	137	40 05	45	137	39 20	45	138
32	42 56	44	136	42 12	44	137	41 28	45	138	40 43	46	138	39 57	46	139
31	43 36	45	137	42 51	46	138	42 05	45	138	41 20	46	139	40 34	47	140
30	44 14	-46	138	43 28	-46	139	42 42	-46	139	41 56	-47	140	41 09	-47	141
29	44 52	47	139	44 05	46	140	43 19	47	140	42 32	48	141	41 44	47	142
28	45 28	46	140	44 42	48	141	43 54	47	142	43 07	48	142	42 19	49	143
27	46 04	47	141	45 17	48	142	44 29	48	143	43 41	49	143	42 52	49	144
26	46 40	49	142	45 51	48	143	45 03	49	144	44 14	49	144	43 25	50	145
25	47 14	-49	143	46 25	-49	144	45 36	-50	145	44 46	-50	146	43 56	-50	146
24	47 47	49	144	46 58	50	145	46 08	50	146	45 18	51	147	44 27	51	147
23	48 20	50	145	47 30	51	146	46 39	51	147	45 48	51	148	44 57	51	149
22	48 51	50	147	48 01	52	147	47 09	51	148	46 18	52	149	45 26	52	150
21	49 22	52	148	48 30	51	149	47 39	53	149	46 46	52	150	45 54	53	151
20	49 51	-52	149	48 59	-52	150	48 07	-53	151	47 14	-53	151	46 21	-53	152
19	50 20	53	151	49 27	53	151	48 34	54	152	47 40	53	153	46 47	54	153
18	50 47	54	152	49 53	53	153	49 00	54	153	48 06	54	154	47 12	55	155
17	51 13	54	153	50 19	54	154	49 25	55	155	48 30	55	155	47 35	55	156
16	51 38	55	155	50 43	55	155	49 48	55	156	48 53	55	157	47 58	56	157
15	52 01	-55	156	51 06	-55	157	50 11	-56	157	49 15	-56	158	48 19	-56	158
14	52 23	55	158	51 28	56	158	50 32	56	159	49 36	57	159	48 39	56	160
13	52 44	56	159	51 48	56	160	50 52	57	160	49 55	57	161	48 58	57	161
12	53 04	57	161	52 07	57	161	51 10	57	162	50 13	57	162	49 16	57	163
11	53 22	57	162	52 25	58	163	51 27	57	163	50 30	58	163	49 32	57	164
10	53 39	-58	164	52 41	-58	164	51 43	-58	165	50 45	-58	165	49 47	-58	165
9	53 54	58	165	52 56	58	166	51 58	59	166	50 59	58	166	50 01	59	167
8	54 08	59	167	53 09	58	167	52 11	59	168	51 12	59	168	50 13	59	168
7	54 20	59	168	53 21	59	169	52 22	59	169	51 23	59	169	50 24	59	170
6	54 30	59	170	53 31	59	170	52 32	59	171	51 33	59	171	50 34	60	171
5	54 39	-59	172	53 40	-59	172	52 41	-60	172	51 41	-59	172	50 42	-60	173
4	54 47	60	173	53 47	59	174	52 48	60	174	51 48	60	174	50 48	59	174
3	54 53	60	175	53 53	60	175	52 53	60	175	51 53	60	175	50 53	59	176
2	54 57	60	177	53 57	60	177	52 57	60	177	51 57	60	177	50 57	60	177
1	54 59	60	178	53 59	60	178	52 59	60	178	51 59	60	179	50 59	60	179
0	55 00	-60	180	54 00	-60	180	53 00	-60	180	52 00	-60	180	51 00	-60	180

LHA	20° Hc	d	Z	21° Hc	d	Z	22° Hc	d	Z	23° Hc	d	Z	24° Hc	d	Z
	° ′	′	°	° ′	′	°	° ′	′	°	° ′	′	°	° ′	′	°
69	11 30	26	117	11 04	27	117	10 37	28	118	10 09	27	119	09 42	27	120
68	12 21	27	117	11 54	28	118	11 26	27	119	10 59	28	120	10 31	28	121
67	13 11	28	117	12 43	27	118	12 16	28	119	11 48	29	120	11 19	28	121
66	14 01	28	118	13 33	28	119	13 05	29	120	12 36	28	121	12 08	29	121
65	14 51	-29	118	14 22	-28	119	13 54	-29	120	13 25	-29	121	12 56	-30	122
64	15 40	29	119	15 11	29	120	14 42	29	121	14 13	30	121	13 43	29	122
63	16 30	30	119	16 00	29	120	15 31	30	121	15 01	30	122	14 31	30	123
62	17 19	30	120	16 49	30	121	16 19	30	122	15 49	31	122	15 18	31	123
61	18 08	31	120	17 37	30	121	17 07	31	122	16 36	31	123	16 05	31	124
60	18 56	-30	121	18 26	-31	122	17 55	-32	123	17 23	-31	123	16 52	-32	124
59	19 45	32	121	19 13	31	122	18 42	32	123	18 1[illegible]	32	124	17 38	32	125
58	20 33	32	122	20 01	32	123	19 29	32	124	18 57	32	124	18 25	33	125
57	21 21	32	122	20 49	33	123	20 16	33	124	19 43	33	125	19 10	33	126
56	22 08	32	123	21 36	33	124	21 03	33	125	20 30	34	125	19 56	34	126
55	22 55	-33	123	22 22	-33	124	21 49	-34	125	21 15	-34	126	20 41	-34	127
54	23 42	33	124	23 09	34	125	22 35	34	126	22 01	35	127	21 26	35	127
53	24 29	34	124	23 55	34	125	23 21	35	126	22 46	35	127	22 11	35	128
52	25 15	34	125	24 41	35	126	24 06	35	127	23 31	36	128	22 55	36	129
51	26 01	35	126	25 26	35	127	24 51	36	127	24 15	36	128	23 39	36	129
50	26 47	-36	126	26 11	-36	127	25 35	-36	128	24 59	-37	129	24 22	-36	130
49	27 32	36	127	26 56	36	128	26 20	37	129	25 43	37	130	25 06	38	130
48	28 17	36	128	27 41	36	128	27 03	37	129	26 25	38	130	25 48	38	131
47	29 02	38	128	28 24	37	129	27 47	38	130	27 09	38	131	26 31	39	132
46	29 46	38	129	29 08	37	130	28 30	39	131	27 51	38	132	27 13	40	132
45	30 30	-39	130	29 51	-38	130	29 12	-39	131	28 33	-39	132	27 54	-40	133
44	31 13	39	130	30 34	38	131	29 55	40	132	29 15	40	133	28 35	40	134
43	31 56	40	131	31 16	40	132	30 36	40	133	29 55	41	134	29 15	40	134
42	32 38	40	132	31 58	41	133	31 17	40	133	30 37	42	134	29 55	41	135
41	33 20	41	133	32 39	41	133	31 58	41	134	31 17	42	135	30 35	42	136
40	34 01	-41	133	33 20	-42	134	32 38	-42	135	31 55	-42	136	31 14	-43	137
39	34 42	42	134	34 00	42	135	33 18	43	136	32 35	43	137	31 52	43	137
38	35 22	42	135	34 40	43	136	33 57	43	137	33 14	44	137	32 30	44	138
37	36 02	43	136	35 19	44	137	34 35	43	137	33 52	44	138	33 08	45	139
36	36 41	44	137	35 57	44	137	35 13	44	138	34 29	45	139	33 44	45	140
35	37 20	-45	137	36 35	-44	138	35 51	-45	139	35 05	-46	140	34 20	-45	141
34	37 58	45	138	37 13	45	139	36 27	45	140	35 42	46	141	34 56	46	142
33	38 35	46	139	37 49	46	140	37 03	46	141	36 17	46	142	35 31	47	142
32	39 11	46	140	38 25	46	141	37 39	47	142	36 52	47	142	36 05	48	143
31	39 47	47	141	39 00	47	142	38 13	47	143	37 25	48	143	36 38	48	144
30	40 22	-47	142	39 35	-48	143	38 47	-48	144	37 59	-48	144	37 11	-49	145
29	40 57	48	143	40 09	48	144	39 20	48	145	38 32	49	145	37 43	49	146
28	41 30	48	144	40 42	48	145	39 53	49	145	39 04	50	145	38 14	50	147
27	42 03	49	145	41 14	50	146	40 24	50	146	39 34	50	147	38 44	50	148
26	42 35	50	146	41 45	50	147	40 55	50	148	40 05	51	148	39 14	51	149
25	43 06	-50	147	42 16	-51	148	41 25	-51	149	40 34	-51	149	39 43	-52	150
24	43 36	51	148	42 45	52	149	41 54	52	150	41 02	51	150	40 11	53	151
23	44 06	52	149	43 14	52	150	42 22	52	151	41 30	52	151	40 38	53	152
22	44 34	52	150	43 42	53	151	42 49	52	152	41 57	53	152	41 04	54	153
21	45 01	52	152	44 09	53	152	43 15	53	153	42 22	53	154	41 29	54	154
20	45 28	-54	153	44 34	-53	153	43 41	-54	154	42 47	-54	155	41 53	-54	155
19	45 53	54	154	44 59	53	155	44 05	54	155	43 11	55	156	42 16	55	156
18	46 17	54	155	45 23	55	156	44 28	55	156	43 33	55	157	42 38	55	157
17	46 40	55	156	45 45	55	157	44 50	55	158	43 55	56	158	42 59	55	159
16	47 02	55	158	46 07	55	158	45 11	56	159	44 15	56	159	43 19	56	160
15	47 23	-56	159	46 27	-55	160	45 31	-56	160	44 35	-57	161	43 38	-56	161
14	47 43	57	160	46 46	56	161	45 50	57	161	44 53	57	162	43 56	57	162
13	48 01	57	162	47 04	57	162	46 07	57	163	45 10	57	163	44 13	57	163
12	48 19	58	163	47 21	57	163	46 24	58	164	45 26	58	164	44 28	57	165
11	48 35	58	164	47 37	58	165	46 39	58	165	45 41	58	165	44 43	58	166
10	48 49	-58	166	47 51	-58	166	46 53	-59	166	45 54	-58	167	44 56	-58	167
9	49 02	58	167	48 04	58	167	47 05	58	168	46 07	59	168	45 08	59	168
8	49 14	58	168	48 16	59	169	47 17	59	169	46 18	59	169	45 19	59	170
7	49 25	59	170	48 26	59	170	47 27	59	170	46 28	59	171	45 29	60	171
6	49 34	59	171	48 35	59	172	47 36	60	172	46 36	59	172	45 37	60	173
5	49 42	-59	173	48 43	-60	173	47 43	-59	173	46 44	-60	173	45 44	-60	174
4	49 49	60	174	48 49	60	174	47 49	60	175	46 49	59	175	45 50	60	175
3	49 54	60	176	48 54	60	176	47 54	60	176	46 54	60	176	45 54	60	175
2	49 57	60	177	48 57	60	177	47 57	60	177	46 57	60	177	45 57	59	177
1	49 59	60	179	48 59	60	179	47 59	60	179	46 59	60	179	45 59	60	179
0	50 00	-60	180	49 00	-60	180	48 00	-60	180	47 00	-60	180	46 00	-60	180

LHA	25° Hc	d	Z	26° Hc	d	Z	27° Hc	d	Z	28° Hc	d	Z	29° Hc	d	Z	LHA
	° ′	′	°	° ′	′	°	° ′	′	°	° ′	′	°	° ′	′	°	
69	09 15	28	121	08 47	27	122	08 20	28	123	07 52	28	124	07 24	28	125	291
68	10 03	28	121	09 35	28	122	09 07	28	123	08 39	29	124	08 10	28	125	292
67	10 51	28	122	10 23	29	123	09 54	29	124	09 25	29	125	08 56	29	125	293
66	11 39	29	122	11 10	29	123	10 41	30	124	10 11	29	125	09 42	30	126	294
65	12 26	-29	123	11 57	-30	124	11 27	-30	125	10 57	-29	125	10 28	-31	126	295
64	13 14	30	123	12 44	30	124	12 14	31	125	11 43	30	126	11 13	31	127	296
63	14 01	31	124	13 30	30	125	13 00	31	125	12 29	31	126	11 58	31	127	297
62	14 47	30	124	14 17	32	125	13 45	31	126	13 14	31	127	12 43	32	128	298
61	15 34	31	125	15 03	32	126	14 31	32	126	13 59	32	127	13 27	32	128	299
60	16 20	-32	125	15 48	-32	126	15 16	-32	127	14 44	-33	128	14 11	-32	129	300
59	17 06	32	126	16 34	33	127	16 01	33	127	15 28	33	128	14 55	33	129	301
58	17 52	33	126	17 19	33	127	16 46	34	128	16 12	33	129	15 39	34	130	302
57	18 37	33	127	18 04	34	128	17 30	34	128	16 56	34	129	16 22	34	130	303
56	19 22	34	127	18 48	34	128	18 14	34	129	17 40	35	130	17 05	35	131	304
55	20 07	-34	128	19 33	-35	129	18 58	-35	130	18 23	-35	130	17 48	-36	131	305
54	20 51	35	128	20 16	35	129	19 41	35	130	19 06	36	131	18 30	36	132	306
53	21 36	36	129	21 00	36	130	20 24	36	131	19 48	36	132	19 12	37	132	307
52	22 19	36	130	21 43	35	130	21 07	37	131	20 30	37	132	19 53	37	133	308
51	23 03	37	130	22 26	37	131	21 49	37	132	21 12	38	133	20 34	37	133	309
50	23 46	-38	131	23 08	-37	132	22 31	-38	132	21 53	-38	133	21 15	-38	134	310
49	24 28	38	131	23 50	38	132	23 12	38	133	22 34	38	134	21 56	39	135	311
48	25 10	38	132	24 32	39	133	23 53	38	134	23 15	40	134	22 35	39	135	312
47	25 52	39	133	25 13	39	133	24 34	39	134	23 55	40	135	23 15	40	136	313
46	26 33	39	133	25 54	40	134	25 14	40	135	24 34	40	136	23 54	41	137	314
45	27 14	-40	134	26 34	-40	135	25 54	-41	136	25 13	-41	136	24 32	-41	137	315
44	27 55	41	135	27 14	41	135	26 33	41	136	25 52	41	137	25 11	42	138	316
43	28 35	42	135	27 53	41	136	27 12	42	137	26 30	42	138	25 48	42	139	317
42	29 14	42	136	28 32	42	137	27 50	42	138	27 08	43	138	26 25	43	139	318
41	29 53	42	137	29 11	43	138	28 28	43	138	27 45	43	139	27 02	44	140	319
40	30 31	-43	137	29 48	-43	138	29 05	-43	139	28 22	-44	140	27 38	-44	141	320
39	31 09	43	138	30 26	44	139	29 42	44	140	28 58	45	141	28 13	44	141	321
38	31 46	44	139	31 02	44	140	30 18	45	141	29 33	45	141	28 48	45	142	322
37	32 23	45	140	31 38	45	141	30 53	45	141	30 08	45	142	29 23	46	143	323
36	32 59	45	141	32 14	46	141	31 28	45	142	30 43	47	143	29 56	46	144	324
35	33 35	-46	141	32 49	-46	142	32 03	-47	143	31 16	-46	144	30 30	-47	144	325
34	34 10	47	142	33 23	47	143	32 36	47	144	31 49	47	145	31 02	47	145	326
33	34 44	47	143	33 57	48	144	33 09	47	145	32 22	48	145	31 34	48	146	327
32	35 17	47	144	34 30	48	145	33 42	48	145	32 54	49	146	32 05	49	147	328
31	35 50	48	145	35 02	49	146	34 13	48	146	33 25	49	147	32 36	50	148	329
30	36 22	-49	146	35 33	-49	147	34 44	-49	147	33 55	-50	148	33 05	-49	149	330
29	36 54	50	147	36 04	49	147	35 15	50	148	34 25	51	149	33 34	50	149	331
28	37 24	50	148	36 34	50	148	35 44	51	149	34 53	50	150	34 03	51	150	332
27	37 54	51	149	37 03	50	149	36 13	51	150	35 22	52	151	34 30	51	151	333
26	38 23	51	150	37 32	52	150	36 40	51	151	35 49	52	152	34 57	52	152	334
25	38 51	-52	151	37 59	-51	151	37 08	-53	152	36 15	-52	152	35 23	-52	153	335
24	39 18	52	152	38 26	52	152	37 34	53	153	36 41	53	153	35 48	53	154	336
23	39 45	53	153	38 52	53	153	37 59	53	154	37 06	53	154	36 13	54	155	337
22	40 10	53	154	39 17	53	154	38 24	54	155	37 30	54	155	36 36	54	156	338
21	40 35	54	155	39 41	54	155	38 47	54	156	37 53	54	156	36 59	55	157	339
20	40 59	-55	156	40 04	-54	156	39 10	-55	157	38 15	-55	157	37 20	-55	158	340
19	41 21	55	157	40 26	55	157	39 31	55	158	38 36	55	158	37 41	55	159	341
18	41 43	55	158	40 48	56	159	39 52	55	159	38 57	56	160	38 01	56	160	342
17	42 04	56	159	41 08	56	160	40 12	56	160	39 16	56	161	38 20	57	161	343
16	42 23	56	160	41 27	56	161	40 31	57	161	39 34	56	162	38 38	57	162	344
15	42 42	-57	161	41 45	-57	162	40 48	-57	162	39 51	-57	163	38 54	-57	163	345
14	42 59	57	163	42 02	57	163	41 05	57	163	40 08	58	164	39 10	57	164	346
13	43 16	58	164	42 18	57	164	41 21	58	165	40 23	58	165	39 25	58	165	347
12	43 31	58	165	42 33	58	165	41 35	58	166	40 37	58	166	39 39	58	166	348
11	43 45	58	166	42 47	59	167	41 48	58	167	40 50	58	167	39 52	59	167	349
10	43 58	-59	167	42 59	-58	168	42 01	-59	168	41 02	-58	168	40 04	-59	169	350
9	44 09	58	169	43 11	59	169	42 12	59	169	41 13	59	169	40 14	59	170	351
8	44 20	59	170	43 21	59	170	42 22	59	170	41 23	59	171	40 24	59	171	352
7	44 29	59	171	43 30	59	171	42 31	59	172	41 32	60	172	40 32	59	172	353
6	44 37	59	172	43 38	59	173	42 39	60	173	41 39	59	173	40 40	60	173	354
5	44 44	-59	174	43 45	-60	174	42 45	-60	174	41 45	-59	174	40 46	-60	174	355
4	44 50	60	175	43 50	60	175	42 50	59	175	41 51	60	175	40 51	60	175	356
3	44 54	59	176	43 55	60	176	42 55	60	176	41 55	60	176	40 55	60	177	357
2	44 58	60	177	43 58	60	178	42 58	60	178	41 58	60	178	40 58	60	178	358
1	44 59	60	179	43 59	60	179	42 59	60	179	41 59	60	179	40 59	60	179	359
0	45 00	-60	180	44 00	-60	180	43 00	-60	180	42 00	-60	180	41 00	-60	180	360

S. Lat. {LHA greater than 180°........Zn=180−Z; LHA less than 180°..........Zn=180+Z

DECLINATION (15°-29°) CONTRARY NAME TO LATITUDE

LAT 20°

Table 36

12
Celestial Navigation without a Sextant

Someday it may happen—you lose your balance and your sextant falls overboard. Or you drop it, hard—so hard that you can't trust it. What do you do? You navigate without it by using the horizon as your sextant. Instead of bringing the sun or moon down to the horizon with the sextant, you wait until the sun or moon has actually reached the horizon at sunset or dawn. At this moment Hs would be 0°. You enter this observation on your worksheet and proceed to reduce it as you would any other observation taken with a sextant. To illustrate, imagine that it is July 3, 1974, and you time the moment the sun's lower edge touches the horizon as 19-35-05. Your watch is 25 seconds slow. Here is the way the worksheet looks:

Watch Time (WT)	=	19-35-05
Fast (−) or Slow (+)	=	+25
Conversion to GMT	=	+4
GMT	=	23-35-30
Hs	=	0°00′
IC	=	00′
Dip	=	−03′
ha	=	−03′
Main Corr.	=	−18′
Additional Corr.	=	+02′
Ho	=	−19′

The additional correction came from page A4 in the *Nautical Almanac* (see Table 37) and is an additional refraction correction caused by high barometer and high temperature. I actually took this sight, and the temperature was 82° F with a barometer of 30.3 inches.

The negative Ho (−19′) indicates that the sun is actually nearly two-thirds of its diameter below the horizon. The effect is caused by refraction, as discussed in Chapter 7.

Turning now to the daily page in the almanac for July 3 (Table 25) and proceeding to the third stage of the worksheet:

ALTITUDE CORRECTION TABLES 0°–10°—SUN, STARS, PLANETS

App. Alt.	OCT.–MAR. SUN Lower Limb	OCT.–MAR. SUN Upper Limb	APR.–SEPT. SUN Lower Limb	APR.–SEPT. SUN Upper Limb	STARS PLANETS
° ′	′	′	′	′	′
0 00	−18·2	−51·7	−18·4	−51·4	−34·5
03	17·5	51·0	17·8	50·8	33·8
06	16·9	50·4	17·1	50·1	33·2
09	16·3	49·8	16·5	49·5	32·6
12	15·7	49·2	15·9	48·9	32·0
15	15·1	48·6	15·3	48·3	31·4
0 18	−14·5	−48·0	−14·8	−47·8	−30·8
21	14·0	47·5	14·2	47·2	30·3
24	13·5	47·0	13·7	46·7	29·8
27	12·9	46·4	13·2	46·2	29·2
30	12·4	45·9	12·7	45·7	28·7
33	11·9	45·4	12·2	45·2	28·2
0 36	−11·5	−45·0	−11·7	−44·7	−27·8
39	11·0	44·5	11·2	44·2	27·3
42	10·5	44·0	10·8	43·8	26·8
45	10·1	43·6	10·3	43·3	26·4
48	9·6	43·1	9·9	42·9	25·9
51	9·2	42·7	9·5	42·5	25·5
0 54	− 8·8	−42·3	− 9·1	−42·1	−25·1
0 57	8·4	41·9	8·7	41·7	24·7
1 00	8·0	41·5	8·3	41·3	24·3
03	7·7	41·2	7·9	40·9	24·0
06	7·3	40·8	7·5	40·5	23·6
09	6·9	40·4	7·2	40·2	23·2
1 12	− 6·6	−40·1	− 6·8	−39·8	−22·9
15	6·2	39·7	6·5	39·5	22·5
18	5·9	39·4	6·2	39·2	22·2
21	5·6	39·1	5·8	38·8	21·9
24	5·3	38·8	5·5	38·5	21·6
27	4·9	38·4	5·2	38·2	21·2
1 30	− 4·6	−38·1	− 4·9	−37·9	−20·9
35	4·2	37·7	4·4	37·4	20·5
40	3·7	37·2	4·0	37·0	20·0
45	3·2	36·7	3·5	36·5	19·5
50	2·8	36·3	3·1	36·1	19·1
1 55	2·4	35·9	2·6	35·6	18·7
2 00	− 2·0	−35·5	− 2·2	−35·2	−18·3
05	1·6	35·1	1·8	34·8	17·9
10	1·2	34·7	1·5	34·5	17·5
15	0·9	34·4	1·1	34·1	17·2
20	0·5	34·0	0·8	33·8	16·8
25	− 0·2	33·7	0·4	33·4	16·5
2 30	+ 0·2	−33·3	− 0·1	−33·1	−16·1
35	0·5	33·0	+ 0·2	32·8	15·8
40	0·8	32·7	0·5	32·5	15·5
45	1·1	32·4	0·8	32·2	15·2
50	1·4	32·1	1·1	31·9	14·9
2 55	1·6	31·9	1·4	31·6	14·7
3 00	+ 1·9	−31·6	+ 1·7	−31·3	−14·4
05	2·2	31·3	1·9	31·1	14·1
10	2·4	31·1	2·1	30·9	13·9
15	2·6	30·9	2·4	30·6	13·7
20	2·9	30·6	2·6	30·4	13·4
25	3·1	30·4	2·9	30·1	13·2
3 30	+ 3·3	−30·2	+ 3·1	−29·9	−13·0

App. Alt.	OCT.–MAR. SUN Lower Limb	OCT.–MAR. SUN Upper Limb	APR.–SEPT. SUN Lower Limb	APR.–SEPT. SUN Upper Limb	STARS PLANETS
° ′	′	′	′	′	′
3 30	+ 3·3	−30·2	+ 3·1	−29·9	−13·0
35	3·6	29·9	3·3	29·7	12·7
40	3·8	29·7	3·5	29·5	12·5
45	4·0	29·5	3·7	29·3	12·3
50	4·2	29·3	3·9	29·1	12·1
3 55	4·4	29·1	4·1	28·9	11·9
4 00	+ 4·5	−29·0	+ 4·3	−28·7	−11·8
05	4·7	28·8	4·5	28·5	11·6
10	4·9	28·6	4·6	28·4	11·4
15	5·1	28·4	4·8	28·2	11·2
20	5·2	28·3	5·0	28·0	11·1
25	5·4	28·1	5·1	27·9	10·9
4 30	+ 5·6	−27·9	+ 5·3	−27·7	−10·7
35	5·7	27·8	5·5	27·5	10·6
40	5·9	27·6	5·6	27·4	10·4
45	6·0	27·5	5·8	27·2	10·3
50	6·2	27·3	5·9	27·1	10·1
4 55	6·3	27·2	6·0	27·0	10·0
5 00	+ 6·4	−27·1	+ 6·2	−26·8	− 9·9
05	6·6	26·9	6·3	26·7	9·7
10	6·7	26·8	6·4	26·6	9·6
15	6·8	26·7	6·6	26·4	9·5
20	6·9	26·6	6·7	26·3	9·4
25	7·1	26·4	6·8	26·2	9·2
5 30	+ 7·2	−26·3	+ 6·9	−26·1	− 9·1
35	7·3	26·2	7·0	26·0	9·0
40	7·4	26·1	7·2	25·8	8·9
45	7·5	26·0	7·3	25·7	8·8
50	7·6	25·9	7·4	25·6	8·7
5 55	7·7	25·8	7·5	25·5	8·6
6 00	+ 7·8	−25·7	+ 7·6	−25·4	− 8·5
10	8·0	25·5	7·8	25·2	8·3
20	8·2	25·3	8·0	25·0	8·1
30	8·4	25·1	8·1	24·9	7·9
40	8·6	24·9	8·3	24·7	7·7
6 50	8·7	24·8	8·5	24·5	7·6
7 00	+ 8·9	−24·6	+ 8·6	−24·4	− 7·4
10	9·1	24·4	8·8	24·2	7·2
20	9·2	24·3	9·0	24·0	7·1
30	9·3	24·2	9·1	23·9	7·0
40	9·5	24·0	9·2	23·8	6·8
7 50	9·6	23·9	9·4	23·6	6·7
8 00	+ 9·7	−23·8	+ 9·5	−23·5	− 6·6
10	9·9	23·6	9·6	23·4	6·4
20	10·0	23·5	9·7	23·3	6·3
30	10·1	23·4	9·8	23·2	6·2
40	10·2	23·3	10·0	23·0	6·1
8 50	10·3	23·2	10·1	22·9	6·0
9 00	+10·4	−23·1	+10·2	−22·8	− 5·9
10	10·5	23·0	10·3	22·7	5·8
20	10·6	22·9	10·4	22·6	5·7
30	10·7	22·8	10·5	22·5	5·6
40	10·8	22·7	10·6	22·4	5·5
9 50	10·9	22·6	10·6	22·4	5·4
10 00	+11·0	−22·5	+10·7	−22·3	− 5·3

Additional corrections for temperature and pressure are given on the following page.
For bubble sextant observations ignore dip and use the star corrections for Sun, planets, and stars.

Table 37

ALTITUDE CORRECTION TABLES—ADDITIONAL CORRECTIONS

ADDITIONAL REFRACTION CORRECTIONS FOR NON-STANDARD CONDITIONS

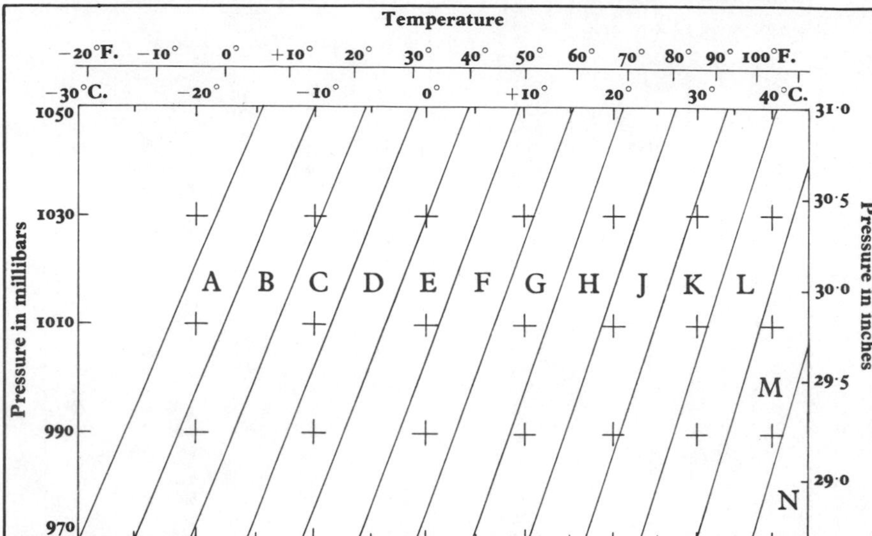

App. Alt.	A	B	C	D	E	F	G	H	J	K	L	M	N	App. Alt.
° ′	′	′	′	′	′	′	′	′	′	′	′	′	′	° ′
0 00	−6·9	−5·7	−4·6	−3·4	−2·3	−1·1	0·0	+1·1	+2·3	+3·4	+4·6	+5·7	+6·9	0 00
0 30	5·2	4·4	3·5	2·6	1·7	0·9	0·0	0·9	1·7	2·6	3·5	4·4	5·2	0 30
1 00	4·3	3·5	2·8	2·1	1·4	0·7	0·0	0·7	1·4	2·1	2·8	3·5	4·3	1 00
1 30	3·5	2·9	2·4	1·8	1·2	0·6	0·0	0·6	1·2	1·8	2·4	2·9	3·5	1 30
2 00	3·0	2·5	2·0	1·5	1·0	0·5	0·0	0·5	1·0	1·5	2·0	2·5	3·0	2 00
2 30	−2·5	−2·1	−1·6	−1·2	−0·8	−0·4	0·0	+0·4	+0·8	+1·2	+1·6	+2·1	+2·5	2 30
3 00	2·2	1·8	1·5	1·1	0·7	0·4	0·0	0·4	0·7	1·1	1·5	1·8	2·2	3 00
3 30	2·0	1·6	1·3	1·0	0·7	0·3	0·0	0·3	0·7	1·0	1·3	1·6	2·0	3 30
4 00	1·8	1·5	1·2	0·9	0·6	0·3	0·0	0·3	0·6	0·9	1·2	1·5	1·8	4 00
4 30	1·6	1·4	1·1	0·8	0·5	0·3	0·0	0·3	0·5	0·8	1·1	1·4	1·6	4 30
5 00	−1·5	−1·3	−1·0	−0·8	−0·5	−0·2	0·0	+0·2	+0·5	+0·8	+1·0	+1·3	+1·5	5 00
6	1·3	1·1	0·9	0·6	0·4	0·2	0·0	0·2	0·4	0·6	0·9	1·1	1·3	6
7	1·1	0·9	0·7	0·6	0·4	0·2	0·0	0·2	0·4	0·6	0·7	0·9	1·1	7
8	1·0	0·8	0·7	0·5	0·3	0·2	0·0	0·2	0·3	0·5	0·7	0·8	1·0	8
9	0·9	0·7	0·6	0·4	0·3	0·1	0·0	0·1	0·3	0·4	0·6	0·7	0·9	9
10 00	−0·8	−0·7	−0·5	−0·4	−0·3	−0·1	0·0	+0·1	+0·3	+0·4	+0·5	+0·7	+0·8	10 00
12	0·7	0·6	0·5	0·3	0·2	0·1	0·0	0·1	0·2	0·3	0·5	0·6	0·7	12
14	0·6	0·5	0·4	0·3	0·2	0·1	0·0	0·1	0·2	0·3	0·4	0·5	0·6	14
16	0·5	0·4	0·3	0·3	0·2	0·1	0·0	0·1	0·2	0·3	0·3	0·4	0·5	16
18	0·4	0·4	0·3	0·2	0·2	0·1	0·0	0·1	0·2	0·2	0·3	0·4	0·4	18
20 00	−0·4	−0·3	−0·3	−0·2	−0·1	−0·1	0·0	+0·1	+0·1	+0·2	+0·3	+0·3	+0·4	20 00
25	0·3	0·3	0·2	0·2	0·1	−0·1	0·0	+0·1	0·1	0·2	0·2	0·3	0·3	25
30	0·3	0·2	0·2	0·1	0·1	0·0	0·0	0·0	0·1	0·1	0·2	0·2	0·3	30
35	0·2	0·2	0·1	0·1	0·1	0·0	0·0	0·0	0·1	0·1	0·1	0·2	0·2	35
40	0·2	0·1	0·1	0·1	−0·1	0·0	0·0	0·0	+0·1	0·1	0·1	0·1	0·2	40
50 00	−0·1	−0·1	−0·1	−0·1	0·0	0·0	0·0	0·0	0·0	+0·1	+0·1	+0·1	+0·1	50 00

The graph is entered with arguments temperature and pressure to find a zone letter; using as arguments this zone letter and apparent altitude (sextant altitude corrected for dip), a correction is taken from the table. This correction is to be applied to the sextant altitude in addition to the corrections for standard conditions (for the Sun, planets and stars from the inside front cover and for the Moon from the inside back cover).

Table 37, continued

SUN

GHA 23 Hr.	= 163° 57′	Dec. 22°56′ N
35 Min. 30 Sec.	= 8° 53′	
GHA	= 171°110′	
GHA	= 172° 50′	
Assumed Long. W.	= 66° 50′	Assumed Lat. = 34° N
LHA	= 106°	

Proceeding to the sight reduction tables (Table 38):

Hc	= −08′	d	= +36′
Corr.	= +34′	Dec. Inc.	= 56′
Hc	= +26′	Corr.	= +34′
Ho	= −19′		
	45′ Away		
Z = 63°		Zn	= 360
			−63
		Zn	= 297°

To realize why the LOP is away from the sun, remember this: Hc is the angle you would have observed had you been at the assumed position at the time of the sight. Then the sun would have been 26′ *above* the horizon. You actually observed it, however, *below* the horizon. The observed altitude was less than the computed (or theoretical) altitude, so you are farther away from the sun, and you are farther away by the total difference between +26′ and −19′ or 45′. All I can say is that when I took this sight, the LOP it produced agreed *exactly* with the distance run since a fix made earlier by advancing sun lines.

Suppose the final result had been this:

Hc	= −26′
Ho	= −30′
	4′ Away

In this case your observation again showed the sun to be lower than calculated based on the assumed position.

Table 38

N. Lat. {LHA greater than 180°....... Zn=Z; LHA less than 180°........ Zn=360−Z}

DECLINATION (15°-29°) SAME NAME AS LATITUDE

LAT 34°

LHA	15° Hc	d	Z	16° Hc	d	Z	17° Hc	d	Z	18° Hc	d	Z	19° Hc	d	Z
	° ′	′	°	° ′	′	°	° ′	′	°	° ′	′	°	° ′	′	°
70	24 45	+31	88	25 16	+30	87	25 46	+30	86	26 16	+29	85	26 45	+29	84
71	23 55	31	88	24 26	30	87	24 56	30	86	25 26	30	85	25 56	29	84
72	23 05	31	87	23 36	31	86	24 07	30	85	24 37	29	84	25 06	30	83
73	22 16	31	87	22 47	30	86	23 17	30	85	23 47	30	84	24 17	29	83
74	21 26	31	86	21 57	31	85	22 28	30	84	22 58	30	83	23 28	29	82
75	20 37	+31	85	21 08	+30	85	21 38	+30	84	22 08	+30	83	22 38	+30	82
76	19 47	31	85	20 18	31	84	20 49	30	83	21 19	30	82	21 49	30	81
77	18 58	31	84	19 29	30	83	19 59	31	83	20 30	30	82	21 00	30	81
78	18 08	31	84	18 39	31	83	19 10	31	82	19 41	30	81	20 11	30	80
79	17 19	31	83	17 50	31	82	18 21	31	82	18 52	30	81	19 22	30	80
80	16 29	+32	83	17 01	+31	82	17 32	+31	81	18 03	+30	80	18 33	+31	79
81	15 40	31	82	16 11	32	81	16 43	31	81	17 14	30	80	17 44	31	79
82	14 51	31	82	15 22	32	81	15 54	31	80	16 25	31	79	16 56	30	78
83	14 01	32	81	14 33	32	80	15 05	31	79	15 36	31	79	16 07	31	78
84	13 12	32	81	13 44	32	80	14 16	31	79	14 47	31	78	15 18	32	77
85	12 23	+32	80	12 55	+32	79	13 27	+32	78	13 59	+31	78	14 30	+31	77
86	11 34	32	80	12 06	32	79	12 38	32	78	13 10	32	77	13 42	31	76
87	10 45	33	79	11 18	32	78	11 50	32	77	12 22	31	77	12 53	32	76
88	09 57	32	79	10 29	32	78	11 01	32	77	11 33	32	76	12 05	32	75
89	09 08	32	78	09 40	33	77	10 13	32	76	10 45	32	75	11 17	32	75
90	08 19	+33	78	08 52	+33	77	09 25	+32	76	09 57	+32	75	10 29	+33	74
91	07 31	33	77	08 04	32	76	08 36	33	75	09 09	33	74	09 42	32	74
92	06 42	33	76	07 15	33	76	07 48	33	75	08 21	33	74	08 54	33	73
93	05 54	33	76	06 27	34	75	07 01	33	74	07 34	32	73	08 06	33	73
94	05 06	33	75	05 39	34	75	06 13	33	74	06 46	33	73	07 19	33	72
95	04 18	+33	75	04 51	+34	74	05 25	+34	73	05 59	+33	72	06 32	+33	72
96	03 30	34	74	04 04	33	73	04 37	34	73	05 11	34	72	05 45	33	71
97	02 42	34	74	03 16	34	73	03 50	34	72	04 24	34	71	04 58	34	71
98	01 54	35	73	02 29	34	72	03 03	34	72	03 37	34	71	04 11	34	70
99	01 07	34	73	01 41	35	72	02 16	34	71	02 50	34	70	03 24	35	69
100	00 20	+34	72	00 54	+35	71	01 29	+34	70	02 03	+35	70	02 38	+35	69
101	−0 28	35	71	00 07	35	71	00 42	35	70	01 17	35	69	01 52	35	68
102	−1 15	35	71	−0 40	35	70	−0 05	35	69	00 30	36	68	01 06	35	68
103	−2 02	36	70	−1 26	35	70	−0 51	35	69	−0 16	36	68	00 20	35	67
104	−2 49	36	70	−2 13	36	69	−1 37	35	68	−1 02	36	67	−0 26	36	67
105	−3 35	+36	69	−2 59	+36	68	−2 23	+35	68	−1 48	+36	67	−1 12	+36	66
106	−4 22	37	69	−3 45	36	68	−3 09	36	67	−2 33	36	66	−1 57	36	65
107	−5 08	37	68	−4 31	36	67	−3 55	36	66	−3 19	37	66	−2 42	36	65
108	−5 54	37	67	−5 17	37	67	−4 40	36	66	−4 04	37	65	−3 27	37	64
109	109			−6 03	37	66	−5 26	37	65	−4 49	37	64	−4 12	37	64
110				110			−6 11	+37	65	−5 34	+38	64	−4 56	+37	63
111							111			−6 18	38	63	−5 40	37	63
112										112			−6 24	38	62

LHA	20° Hc	d	Z	21° Hc	d	Z	22° Hc	d	Z	23° Hc	d	Z	24° Hc	d	Z
	° ′	′	°	° ′	′	°	° ′	′	°	° ′	′	°	° ′	′	°
70	27 14	+29	83	27 43	+28	82	28 11	+28	81	28 39	+28	80	29 07	+27	79
71	26 25	29	83	26 54	28	82	27 22	28	81	27 50	28	80	28 18	27	79
72	25 36	29	82	26 05	28	81	26 33	28	80	27 01	28	79	27 29	27	78
73	24 46	29	82	25 15	29	81	25 44	28	80	26 12	28	79	26 40	28	78
74	23 57	29	81	24 26	29	80	24 55	29	79	25 24	28	78	25 52	27	77
75	23 08	+29	81	23 37	+29	80	24 06	+29	79	24 35	+28	78	25 03	+28	77
76	22 19	30	80	22 49	29	79	23 18	28	78	23 46	29	77	24 15	28	77
77	21 30	30	80	22 00	29	79	22 29	29	78	22 58	29	77	23 27	28	76
78	20 41	30	79	21 11	29	78	21 40	30	77	22 10	28	77	22 38	29	76
79	19 52	30	79	20 22	30	78	20 52	29	77	21 21	29	76	21 50	29	75
80	19 04	+30	78	19 34	+29	77	20 03	+30	76	20 33	+29	76	21 02	+29	75
81	18 15	30	78	18 45	30	77	19 15	30	76	19 45	29	75	20 14	29	74
82	17 26	31	77	17 57	30	76	18 27	30	75	18 57	30	75	19 27	29	74
83	16 38	30	77	17 08	31	76	17 39	30	75	18 09	30	75	18 39	30	73
84	15 50	30	76	16 20	31	75	16 51	30	75	17 21	30	74	17 51	30	73
85	15 01	+31	76	15 32	+31	75	16 03	+31	74	16 34	+30	73	17 04	+30	72
86	14 13	31	75	14 44	31	74	15 15	31	74	15 46	31	73	16 17	30	72
87	13 25	31	75	13 56	32	74	14 28	31	73	14 59	30	72	15 29	31	71
88	12 37	32	74	13 09	31	73	13 40	31	73	14 11	31	72	14 42	31	71
89	11 49	32	74	12 21	32	73	12 53	31	72	13 24	32	71	13 56	31	70
90	11 02	+32	73	11 34	+32	72	12 06	+31	72	12 37	+32	71	13 09	+31	70
91	10 14	32	73	10 46	32	72	11 18	32	71	11 50	32	70	12 22	32	69
92	09 27	32	72	09 59	33	71	10 32	31	71	11 04	32	70	11 36	32	69
93	08 39	33	72	09 12	33	71	09 45	32	70	10 17	33	69	10 50	32	68
94	07 52	33	71	08 25	33	70	08 58	33	70	09 31	32	69	10 03	33	68
95	07 05	+33	71	07 38	+34	70	08 12	+33	69	08 45	+32	68	09 17	+33	67
96	06 18	34	70	06 52	33	69	07 25	33	69	07 58	34	68	08 32	33	67
97	05 32	33	70	06 05	34	69	06 39	34	68	07 13	33	67	07 46	33	66
98	04 45	34	69	05 19	34	68	05 53	34	67	06 27	34	67	07 01	33	66
99	03 59	34	68	04 33	34	68	05 07	34	67	05 41	34	66	06 15	34	65
100	03 13	+34	68	03 47	+35	67	04 22	+34	66	04 56	+34	65	05 30	+35	65
101	02 27	34	67	03 01	35	67	03 36	35	66	04 11	34	65	04 45	35	64
102	01 41	35	67	02 16	35	66	02 51	35	65	03 26	35	64	04 01	35	64
103	00 55	35	66	01 30	36	65	02 06	35	65	02 41	35	64	03 16	36	63
104	00 10	35	66	00 45	36	65	01 21	35	64	01 56	36	63	02 32	36	63
105	−0 36	+36	65	00 00	+36	64	00 36	+36	64	01 12	+36	63	01 48	+36	62
106	−1 21	37	65	−0 44	36	64	−0 08	36	63	00 28	36	62	01 04	36	61
107	−2 06	37	64	−1 29	37	63	−0 52	36	62	−0 16	37	62	00 21	36	61
108	−2 50	37	63	−2 13	37	63	−1 36	36	62	−1 00	37	61	−0 23	37	60
109	−3 35	38	63	−2 57	37	62	−2 20	37	61	−1 43	37	61	−1 06	37	60
110	−4 19	+38	62	−3 41	+37	62	−3 04	+38	61	−2 26	+37	60	−1 49	+38	59
111	−5 03	38	62	−4 25	38	61	−3 47	38	60	−3 09	38	60	−2 31	38	59
112	−5 46	38	61	−5 08	38	60	−4 30	38	60	−3 52	38	59	−3 14	39	58
113	113			−5 51	38	60	−5 13	39	59	−4 34	38	58	−3 56	39	57
114				114			−5 55	39	58	−5 16	39	58	−4 37	39	57
115							115			−5 58	+39	57	−5 19	+39	56
116										116			−6 00	40	56
117													117		

LHA	25° Hc	d	Z	26° Hc	d	Z	27° Hc	d	Z	28° Hc	d	Z	29° Hc	d	Z	LHA
	° ′	′	°	° ′	′	°	° ′	′	°	° ′	′	°	° ′	′	°	
70	29 34	+26	78	30 00	+26	77	30 26	+25	76	30 51	+25	75	31 16	+25	74	290
71	28 45	27	78	29 12	26	77	29 38	25	76	30 03	26	75	30 29	24	74	289
72	27 56	27	77	28 23	27	76	28 50	25	75	29 15	26	74	29 41	25	73	288
73	27 08	27	77	27 35	26	76	28 01	27	75	28 28	25	74	28 53	26	73	287
74	26 19	28	76	26 47	27	75	27 14	26	74	27 40	26	73	28 06	25	72	286
75	25 31	+28	76	25 59	+27	75	26 26	+26	74	26 52	+27	73	27 19	+25	72	285
76	24 43	28	76	25 11	27	74	25 38	27	74	26 05	26	72	26 31	26	72	284
77	23 55	28	75	24 23	27	74	24 50	27	73	25 17	27	72	25 44	27	71	283
78	23 07	28	75	23 35	28	73	24 03	27	73	24 30	27	71	24 57	27	71	282
79	22 19	28	74	22 47	28	73	23 15	28	72	23 43	27	71	24 10	27	70	281
80	21 31	+29	74	22 00	+28	72	22 28	+28	72	22 56	+28	71	23 24	+27	70	280
81	20 43	29	73	21 12	29	72	21 41	28	71	22 09	28	70	22 37	27	69	279
82	19 56	29	73	20 25	29	71	20 54	28	71	21 22	28	70	21 50	28	69	278
83	19 09	29	72	19 38	29	71	20 07	29	70	20 36	28	69	21 04	28	69	277
84	18 21	30	72	18 51	29	70	19 20	29	70	19 49	29	69	20 18	28	68	276
85	17 34	+30	71	18 04	+29	70	18 33	+30	69	19 03	+29	69	19 32	+29	68	275
86	16 47	30	71	17 17	30	69	17 47	30	69	18 17	29	68	18 46	29	67	274
87	16 00	30	70	16 30	31	69	17 01	30	68	17 31	29	68	18 00	30	67	273
88	15 13	31	70	15 44	30	68	16 14	31	68	16 45	30	67	17 15	29	66	272
89	14 27	31	69	14 58	30	68	15 28	31	67	15 59	30	67	16 29	30	66	271
90	13 40	+31	68	14 11	+31	67	14 42	+31	67	15 13	+31	66	15 44	+30	65	270
91	12 54	31	68	13 25	32	67	13 57	31	66	14 28	31	66	14 59	30	65	269
92	12 08	32	67	12 40	31	66	13 11	32	66	13 43	31	65	14 14	31	64	268
93	11 22	32	67	11 54	32	66	12 26	31	65	12 57	32	65	13 29	31	64	267
94	10 36	32	66	11 08	32	65	11 40	33	65	12 13	31	64	12 44	32	63	266
95	09 50	+33	66	10 23	+32	65	10 55	+33	64	11 28	+32	64	12 00	+32	63	265
96	09 05	33	65	09 38	33	64	10 11	32	64	10 43	33	63	11 16	32	63	264
97	08 19	34	65	08 53	33	64	09 26	33	63	09 59	33	63	10 32	33	62	263
98	07 34	34	64	08 08	33	63	08 41	34	63	09 15	33	62	09 48	33	62	262
99	06 49	34	64	07 23	34	63	07 57	34	62	08 31	33	62	09 04	33	61	261
100	06 05	+34	63	06 39	+34	62	07 13	+34	62	07 47	+34	61	08 21	+34	61	260
101	05 20	35	63	05 55	34	62	06 29	34	61	07 03	35	61	07 38	34	60	259
102	04 36	35	62	05 11	34	61	05 45	35	61	06 20	35	60	06 55	34	60	258
103	03 52	35	62	04 27	35	61	05 02	35	60	05 37	35	60	06 12	35	59	257
104	03 08	35	61	03 43	36	60	04 19	35	60	04 54	35	59	05 29	36	59	256
105	02 24	+36	61	03 00	+36	60	03 36	+35	59	04 11	+36	59	04 47	+36	58	255
106	01 40	37	60	02 17	36	59	02 53	36	59	03 29	36	58	04 05	36	57	254
107	00 57	37	60	01 34	36	59	02 10	37	58	02 47	36	58	03 23	37	57	253
108	00 14	37	59	00 51	37	58	01 28	37	58	02 05	37	57	02 42	37	56	252
109	−0 29	38	59	00 09	37	58	00 46	37	57	01 23	37	57	02 00	38	56	251
110	−1 11	+38	58	−0 33	+37	57	00 04	+38	57	00 42	+37	56	01 19	+38	55	250
111	−1 53	38	58	−1 15	38	57	−0 37	38	56	00 01	38	56	00 39	38	55	249
112	−2 35	38	57	−1 57	38	56	−1 19	39	56	−0 40	38	55	−0 02	39	54	248
113	−3 17	39	57	−2 38	38	56	−2 00	39	55	−1 21	39	54	−0 42	39	54	247
114	−3 58	39	56	−3 19	39	55	−2 40	39	55	−2 01	39	54	−1 22	39	53	246
115	−4 40	+40	55	−4 00	+39	55	−3 21	+40	54	−2 41	+40	53	−2 01	+39	52	245
116	−5 20	39	55	−4 41	40	54	−4 01	40	53	−3 21	40	53	−2 41	40	52	244
117	−6 01	40	54	−5 21	41	54	−4 40	40	53	−4 00	40	52	−3 20	41	51	243
118	118			−6 01	41	53	−5 20	41	52	−4 39	41	51	−3 58	41	51	242
119				119			−5 59	41	52	−5 18	41	51	−4 37	42	50	241
120							120			−5 56	+41	50	−5 15	+42	50	240
121										121			−5 53	42	49	239

LHA	15° Hc	d	Z	16° Hc	d	Z	17° Hc	d	Z	18° Hc	d	Z	19° Hc	d	Z
87	−5 54	33	104	−6 27	34	105	273								
86	−5 06	33	105	−5 39	34	105	−6 13	33	106	274					
85	−4 18	−33	105	−4 51	−34	106	−5 25	−34	107	−5 59	−33	108	−6 32	−33	108
84	−3 30	34	106	−4 04	33	107	−4 37	34	107	−5 11	34	108	−5 45	33	109
83	−2 42	34	106	−3 16	34	107	−3 50	34	108	−4 24	34	109	−4 58	34	110
82	−1 54	35	107	−2 29	34	108	−3 03	34	108	−3 37	34	109	−4 11	34	110
81	−1 07	34	107	−1 41	35	108	−2 16	34	109	−2 50	34	110	−3 24	35	111
80	−0 20	−34	108	−0 54	−35	109	−1 29	−34	110	−2 03	−35	110	−2 38	−35	111
79	00 28	35	109	−0 07	35	109	−0 42	35	110	−1 17	35	111	−1 52	35	112
78	01 15	35	109	00 40	35	110	00 05	35	111	−0 30	36	112	−1 06	35	112
77	02 02	36	110	01 26	35	110	00 51	35	111	00 16	36	112	−0 20	35	113
76	02 49	36	110	02 13	36	111	01 37	35	112	01 02	36	113	00 26	36	113
75	03 35	−36	111	02 59	−36	111	02 23	−35	112	01 48	−36	113	01 12	−36	114
74	04 22	37	111	03 45	36	112	03 09	36	113	02 33	36	114	01 57	36	115
73	05 08	37	112	04 31	36	113	03 55	36	114	03 19	37	114	02 42	36	115
72	05 54	37	113	05 17	37	113	04 40	36	114	04 04	37	115	03 27	37	116
71	06 40	37	113	06 03	37	114	05 26	37	115	04 49	37	116	04 12	37	116
70	07 25	−37	114	06 48	−37	115	06 11	−37	115	05 34	−38	116	04 56	−37	117

LHA	20° Hc	d	Z	21° Hc	d	Z	22° Hc	d	Z	23° Hc	d	Z	24° Hc	d	Z
85	275														
84	−6 18	34	110	276											
83	−5 32	33	110	−6 05	34	111	277								
82	−4 45	34	111	−5 19	34	112	−5 53	34	113	−6 27	34	114	278		
81	−3 59	34	112	−4 33	34	112	−5 07	34	113	−5 41	34	114	−6 15	34	115
80	−3 13	−34	112	−3 47	−35	113	−4 22	−34	114	−4 56	−34	115	−5 30	−35	115
79	−2 27	34	113	−3 01	35	113	−3 36	35	114	−4 11	34	115	−4 45	35	116
78	−1 41	35	113	−2 16	35	114	−2 51	35	115	−3 26	35	116	−4 01	35	116
77	−0 55	35	114	−1 30	36	115	−2 06	35	115	−2 41	35	116	−3 16	36	117
76	−0 10	35	114	−0 45	36	115	−1 21	35	116	−1 56	36	117	−2 32	36	117
75	00 36	−36	115	00 00	−36	116	−0 36	−36	116	−1 12	−36	117	−1 48	−36	118
74	01 21	37	115	00 44	36	116	00 08	36	117	−0 28	36	118	−1 04	36	119
73	02 06	37	116	01 29	37	117	00 52	36	118	00 16	37	118	−0 21	36	119
72	02 50	37	117	02 13	37	117	01 36	36	118	01 00	37	119	00 23	37	120
71	03 35	38	117	02 57	37	118	02 20	37	119	01 43	37	119	01 06	37	120
70	04 19	−38	118	03 41	−37	118	03 04	−38	119	02 26	−37	120	01 49	−38	121

LHA	25° Hc	d	Z	26° Hc	d	Z	27° Hc	d	Z	28° Hc	d	Z	29° Hc	d	Z	LHA
81	279															
80	−6 05	−34	116	280												
79	−5 20	35	117	−5 55	34	117	281									
78	−4 36	35	117	−5 11	34	118	−5 45	35	119	−6 20	35	120	282			
77	−3 52	35	118	−4 27	35	119	−5 02	35	119	−5 37	35	120	−6 12	35	121	283
76	−3 08	35	118	−3 43	36	119	−4 19	35	120	−4 54	35	121	−5 29	36	122	284
75	−2 24	−36	119	−3 00	−36	120	−3 36	−35	120	−4 11	−36	121	−4 47	−36	122	285
74	−1 40	37	119	−2 17	36	120	−2 53	36	121	−3 29	36	122	−4 05	36	123	286
73	−0 57	37	120	−1 34	36	121	−2 10	37	121	−2 47	36	122	−3 23	37	123	287
72	−0 14	37	120	−0 51	37	121	−1 28	37	122	−2 05	37	123	−2 42	37	124	288
71	00 29	38	121	−0 09	37	122	−0 46	37	123	−1 23	37	123	−2 00	38	124	289
70	01 11	−38	122	00 33	−37	122	−0 04	−38	123	−0 42	−37	124	−1 19	−38	125	290

LAT 34°

Hc	=	−26′
Ho	=	− 6′
		20′ Toward

Your observation shows the sun to be higher (less negative) than it would be at the assumed position; therefore, you are closer to the sun.

It is fascinating to speculate that if the moon were in the right position (quarter to half full) a fix could be made by taking a horizon sight of the moon, then the sun, and advancing the LOP for the distance run. It is also possible that Venus could be used for this, because it is so bright. I haven't been at sea at a time when the proper conditions prevailed, but I would be very interested to hear from any reader who has the opportunity to try these possibilities.

13 Celestial Navigation without a Sextant or Tables

You now know that it possible to look up the point on earth that is, at any moment, directly beneath any star in the sky. To put it another way, at any moment every star is directly *over* some definite point on earth. If you happen to want to reach one of those points, all you have to do is aim your boat at the star at the moment it is over your destination. You will then be pointed at your destination, so note the compass heading. Repeat this process every night, and sail the compass course during the day, and you will automatically sail the shortest course to your destination—the great circle route.

This procedure is useful only on long voyages, and you can see why if you remember what you know about zenith distance. Suppose, for example, that there was a star over Bermuda and you were in Newport, R.I. The distance from Newport to Bermuda is a little over 600 nautical miles. This means that the zenith distance would be 10° (1° = 60 nautical miles). From the deck of your boat, then, this star would be only 10° from the vertical. It is difficult to aim a boat accurately with the object so high in the sky—after all, the altitude of this star would be 80°! For accurate aiming, the star needs to be lower, which means you need to be farther away.

Suppose you were off the west coast of Africa and headed for Bermuda. In this case, the distance is around 2,400 miles, and the zenith distance is about 40°, which means the star would be about 50° above the horizon and would be easy to aim at. You could sail about two-thirds of the trip before you needed to pick up a sextant. The Polynesians learned by heart what stars were over which islands at any given time of the year (they carried their *Nautical Almanac*s in their heads) and used this method for centuries. According to a recent book on the subject, they still do.

To illustrate, suppose you were leaving the Canary Islands to sail to the Virgin Islands on July 16, 1974. The first thing to do is find those stars that have the same or nearly the same latitude as your destination. To do this, look at the star chart in the back of the *Nautical Almanac* or the *Air Almanac* (the *Air Almanac*, by the way, has the best). The latitude of the Virgins is about 18°30′. Looking along the 18° declination line on the chart of northern stars (Table 39), and referring to the list of stars in Table 41, you find that Arcturus (a nice, bright star) has a declination of 19°19′ north, which means that it passes 34 nautical miles north of the northernmost of the Virgins (Anegada). It would be reasonable to aim at this point, since you are going to go to sextant observations about 1,000 miles out and head for the more southern of the Virgin Islands. (You don't want to make a landfall at Anegada; it is very low and hard to see, and many vessels have been wrecked there.)

The longitude of the Virgins is about 64° west. You have learned that the GHA of a star is the GHA of Aries plus the sidereal hour angle of the star:

GHA Star = GHA Aries + SHA Star

You want to know when the GHA of the star is the same as the longitude of the Virgins:

64° = GHA Aries + SHA Star

From the almanac (Table 41), the SHA of Arcturus is 146°22′. Therefore:

64° = GHA Aries + 146°22′
64° − 146°22′ = GHA Aries

To make this process simple, let's make the longitude 64°22′ W, which passes right down the middle of the Virgin Islands. First add 360° so you can perform the subtraction. Then: 64°22′ + 360° − 146°22′ = 278°. So when GHA Aries = 278°, Arcturus is over a spot on the earth at 19°19′ N, 64°22′ W, or 34 nautical miles north of Anegada.

Looking in the *Nautical Almanac* for July 16, 1974 (Table 41), you can see that Arcturus is going to be over that spot around 11 P.M. (2300 hours), GMT, which is also local time in the Canary Islands.

To find the exact time, note that 2300 hours GMT is 279°24′ Aries GHA. The difference, 1°24′ of arc, equals 5 minutes and 36 seconds of

STAR CHARTS

NORTHERN STARS

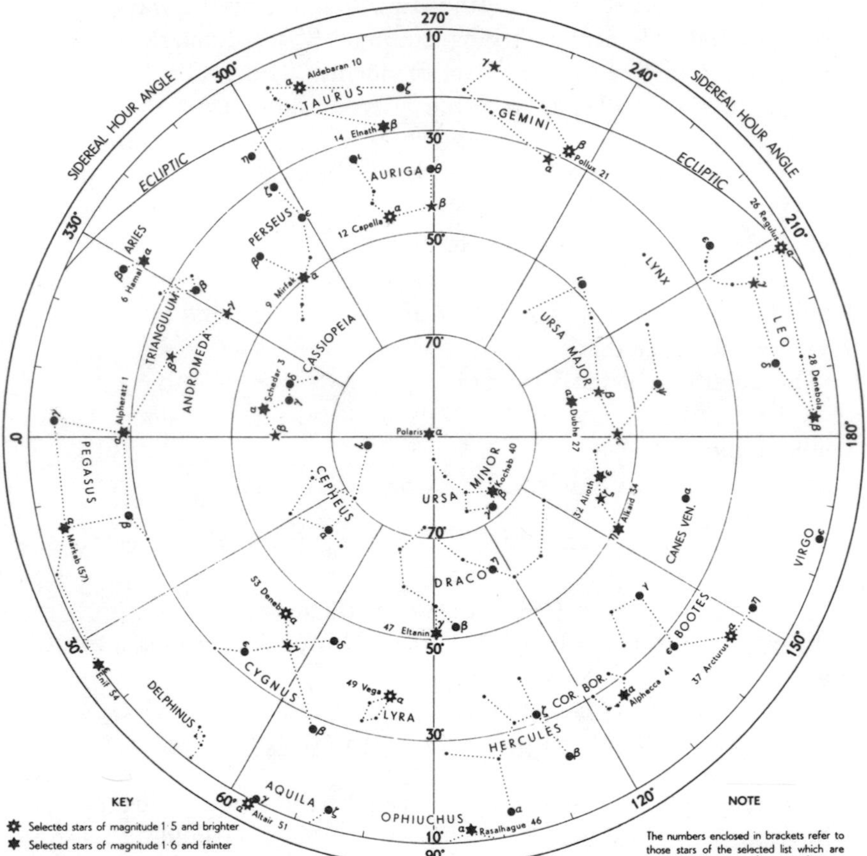

KEY

- Selected stars of magnitude 1·5 and brighter
- Selected stars of magnitude 1·6 and fainter
- Other tabulated stars of magnitude 2·5 and brighter
- Other tabulated stars of magnitude 2·6 and fainter
- Untabulated stars

NOTE

The numbers enclosed in brackets refer to those stars of the selected list which are not used in H.O. 249 (A.P. 3270).

EQUATORIAL STARS (S.H.A. 0° to 180°)

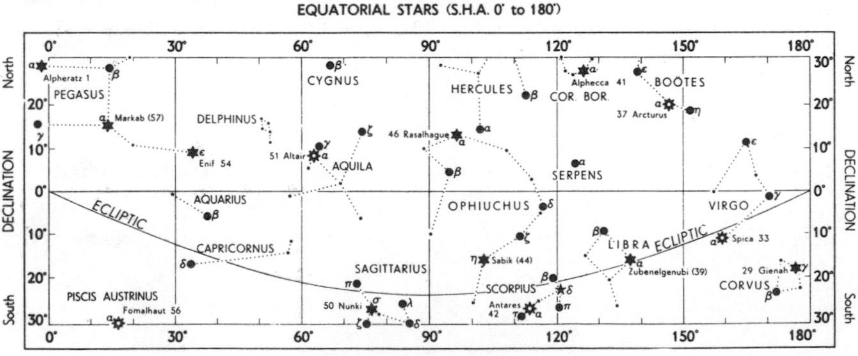

Table 39

STAR CHARTS

SOUTHERN STARS

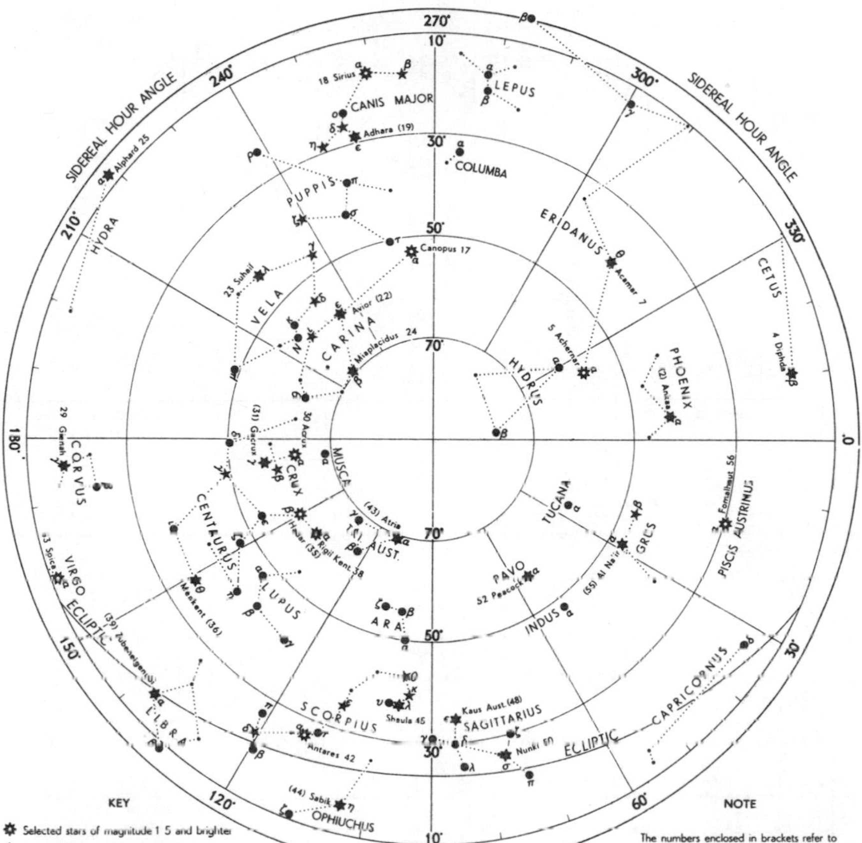

KEY

- Selected stars of magnitude 1·5 and brighter
- Selected stars of magnitude 1·6 and fainter
- Other tabulated stars of magnitude 2·5 and brighter
- Other tabulated stars of magnitude 2·6 and fainter
- Untabulated stars

NOTE

The numbers enclosed in brackets refer to those stars of the selected list which are not used in H.O. 249 (A.P. 3270)

EQUATORIAL STARS (S.H.A. 180° to 360°)

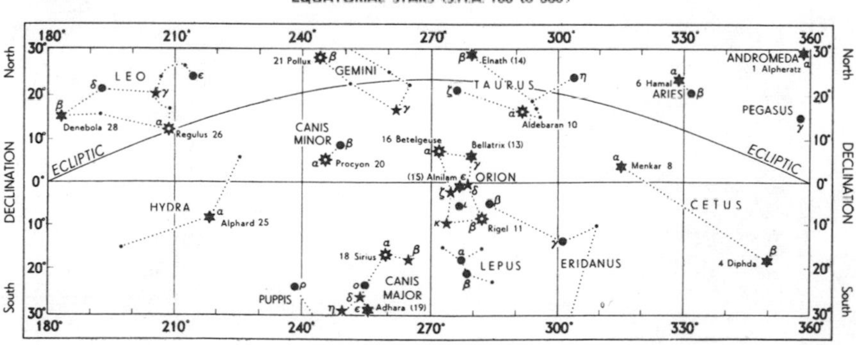

Table 40

1974 JULY 15, 16, 17 (MON., TUES., WED.)

G.M.T.		ARIES	VENUS −3·4		MARS +2·0		JUPITER −2·2		SATURN +0·3		STARS		
		G.H.A.	G.H.A.	Dec.	G.H.A.	Dec.	G.H.A.	Dec.	G.H.A.	Dec.	Name	S.H.A.	Dec.
d	h	° ′	° ′	° ′	° ′	° ′	° ′	° ′	° ′	° ′		° ′	° ′
15	00	292 28·0	210 28·6	N22 09·9	147 33·5	N15 14·1	303 12·4	S 6 00·0	191 24·6	N22 29·2	Acamar	315 40·6	S 40 24·1
	01	307 30·4	225 27·9	10·2	162 34·5	13·6	318 14·9	00·0	206 26·8	29·1	Achernar	335 48·4	S 57 21·5
	02	322 32·9	240 27·1	10·4	177 35·4	13·0	333 17·4	00·1	221 28·9	29·1	Acrux	173 42·2	S 62 57·9
	03	337 35·3	255 26·4	· · 10·7	192 36·4	· · 12·5	348 19·9	· · 00·1	236 31·0	· · 29·1	Adhara	255 35·8	S 28 56·2
	04	352 37·8	270 25·6	11·0	207 37·3	12·0	3 22·4	00·1	251 33·1	29·1	Aldebaran	291 23·1	N 16 27·6
	05	7 40·3	285 24·9	11·2	222 38·3	11·5	18 24·9	00·2	266 35·2	29·1			
	06	22 42·7	300 24·1	N22 11·5	237 39·2	N15 11·0	33 27·5	S 6 00·2	281 37·3	N22 29·0	Alioth	166 46·2	N 56 06·1
	07	37 45·2	315 23·4	11·8	252 40·2	10·4	48 30·0	00·2	296 39·5	29·0	Alkaid	153 21·7	N 49 26·6
	08	52 47·7	330 22·6	12·0	267 41·1	09·9	63 32·5	00·3	311 41·6	29·0	Al Na'ir	28 19·8	S 47 04·8
M	09	67 50·1	345 21·9	· · 12·3	282 42·1	· · 09·4	78 35·0	· · 00·3	326 43·7	· · 29·0	Alnilam	276 16·2	S 1 13·0
O	10	82 52·6	0 21·1	12·6	297 43·0	08·9	93 37·5	00·3	341 45·8	28·9	Alphard	218 25·0	S 8 32·9
N	11	97 55·1	15 20·4	12·8	312 43·9	08·4	108 40·0	00·4	356 47·9	28·9			
D	12	112 57·5	30 19·6	N22 13·1	327 44·9	N15 07·8	123 42·6	S 6 00·4	11 50·1	N22 28·9	Alphecca	126 35·5	N 26 48·1
A	13	128 00·0	45 18·8	13·3	342 45·8	07·3	138 45·1	00·4	26 52·2	28·9	Alpheratz	358 13·6	N 28 57·0
Y	14	143 02·5	60 18·1	13·6	357 46·8	06·8	153 47·6	00·5	41 54·3	28·9	Altair	62 36·3	N 8 48·1
	15	158 04·9	75 17·3	· · 13·9	12 47·7	· · 06·3	168 50·1	· · 00·5	56 56·4	· · 28·8	Ankaa	353 44·2	S 42 26·3
	16	173 07·4	90 16·6	14·1	27 48·7	05·8	183 52·6	00·6	71 58·5	28·8	Antares	113 01·8	S 26 22·7
	17	188 09·8	105 15·8	14·4	42 49·6	05·2	198 55·2	00·6	87 00·6	28·8			
	18	203 12·3	120 15·1	N22 14·6	57 50·6	N15 04·7	213 57·7	S 6 00·6	102 02·8	N22 28·8	Arcturus	146 22·2	N 19 18·9
	19	218 14·8	135 14·3	14·9	72 51·5	04·2	229 00·2	00·7	117 04·9	28·8	Atria	108 29·4	S 68 59·2
	20	233 17·2	150 13·5	15·1	87 52·5	03·7	244 02·7	00·7	132 07·0	28·7	Avior	234 30·6	S 59 25·8
	21	248 19·7	165 12·8	· · 15·4	102 53·4	· · 03·1	259 05·2	· · 00·7	147 09·1	· · 28·7	Bellatrix	279 03·6	N 6 19·7
	22	263 22·2	180 12·0	15·6	117 54·4	02·6	274 07·8	00·8	162 11·2	28·7	Betelgeuse	271 33·1	N 7 24·2
	23	278 24·6	195 11·3	15·9	132 55·3	02·1	289 10·3	00·8	177 13·4	28·7			
16	00	293 27·1	210 10·5	N22 16·1	147 56·3	N15 01·6	304 12·8	S 6 00·9	192 15·5	N22 28·6	Canopus	264 09·6	S 52 40·8
	01	308 29·6	225 09·8	16·4	162 57·2	01·1	319 15·3	00·9	207 17·6	28·6	Capella	281 17·9	N 45 58·3
	02	323 32·0	240 09·0	16·6	177 58·2	00·5	334 17·9	00·9	222 19·7	28·6	Deneb	49 50·9	N 45 11·4
	03	338 34·5	255 08·2	· · 16·9	192 59·1	15 00·0	349 20·4	· · 01·0	237 21·8	· · 28·6	Denebola	183 03·5	N 14 42·9
	04	353 37·0	270 07·5	17·1	208 00·1	14 59·5	4 22·9	01·0	252 23·9	28·6	Diphda	349 25·0	S 18 07·3
	05	8 39·4	285 06·7	17·4	223 01·0	59·0	19 25·4	01·0	267 26·1	28·5			
	06	23 41·9	300 06·0	N22 17·6	238 02·0	N14 58·4	34 27·9	S 6 01·1	282 28·2	N22 28·5	Dubhe	194 27·6	N 61 53·5
	07	38 44·3	315 05·2	17·8	253 02·9	57·9	49 30·5	01·1	297 30·3	28·5	Elnath	278 49·8	N 28 35·2
T	08	53 46·8	330 04·4	18·1	268 03·9	57·4	64 33·0	01·2	312 32·4	28·5	Eltanin	90 59·2	N 51 29·7
U	09	68 49·3	345 03·7	· · 18·3	283 04·8	· · 56·9	79 35·5	· · 01·2	327 34·5	· · 28·4	Enif	34 15·4	N 9 45·6
E	10	83 51·7	0 02·9	18·6	298 05·8	56·3	94 38·0	01·2	342 36·7	28·4	Fomalhaut	15 55·8	S 29 45·2
S	11	98 54·2	15 02·1	18·8	313 06·7	55·8	109 40·6	01·3	357 38·8	28·4			
D	12	113 56·7	30 01·4	N22 19·0	328 07·7	N14 55·3	124 43·1	S 6 01·3	12 40·9	N22 28·4	Gacrux	172 33·7	S 56 58·6
A	13	128 59·1	45 00·6	19·3	343 08·6	54·8	139 45·6	01·4	27 43·0	28·4	Gienah	176 22·4	S 17 24·2
Y	14	144 01·6	59 59·9	19·5	358 09·6	54·2	154 48·1	01·4	42 45·1	28·3	Hadar	149 29·3	S 60 15·4
	15	159 04·1	74 59·1	· · 19·7	13 10·5	· · 53·7	169 50·7	· · 01·4	57 47·2	· · 28·3	Hamal	328 33·8	N 23 20·6
	16	174 06·5	89 58·3	20·0	28 11·5	53·2	184 53·2	01·5	72 49·4	28·3	Kaus Aust.	84 22·0	S 34 23·9
	17	189 09·0	104 57·6	20·2	43 12·4	52·7	199 55·7	01·5	87 51·5	28·3			
	18	204 11·4	119 56·8	N22 20·4	58 13·4	N14 52·1	214 58·3	S 6 01·6	102 53·6	N22 28·3	Kochab	137 18·2	N 74 15·8
	19	219 13·9	134 56·0	20·6	73 14·3	51·6	230 00·8	01·6	117 55·7	28·2	Markab	14 07·2	N 15 04·2
	20	234 16·4	149 55·3	20·9	88 15·3	51·1	245 03·3	01·6	132 57·8	28·2	Menkar	314 45·7	N 3 59·5
	21	249 18·8	164 54·5	· · 21·1	103 16·2	· · 50·6	260 05·8	· · 01·7	148 00·0	· · 28·2	Menkent	148 41·9	S 36 15·0
	22	264 21·3	179 53·7	21·3	118 17·2	50·0	275 08·4	01·7	163 02·1	28·2	Miaplacidus	221 46·8	S 69 37·0
	23	279 23·8	194 53·0	21·5	133 18·1	49·5	290 10·9	01·8	178 04·2	28·1			
17	00	294 26·2	209 52·2	N22 21·8	148 19·1	N14 49·0	305 13·4	S 6 01·8	193 06·3	N22 28·1	Mirfak	309 22·4	N 49 46·2
	01	309 28·7	224 51·5	22·0	163 20·0	48·5	320 16·0	01·8	208 08·4	28·1	Nunki	76 34·0	S 26 19·7
	02	324 31·2	239 50·7	22·2	178 21·0	47·9	335 18·5	01·9	223 10·6	28·1	Peacock	54 04·4	S 56 48·9
	03	339 33·6	254 49·9	· · 22·4	193 21·9	· · 47·4	350 21·0	· · 01·9	238 12·7	· · 28·1	Pollux	244 03·7	N 28 05·3
	04	354 36·1	269 49·2	22·6	208 22·9	46·9	5 23·5	02·0	253 14·8	28·0	Procyon	245 30·5	N 5 17·5
	05	9 38·6	284 48·4	22·9	223 23·8	46·3	20 26·1	02·0	268 16·9	28·0			
	06	24 41·0	299 47·6	N22 23·1	238 24·8	N14 45·8	35 28·6	S 6 02·0	283 19·0	N22 28·0	Rasalhague	96 33·2	N 12 34·8
W	07	39 43·5	314 46·9	23·3	253 25·7	45·3	50 31·1	02·1	298 21·1	28·0	Regulus	208 14·7	N 12 05·5
E	08	54 45·9	329 46·1	23·5	268 26·7	44·8	65 33·7	02·1	313 23·3	27·9	Rigel	281 40·3	S 8 13·7
D	09	69 48·4	344 45·3	· · 23·7	283 27·6	· · 44·2	80 36·2	· · 02·2	328 25·4	· · 27·9	Rigil Kent.	140 31·4	S 60 44·1
N	10	84 50·9	359 44·5	23·9	298 28·6	43·7	95 38·7	02·2	343 27·5	27·9	Sabik	102 45·7	S 15 41·7
E	11	99 53·3	14 43·8	24·1	313 29·5	43·2	110 41·3	02·3	358 29·6	27·9			
S	12	114 55·8	29 43·0	N22 24·4	328 30·5	N14 42·6	125 43·8	S 6 02·3	13 31·7	N22 27·9	Schedar	350 13·7	N 56 23·7
D	13	129 58·3	44 42·2	24·6	343 31·4	42·1	140 46·3	02·3	28 33·9	27·8	Shaula	97 01·1	S 37 05·2
A	14	145 00·7	59 41·5	24·8	358 32·4	41·6	155 48·9	02·4	43 36·0	27·8	Sirius	258 59·8	S 16 40·8
Y	15	160 03·2	74 40·7	· · 25·0	13 33·3	· · 41·1	170 51·4	· · 02·4	58 38·1	· · 27·8	Spica	159 02·0	S 11 01·8
	16	175 05·7	89 39·9	25·2	28 34·3	40·5	185 53·9	02·5	73 40·2	27·8	Suhail	223 14·3	S 43 19·9
	17	190 08·1	104 39·2	25·4	43 35·2	40·0	200 56·5	02·5	88 42·3	27·7			
	18	205 10·6	119 38·4	N22 25·6	58 36·2	N14 39·5	215 59·0	S 6 02·6	103 44·5	N22 27·7	Vega	80 58·3	N 38 45·7
	19	220 13·1	134 37·6	25·8	73 37·1	38·9	231 01·5	02·6	118 46·6	27·7	Zuben'ubi	137 37·6	S 15 56·3
	20	235 15·5	149 36·9	26·0	88 38·1	38·4	246 04·1	02·6	133 48·7	27·7		S.H.A.	Mer. Pass.
	21	250 18·0	164 36·1	· · 26·2	103 39·1	· · 37·9	261 06·6	· · 02·7	148 50·8	· · 27·7		° ′	h m
	22	265 20·4	179 35·3	26·4	118 40·0	37·4	276 09·1	02·7	163 52·9	27·6	Venus	276 43·4	10 00
	23	280 22·9	194 34·5	26·6	133 41·0	36·8	291 11·7	02·8	178 55·1	27·6	Mars	214 29·2	14 07
Mer. Pass.		h m 4 25·5	v −0·8	d 0·2	v 0·9	d 0·5	v 2·5	d 0·0	v 2·1	d 0·0	Jupiter	10 45·7	3 43
											Saturn	258 48·4	11 09

Table 41

time (from Table 26 on page 103). So, subtracting 5 minutes and 36 seconds from 2300 hours, at 22-54-24 hours in the Canaries, you can look west at Arcturus and know that it is directly over your destination. And it will be over that spot 4 minutes earlier every night of your voyage, because the stars rise 4 minutes earlier every day.

14
Celestial Navigation without a Chronometer

You may have heard or read that it is not possible to determine longitude without an accurate knowledge of GMT. Actually, there are two methods of finding longitude without a chronometer. If you have read Joshua Slocum's *Sailing Alone Around the World*, you will know that he determined his longitude by a method known as *lunar distance*. This method was developed in the early 1800s and basically consists in measuring with the sextant the angle ("distance") between the moon and a star or planet. Since the GHA and declination of the moon change relatively rapidly, this "distance" is unique, and until 1914 the almanac included a table of lunar distances from various stars and the respective GMTs. The distance measured with the sextant was compared with the data in the almanac to find the GMT of the observation.

Unfortunately, this method involved a lot of very tedious arithmetic so, with the advent of time signals, it fell out of use and the distance table was dropped from the almanac.

A more straightforward method was discovered by Sir Francis Chichester, and he mentions it in his book *Along the Clipper Way*. Here is his explanation in his own words: "Make a simultaneous observation of the sun and the moon for altitude when the moon is nearly east or west of the ship. From this observation compute a sun-moon fix in the ordinary way, using a guessed-at Greenwich Mean Time. Now compute a second fix from the same observation but using a Greenwich Mean Time which differs from the first one by half an hour or an hour. Now establish the latitude by a meridian altitude [noon sight] of the sun or any other heavenly body as it crosses the meridian. This observation does not require accurate time. Now join the two sun-moon fixes and the point where the line joining them, produced if necessary, cuts the known latitude must be the correct longitude at the time of the observation."[1]

[1]Quoted with permission from the publisher, Coward, McCann & Geoghegan, Inc.

To understand what Sir Francis is saying, you must turn again to zenith distance. You know that the zenith distance is your distance from the GP of the body. Look at Figure 14.1. You can see that the same zenith distance could be measured anywhere on a circle around the GP. The radius would be the zenith distance. You could be anywhere on the perimeter of that circle. Likewise, if you take simultaneous observations of two bodies, you are where the circles intersect (Figure 14.2). You can see that they intersect in two places. The distances involved are usually so great that there is little doubt about which intersection you are at. After all, you know whether you are in the North Atlantic or South Atlantic ocean.

Suppose you take a simultaneous observation of a star and a planet and do not know the time. You guess at the time and plot the fix (Figure 14.3) and calculate it again for 1 hour later. The fix has moved west about 15° in a parallel line. This is because neither the star nor the planet has changed declination very much in 1 hour.

The moon, on the other hand, is often changing declination very rapidly (as much as 14′ in an hour). Therefore, if you made a plot of sun-moon fixes when the moon was moving, say, south at 14′ per hour, you would get a diagonal line (Figure 14.4). If you had taken a noon sight of the sun and knew your latitude was 45° north when you took the simultaneous observation of sun and moon, you have to be where that fix coincides with the latitude line. The line between the fixes represents a continuous string of spots at which you could have been. Since you must be at one of those spots and also on the latitude by noon sight, you must be where the lines cross.

If your DR is really off, your guess of times might result in fixes that do not straddle your position. This is what Sir Francis was thinking of when he wrote "where the line joining them [the fixes based on the guessed GMT], produced if necessary, cuts the known latitude" (Figure 14.5). To produce a line means to extend it.

In using this method, you have to contend with two major difficulties. First, in an hour's time the sun and moon traverse 15° of longitude, which means you are working to a small scale. Second, the declination of the moon is not always changing at a very high rate—it gets as low as 0.2′ per hour, which means that the line of fixes would

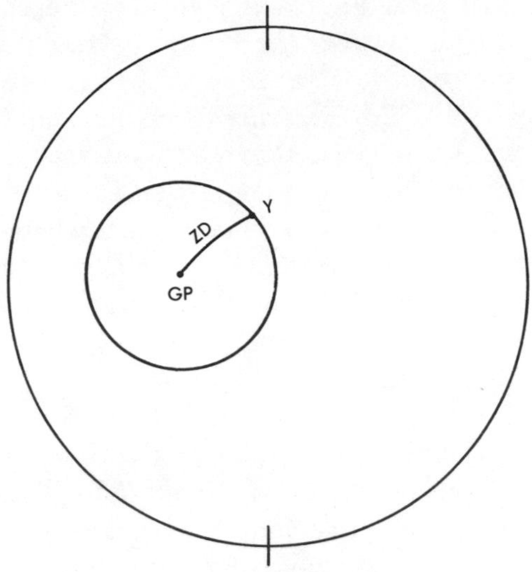

Fig. 14.1

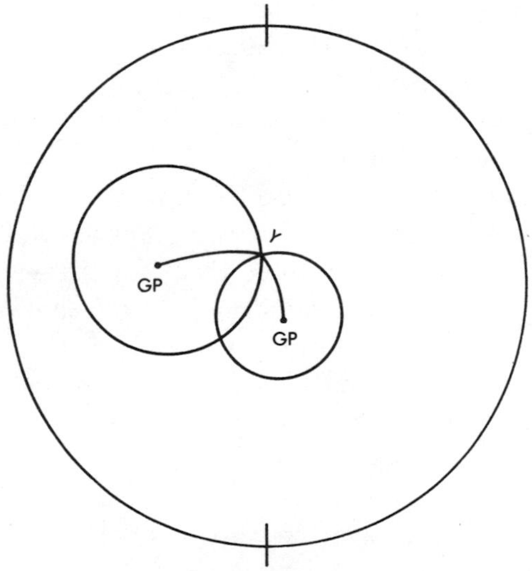

Fig. 14.2

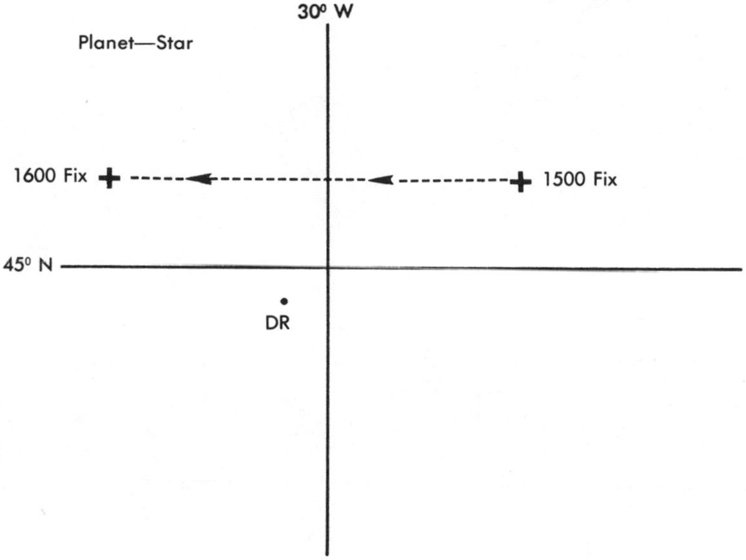

Fig. 14.3

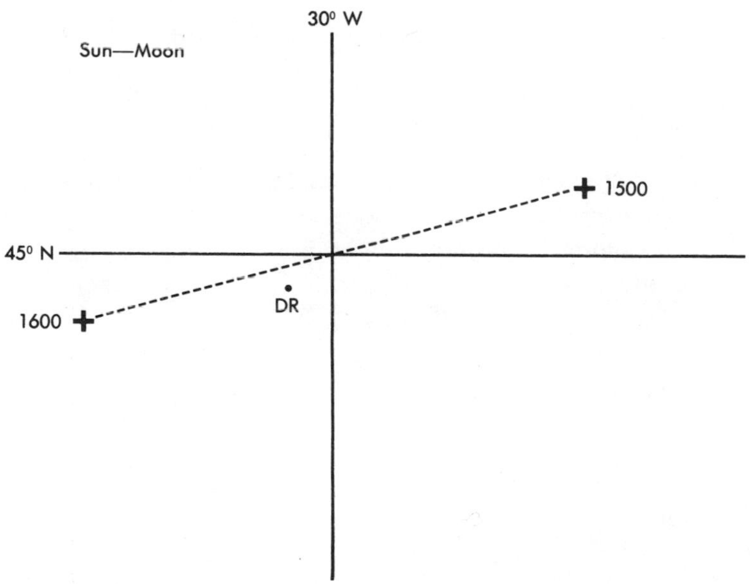

Fig. 14.4

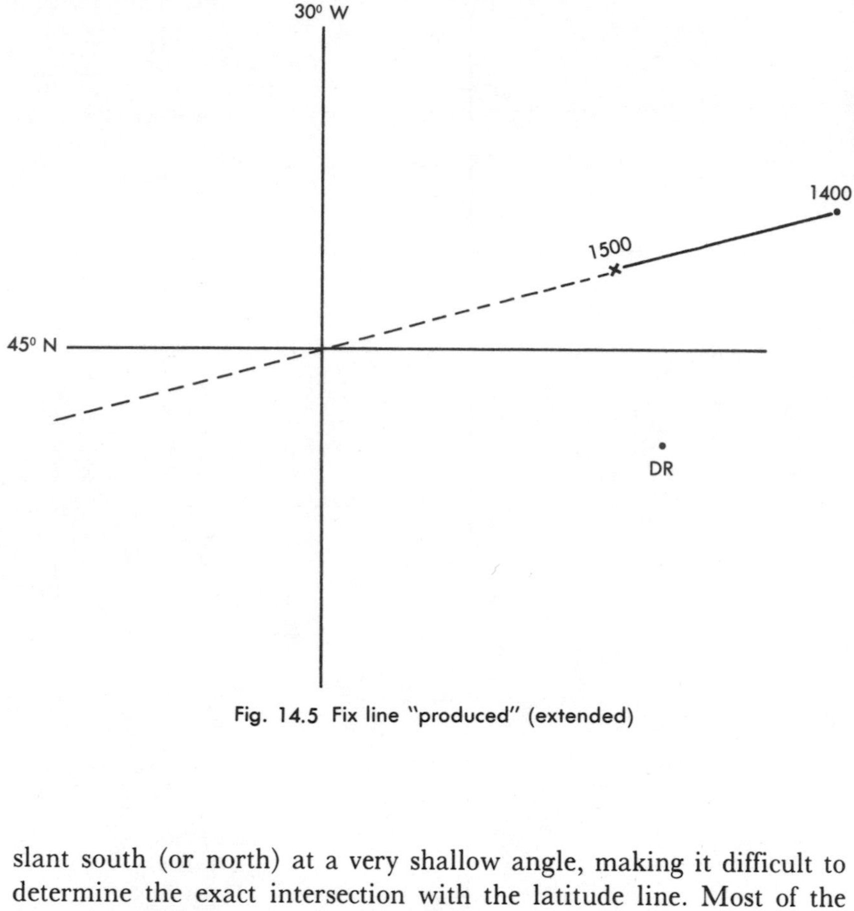

Fig. 14.5 Fix line "produced" (extended)

slant south (or north) at a very shallow angle, making it difficult to determine the exact intersection with the latitude line. Most of the time, though, the rate of change of declination is in the range of 5′ to 10′ per hour.

Here is an example, using Table 42, 42a, and H.O. 229 (Tables 42b, c, d, e, and f):

6-16-75 1530 Hours GMT (Guess)

Ho Moon = 19°00′
Ho Sun = 57°30′
Latitude = 45°N (By noon sight advanced to moment of simultaneous observation)

1975 JUNE 15, 16, 17 (SUN., MON., TUES.)

	G.M.T.	SUN G.H.A.	SUN Dec.	MOON G.H.A.	v	MOON Dec.	d	H.P.
	d h	° ′	° ′	° ′	′	° ′	′	′
	15 00	179 57.1	N23 16.3	110 50.3	10.0	N 5 58.7	12.5	59.4
	01	194 57.0	16.4	125 19.3	9.9	5 46.2	12.6	59.4
	02	209 56.8	16.5	139 48.2	10.0	5 33.6	12.5	59.4
	03	224 56.7	·· 16.6	154 17.2	10.1	5 21.1	12.6	59.4
	04	239 56.6	16.8	168 46.3	10.0	5 08.5	12.7	59.4
	05	254 56.4	16.9	183 15.3	10.1	4 55.8	12.6	59.4
	06	269 56.3	N23 17.0	197 44.4	10.0	N 4 43.2	12.7	59.4
	07	284 56.2	17.1	212 13.4	10.2	4 30.5	12.7	59.4
	08	299 56.0	17.2	226 42.6	10.1	4 17.8	12.7	59.4
S	09	314 55.9	·· 17.3	241 11.7	10.1	4 05.1	12.8	59.4
U	10	329 55.8	17.5	255 40.8	10.2	3 52.3	12.8	59.4
N	11	344 55.6	17.6	270 10.0	10.2	3 39.5	12.8	59.4
D	12	359 55.5	N23 17.7	284 39.2	10.2	N 3 26.7	12.8	59.4
A	13	14 55.4	17.8	299 08.4	10.2	3 13.9	12.8	59.4
Y	14	29 55.2	17.9	313 37.6	10.3	3 01.1	12.8	59.4
	15	44 55.1	·· 18.0	328 06.9	10.3	2 48.3	12.9	59.4
	16	59 55.0	18.1	342 36.2	10.2	2 35.4	12.8	59.4
	17	74 54.8	18.2	357 05.4	10.3	2 22.6	12.9	59.4
	18	89 54.7	N23 18.3	11 34.7	10.3	N 2 09.7	12.9	59.4
	19	104 54.6	18.5	26 04.0	10.4	1 56.8	12.9	59.4
	20	119 54.4	18.6	40 33.4	10.3	1 43.9	12.9	59.4
	21	134 54.3	·· 18.7	55 02.7	10.3	1 31.0	12.9	59.4
	22	149 54.2	18.8	69 32.0	10.4	1 18.1	12.9	59.4
	23	164 54.0	18.9	84 01.4	10.4	1 05.2	12.9	59.4
	16 00	179 53.9	N23 19.0	98 30.8	10.3	N 0 52.3	13.0	59.4
	01	194 53.8	19.1	113 00.1	10.4	0 39.3	12.9	59.4
	02	209 53.6	19.2	127 29.5	10.4	0 26.4	12.9	59.4
	03	224 53.5	·· 19.3	141 58.9	10.4	0 13.5	12.9	59.3
	04	239 53.4	19.4	156 28.3	10.4	N 0 00.6	12.9	59.3
	05	254 53.2	19.5	170 57.7	10.4	S 0 12.3	13.0	59.3
	06	269 53.1	N23 19.6	185 27.1	10.4	S 0 25.3	12.9	59.3
	07	284 53.0	19.7	199 56.5	10.4	0 38.2	12.9	59.3
	08	299 52.8	19.8	214 25.9	10.5	0 51.1	12.9	59.3
M	09	314 52.7	·· 19.9	228 55.4	10.4	1 04.0	12.9	59.3
O	10	329 52.6	20.0	243 24.8	10.4	1 16.9	12.8	59.3
N	11	344 52.4	20.1	257 54.2	10.4	1 29.7	12.9	59.3
D	12	359 52.3	N23 20.2	272 23.6	10.4	S 1 42.6	12.9	59.3
A	13	14 52.2	20.3	286 53.0	10.5	1 55.5	12.8	59.3
Y	14	29 52.0	20.4	301 22.5	10.4	2 08.3	12.8	59.3
	15	44 51.9	·· 20.5	315 51.9	10.4	2 21.1	12.9	59.3
	16	59 51.8	20.5	330 21.3	10.4	2 34.0	12.8	59.3
	17	74 51.6	20.6	344 50.7	10.4	2 46.8	12.7	59.3
	18	89 51.5	N23 20.7	359 20.1	10.4	S 2 59.5	12.8	59.3
	19	104 51.4	20.8	13 49.5	10.4	3 12.3	12.7	59.3
	20	119 51.2	20.9	28 18.9	10.4	3 25.0	12.7	59.2
	21	134 51.1	·· 21.0	42 48.3	10.4	3 37.7	12.7	59.2
	22	149 51.0	21.1	57 17.7	10.4	3 50.4	12.7	59.2
	23	164 50.8	21.2	71 47.1	10.3	4 03.1	12.7	59.2
	17 00	179 50.7	N23 21.3	86 16.4	10.4	S 4 15.8	12.6	59.2
	01	194 50.5	21.3	100 45.8	10.3	4 28.4	12.6	59.2
	02	209 50.4	21.4	115 15.1	10.3	4 41.0	12.5	59.2
	03	224 50.3	·· 21.5	129 44.4	10.4	4 53.5	12.5	59.2
	04	239 50.1	21.6	144 13.8	10.3	5 06.0	12.5	59.2
	05	254 50.0	21.7	158 43.1	10.3	5 18.5	12.5	59.2
	06	269 49.9	N23 21.8	173 12.4	10.2	S 5 31.0	12.4	59.2
	07	284 49.7	21.8	187 41.6	10.3	5 43.4	12.4	59.2
	08	299 49.6	21.9	202 10.9	10.2	5 55.8	12.4	59.2
T	09	314 49.5	·· 22.0	216 40.1	10.3	6 08.2	12.3	59.1
U	10	329 49.3	22.1	231 09.4	10.2	6 20.5	12.3	59.1
E	11	344 49.2	22.2	245 38.6	10.2	6 32.8	12.2	59.1
S	12	359 49.1	N23 22.2	260 07.8	10.2	S 6 45.0	12.2	59.1
D	13	14 48.9	22.3	274 37.0	10.1	6 57.2	12.2	59.1
A	14	29 48.8	22.4	289 06.1	10.2	7 09.4	12.1	59.1
Y	15	44 48.7	·· 22.5	303 35.3	10.1	7 21.5	12.1	59.1
	16	59 48.5	22.5	318 04.4	10.1	7 33.6	12.0	59.1
	17	74 48.4	22.6	332 33.5	10.0	7 45.6	12.0	59.1
	18	89 48.3	N23 22.7	347 02.5	10.1	S 7 57.6	11.9	59.1
	19	104 48.1	22.8	1 31.6	10.0	8 09.5	11.9	59.0
	20	119 48.0	22.8	16 00.6	10.0	8 21.4	11.8	59.0
	21	134 47.9	·· 22.9	30 29.6	10.0	8 33.2	11.8	59.0
	22	149 47.7	23.0	44 58.6	10.0	8 45.0	11.8	59.0
	23	164 47.6	23.1	59 27.6	9.9	8 56.8	11.6	59.0
		S.D. 15.8	*d* 0.1	S.D.	16.2	16.2		16.1

Lat.	Twilight Naut.	Twilight Civil	Sunrise	Moonrise 15	Moonrise 16	Moonrise 17	Moonrise 18
°	h m	h m	h m	h m	h m	h m	h m
N 72	▭	▭	▭	10 12	12 15	14 18	16 28
N 70	▭	▭	▭	10 18	12 12	14 07	16 05
68	▭	▭	▭	10 22	12 10	13 58	15 47
66	▭	▭	▭	10 26	12 08	13 50	15 33
64	////	////	01 33	10 29	12 06	13 44	15 21
62	////	////	02 10	10 31	12 05	13 38	15 11
60	////	00 53	02 36	10 34	12 04	13 34	15 03
N 58	////	01 41	02 56	10 36	12 03	13 29	14 56
56	////	02 11	03 13	10 37	12 02	13 26	14 49
54	00 48	02 33	03 27	10 39	12 01	13 22	14 43
52	01 33	02 51	03 39	10 41	12 00	13 19	14 38
50	02 01	03 06	03 50	10 42	12 00	13 17	14 34
45	02 46	03 35	04 12	10 45	11 58	13 11	14 23
N 40	03 16	03 58	04 30	10 47	11 57	13 06	14 15
35	03 39	04 16	04 45	10 49	11 56	13 02	14 08
30	03 58	04 31	04 58	10 51	11 55	12 58	14 02
20	04 27	04 56	05 20	10 54	11 53	12 52	13 51
N 10	04 49	05 16	05 39	10 57	11 52	12 47	13 42
0	05 08	05 34	05 57	11 00	11 51	12 42	13 33
S 10	05 25	05 51	06 14	11 03	11 50	12 36	13 24
20	05 41	06 09	06 33	11 05	11 48	12 31	13 15
30	05 58	06 27	06 54	11 09	11 47	12 25	13 05
35	06 06	06 38	07 06	11 11	11 46	12 22	12 59
40	06 16	06 50	07 20	11 13	11 45	12 18	12 52
45	06 26	07 03	07 37	11 15	11 44	12 13	12 45
S 50	06 38	07 19	07 58	11 18	11 43	12 08	12 35
52	06 43	07 27	08 07	11 19	11 42	12 06	12 31
54	06 40	07 35	08 18	11 21	11 42	12 03	12 26
56	06 54	07 44	08 31	11 22	11 41	12 00	12 21
58	07 01	07 54	08 46	11 24	11 40	11 57	12 15
S 60	07 09	08 06	09 03	11 26	11 40	11 53	12 09

Lat.	Sunset	Twilight Civil	Twilight Naut.	Moonset 15	Moonset 16	Moonset 17	Moonset 18
°	h m	h m	h m	h m	h m	h m	h m
N 72	▭	▭	▭	23 35	23 20	23 04	22 43
N 70	▭	▭	▭	23 34	23 26	23 18	23 08
68	▭	▭	▭	23 33	23 31	23 29	23 27
66	▭	▭	▭	23 32	23 35	23 38	23 42
64	22 29	////	////	23 32	23 38	23 46	23 55
62	21 51	////	////	23 31	23 41	23 52	24 06
60	21 25	23 09	////	23 30	23 44	23 58	24 15
N 58	21 05	22 20	////	23 30	23 46	24 03	00 03
56	20 48	21 51	////	23 30	23 48	24 08	00 08
54	20 34	21 28	23 14	23 29	23 50	24 12	00 12
52	20 22	21 11	22 28	23 29	23 52	24 16	00 16
50	20 11	20 55	22 01	23 29	23 53	24 19	00 19
45	19 49	20 26	21 16	23 28	23 57	24 27	00 27
N 40	19 30	20 03	20 45	23 27	24 00	00 00	00 33
35	19 16	19 45	20 22	23 27	24 02	00 02	00 39
30	19 03	19 30	20 03	23 26	24 04	00 04	00 43
20	18 41	19 05	19 34	23 26	24 08	00 08	00 52
N 10	18 22	18 45	19 12	23 25	24 12	00 12	00 59
0	18 04	18 27	18 53	23 24	24 15	00 15	01 06
S 10	17 47	18 10	18 36	23 23	24 18	00 18	01 13
20	17 28	17 52	18 20	23 23	24 21	00 21	01 20
30	17 07	17 33	18 03	23 22	24 25	00 25	01 29
35	16 55	17 23	17 55	23 21	24 27	00 27	01 33
40	16 41	17 11	17 45	23 20	24 30	00 30	01 39
45	16 24	16 58	17 35	23 20	24 33	00 33	01 45
S 50	16 03	16 42	17 23	23 19	24 36	00 36	01 53
52	15 53	16 34	17 18	23 18	24 38	00 38	01 57
54	15 42	16 26	17 12	23 18	24 40	00 40	02 01
56	15 30	16 17	17 06	23 17	24 41	00 41	02 05
58	15 15	16 07	16 59	23 17	24 44	00 44	02 10
S 60	14 58	15 55	16 52	23 16	24 46	00 46	02 16

Day	SUN Eqn. of Time 00^h	SUN Eqn. of Time 12^h	SUN Mer. Pass.	MOON Mer. Pass. Upper	MOON Mer. Pass. Lower	Age	Phase
	m s	m s	h m	h m	h m	d	
15	00 11	00 18	12 00	17 12	04 47	06	
16	00 24	00 31	12 01	18 03	05 37	07	◐
17	00 37	00 43	12 01	18 54	06 28	08	

Table 42

30ᵐ INCREMENTS AND CORRECTIONS 31ᵐ

30ᵐ	SUN PLANETS	ARIES	MOON	v or d	Corrⁿ	v or d	Corrⁿ	v or d	Corrⁿ
s	° ′	° ′	° ′	′	′	′	′	′	′
00	7 30·0	7 31·2	7 09·5	0·0	0·0	6·0	3·1	12·0	6·1
01	7 30·3	7 31·5	7 09·7	0·1	0·1	6·1	3·1	12·1	6·2
02	7 30·5	7 31·7	7 10·0	0·2	0·1	6·2	3·2	12·2	6·2
03	7 30·8	7 32·0	7 10·2	0·3	0·2	6·3	3·2	12·3	6·3
04	7 31·0	7 32·2	7 10·5	0·4	0·2	6·4	3·3	12·4	6·3
05	7 31·3	7 32·5	7 10·7	0·5	0·3	6·5	3·3	12·5	6·4
06	7 31·5	7 32·7	7 10·9	0·6	0·3	6·6	3·4	12·6	6·4
07	7 31·8	7 33·0	7 11·2	0·7	0·4	6·7	3·4	12·7	6·5
08	7 32·0	7 33·2	7 11·4	0·8	0·4	6·8	3·5	12·8	6·5
09	7 32·3	7 33·5	7 11·6	0·9	0·5	6·9	3·5	12·9	6·6
10	7 32·5	7 33·7	7 11·9	1·0	0·5	7·0	3·6	13·0	6·6
11	7 32·8	7 34·0	7 12·1	1·1	0·6	7·1	3·6	13·1	6·7
12	7 33·0	7 34·2	7 12·4	1·2	0·6	7·2	3·7	13·2	6·7
13	7 33·3	7 34·5	7 12·6	1·3	0·7	7·3	3·7	13·3	6·8
14	7 33·5	7 34·7	7 12·8	1·4	0·7	7·4	3·8	13·4	6·8
15	7 33·8	7 35·0	7 13·1	1·5	0·8	7·5	3·8	13·5	6·9
16	7 34·0	7 35·2	7 13·3	1·6	0·8	7·6	3·9	13·6	6·9
17	7 34·3	7 35·5	7 13·6	1·7	0·9	7·7	3·9	13·7	7·0
18	7 34·5	7 35·7	7 13·8	1·8	0·9	7·8	4·0	13·8	7·0
19	7 34·8	7 36·0	7 14·0	1·9	1·0	7·9	4·0	13·9	7·1
20	7 35·0	7 36·2	7 14·3	2·0	1·0	8·0	4·1	14·0	7·1
21	7 35·3	7 36·5	7 14·5	2·1	1·1	8·1	4·1	14·1	7·2
22	7 35·5	7 36·7	7 14·7	2·2	1·1	8·2	4·2	14·2	7·2
23	7 35·8	7 37·0	7 15·0	2·3	1·2	8·3	4·2	14·3	7·3
24	7 36·0	7 37·2	7 15·2	2·4	1·2	8·4	4·3	14·4	7·3
25	7 36·3	7 37·5	7 15·5	2·5	1·3	8·5	4·3	14·5	7·4
26	7 36·5	7 37·7	7 15·7	2·6	1·3	8·6	4·4	14·6	7·4
27	7 36·8	7 38·0	7 15·9	2·7	1·4	8·7	4·4	14·7	7·5
28	7 37·0	7 38·3	7 16·2	2·8	1·4	8·8	4·5	14·8	7·5
29	7 37·3	7 38·5	7 16·4	2·9	1·5	8·9	4·5	14·9	7·6
30	7 37·5	7 38·8	7 16·7	3·0	1·5	9·0	4·6	15·0	7·6
31	7 37·8	7 39·0	7 16·9	3·1	1·6	9·1	4·6	15·1	7·7
32	7 38·0	7 39·3	7 17·1	3·2	1·6	9·2	4·7	15·2	7·7
33	7 38·3	7 39·5	7 17·4	3·3	1·7	9·3	4·7	15·3	7·8
34	7 38·5	7 39·8	7 17·6	3·4	1·7	9·4	4·8	15·4	7·8
35	7 38·8	7 40·0	7 17·9	3·5	1·8	9·5	4·8	15·5	7·9
36	7 39·0	7 40·3	7 18·1	3·6	1·8	9·6	4·9	15·6	7·9
37	7 39·3	7 40·5	7 18·3	3·7	1·9	9·7	4·9	15·7	8·0
38	7 39·5	7 40·8	7 18·6	3·8	1·9	9·8	5·0	15·8	8·0
39	7 39·8	7 41·0	7 18·8	3·9	2·0	9·9	5·0	15·9	8·1
40	7 40·0	7 41·3	7 19·0	4·0	2·0	10·0	5·1	16·0	8·1
41	7 40·3	7 41·5	7 19·3	4·1	2·1	10·1	5·1	16·1	8·2
42	7 40·5	7 41·8	7 19·5	4·2	2·1	10·2	5·2	16·2	8·2
43	7 40·8	7 42·0	7 19·8	4·3	2·2	10·3	5·2	16·3	8·3
44	7 41·0	7 42·3	7 20·0	4·4	2·2	10·4	5·3	16·4	8·3
45	7 41·3	7 42·5	7 20·2	4·5	2·3	10·5	5·3	16·5	8·4
46	7 41·5	7 42·8	7 20·5	4·6	2·3	10·6	5·4	16·6	8·4
47	7 41·8	7 43·0	7 20·7	4·7	2·4	10·7	5·4	16·7	8·5
48	7 42·0	7 43·3	7 21·0	4·8	2·4	10·8	5·5	16·8	8·5
49	7 42·3	7 43·5	7 21·2	4·9	2·5	10·9	5·5	16·9	8·6
50	7 42·5	7 43·8	7 21·4	5·0	2·5	11·0	5·6	17·0	8·6
51	7 42·8	7 44·0	7 21·7	5·1	2·6	11·1	5·6	17·1	8·7
52	7 43·0	7 44·3	7 21·9	5·2	2·6	11·2	5·7	17·2	8·7
53	7 43·3	7 44·5	7 22·1	5·3	2·7	11·3	5·7	17·3	8·8
54	7 43·5	7 44·8	7 22·4	5·4	2·7	11·4	5·8	17·4	8·8
55	7 43·8	7 45·0	7 22·6	5·5	2·8	11·5	5·8	17·5	8·9
56	7 44·0	7 45·3	7 22·9	5·6	2·8	11·6	5·9	17·6	8·9
57	7 44·3	7 45·5	7 23·1	5·7	2·9	11·7	5·9	17·7	9·0
58	7 44·5	7 45·8	7 23·3	5·8	2·9	11·8	6·0	17·8	9·0
59	7 44·8	7 46·0	7 23·6	5·9	3·0	11·9	6·0	17·9	9·1
60	7 45·0	7 46·3	7 23·8	6·0	3·1	12·0	6·1	18·0	9·2

31ᵐ	SUN PLANETS	ARIES	MOON	v or d	Corrⁿ	v or d	Corrⁿ	v or d	Corrⁿ
s	° ′	° ′	° ′	′	′	′	′	′	′
00	7 45·0	7 46·3	7 23·8	0·0	0·0	6·0	3·2	12·0	6·3
01	7 45·3	7 46·5	7 24·1	0·1	0·1	6·1	3·2	12·1	6·4
02	7 45·5	7 46·8	7 24·3	0·2	0·1	6·2	3·3	12·2	6·4
03	7 45·8	7 47·0	7 24·5	0·3	0·2	6·3	3·3	12·3	6·5
04	7 46·0	7 47·3	7 24·8	0·4	0·2	6·4	3·4	12·4	6·5
05	7 46·3	7 47·5	7 25·0	0·5	0·3	6·5	3·4	12·5	6·6
06	7 46·5	7 47·8	7 25·2	0·6	0·3	6·6	3·5	12·6	6·6
07	7 46·8	7 48·0	7 25·5	0·7	0·4	6·7	3·5	12·7	6·7
08	7 47·0	7 48·3	7 25·7	0·8	0·4	6·8	3·6	12·8	6·7
09	7 47·3	7 48·5	7 26·0	0·9	0·5	6·9	3·6	12·9	6·8
10	7 47·5	7 48·8	7 26·2	1·0	0·5	7·0	3·7	13·0	6·8
11	7 47·8	7 49·0	7 26·4	1·1	0·6	7·1	3·7	13·1	6·9
12	7 48·0	7 49·3	7 26·7	1·2	0·6	7·2	3·8	13·2	6·9
13	7 48·3	7 49·5	7 26·9	1·3	0·7	7·3	3·8	13·3	7·0
14	7 48·5	7 49·8	7 27·2	1·4	0·7	7·4	3·9	13·4	7·0
15	7 48·8	7 50·0	7 27·4	1·5	0·8	7·5	3·9	13·5	7·1
16	7 49·0	7 50·3	7 27·6	1·6	0·8	7·6	4·0	13·6	7·1
17	7 49·3	7 50·5	7 27·9	1·7	0·9	7·7	4·0	13·7	7·2
18	7 49·5	7 50·8	7 28·1	1·8	0·9	7·8	4·1	13·8	7·2
19	7 49·8	7 51·0	7 28·4	1·9	1·0	7·9	4·1	13·9	7·3
20	7 50·0	7 51·3	7 28·6	2·0	1·1	8·0	4·2	14·0	7·4
21	7 50·3	7 51·5	7 28·8	2·1	1·1	8·1	4·3	14·1	7·4
22	7 50·5	7 51·8	7 29·1	2·2	1·2	8·2	4·3	14·2	7·5
23	7 50·8	7 52·0	7 29·3	2·3	1·2	8·3	4·4	14·3	7·5
24	7 51·0	7 52·3	7 29·5	2·4	1·3	8·4	4·4	14·4	7·6
25	7 51·3	7 52·5	7 29·8	2·5	1·3	8·5	4·5	14·5	7·6
26	7 51·5	7 52·8	7 30·0	2·6	1·4	8·6	4·5	14·6	7·7
27	7 51·8	7 53·0	7 30·3	2·7	1·4	8·7	4·6	14·7	7·7
28	7 52·0	7 53·3	7 30·5	2·8	1·5	8·8	4·6	14·8	7·8
29	7 52·3	7 53·5	7 30·7	2·9	1·5	8·9	4·7	14·9	7·8
30	7 52·5	7 53·8	7 31·0	3·0	1·6	9·0	4·7	15·0	7·9
31	7 52·8	7 54·0	7 31·2	3·1	1·6	9·1	4·8	15·1	7·9
32	7 53·0	7 54·3	7 31·5	3·2	1·7	9·2	4·8	15·2	8·0
33	7 53·3	7 54·5	7 31·7	3·3	1·7	9·3	4·9	15·3	8·0
34	7 53·5	7 54·8	7 31·9	3·4	1·8	9·4	4·9	15·4	8·1
35	7 53·8	7 55·0	7 32·2	3·5	1·8	9·5	5·0	15·5	8·1
36	7 54·0	7 55·3	7 32·4	3·6	1·9	9·6	5·0	15·6	8·2
37	7 54·3	7 55·5	7 32·6	3·7	1·9	9·7	5·1	15·7	8·2
38	7 54·5	7 55·8	7 32·9	3·8	2·0	9·8	5·1	15·8	8·3
39	7 54·8	7 56·0	7 33·1	3·9	2·0	9·9	5·2	15·9	8 3
40	7 55·0	7 56·3	7 33·4	4·0	2·1	10·0	5·3	16·0	8·4
41	7 55·3	7 56·6	7 33·6	4·1	2·2	10·1	5·3	16·1	8·5
42	7 55·5	7 56·8	7 33·8	4·2	2·2	10·2	5·4	16·2	8·5
43	7 55·8	7 57·1	7 34·1	4·3	2·3	10·3	5·4	16·3	8·6
44	7 56·0	7 57·3	7 34·3	4·4	2·3	10·4	5·5	16·4	8·6
45	7 56·3	7 57·6	7 34·6	4·5	2·4	10·5	5·5	16·5	8·7
46	7 56·5	7 57·8	7 34·8	4·6	2·4	10·6	5·6	16·6	8·7
47	7 56·8	7 58·1	7 35·0	4·7	2·5	10·7	5·6	16·7	8·8
48	7 57·0	7 58·3	7 35·3	4·8	2·5	10·8	5·7	16·8	8·8
49	7 57·3	7 58·6	7 35·5	4·9	2·6	10·9	5·7	16·9	8·9
50	7 57·5	7 58·8	7 35·7	5·0	2·6	11·0	5·8	17·0	8·9
51	7 57·8	7 59·1	7 36·0	5·1	2·7	11·1	5·8	17·1	9·0
52	7 58·0	7 59·3	7 36·2	5·2	2·7	11·2	5·9	17·2	9·0
53	7 58·3	7 59·6	7 36·5	5·3	2·8	11·3	5·9	17·3	9·1
54	7 58·5	7 59·8	7 36·7	5·4	2·8	11·4	6·0	17·4	9·1
55	7 58·8	8 00·1	7 36·9	5·5	2·9	11·5	6·0	17·5	9·2
56	7 59·0	8 00·3	7 37·2	5·6	2·9	11·6	6·1	17·6	9·2
57	7 59·3	8 00·6	7 37·4	5·7	3·0	11·7	6·1	17·7	9·3
58	7 59·5	8 00·8	7 37·7	5·8	3·0	11·8	6·2	17·8	9·3
59	7 59·8	8 01·1	7 37·9	5·9	3·1	11·9	6·2	17·9	9·4
60	8 00·0	8 01·3	7 38·1	6·0	3·2	12·0	6·3	18·0	9·5

Table 42A

LATITUDE CONTRARY NAME TO DECLINATION — L.H.A. 59°, 301°

Dec.	38° Hc	d	Z	39° Hc	d	Z	40° Hc	d	Z	41° Hc	d	Z	Dec.
°	° ′	′	°	° ′	′	°	° ′	′	°	° ′	′	°	°
0	23 56.7	−40.6	110.3	23 35.7	−41.4	110.7	23 14.2	−42.0	111.1	22 52.4	−42.8	111.5	0
1	23 16.1	40.7	111.1	22 54.3	41.5	111.5	22 32.2	42.3	111.9	22 09.6	43.0	112.3	1
2	22 35.4	41.1	111.9	22 12.8	41.8	112.3	21 49.9	42.6	112.7	21 26.6	43.3	113.0	2
3	21 54.3	41.2	112.7	21 31.0	42.0	113.1	21 07.3	42.7	113.4	20 43.3	43.4	113.8	3
4	21 13.1	41.5	113.5	20 49.0	42.2	113.8	20 24.6	42.9	114.2	19 59.9	43.6	114.5	4
5	20 31.6	−41.6	114.2	20 06.8	−42.3	114.6	19 41.7	−43.1	114.9	19 16.3	−43.8	115.2	5
6	19 50.0	41.9	115.0	19 24.5	42.6	115.3	18 58.6	43.2	115.6	18 32.5	43.9	116.0	6
7	19 08.1	42.0	115.8	18 41.9	42.8	116.1	18 15.4	43.5	116.4	17 48.6	44.1	116.7	7
8	18 26.1	42.2	116.5	17 59.1	42.9	116.8	17 31.9	43.5	117.1	17 04.5	44.3	117.4	8
9	17 43.9	42.4	117.3	17 16.2	43.0	117.6	16 48.4	43.8	117.8	16 20.2	44.4	118.1	9
10	17 01.5	−42.6	118.0	16 33.2	−43.3	118.3	16 04.6	−43.9	118.5	15 35.8	−44.5	118.8	10
11	16 18.9	42.7	118.8	15 49.9	43.3	119.0	15 20.7	44.0	119.2	14 51.3	44.6	119.5	11
12	15 36.2	42.8	119.5	15 06.6	43.5	119.7	14 36.7	44.1	119.9	14 06.7	44.8	120.2	12
13	14 53.4	43.0	120.2	14 23.1	43.6	120.4	13 52.6	44.2	120.6	13 21.9	44.8	120.9	13
14	14 10.4	43.1	120.9	13 39.5	43.8	121.1	13 08.4	44.4	121.3	12 37.1	45.0	121.5	14
15	13 27.3	−43.2	121.6	12 55.7	−43.8	121.8	12 24.0	−44.5	122.0	11 52.1	−45.1	122.2	15
16	12 44.1	43.4	122.4	12 11.9	44.0	122.5	11 39.5	44.5	122.7	11 07.0	45.1	122.9	16
17	12 00.7	43.4	123.1	11 27.9	44.0	123.2	10 55.0	44.7	123.4	10 21.9	45.3	123.6	17
18	11 17.3	43.6	123.8	10 43.9	44.2	123.9	10 10.3	44.7	124.1	9 36.6	45.3	124.2	18
19	10 33.7	43.6	124.5	9 59.7	44.2	124.6	9 25.6	44.8	124.8	8 51.3	45.4	124.9	19
20	9 50.1	−43.7	125.2	9 15.5	−44.3	125.3	8 40.8	−44.9	125.4	8 05.9	−45.4	125.6	20
21	9 06.4	43.8	125.9	8 31.2	44.4	126.0	7 55.9	45.0	126.1	7 20.5	45.5	126.2	21
22	8 22.6	43.9	126.6	7 46.8	44.5	126.7	7 10.9	45.0	126.8	6 35.0	45.6	126.9	22
23	7 38.7	43.9	127.2	7 02.3	44.5	127.3	6 25.9	45.1	127.4	5 49.4	45.6	127.5	23
24	6 54.8	44.0	127.9	6 17.8	44.5	128.0	5 40.8	45.1	128.1	5 03.8	45.6	128.2	24
25	6 10.8	−44.1	128.6	5 33.3	−44.6	128.7	4 55.7	−45.1	128.8	4 18.2	−45.7	128.8	25
26	5 26.7	44.1	129.3	4 48.7	44.7	129.4	4 10.6	45.2	129.4	3 32.5	45.7	129.5	26
27	4 42.6	44.1	130.0	4 04.0	44.6	130.0	3 25.4	45.2	130.1	2 46.8	45.8	130.1	27
28	3 58.5	44.2	130.7	3 19.4	44.7	130.7	2 40.2	45.2	130.7	2 01.0	45.7	130.8	28
29	3 14.3	44.2	131.3	2 34.7	44.8	131.4	1 55.0	45.2	131.4	1 15.3	45.7	131.4	29
30	2 30.1	−44.2	132.0	1 49.9	−44.7	132.0	1 09.8	−45.3	132.1	0 29.6	−45.8	132.1	30
31	1 45.9	44.2	132.7	1 05.2	44.7	132.7	0 24.5	−45.2	132.7	0 16.2	+45.7	47.3	31
32	1 01.7	44.3	133.4	0 20.5	−44.8	133.4	0 20.7	+45.3	46.6	1 01.9	45.8	46.6	32
33	0 17.4	−44.2	134.0	0 24.3	+44.7	46.0	1 06.0	45.2	46.0	1 47.7	45.7	46.0	33
34	0 26.8	+44.3	45.3	1 09.0	44.8	45.3	1 51.2	45.3	45.3	2 33.4	45.7	45.3	34
35	1 11.1	+44.2	44.6	1 53.8	+44.7	44.6	2 36.5	+45.2	44.7	3 19.1	+45.7	44.7	35
36	1 55.3	44.2	43.9	2 38.5	44.7	44.0	3 21.7	45.1	44.0	4 04.8	45.7	44.0	36
37	2 39.5	44.2	43.3	3 23.2	44.7	43.3	4 06.8	45.2	43.3	4 50.5	45.6	43.4	37
38	3 23.7	44.2	42.6	4 07.9	44.6	42.6	4 52.0	45.1	42.7	5 36.1	45.6	42.7	38
39	4 07.9	44.1	41.9	4 52.5	44.6	42.0	5 37.1	45.1	42.0	6 21.7	45.5	42.1	39
40	4 52.0	+44.1	41.2	5 37.1	+44.6	41.3	6 22.2	+45.0	41.4	7 07.2	+45.4	41.4	40
41	5 36.1	44.0	40.5	6 21.7	44.5	40.6	7 07.2	44.9	40.7	7 52.6	45.5	40.8	41
42	6 20.1	44.0	39.9	7 06.2	44.4	39.9	7 52.1	44.9	40.0	8 38.1	45.3	40.1	42
43	7 04.1	43.9	39.2	7 50.6	44.4	39.3	8 37.0	44.8	39.3	9 23.4	45.3	39.5	43
44	7 48.0	43.9	38.5	8 35.0	44.3	38.6	9 21.8	44.8	38.7	10 08.7	45.1	38.8	44
45	8 31.9	+43.8	37.8	9 19.3	+44.2	37.9	10 06.6	+44.7	38.0	10 53.8	+45.1	38.1	45
46	9 15.7	43.7	37.1	10 03.5	44.1	37.2	10 51.3	44.5	37.3	11 38.9	45.0	37.4	46
47	9 59.4	43.6	36.4	10 47.6	44.1	36.5	11 35.8	44.5	36.6	12 23.9	44.9	36.8	47
48	10 43.0	43.5	35.7	11 31.7	43.9	35.8	12 20.3	44.4	36.0	13 08.8	44.8	36.1	48
49	11 26.5	43.5	35.0	12 15.6	43.9	35.1	13 04.7	44.2	35.3	13 53.6	44.7	35.4	49
50	12 10.0	+43.3	34.3	12 59.5	+43.7	34.4	13 48.9	+44.2	34.6	14 38.3	+44.6	34.7	50
51	12 53.3	43.2	33.6	13 43.2	43.6	33.7	14 33.1	44.0	33.9	15 22.9	44.4	34.0	51
52	13 36.5	43.1	32.9	14 26.8	43.5	33.0	15 17.1	43.9	33.2	16 07.3	44.3	33.3	52
53	14 19.6	42.9	32.2	15 10.3	43.4	32.3	16 01.0	43.7	32.5	16 51.6	44.1	32.6	53
54	15 02.5	42.8	31.4	15 53.7	43.2	31.6	16 44.7	43.6	31.7	17 35.7	44.0	31.9	54
55	15 45.3	+42.7	30.7	16 36.9	+43.0	30.9	17 28.3	+43.5	31.0	18 19.7	+43.8	31.2	55
56	16 28.0	42.5	30.0	17 19.9	42.9	30.1	18 11.8	43.2	30.3	19 03.5	43.7	30.5	56
57	17 10.5	42.4	29.3	18 02.8	42.7	29.4	18 55.0	43.1	29.6	19 47.2	43.5	29.7	57
58	17 52.9	42.1	28.5	18 45.5	42.6	28.7	19 38.1	43.0	28.8	20 30.7	43.3	29.0	58
59	18 35.0	42.0	27.8	19 28.1	42.4	27.9	20 21.1	42.7	28.1	21 14.0	43.1	28.3	59
60	19 17.0	+41.9	27.0	20 10.5	+42.1	27.2	21 03.8	+42.5	27.3	21 57.1	+42.8	27.5	60
61	19 58.9	41.6	26.2	20 52.6	42.0	26.4	21 46.3	42.4	26.6	22 39.9	42.7	26.8	61
62	20 40.5	41.4	25.5	21 34.6	41.8	25.6	22 28.7	42.1	25.8	23 22.6	42.5	26.0	62
63	21 21.9	41.2	24.7	22 16.4	41.5	24.9	23 10.8	41.8	25.0	24 05.1	42.2	25.2	63
64	22 03.1	41.0	23.9	22 57.9	41.3	24.1	23 52.6	41.7	24.3	24 47.3	42.0	24.4	64
65	22 44.1	+40.7	23.1	23 39.2	+41.1	23.3	24 34.3	+41.4	23.5	25 29.3	+41.7	23.7	65
66	23 24.8	40.5	22.3	24 20.3	40.8	22.5	25 15.7	41.1	22.7	26 11.0	41.4	22.9	66
67	24 05.3	40.2	21.5	25 01.1	40.5	21.7	25 56.8	40.8	21.9	26 52.4	41.2	22.1	67
68	24 45.5	40.0	20.7	25 41.6	40.3	20.9	26 37.6	40.6	21.1	27 33.6	40.9	21.2	68
69	25 25.5	39.7	19.9	26 21.9	40.0	20.1	27 18.2	40.3	20.2	28 14.5	40.5	20.4	69
70	26 05.2	+39.4	19.1	27 01.9	+39.6	19.2	27 58.5	+39.9	19.4	28 55.0	+40.3	19.6	70
71	26 44.6	39.1	18.2	27 41.5	39.4	18.4	28 38.4	39.7	18.5	29 35.3	39.9	18.7	71
72	27 23.7	38.7	17.4	28 20.9	39.1	17.5	29 18.1	39.3	17.7	30 15.2	39.6	17.9	72
73	28 02.4	38.5	16.5	29 00.0	38.7	16.7	29 57.4	39.0	16.8	30 54.8	39.3	17.0	73
74	28 40.9	38.1	15.6	29 38.7	38.3	15.8	30 36.4	38.6	15.9	31 34.1	38.8	16.1	74
75	29 19.0	+37.8	14.7	30 17.0	+38.0	14.9	31 15.0	+38.2	15.0	32 12.9	+38.5	15.2	75
76	29 56.8	37.4	13.8	30 55.0	37.6	14.0	31 53.2	37.9	14.1	32 51.4	38.1	14.3	76
77	30 34.2	37.0	12.9	31 32.6	37.3	13.1	32 31.1	37.4	13.2	33 29.5	37.6	13.4	77
78	31 11.2	36.6	12.0	32 09.9	36.8	12.2	33 08.5	37.0	12.3	34 07.1	37.2	12.4	78
79	31 47.8	36.2	11.1	32 46.7	36.4	11.2	33 45.5	36.6	11.3	34 44.3	36.8	11.5	79
80	32 24.0	+35.8	10.2	33 23.1	+35.9	10.3	34 22.1	+36.1	10.4	35 21.1	+36.3	10.5	80
81	32 59.8	35.4	9.2	33 59.0	35.5	9.3	34 58.2	35.7	9.4	35 57.4	35.8	9.5	81
82	33 35.2	34.8	8.2	34 34.5	35.1	8.3	35 33.9	35.2	8.4	36 33.2	35.3	8.5	82
83	34 10.0	34.5	7.3	35 09.6	34.5	7.3	36 09.1	34.6	7.4	37 08.5	34.8	7.5	83
84	34 44.5	33.9	6.3	35 44.1	34.0	6.3	36 43.7	34.2	6.4	37 43.3	34.3	6.5	84
85	35 18.4	+33.4	5.3	36 18.1	+33.5	5.3	37 17.9	+33.6	5.4	38 17.6	+33.7	5.5	85
86	35 51.8	32.9	4.2	36 51.6	33.0	4.3	37 51.5	33.0	4.3	38 51.3	33.1	4.4	86
87	36 24.7	32.3	3.2	37 24.6	32.4	3.2	38 24.5	32.4	3.3	39 24.4	32.5	3.3	87
88	36 57.0	31.8	2.1	37 57.0	31.8	2.2	38 56.9	31.9	2.2	39 56.9	31.9	2.2	88
89	37 28.8	31.2	1.1	38 28.8	31.2	1.1	39 28.8	31.2	1.1	40 28.8	31.2	1.1	89
90	38 00.0	+30.6	0.0	39 00.0	+30.6	0.0	40 00.0	+30.6	0.0	41 00.0	+30.6	0.0	90

Dec.	42° Hc	d	Z	43° Hc	d	Z	44° Hc	d	Z	45° Hc	d	Z	Dec.
°	° ′	′	°	° ′	′	°	° ′	′	°	° ′	′	°	°
0	22 30.2	−43.5	111.9	22 07.7	−44.3	112.3	21 44.7	−44.9	112.7	21 21.5	−45.7	113.0	0
1	21 46.7	43.8	112.6	21 23.4	44.5	113.0	20 59.8	45.2	113.4	20 35.8	45.8	113.7	1
2	21 02.9	43.9	113.4	20 38.9	44.6	113.7	20 14.6	45.3	114.1	19 50.0	46.0	114.4	2
3	20 19.0	44.1	114.1	19 54.3	44.8	114.4	19 29.3	45.4	114.8	19 04.0	46.1	115.1	3
4	19 34.9	44.3	114.8	19 09.5	44.9	115.1	18 43.9	45.6	115.5	18 17.9	46.2	115.8	4
5	18 50.6	−44.5	115.5	18 24.6	−45.2	115.8	17 58.3	−45.8	116.1	17 31.7	−46.4	116.4	5
6	18 06.1	44.6	116.3	17 39.4	45.2	116.5	17 12.5	45.9	116.8	16 45.3	46.5	117.1	6
7	17 21.5	44.7	117.0	16 54.2	45.4	117.2	16 26.6	46.0	117.5	15 58.8	46.6	117.8	7
8	16 36.8	44.9	117.7	16 08.8	45.5	117.9	15 40.6	46.2	118.2	15 12.2	46.8	118.4	8
9	15 51.9	45.1	118.3	15 23.3	45.7	118.6	14 54.4	46.2	118.8	14 25.4	46.8	119.1	9
10	15 06.8	−45.1	119.0	14 37.6	−45.7	119.3	14 08.2	−46.4	119.5	13 38.6	−47.0	119.7	10
11	14 21.7	45.3	119.7	13 51.9	45.9	119.9	13 21.8	46.4	120.1	12 51.6	47.0	120.3	11
12	13 36.4	45.3	120.4	13 06.0	46.0	120.6	12 35.4	46.6	120.8	12 04.6	47.2	121.0	12
13	12 51.1	45.5	121.1	12 20.0	46.0	121.2	11 48.8	46.6	121.4	11 17.4	47.2	121.6	13
14	12 05.6	45.6	121.7	11 34.0	46.2	121.9	11 02.2	46.7	122.1	10 30.2	47.2	122.2	14
15	11 20.0	−45.6	122.4	10 47.8	−46.2	122.6	10 15.5	−46.8	122.7	9 43.0	−47.4	122.9	15
16	10 34.4	45.8	123.1	10 01.6	46.3	123.2	9 28.7	46.9	123.3	8 55.6	47.4	123.5	16
17	9 48.6	45.8	123.7	9 15.3	46.4	123.8	8 41.8	46.9	124.0	8 08.2	47.5	124.1	17
18	9 02.8	45.9	124.4	8 28.9	46.4	124.5	7 54.9	47.0	124.6	7 20.7	47.5	124.7	18
19	8 16.9	45.9	125.0	7 42.5	46.5	125.1	7 07.9	47.1	125.2	6 33.2	47.5	125.3	19
20	7 31.0	−46.0	125.7	6 56.0	−46.6	125.8	6 20.8	−47.0	125.9	5 45.7	−47.6	125.9	20
21	6 45.0	46.1	126.3	6 09.4	46.6	126.4	5 33.8	47.2	126.5	4 58.1	47.7	126.6	21
22	5 58.9	46.1	127.0	5 22.8	46.6	127.0	4 46.6	47.1	127.1	4 10.4	47.6	127.2	22
23	5 12.8	46.1	127.6	4 36.2	46.7	127.7	3 59.5	47.2	127.7	3 22.8	47.7	127.8	23
24	4 26.7	46.2	128.2	3 49.5	46.7	128.3	3 12.3	47.2	128.3	2 35.1	47.7	128.4	24
25	3 40.5	−46.2	128.9	3 02.8	−46.7	128.9	2 25.1	−47.2	129.0	1 47.4	−47.7	129.0	25
26	2 54.3	46.2	129.5	2 16.1	46.7	129.6	1 37.9	47.2	129.6	0 59.7	47.8	129.6	26
27	2 08.1	46.2	130.2	1 29.4	46.8	130.2	0 50.7	47.3	130.2	0 11.9	−47.7	130.2	27
28	1 21.9	46.3	130.8	0 42.6	−46.7	130.8	0 03.4	−47.2	130.8	0 35.0	+47.7	49.2	28
29	0 35.6	−46.2	131.4	0 04.1	+46.7	48.6	0 43.8	+47.2	48.6	1 23.5	47.7	48.6	29
30	0 10.6	+46.3	47.9	0 50.8	+46.8	47.9	1 31.0	+47.3	48.0	2 11.2	+47.7	48.0	30
31	0 56.9	46.2	47.3	1 37.6	46.7	47.3	2 18.3	47.2	47.3	2 58.9	47.7	47.4	31
32	1 43.1	46.3	46.7	2 24.3	46.7	46.7	3 05.5	47.1	46.7	3 46.6	47.6	46.8	32
33	2 29.4	46.2	46.0	3 11.0	46.7	46.1	3 52.6	47.2	46.1	4 34.2	47.6	46.2	33
34	3 15.6	46.2	45.4	3 57.7	46.7	45.4	4 39.8	47.1	45.5	5 21.8	47.6	45.5	34
35	4 01.8	+46.1	44.7	4 44.4	+46.6	44.8	5 26.9	+47.1	44.9	6 09.4	+47.6	44.9	35
36	4 47.9	46.1	44.1	5 31.0	46.6	44.2	6 14.0	47.0	44.2	6 57.0	47.5	44.3	36
37	5 34.0	46.1	43.5	6 17.6	46.5	43.5	7 01.0	47.0	43.6	7 44.5	47.4	43.7	37
38	6 20.1	46.1	42.8	7 04.1	46.5	42.9	7 48.0	47.0	43.0	8 31.9	47.4	43.1	38
39	7 06.2	45.9	42.2	7 50.6	46.4	42.3	8 35.0	46.8	42.4	9 19.3	47.3	42.5	39
40	7 52.1	+46.0	41.5	8 37.0	+46.4	41.6	9 21.8	+46.9	41.7	10 06.6	+47.2	41.8	40
41	8 38.1	45.8	40.9	9 23.4	46.3	41.0	10 08.7	46.7	41.1	10 53.8	47.2	41.2	41
42	9 23.9	45.8	40.2	10 09.7	46.2	40.3	10 55.4	46.6	40.4	11 41.0	47.1	40.6	42
43	10 09.7	45.7	39.6	10 55.9	46.1	39.7	11 42.0	46.6	39.8	12 28.1	47.0	39.9	43
44	10 55.4	45.6	38.9	11 42.0	46.1	39.0	12 28.6	46.5	39.2	13 15.1	46.9	39.3	44
45	11 41.0	+45.5	38.2	12 28.1	+45.9	38.4	13 15.1	+46.4	38.5	14 02.0	+46.8	38.7	45
46	12 26.5	45.5	37.6	13 14.0	45.9	37.7	14 01.5	46.2	37.9	14 48.8	46.7	38.0	46
47	13 12.0	45.3	36.9	13 59.9	45.7	37.0	14 47.7	46.2	37.2	15 35.5	46.5	37.4	47
48	13 57.3	45.2	36.2	14 45.6	45.7	36.4	15 33.9	46.0	36.5	16 22.0	46.5	36.7	48
49	14 42.5	45.1	35.5	15 31.3	45.5	35.7	16 19.9	45.9	35.9	17 08.5	46.3	36.1	49
50	15 27.6	+44.9	34.9	16 16.8	+45.3	35.0	17 05.8	+45.8	35.2	17 54.8	+46.2	35.4	50
51	16 12.5	44.9	34.2	17 02.1	45.3	34.3	17 51.6	45.6	34.5	18 41.0	46.0	34.7	51
52	16 57.4	44.7	33.5	17 47.4	45.0	33.7	18 37.2	45.5	33.8	19 27.0	45.9	34.0	52
53	17 42.1	44.5	32.8	18 32.4	45.0	33.0	19 22.7	45.4	33.2	20 12.9	45.7	33.3	53
54	18 26.6	44.4	32.1	19 17.4	44.8	32.3	20 08.1	45.1	32.5	20 58.6	45.6	32.7	54
55	19 11.0	+44.2	31.4	20 02.2	+44.6	31.6	20 53.2	+45.0	31.8	21 44.2	+45.4	32.0	55
56	19 55.2	44.0	30.7	20 46.8	44.4	30.8	21 38.2	44.8	31.0	22 29.6	45.2	31.3	56
57	20 39.2	43.9	29.9	21 31.2	44.2	30.1	22 23.0	44.6	30.3	23 14.8	45.0	30.5	57
58	21 23.1	43.7	29.2	22 15.4	44.1	29.4	23 07.6	44.5	29.6	23 59.8	44.7	29.8	58
59	22 06.8	43.4	28.5	22 59.5	43.8	28.7	23 52.1	44.2	28.9	24 44.5	44.6	29.1	59
60	22 50.2	+43.3	27.7	23 43.3	+43.6	27.9	24 36.3	+43.9	28.1	25 29.1	+44.4	28.3	60
61	23 33.5	43.0	27.0	24 26.9	43.4	27.2	25 20.2	43.8	27.4	26 13.5	44.1	27.6	61
62	24 16.5	42.8	26.2	25 10.3	43.2	26.4	26 04.0	43.5	26.6	26 57.6	43.8	26.8	62
63	24 59.3	42.6	25.4	25 53.5	42.9	25.6	26 47.5	43.3	25.8	27 41.4	43.7	26.1	63
64	25 41.9	42.3	24.6	26 36.4	42.6	24.9	27 30.8	43.0	25.1	28 25.1	43.3	25.3	64
65	26 24.2	+42.0	23.9	27 19.0	+42.4	24.1	28 13.8	+42.7	24.3	29 08.4	+43.1	24.5	65
66	27 06.2	41.8	23.1	28 01.4	42.1	23.3	28 56.5	42.4	23.5	29 51.5	42.7	23.7	66
67	27 48.0	41.5	22.2	28 43.5	41.8	22.5	29 38.9	42.1	22.7	30 34.2	42.5	22.9	67
68	28 29.5	41.2	21.4	29 25.3	41.5	21.6	30 21.0	41.9	21.8	31 16.7	42.1	22.1	68
69	29 10.7	40.8	20.6	30 06.8	41.2	20.8	31 02.9	41.4	21.0	31 58.8	41.8	21.2	69
70	29 51.5	+40.6	19.8	30 48.0	+40.8	20.0	31 44.3	+41.2	20.2	32 40.6	+41.5	20.4	70
71	30 32.1	40.2	18.9	31 28.8	40.5	19.1	32 25.5	40.8	19.3	33 22.1	41.1	19.5	71
72	31 12.3	39.9	18.0	32 09.3	40.2	18.2	33 06.3	40.4	18.4	34 03.2	40.7	18.6	72
73	31 52.2	39.5	17.2	32 49.5	39.7	17.4	33 46.7	40.1	17.5	34 43.9	40.3	17.8	73
74	32 31.7	39.1	16.3	33 29.2	39.4	16.5	34 26.8	39.6	16.6	35 24.2	39.9	16.8	74
75	33 10.8	+38.7	15.4	34 08.6	+39.0	15.5	35 06.4	+39.2	15.7	36 04.1	+39.5	15.9	75
76	33 49.5	38.3	14.5	34 47.6	38.5	14.6	35 45.6	38.8	14.8	36 43.6	39.0	15.0	76
77	34 27.8	37.9	13.5	35 26.1	38.1	13.7	36 24.4	38.3	13.9	37 22.6	38.6	14.0	77
78	35 05.7	37.4	12.6	36 04.2	37.7	12.7	37 02.7	37.9	12.9	38 01.2	38.1	13.1	78
79	35 43.1	37.0	11.6	36 41.9	37.1	11.8	37 40.6	37.4	11.9	38 39.3	37.6	12.1	79
80	36 20.1	+36.5	10.6	37 19.0	+36.7	10.8	38 18.0	+36.8	10.9	39 16.9	+37.0	11.1	80
81	36 56.6	36.0	9.7	37 55.7	36.2	9.8	38 54.8	36.3	9.9	39 53.9	36.5	10.1	81
82	37 32.6	35.4	8.7	38 31.9	35.6	8.8	39 31.1	35.8	8.9	40 30.4	36.0	9.0	82
83	38 08.0	34.9	7.6	39 07.5	35.0	7.7	40 06.9	35.2	7.9	41 06.4	35.3	8.0	83
84	38 42.9	34.4	6.6	39 42.5	34.5	6.7	40 42.1	34.6	6.8	41 41.7	34.7	6.9	84
85	39 17.3	+33.8	5.5	40 17.0	+33.9	5.6	41 16.7	+34.0	5.7	42 16.4	+34.1	5.8	85
86	39 51.1	33.2	4.5	40 50.9	33.3	4.5	41 50.7	33.4	4.6	42 50.5	33.5	4.7	86
87	40 24.3	32.5	3.4	41 24.2	32.6	3.4	42 24.1	32.6	3.5	43 24.0	32.7	3.5	87
88	40 56.8	32.0	2.3	41 56.8	31.9	2.3	42 56.7	32.0	2.3	43 56.7	32.0	2.4	88
89	41 28.8	31.2	1.1	42 28.7	31.3	1.2	43 28.7	31.3	1.2	44 28.7	31.3	1.2	89
90	42 00.0	+30.6	0.0	43 00.0	+30.5	0.0	44 00.0	+30.5	0.0	45 00.0	+30.5	0.0	90

S. Lat. { L.H.A. greater than 180°......Zn=180°−Z; L.H.A. less than 180°............Zn=180°+Z

LATITUDE SAME NAME AS DECLINATION — L.H.A. 121°, 239°

Table 42B

30°, 330° L.H.A. LATITUDE SAME NAME AS DECLINATION

N. Lat. {L.H.A. greater than 180°......Zn=Z; L.H.A. less than 180°............Zn=360°−

Dec.	38°			39°			40°			41°			42°			43°			44°			45°			Dec.
	Hc	d	Z	Hc	d	Z	Hc	d	Z	Hc	d	Z	Hc	d	Z	Hc	d	Z	Hc	d	Z	Hc	d	Z	
°	° ′	′	°	° ′	′	°	° ′	′	°	° ′	′	°	° ′	′	°	° ′	′	°	° ′	′	°	° ′	′	°	°
0	43 02.1	+50.4	136.8	42 18.1	+50.9	137.5	41 33.6	+51.5	138.1	40 48.8	+51.9	138.7	40 03.6	+52.3	139.2	39 18.0	+52.7	139.8	38 32.0	+53.2	140.3	37 45.7	+53.6	140.8	0
1	43 52.5	50.0	136.1	43 09.0	50.6	136.7	42 25.1	51.1	137.4	41 40.7	51.7	138.0	40 55.9	52.1	138.6	40 10.7	52.6	139.1	39 25.2	53.0	139.7	38 39.3	53.4	140.2	1
2	44 42.5	49.8	135.3	43 59.6	50.4	136.0	43 16.2	50.9	136.7	42 32.4	51.4	137.3	41 48.0	51.9	137.9	41 03.3	52.4	138.5	40 18.2	52.8	139.1	39 32.7	53.2	139.6	2
3	45 32.3	49.5	134.5	44 50.0	50.0	135.2	44 07.1	50.6	135.9	43 23.8	51.1	136.6	42 39.9	51.7	137.2	41 55.7	52.1	137.8	41 11.0	52.6	138.4	40 25.9	53.0	139.0	3
4	46 21.8	49.0	133.7	45 40.0	49.7	134.5	44 57.7	50.3	135.2	44 14.9	50.9	135.9	43 31.6	51.4	136.5	42 47.8	51.9	137.2	42 03.6	52.4	137.8	41 18.9	52.9	138.4	4
5	47 10.8	+48.7	132.9	46 29.7	+49.4	133.7	45 48.0	+50.0	134.4	45 05.8	+50.5	135.1	44 23.0	+51.1	135.8	43 39.7	+51.7	136.5	42 56.0	+52.1	137.1	42 11.8	+52.6	137.8	5
6	47 59.5	48.3	132.0	47 19.1	48.9	132.8	46 38.0	49.6	133.6	45 56.3	50.2	134.4	45 14.1	50.8	135.1	44 31.4	51.3	135.8	43 48.1	51.9	136.5	43 04.4	52.4	137.1	6
7	48 47.8	47.9	131.1	48 08.0	48.6	132.0	47 27.6	49.3	132.8	46 46.5	49.9	133.6	46 04.9	50.5	134.3	45 22.7	51.1	135.0	44 40.0	51.6	135.8	43 56.8	52.1	136.4	7
8	49 35.7	47.4	130.2	48 56.6	48.2	131.1	48 16.9	48.8	131.9	47 36.4	49.6	132.7	46 55.4	50.2	133.5	46 13.8	50.8	134.3	45 31.6	51.4	135.0	44 48.9	51.9	135.7	8
9	50 23.1	46.9	129.2	49 44.8	47.7	130.2	49 05.7	48.5	131.0	48 26.0	49.1	131.9	47 45.6	49.8	132.7	47 04.6	50.4	133.5	46 23.0	51.0	134.3	45 40.8	51.6	135.0	9
10	51 10.0	+46.4	128.3	50 32.5	+47.2	129.2	49 54.2	+48.0	130.1	49 15.1	+48.8	131.0	48 35.4	+49.5	131.9	47 55.0	+50.1	132.7	47 14.0	+50.7	133.5	46 32.4	+51.3	134.3	10
11	51 56.4	45.9	127.2	51 19.7	46.7	128.2	50 42.2	47.5	129.2	50 03.9	48.3	130.1	49 24.9	49.0	131.0	48 45.1	49.8	131.9	48 04.7	50.4	132.7	47 23.7	51.0	133.5	11
12	52 42.3	45.3	126.2	52 06.4	46.2	127.2	51 29.7	47.1	128.2	50 52.2	47.9	129.2	50 13.9	48.6	130.1	49 34.9	49.3	131.0	48 55.1	50.1	131.9	48 14.7	50.7	132.7	12
13	53 27.6	44.7	125.1	52 52.6	45.7	126.2	52 16.8	46.5	127.2	51 40.1	47.3	128.2	51 02.5	48.2	129.2	50 24.2	48.9	130.1	49 45.2	49.6	131.1	49 05.4	50.3	131.9	13
14	54 12.3	44.0	124.0	53 38.3	45.0	125.1	53 03.3	45.9	126.2	52 27.4	46.9	127.2	51 50.7	47.7	128.3	51 13.1	48.5	129.2	50 34.8	49.3	130.2	49 55.7	50.0	131.1	14
15	54 56.3	+43.3	122.8	54 23.3	+44.3	124.0	53 49.2	+45.4	125.1	53 14.3	+46.3	126.2	52 38.4	+47.2	127.3	52 01.6	+48.0	128.3	51 24.1	+48.8	129.3	50 45.7	+49.5	130.2	15
16	55 39.6	42.5	121.6	55 07.6	43.7	122.8	54 34.6	44.7	124.0	54 00.6	45.7	125.1	53 25.6	46.6	126.2	52 49.6	47.6	127.3	52 12.9	48.3	128.3	51 35.2	49.2	129.3	16
17	56 22.1	41.8	120.3	55 51.3	42.9	121.6	55 19.3	44.0	122.8	54 46.3	45.0	124.0	54 12.2	46.0	125.2	53 37.2	46.9	126.3	53 01.2	47.8	127.4	52 24.4	48.6	128.4	17
18	57 03.9	40.9	119.0	56 34.2	42.1	120.3	56 03.3	43.3	121.6	55 31.3	44.4	122.9	54 58.2	45.4	124.1	54 24.1	46.4	125.2	53 49.0	47.3	126.3	53 13.0	48.2	127.4	18
19	57 44.8	40.0	117.6	57 16.3	41.3	119.0	56 46.6	42.5	120.4	56 15.7	43.6	121.7	55 43.6	44.8	122.9	55 10.5	45.8	124.1	54 36.3	46.8	125.3	54 01.2	47.6	126.4	19
20	58 24.8	+39.0	116.2	57 57.6	+40.4	117.7	57 29.1	+41.7	119.1	56 59.3	+42.9	120.4	56 28.4	+44.0	121.7	55 56.3	+45.1	123.0	55 23.1	+46.1	124.2	54 48.8	+47.1	125.4	20
21	59 03.8	38.1	114.8	58 38.0	39.4	116.3	58 10.8	40.8	117.7	57 42.2	42.1	119.1	57 12.4	43.3	120.5	56 41.4	44.4	121.8	56 09.2	45.5	123.1	55 35.9	46.5	124.3	21
22	59 41.9	36.9	113.2	59 17.4	38.4	114.8	58 51.6	39.8	116.3	58 24.3	41.1	117.8	57 55.7	42.4	119.2	57 25.8	43.6	120.6	56 54.7	44.8	121.9	56 22.4	45.9	123.2	22
23	60 18.8	35.7	111.7	59 55.8	37.4	113.3	59 31.4	38.8	114.8	59 05.4	40.3	116.4	58 38.1	41.6	117.8	58 09.4	42.9	119.3	57 39.5	44.0	120.6	57 08.3	45.1	122.0	23
24	60 54.5	34.6	110.0	60 33.2	36.1	111.7	60 10.2	37.7	113.3	59 45.7	39.2	114.9	59 19.7	40.6	116.4	58 52.3	41.9	117.9	58 23.5	43.2	119.4	57 53.4	44.5	120.8	24
25	61 29.1	+33.2	108.3	61 09.3	+35.0	110.1	60 47.9	+36.6	111.7	60 24.9	+38.1	113.4	60 00.3	+39.6	115.0	59 34.2	+41.1	116.5	59 06.7	+42.4	118.0	58 37.9	+43.6	119.5	25
26	62 02.3	31.8	106.6	61 44.3	33.6	108.4	61 24.5	35.3	110.1	61 03.0	37.0	111.8	60 39.9	38.6	113.5	60 15.3	40.0	115.1	59 49.1	41.5	116.6	59 21.5	42.8	118.1	26
27	62 34.1	30.3	104.8	62 17.9	32.2	106.6	61 59.8	34.1	108.4	61 40.0	35.8	110.2	61 18.5	37.4	111.9	60 55.3	39.0	113.6	60 30.6	40.4	115.2	60 04.3	41.8	116.8	27
28	63 04.4	28.8	102.9	62 50.1	30.8	104.8	62 33.9	32.6	106.6	62 15.8	34.4	108.5	61 55.9	36.2	110.2	61 34.3	37.8	112.0	61 11.0	39.4	113.7	60 46.1	40.9	115.3	28
29	63 33.2	27.2	100.9	63 20.9	29.1	102.9	63 06.5	31.2	104.8	62 50.2	33.1	106.7	62 32.1	34.9	108.5	62 12.1	36.6	110.3	61 50.4	38.3	112.1	61 27.0	39.9	113.8	29
30	64 00.4	+25.3	98.9	63 50.0	+27.6	100.9	63 37.7	+29.6	102.9	63 23.3	+31.6	104.8	63 07.0	+33.4	106.7	62 48.7	+35.3	108.6	62 28.7	+37.0	110.4	62 06.9	+38.7	112.2	30
31	64 25.7	23.6	96.8	64 17.6	25.7	98.9	64 07.3	27.9	100.9	63 54.9	30.0	102.9	63 40.4	32.0	104.9	63 24.0	34.0	106.8	63 05.7	35.8	108.7	62 45.6	37.5	110.6	31
32	64 49.3	21.7	94.7	64 43.3	24.0	96.8	64 35.2	26.1	98.9	64 24.9	28.3	100.9	64 12.4	30.4	103.0	63 58.0	32.4	105.0	63 41.5	34.3	106.9	63 23.1	36.1	108.8	32
33	65 11.0	19.7·	92.5	65 07.3	22.0·	94.6	65 01.3	24.4	96.8	64 53.2	26.5	98.9	64 42.8	28.8	101.0	64 30.4	30.8	103.0	64 15.8	32.9	105.1	63 59.2	34.8	107.0	33
34	65 30.7	17.5·	90.2	65 29.3	20.0·	92.4	65 25.7	22.3·	94.6	65 19.7	24.7	96.7	65 11.6	26.9	98.9	65 01.2	29.2	101.0	64 48.7	31.2	103.1	64 34.0	33.3	105.2	34
35	65 48.2	+15.5·	87.9	65 49.3	+17.9·	90.1	65 48.0	+20.4·	92.3	65 44.4	+22.8·	94.5	65 38.5	+25.1	96.7	65 30.4	+27.3	98.9	65 19.9	-29.6	101.1	65 07.3	+31.7	103.2	35
36	66 03.7	13.2·	85.5	66 07.2	15.7·	87.8	66 08.4	18.2·	90.0	66 07.2	20.7·	92.3	66 03.6	23.1·	94.5	65 57.7	25.5	96.8	65 49.5	27.8	99.0	65 39.0	30.0	101.2	36
37	66 16.9	10.9·	83.1	66 22.9	13.5·	85.4	66 26.6	16.0·	87.6	66 27.9	18.5·	89.9	66 26.7	21.1·	92.2	66 23.2	23.5·	94.5	66 17.3	25.9·	96.8	66 09.0	28.2	99.0	37
38	66 27.8	8.6·	80.6	66 36.4	11.2·	82.9	66 42.6	13.7·	85.2	66 46.4	16.3·	87.5	66 47.8	18.8·	89.9	66 46.7	21.3·	92.2	66 43.2	23.8·	94.5	66 37.2	26.3·	96.8	38
39	66 36.4	6.2·	78.1	66 47.6	8.7·	80.4	66 56.3	11.4·	82.7	67 02.7	14.0·	85.1	67 06.6	16.6·	87.4	67 08.0	19.2·	89.8	67 07.0	21.7·	92.2	67 03.5	24.2·	94.5	39
40	66 42.6	+3.8·	75.6	66 56.3	+6.4·	77.9	67 07.7	+9.0·	80.2	67 16.7	+11.6·	82.6	67 23.2	+14.3·	85.0	67 27.2	+16.9·	87.4	67 28.7	+19.5·	89.8	67 27.7	+22.1·	92.2	40
41	66 46.4	+1.4·	73.1	67 02.7	3.9·	75.4	67 16.7	6.5·	77.7	67 28.3	9.2·	80.0	67 37.5	11.8·	82.4	67 44.1	14.5·	84.8	67 48.2	17.2·	87.3	67 49.8	19.8·	89.7	41
42	66 47.8	-1.1·	70.6	67 06.6	+1.4·	72.8	67 23.2	4.0·	75.1	67 37.5	6.6·	77.4	67 49.3	9.3·	79.8	67 58.6	12.0·	82.3	68 05.4	14.8·	84.7	68 09.6	17.5·	87.2	42
43	66 46.7	3.5·	68.0	67 08.0	-1.0·	70.2	67 27.2	+1.5·	72.5	67 44.1	4.1·	74.8	67 58.6	6.8·	77.2	68 10.6	9.6·	79.6	68 20.2	12.2·	82.1	68 27.1	15.0·	84.6	43
44	66 43.2	6.0·	65.5	67 07.0	3.5·	67.7	67 28.7	-1.0·	69.9	67 48.2	+1.6·	72.2	68 05.4	4.2·	74.6	68 20.2	6.9·	77.0	68 32.4	9.7·	79.5	68 42.1	12.5·	82.0	44
45	66 37.2	-8.3·	63.0	67 03.5	-6.0·	65.1	67 27.7	-3.5·	67.3	67 49.8	-1.0·	69.5	68 09.6	+1.6·	71.9	68 27.1	+4.3·	74.3	68 42.1	+7.0·	76.7	68 54.6	+9.8·	79.3	45
46	66 28.9	10.7·	60.5	66 57.5	8.4·	62.5	67 24.2	6.1·	64.7	67 48.8	3.6·	66.9	68 11.2	-1.0·	69.2	68 31.4	+1.6·	71.6	68 49.1	4.4·	74.0	69 04.4	7.2·	76.5	46
47	66 18.2	13.0·	58.0	66 49.1	10.8·	60.0	67 18.1	8.5·	62.1	67 45.2	6.1·	64.3	68 10.2	3.6·	66.5	68 33.0	-1.0·	68.8	68 53.5	+1.7·	71.2	69 11.6	4.4·	73.7	47
48	66 05.2	15.1·	55.6	66 38.3	13.1·	57.5	67 09.6	10.9·	59.5	67 39.1	8.6·	61.6	68 06.6	6.2·	63.8	68 32.0	3.7·	66.1	68 55.2	-1.1·	68.5	69 16.0	+1.7·	70.9	48
49	65 50.1	17.4·	53.3	66 25.2	15.4·	55.1	66 58.7	13.3·	57.0	67 30.5	11.1·	59.0	68 00.4	8.8·	61.2	68 28.3	6.3·	63.4	68 54.1	3.7·	65.7	69 17.7	-1.1·	68.1	49
50	65 32.7	-19.4·	50.9	66 09.8	-17.6·	52.7	66 45.4	-15.6·	54.5	67 19.4	-13.5·	56.5	67 51.6	-11.2·	58.5	68 22.0	-8.9·	60.7	68 50.4	-6.5·	62.9	69 16.6	-3.8·	65.3	50
51	65 13.3	21.5·	48.7	65 52.2	19.6·	50.3	66 29.8	17.8·	52.1	67 05.9	15.8·	54.0	67 40.4	13.7·	55.9	68 13.1	11.4·	58.0	68 43.9	9.0·	60.2	69 12.8	6.6·	62.5	51
52	64 51.8	23.4	46.4	65 32.6	21.8·	48.0	66 12.0	19.9·	49.7	66 50.1	18.1·	51.5	67 26.7	16.1·	53.4	68 01.7	14.0·	55.4	68 34.9	11.7·	57.5	69 06.2	9.3·	59.7	52
53	64 28.4	25.2	44.3	65 10.8	23.6	45.8	65 52.1	22.0·	47.4	66 32.0	20.2·	49.1	67 10.6	18.3·	50.9	67 47.7	16.3·	52.8	68 23.2	14.2·	54.8	68 56.9	11.9·	56.9	53
54	64 03.2	26.9	42.2	64 47.2	25.5	43.6	65 30.1	24.0	45.1	66 11.8	22.3·	46.7	66 52.3	20.5·	48.4	67 31.4	18.7·	50.2	68 09.0	16.6·	52.2	68 45.0	14.5·	54.2	54
55	63 36.3	-28.6	40.2	64 21.7	-27.3	41.5	65 06.1	-25.8	42.9	65 49.5	-24.3	44.5	66 31.8	-22.7·	46.1	67 12.7	-20.8·	47.8	67 52.4	-19.0·	49.6	68 30.5	-17.0·	51.5	55
56	63 07.7	30.2	38.2	63 54.4	28.9	39.5	64 40.3	27.6	40.8	65 25.2	26.1	42.2	66 09.1	24.6	43.8	66 51.9	23.0·	45.4	67 33.4	21.2·	47.1	68 13.5	19.3·	48.9	56
57	62 37.5	31.6	36.3	63 25.5	30.5	37.5	64 12.7	29.3	38.8	64 59.1	28.0	40.1	65 44.5	26.6	41.5	66 28.9	25.0·	43.0	67 12.2	23.4·	44.7	67 54.2	21.6·	46.4	57
58	62 05.9	33.1	34.5	62 55.0	32.0	35.6	63 43.4	30.8	36.8	64 31.1	29.6	38.0	65 17.9	28.3	39.3	66 03.9	27.0	40.8	66 48.8	25.4·	42.3	67 32.6	23.8·	43.9	58
59	61 32.8	34.4	32.7	62 23.0	33.4	33.7	63 12.6	32.4	34.8	64 01.5	31.3	36.0	64 49.6	30.0	37.3	65 36.9	28.7	38.6	66 23.4	27.4	40.0	67 08.8	25.8·	41.5	59
60	60 58.4	-35.6	31.0	61 49.6	-34.7	32.0	62 40.2	-33.8	33.0	63 30.2	-32.7	34.1	64 19.6	-31.7	35.2	65 08.2	-30.5	36.5	65 56.0	-29.1	37.8	66 43.0	-27.8	39.2	60
61	60 22.8	36.8	29.4	61 14.9	36.0	30.3	62 06.4	35.1	31.2	62 57.5	34.2	32.2	63 47.9	33.1	33.3	64 37.7	32.0	34.5	65 26.9	30.9	35.7	66 15.2	29.6	37.0	61
62	59 46.0	37.9	27.8	60 38.9	37.2	28.6	61 31.3	36.3	29.5	62 23.3	35.5	30.4	63 14.8	34.6	31.4	64 05.7	33.6	32.5	64 56.0	32.5	33.6	65 45.6	31.4	34.9	62
63	59 08.1	38.9	26.3	60 01.7	38.2	27.0	60 55.0	37.5	27.8	61 47.8	36.7	28.7	62 40.2	35.9	29.6	63 32.1	35.0	30.6	64 23.5	34.0	31.7	65 14.2	32.9	32.8	63
64	58 29.2	39.9	24.8	59 23.5	39.3	25.5	60 17.5	38.6	26.2	61 11.1	37.9	27.0	62 04.3	37.1	27.9	62 57.1	36.3	28.8	63 49.5	35.5	29.8	64 41.3	34.5	30.8	64
65	57 49.3	-40.8	23.4	58 44.2	-40.2	24.0	59 38.9	-39.7	24.7	60 33.2	-39.0	25.5	61 27.2	-38.3	26.2	62 20.8	-37.5	27.1	63 14.0	-36.7	28.0	64 06.8	-35.9	28.9	65
66	57 08.5	41.7	22.0	58 04.0	41.2	22.6	58 59.2	40.6	23.2	59 54.2	40.0	23.9	60 48.9	39.4	24.6	61 43.3	38.8	25.4	62 37.3	38.0	26.2	63 30.9	37.2	27.1	66
67	56 26.8	42.5	20.7	57 22.8	42.0	21.2	58 18.6	41.5	21.8	59 14.2	41.0	22.5	60 09.5	40.4	23.1	61 04.5	39.7	23.8	61 59.3	39.2	24.6	62 53.7	38.5	25.4	67
68	55 44.3	43.2	19.4	56 40.8	42.8	19.9	57 37.1	42.3	20.5	58 33.2	41.8	21.0	59 29.1	41.4	21.6	60 24.8	40.9	22.3	61 20.1	40.2	23.0	62 15.2	39.6	23.7	68
69	55 01.1	43.9	18.2	55 58.0	43.5	18.7	56 54.8	43.2	19.2	57 51.4	42.8	19.7	58 47.7	42.2	20.2	59 43.9	41.7	20.8	60 39.9	41.2	21.5	61 35.6	40.6	22.1	69
70	54 17.2	-44.6	17.0	55 14.5	-44.3	17.5	56 11.6	-43.9	17.9	57 08.6	-43.4	18.4	58 05.5	-43.1	18.9	59 02.2	-42.7	19.4	59 58.7	-42.2	20.0	60 55.0	-41.7	20.6	70
71	53 32.6	45.3	15.9	54 30.2	44.9	16.3	55 27.7	44.5	16.7	56 25.2	44.3	17.1	57 22.4	43.8	17.6	58 19.5	43.4	18.1	59 16.5	43.0	18.6	60 13.3	42.6	19.1	71
72	52 47.3	45.8	14.8	53 45.3	45.5	15.1	54 43.2	45.3	15.5	55 40.9	44.9	15.9	56 38.6	44.6	16.3	57 36.1	44.2	16.8	58 33.5	43.9	17.2	59 30.7	43.5	17.7	72
73	52 01.5	46.3	13.7	52 59.8	46.1	14.1	53 57.9	45.8	14.4	54 56.0	45.5	14.7	55 54.0	45.3	15.1	56 51.9	45.0	15.5	57 49.6	44.6	15.9	58 47.2	44.2	16.4	73
74	51 15.2	46.9	12.7	52 13.7	46.7	13.0	53 12.1	46.4	13.3	54 10.5	46.2	13.6	55 08.7	45.9	14.0	56 06.9	45.6	14.3	57 05.0	45.3	14.7	58 03.0	45.0	15.1	74
75	50 28.3	-47.4	11.7	51 27.0	-47.1	12.0	52 25.7	-47.0	12.3	53 24.3	-46.7	12.5	54 22.8	-46.5	12.8	55 21.3	-46.3	13.2	56 19.7	-46.0	13.5	57 18.0	-45.7	13.9	75
76	49 40.9	47.8	10.8	50 39.9	47.7	11.0	51 38.7	47.4	11.2	52 37.6	47.3	11.5	53 36.3	47.0	11.8	54 35.0	46.8	12.0	55 33.7	46.6	12.4	56 32.3	46.4	12.7	76
77	48 53.1	48.2	9.8	49 52.2	48.1	10.1	50 51.3	47.9	10.3	51 50.3	47.7	10.5	52 49.3	47.6	10.7	53 48.2	47.4	11.0	54 47.1	47.2	11.2	55 45.9	47.0	11.5	77
78	48 04.9	48.7	9.0	49 04.1	48.5	9.1	50 03.4	48.4	9.3	51 02.6	48.3	9.5	52 01.7	48.1	9.7	53 00.8	47.9	10.0	53 59.9	47.7	10.2	54 58.9	47.5	10.4	78
79	47 16.2	49.0	8.1	48 15.6	48.9	8.2	49 15.0	48.8	8.4	50 14.3	48.6	8.6	51 13.6	48.5	8.8	52 12.9	48.3	9.0	53 12.2	48.2	9.2	54 11.4	48.1	9.4	79
80	46 27.2	-49.4	7.2	47 26.7	-49.3	7.4	48 26.2	-49.2	7.5	49 25.7	-49.1	7.7	50 25.1	-48.9	7.8	51 24.6	-48.9	8.0	52 24.0	-48.7	8.2	53 23.3	-48.5	8.4	80
81	45 37.8	49.7	6.4	46 37.4	49.6	6.5	47 37.0	49.5	6.7	48 36.6	49.5	6.8	49 36.2	49.4	6.9	50 35.7	49.2	7.1	51 35.3	49.2	7.2	52 34.8	49.1	7.4	81
82	44 48.1	50.1	5.6	45 47.8	50.0	5.7	46 47.5	49.9	5.8	47 47.1	49.8	5.9	48 46.8	49.7	6.1	49 46.5	49.7	6.2	50 46.1	49.5	6.3	51 45.7	49.4	6.5	82
83	43 58.0	50.4	4.9	44 57.8	50.3	4.9	45 57.6	50.3	5.0	46 57.3	50.1	5.1	47 57.1	50.1	5.2	48 56.8	50.0	5.3	49 56.6	50.0	5.4	50 56.3	49.9	5.5	83
84	43 07.6	50.6	4.1	44 07.5	50.6	4.2	45 07.3	50.5	4.2	46 07.2	50.5	4.3	47 07.0	50.4	4.4	48 06.8	50.4	4.5	49 06.6	50.3	4.6	50 06.4	50.2	4.7	84
85	42 17.0	-50.9	3.4	43 16.9	-50.9	3.4	44 16.8	-50.9	3.5	45 16.7	-50.8	3.6	46 16.6	-50.8	3.6	47 16.4	-50.7	3.7	48 16.3	-50.6	3.8	49 16.2	-50.6	3.8	85
86	41 26.1	51.2	2.7	42 26.0	51.1	2.7	43 25.9	51.1	2.8	44 25.9	51.1	2.8	45 25.8	51.0	2.8	46 25.7	51.0	2.9	47 25.7	51.0	3.0	48 25.6	51.0	3.0	86
87	40 34.9	51.4	2.0	41 34.9	51.4	2.0	42 34.8	51.3	2.0	43 34.8	51.3	2.1	44 34.8	51.4	2.1	45 34.7	51.3	2.1	46 34.7	51.3	2.2	47 34.6	51.2	2.2	87
88	39 43.5	51.6	1.3	40 43.5	51.6	1.3	41 43.5	51.7	1.3	42 43.5	51.7	1.4	43 43.4	51.6	1.4	44 43.4	51.6	1.4	45 43.4	51.6	1.4	46 43.4	51.6	1.5	88
89	38 51.9	51.9	0.6	39 51.9	51.9	0.7	40 51.8	51.8	0.7	41 51.8	51.8	0.7	42 51.8	51.8	0.7	43 51.8	51.8	0.7	44 51.8	51.8	0.7	45 51.8	51.8	0.7	89
90	38 00.0	-52.1	0.0	39 00.0	-52.1	0.0	40 00.0	-52.1	0.0	41 00.0	-52.1	0.0	42 00.0	-52.1	0.0	43 00.0	-52.1	0.0	44 00.0	-52.1	0.0	45 00.0	-52.1	0.0	90
	38°			39°			40°			41°			42°			43°			44°			45°			

30°, 330° L.H.A. LATITUDE SAME NAME AS DECLINATION

Table 42C

LATITUDE CONTRARY NAME TO DECLINATION — L.H.A. 60°, 300°

Dec.	38° Hc	d	Z	39° Hc	d	Z	40° Hc	d	Z	41° Hc	d	Z	42° Hc	d	Z	43° Hc	d	Z	44° Hc	d	Z	45° Hc	d	Z	Dec.
°	° ′	′	°	° ′	′	°	° ′	′	°	° ′	′	°	° ′	′	°	° ′	′	°	° ′	′	°	° ′	′	°	°
0	23 12.2	-40.3	109.6	22 51.9	-41.1	110.0	22 31.3	-41.9	110.4	22 10.2	-42.6	110.7	21 48.8	-43.4	111.1	21 27.0	-44.1	111.5	21 04.8	-44.8	111.9	20 42.3	-45.5	112.2	0
1	22 31.9	40.5	110.4	22 10.8	41.3	110.8	21 49.4	42.1	111.1	21 27.6	42.8	111.5	21 05.4	43.5	111.9	20 42.9	44.2	112.2	20 20.0	44.9	112.6	19 56.8	45.6	112.9	1
2	21 51.4	40.8	111.2	21 29.5	41.5	111.5	21 07.3	42.3	111.9	20 44.8	43.0	112.3	20 21.9	43.7	112.6	19 58.7	44.5	112.9	19 35.1	45.1	113.3	19 11.2	45.7	113.6	2
3	21 10.6	41.0	112.0	20 48.0	41.8	112.3	20 25.0	42.4	112.7	20 01.8	43.2	113.0	19 38.2	43.9	113.3	19 14.2	44.5	113.7	18 50.0	45.2	114.0	18 25.5	45.9	114.3	3
4	20 29.6	41.2	112.7	20 06.2	41.9	113.1	19 42.6	42.7	113.4	19 18.6	43.4	113.7	18 54.3	44.1	114.1	18 29.7	44.8	114.4	18 04.8	45.4	114.7	17 39.6	46.0	115.0	4
5	19 48.4	-41.4	113.5	19 24.3	-42.1	113.8	18 59.9	-42.8	114.2	18 35.2	-43.5	114.5	18 10.2	-44.2	114.8	17 44.9	-44.9	115.1	17 19.4	-45.6	115.3	16 53.6	-46.2	115.6	5
6	19 07.0	41.6	114.3	18 42.2	42.3	114.6	18 17.1	43.0	114.9	17 51.7	43.7	115.2	17 26.0	44.4	115.5	17 00.0	45.0	115.8	16 33.8	45.6	116.0	16 07.4	46.3	116.3	6
7	18 25.4	41.8	115.0	17 59.9	42.5	115.3	17 34.1	43.2	115.6	17 08.0	43.9	115.9	16 41.6	44.5	116.2	16 15.0	45.1	116.4	15 48.2	45.8	116.7	15 21.1	46.4	117.0	7
8	17 43.6	41.9	115.8	17 17.4	42.6	116.1	16 50.9	43.3	116.4	16 24.1	44.0	116.6	15 57.1	44.6	116.9	15 29.9	45.3	117.1	15 02.4	45.9	117.4	14 34.7	46.5	117.6	8
9	17 01.7	42.1	116.5	16 34.8	42.8	116.8	16 07.6	43.5	117.1	15 40.1	44.1	117.3	15 12.5	44.8	117.6	14 44.6	45.4	117.8	14 16.5	46.0	118.0	13 48.2	46.6	118.3	9
10	16 19.6	-42.3	117.3	15 52.0	-43.0	117.5	15 24.1	-43.6	117.8	14 56.0	-44.2	118.0	14 27.7	-44.9	118.3	13 59.2	-45.5	118.5	13 30.5	-46.1	118.7	13 01.6	-46.7	118.9	10
11	15 37.3	42.4	118.0	15 09.0	43.1	118.3	14 40.5	43.7	118.5	14 11.8	44.4	118.7	13 42.8	45.0	118.9	13 13.7	45.6	119.2	12 44.4	46.3	119.4	12 14.9	46.9	119.6	11
12	14 54.9	42.5	118.8	14 25.9	43.2	119.0	13 56.8	43.9	119.2	13 27.4	44.5	119.4	12 57.8	45.1	119.6	12 28.1	45.7	119.8	11 58.1	46.3	120.0	11 28.0	46.8	120.2	12
13	14 12.4	42.7	119.5	13 42.7	43.3	119.7	13 12.9	44.0	119.9	12 42.9	44.6	120.1	12 12.7	45.2	120.3	11 42.4	45.8	120.5	11 11.8	46.4	120.7	10 41.2	47.0	120.8	13
14	13 29.7	42.8	120.2	12 59.4	43.4	120.4	12 28.9	44.0	120.6	11 58.3	44.7	120.8	11 27.5	45.3	121.0	10 56.6	45.9	121.1	10 25.4	46.4	121.3	9 54.2	47.0	121.5	14
15	12 46.9	-42.9	120.9	12 16.0	-43.6	121.1	11 44.9	-44.2	121.3	11 13.6	-44.8	121.5	10 42.2	-45.4	121.6	10 10.7	-46.0	121.8	9 39.0	-46.6	121.9	9 07.2	-47.1	122.1	15
16	12 04.0	43.1	121.6	11 32.4	43.6	121.8	11 00.7	44.3	122.0	10 28.8	44.8	122.2	9 56.8	45.4	122.3	9 24.7	46.0	122.5	8 52.4	46.6	122.6	8 20.1	47.2	122.7	16
17	11 20.9	43.1	122.4	10 48.8	43.8	122.5	10 16.4	44.3	122.7	9 44.0	45.0	122.8	9 11.4	45.6	123.0	8 38.7	46.1	123.1	8 05.8	46.6	123.2	7 32.9	47.2	123.3	17
18	10 37.8	43.2	123.1	10 05.0	43.8	123.2	9 32.1	44.5	123.4	8 59.0	45.0	123.5	8 25.8	45.6	123.6	7 52.6	46.2	123.7	7 19.2	46.7	123.9	6 45.7	47.3	124.0	18
19	9 54.6	43.3	123.8	9 21.2	43.9	123.9	8 47.6	44.5	124.0	8 14.0	45.1	124.2	7 40.2	45.6	124.3	7 06.4	46.2	124.4	6 32.5	46.8	124.5	5 58.4	47.2	124.6	19
20	9 11.3	-43.4	124.5	8 37.3	-44.0	124.6	8 03.1	-44.5	124.7	7 28.9	-45.1	124.8	6 54.6	-45.7	124.9	6 20.2	-46.3	125.0	5 45.7	-46.8	125.1	5 11.2	-47.4	125.2	20
21	8 27.9	43.5	125.2	7 53.3	44.1	125.3	7 18.6	44.7	125.4	6 43.8	45.2	125.5	6 08.9	45.8	125.6	5 33.9	46.3	125.7	4 58.9	46.8	125.8	4 23.8	47.3	125.8	21
22	7 44.4	43.5	125.9	7 09.2	44.1	126.0	6 33.9	44.7	126.1	5 58.6	45.3	126.2	5 23.1	45.8	126.2	4 47.6	46.3	126.3	4 12.1	46.9	126.4	3 36.5	47.4	126.4	22
23	7 00.9	43.6	126.6	6 25.1	44.2	126.7	5 49.2	44.7	126.7	5 13.3	45.3	126.8	4 37.3	45.8	126.9	4 01.3	46.4	127.0	3 25.2	46.9	127.0	2 49.1	47.4	127.0	23
24	6 17.3	43.7	127.3	5 40.9	44.2	127.3	5 04.5	44.8	127.4	4 28.0	45.3	127.5	3 51.5	45.9	127.5	3 14.9	46.4	127.6	2 38.3	46.9	127.6	2 01.7	47.5	127.7	24
25	5 33.6	-43.7	127.9	4 56.7	-44.2	128.0	4 19.7	-44.8	128.1	3 42.7	-45.3	128.1	3 05.6	-45.9	128.2	2 28.5	-46.4	128.2	1 51.4	-46.9	128.3	1 14.2	-47.4	128.3	25
26	4 49.9	43.7	128.6	4 12.5	44.3	128.7	3 34.9	44.8	128.7	2 57.4	45.4	128.8	2 19.7	45.9	128.8	1 42.1	46.4	128.9	1 04.5	47.0	128.9	0 26.8	-47.4	128.9	26
27	4 06.2	43.8	129.3	3 28.2	44.4	129.4	2 50.1	44.9	129.4	2 12.0	45.4	129.4	1 33.8	45.9	129.5	0 55.7	46.4	129.5	0 17.5	-46.9	129.5	0 20.6	+47.5	50.5	27
28	3 22.4	43.8	130.0	2 43.8	44.3	130.0	2 05.2	44.9	130.1	1 26.6	45.4	130.1	0 47.9	45.9	130.1	0 09.3	-46.5	130.1	0 29.4	+46.9	49.9	1 00.1	47.4	49.9	28
29	2 38.6	43.8	130.7	1 59.5	44.4	130.7	1 20.3	44.8	130.7	0 41.2	-45.4	130.8	0 02.0	-45.9	130.8	0 37.2	+46.4	49.2	1 16.3	47.0	49.3	1 55.5	47.4	49.3	29
30	1 54.8	-43.8	131.4	1 15.1	-44.3	131.4	0 35.5	-44.9	131.4	0 04.2	+45.4	48.6	0 43.9	+45.9	48.6	1 23.6	+46.4	48.6	2 03.3	+46.9	48.6	2 42.9	+47.4	48.7	30
31	1 11.0	43.9	132.1	0 30.8	-44.4	132.1	0 09.4	+44.9	47.9	0 49.6	45.4	47.9	1 29.8	45.9	48.0	2 10.0	46.4	48.0	2 50.2	46.9	48.0	3 30.3	47.4	48.0	31
32	0 27.1	-43.9	132.7	0 13.6	+44.4	47.3	0 54.3	44.9	47.3	1 35.0	45.4	47.3	2 15.7	45.9	47.3	2 56.4	46.4	47.3	3 37.1	46.8	47.4	4 17.7	47.3	47.4	32
33	0 16.8	+43.8	46.6	0 58.0	44.4	46.6	1 39.2	44.9	46.6	2 20.4	45.4	46.6	3 01.6	45.9	46.7	3 42.8	46.3	46.7	4 23.9	46.8	46.8	5 05.0	47.3	46.8	33
34	1 00.6	43.8	45.9	1 42.4	44.3	45.9	2 24.1	44.8	45.9	3 05.8	45.4	46.0	3 47.5	45.8	46.0	4 29.1	46.4	46.1	5 10.7	46.8	46.1	5 52.3	47.3	46.2	34
35	1 44.4	+43.9	45.2	2 26.7	+44.3	45.2	3 08.9	+44.9	45.3	3 51.2	+45.3	45.3	4 33.3	+45.8	45.4	5 15.5	+46.2	45.4	5 57.5	+46.8	45.5	6 39.6	+47.2	45.6	35
36	2 28.3	43.8	44.5	3 11.0	44.3	44.6	3 53.8	44.8	44.6	4 36.5	45.3	44.7	5 19.1	45.8	44.7	6 01.7	46.3	44.8	6 44.3	46.7	44.9	7 26.8	47.1	45.0	36
37	3 12.1	43.8	43.8	3 55.3	44.3	43.9	4 38.6	44.7	43.9	5 21.8	45.2	44.0	6 04.9	45.7	44.1	6 48.0	46.1	44.1	7 31.0	46.6	44.2	8 13.9	47.1	44.3	37
38	3 55.9	43.7	43.2	4 39.6	44.3	43.2	5 23.3	44.7	43.3	6 07.0	45.2	43.3	6 50.6	45.7	43.4	7 34.1	46.2	43.5	8 17.6	46.6	43.6	9 01.0	47.1	43.7	38
39	4 39.6	43.7	42.5	5 23.9	44.1	42.5	6 08.0	44.7	42.6	6 52.2	45.1	42.7	7 36.3	45.6	42.8	8 20.3	46.0	42.9	9 04.2	46.5	43.0	9 48.1	47.0	43.1	39
40	5 23.3	+43.7	41.8	6 08.0	+44.2	41.9	6 52.7	+44.6	41.9	7 37.3	+45.1	42.0	8 21.9	+45.5	42.1	9 06.3	+46.0	42.2	9 50.7	+46.5	42.3	10 35.1	+46.9	42.4	40
41	6 07.0	43.6	41.1	6 52.2	44.1	41.2	7 37.3	44.6	41.3	8 22.4	45.0	41.3	9 07.4	45.5	41.5	9 52.3	46.0	41.6	10 37.2	46.4	41.7	11 22.0	46.8	41.8	41
42	6 50.6	43.5	40.4	7 36.3	44.0	40.5	8 21.9	44.4	40.6	9 07.4	44.9	40.7	9 52.9	45.4	40.8	10 38.3	45.8	40.9	11 23.6	46.2	41.0	12 08.8	46.7	41.2	42
43	7 34.1	43.5	39.7	8 20.3	43.9	39.8	9 06.3	44.4	39.9	9 52.3	44.9	40.0	10 38.3	45.3	40.1	11 24.1	45.7	40.2	12 09.8	46.2	40.4	12 55.5	46.6	40.5	43
44	8 17.6	43.4	39.0	9 04.2	43.9	39.1	9 50.7	44.4	39.2	10 37.2	44.8	39.3	11 23.6	45.2	39.5	12 09.8	45.7	39.6	12 56.0	46.1	39.7	13 42.1	46.6	39.9	44
45	9 01.0	+43.4	38.3	9 48.1	+43.8	38.4	10 35.1	+44.2	38.5	11 22.0	+44.6	38.7	12 08.8	+45.1	38.8	12 55.5	+45.5	38.9	13 42.1	+46.0	39.1	14 28.7	+46.4	39.2	45
46	9 44.4	43.2	37.6	10 31.9	43.7	37.7	11 19.3	44.1	37.8	12 06.6	44.6	38.0	12 53.9	45.0	38.1	13 41.0	45.5	38.3	14 28.1	45.9	38.4	15 15.1	46.3	38.6	46
47	10 27.6	43.2	36.9	11 15.6	43.6	37.0	12 03.4	44.0	37.2	12 51.2	44.5	37.3	13 38.9	44.9	37.4	14 26.5	45.3	37.6	15 14.0	45.7	37.7	16 01.4	46.2	37.9	47
48	11 10.8	43.0	36.2	11 59.2	43.4	36.3	12 47.4	44.0	36.5	13 35.7	44.3	36.6	14 23.8	44.8	36.7	15 11.8	45.2	36.9	15 59.7	45.7	37.1	16 47.6	46.0	37.3	48
49	11 53.8	43.0	35.5	12 42.8	43.4	35.6	13 31.4	43.6	35.8	14 20.0	44.0	35.9	15 08.6	44.6	36.1	15 57.0	45.1	36.2	16 45.4	45.5	36.4	17 33.6	45.9	36.6	49
50	12 36.8	+42.8	34.8	13 26.0	+43.3	34.9	14 15.2	+43.7	35.1	15 04.2	+44.1	35.2	15 53.2	+44.5	35.4	16 42.1	+44.9	35.5	17 30.9	+45.3	35.7	18 19.5	+45.8	35.9	50
51	13 19.6	42.7	34.1	14 09.3	43.1	34.2	14 58.9	43.5	34.3	15 48.3	44.0	34.5	16 37.7	44.4	34.7	17 27.0	44.8	34.8	18 16.2	45.2	35.0	19 05.3	45.6	35.2	51
52	14 02.3	42.6	33.3	14 52.4	43.0	33.5	15 42.4	43.4	33.6	16 32.3	43.8	33.8	17 22.1	44.3	34.0	18 11.8	44.7	34.1	19 01.4	45.1	34.3	19 50.9	45.5	34.5	52
53	14 44.9	42.4	32.6	15 35.4	42.8	32.8	16 25.8	43.3	32.9	17 16.1	43.7	33.1	18 06.4	44.0	33.3	18 56.5	44.5	33.4	19 46.5	44.9	33.6	20 36.4	45.3	33.8	53
54	15 27.3	42.3	31.9	16 18.2	42.7	32.0	17 09.1	43.1	32.2	17 59.8	43.5	32.4	18 50.4	43.9	32.5	19 41.0	44.3	32.7	20 31.4	44.7	32.9	21 21.7	45.1	33.1	54
55	16 09.6	+42.2	31.1	17 00.9	+42.6	31.3	17 52.2	+42.9	31.5	18 43.3	+43.3	31.6	19 34.3	+43.8	31.8	20 25.3	+44.1	32.0	21 16.1	+44.5	32.2	22 06.8	+44.9	32.4	55
56	16 51.8	42.0	30.4	17 43.5	42.4	30.6	18 35.1	42.8	30.7	19 26.6	43.2	30.9	20 18.1	43.5	31.1	21 09.4	43.9	31.3	22 00.6	44.4	31.5	22 51.7	44.8	31.7	56
57	17 33.8	41.8	29.7	18 25.9	42.2	29.8	19 17.9	42.6	30.0	20 09.8	43.0	30.2	21 01.6	43.4	30.4	21 53.3	43.8	30.6	22 45.0	44.1	30.8	23 36.5	44.5	31.0	57
58	18 15.6	41.6	28.9	19 08.1	42.0	29.1	20 00.5	42.3	29.2	20 52.8	42.7	29.4	21 45.0	43.1	29.6	22 37.1	43.5	29.8	23 29.1	43.9	30.0	24 21.0	44.3	30.2	58
59	18 57.2	41.4	28.1	19 50.1	41.8	28.3	20 42.8	42.2	28.5	21 35.5	42.6	28.7	22 28.1	43.0	28.9	23 20.6	43.4	29.1	24 13.0	43.7	29.3	25 05.3	44.1	29.5	59
60	19 38.6	+41.3	27.4	20 31.9	+41.6	27.5	21 25.0	+42.0	27.7	22 18.1	+42.4	27.9	23 11.1	+42.7	28.1	24 04.0	+43.1	28.3	24 56.7	+43.5	28.5	25 49.4	+43.8	28.8	60
61	20 19.9	41.0	26.6	21 13.5	41.4	26.8	22 07.0	41.8	26.9	23 00.5	42.1	27.1	23 53.8	42.5	27.3	24 47.1	42.0	27.5	25 40.2	43.3	27.8	26 33.2	43.7	28.0	61
62	21 00.9	40.9	25.8	21 54.9	41.2	26.0	22 48.8	41.5	26.2	23 42.6	41.9	26.4	24 36.3	42.3	26.6	25 29.9	42.7	26.8	26 23.5	42.9	27.0	27 16.9	43.3	27.2	62
63	21 41.8	40.6	25.0	22 36.1	40.9	25.2	23 30.3	41.3	25.4	24 24.5	41.7	25.6	25 18.6	42.0	25.8	26 12.6	42.3	26.0	27 06.4	42.8	26.2	28 00.2	43.1	26.4	63
64	22 22.4	40.3	24.2	23 17.0	40.8	24.4	24 11.6	41.1	24.6	25 06.2	41.4	24.8	26 00.6	41.7	25.0	26 54.9	42.1	25.2	27 49.2	42.4	25.4	28 43.3	42.8	25.7	64
65	23 02.7	+40.2	23.4	23 57.8	+40.4	23.6	24 52.7	+40.8	23.8	25 47.6	+41.1	24.0	26 42.3	+41.5	24.2	27 37.0	+41.8	24.4	28 31.6	+42.2	24.6	29 26.1	+42.5	24.9	65
66	23 42.9	39.8	22.6	24 38.2	40.2	22.8	25 33.5	40.5	23.0	26 28.7	40.8	23.2	27 23.8	41.2	23.4	28 18.8	41.6	23.6	29 13.8	41.8	23.8	30 00.6	42.2	24.0	66
67	24 22.7	39.6	21.8	25 18.4	39.9	22.0	26 14.0	40.3	22.2	27 09.5	40.6	22.4	28 05.0	40.9	22.6	29 00.4	41.2	22.8	29 55.6	41.6	23.0	30 50.8	41.9	23.2	67
68	25 02.3	39.4	21.0	25 58.3	39.7	21.2	26 54.3	39.9	21.3	27 50.1	40.3	21.5	28 45.9	40.6	21.7	29 41.6	40.9	21.9	30 37.2	41.2	22.1	31 32.7	41.6	22.4	68
69	25 41.7	39.0	20.1	26 38.0	39.3	20.3	27 34.2	39.6	20.5	28 30.4	39.9	20.7	29 26.5	40.2	20.9	30 22.5	40.6	21.1	31 18.4	40.9	21.3	32 14.3	41.2	21.5	69
70	26 20.7	+38.7	19.3	27 17.3	+39.0	19.5	28 13.8	+39.4	19.6	29 10.3	+39.6	19.8	30 06.7	+39.9	20.0	31 03.1	+40.2	20.2	31 59.3	+40.6	20.4	32 55.5	+40.9	20.7	70
71	26 59.4	38.5	18.4	27 56.3	38.7	18.6	28 53.2	38.9	18.8	29 49.9	39.3	19.0	30 46.6	39.6	19.2	31 43.3	39.9	19.4	32 39.9	40.1	19.6	33 36.4	40.4	19.8	71
72	27 37.9	38.1	17.6	28 35.0	38.4	17.7	29 32.1	38.7	17.9	30 29.2	38.9	18.1	31 26.2	39.2	18.3	32 23.2	39.4	18.5	33 20.0	39.8	18.7	34 16.8	40.1	18.9	72
73	28 16.0	37.7	16.7	29 13.4	38.0	16.9	30 10.8	38.3	17.0	31 08.1	38.6	17.2	32 05.4	38.8	17.4	33 02.6	39.2	17.6	33 59.8	39.4	17.8	34 56.9	39.7	18.0	73
74	28 53.7	37.4	15.8	29 51.4	37.7	16.0	30 49.1	37.9	16.1	31 46.7	38.2	16.3	32 44.2	38.5	16.5	33 41.8	38.7	16.7	34 39.2	39.0	16.9	35 36.6	39.2	17.1	74
75	29 31.1	+37.1	14.9	30 29.1	+37.3	15.1	31 27.0	+37.5	15.2	32 24.9	+37.7	15.4	33 22.7	+38.0	15.6	34 20.5	+38.2	15.8	35 18.2	+38.5	15.9	36 15.8	+38.8	16.1	75
76	30 08.2	36.6	14.0	31 06.4	36.9	14.2	32 04.5	37.1	14.3	33 02.6	37.4	14.5	34 00.7	37.6	14.6	34 58.7	37.9	14.8	35 56.7	38.1	15.0	36 54.6	38.4	15.2	76
77	30 44.8	36.3	13.1	31 43.3	36.4	13.2	32 41.6	36.7	13.4	33 40.0	36.9	13.5	34 38.3	37.2	13.7	35 36.6	37.4	13.9	36 34.8	37.6	14.0	37 33.0	37.9	14.2	77
78	31 21.1	35.9	12.2	32 19.7	36.1	12.3	33 18.3	36.3	12.4	34 16.9	36.5	12.6	35 15.5	36.7	12.7	36 14.0	36.9	12.9	37 12.4	37.2	13.1	38 10.9	37.3	13.2	78
79	31 57.0	35.4	11.2	32 55.8	35.6	11.4	33 54.6	35.8	11.5	34 53.4	36.0	11.6	35 52.2	36.2	11.8	36 50.9	36.4	11.9	37 49.6	36.6	12.1	38 48.2	36.9	12.2	79
80	32 32.4	+35.0	10.3	33 31.4	+35.2	10.4	34 30.4	+35.4	10.5	35 29.4	+35.6	10.6	36 28.4	+35.7	10.8	37 27.3	+35.9	10.9	38 26.2	+36.1	11.1	39 25.1	+36.3	11.2	80
81	33 07.4	34.6	9.3	34 06.6	34.7	9.4	35 05.8	34.9	9.5	36 05.0	35.0	9.7	37 04.1	35.2	9.8	38 03.2	35.4	9.9	39 02.3	35.6	10.0	40 01.4	35.7	10.2	81
82	33 42.0	34.0	8.3	34 41.3	34.2	8.4	35 40.7	34.3	8.5	36 40.0	34.5	8.6	37 39.3	34.7	8.8	38 38.6	34.8	8.9	39 37.9	35.0	9.0	40 37.1	35.2	9.1	82
83	34 16.0	33.6	7.3	35 15.5	33.8	7.4	36 15.0	33.9	7.5	37 14.5	34.0	7.6	38 14.0	34.1	7.7	39 13.4	34.3	7.8	40 12.9	34.4	7.9	41 12.3	34.5	8.1	83
84	34 49.6	33.1	6.3	35 49.3	33.2	6.4	36 48.9	33.3	6.5	37 48.5	33.4	6.6	38 48.1	33.5	6.7	39 47.7	33.6	6.8	40 47.3	33.7	6.9	41 46.8	33.9	7.0	84
85	35 22.7	+32.6	5.3	36 22.5	+32.6	5.4	37 22.2	+32.8	5.4	38 21.9	+32.9	5.5	39 21.6	+33.0	5.6	40 21.3	+33.1	5.7	41 21.0	+33.2	5.8	42 20.7	+33.3	5.9	85
86	35 55.3	32.0	4.3	36 55.1	32.1	4.3	37 55.0	32.1	4.4	38 54.8	32.2	4.5	39 54.6	32.3	4.5	40 54.4	32.4	4.6	41 54.2	32.5	4.7	42 54.0	32.6	4.7	86
87	36 27.3	31.5	3.2	37 27.2	31.6	3.3	38 27.1	31.6	3.3	39 27.0	31.7	3.4	40 26.9	31.7	3.4	41 26.8	31.8	3.5	42 26.7	31.8	3.5	43 26.6	31.9	3.6	87
88	36 58.8	30.9	2.2	37 58.8	30.9	2.2	38 58.7	31.0	2.2	39 58.7	31.0	2.3	40 58.6	31.1	2.3	41 58.6	31.0	2.3	42 58.5	31.1	2.4	43 58.5	31.1	2.4	88
89	37 29.7	30.3	1.1	38 29.7	30.3	1.1	39 29.7	30.3	1.1	40 29.7	30.3	1.1	41 29.7	30.3	1.2	42 29.6	30.4	1.2	43 29.6	30.4	1.2	44 29.6	30.4	1.2	89
90	38 00.0	+29.7	0.0	39 00.0	+29.7	0.0	40 00.0	+29.7	0.0	41 00.0	+29.7	0.0	42 00.0	+29.6	0.0	43 00.0	+29.6	0.0	44 00.0	+29.6	0.0	45 00.0	+29.6	0.0	90
	38°			39°			40°			41°			42°			43°			44°			45°			

S. Lat. { L.H.A. greater than 180°......Zn=180°−Z; L.H.A. less than 180°............Zn=180°+Z

LATITUDE SAME NAME AS DECLINATION — L.H.A. 120°, 240°

Table 42D

INTERPOLATION TABLE

Dec. Inc.	Tens 10'	20'	30'	40'	50'	Decimals	Units 0'	1'	2'	3'	4'	5'	6'	7'	8'	9'
16.0	2.6	5.3	8.0	10.6	13.3	.0	0.0	0.3	0.5	0.8	1.1	1.4	1.6	1.9	2.2	2.5
16.1	2.7	5.3	8.0	10.7	13.4	.1	0.0	0.3	0.6	0.9	1.1	1.4	1.7	2.0	2.2	2.5
16.2	2.7	5.4	8.1	10.8	13.5	.2	0.1	0.3	0.6	0.9	1.2	1.4	1.7	2.0	2.3	2.5
16.3	2.7	5.4	8.1	10.9	13.6	.3	0.1	0.4	0.6	0.9	1.2	1.5	1.7	2.0	2.3	2.6
16.4	2.7	5.5	8.2	10.9	13.7	.4	0.1	0.4	0.7	0.9	1.2	1.5	1.8	2.0	2.3	2.6
16.5	2.8	5.5	8.3	11.0	13.8	.5	0.1	0.4	0.7	1.0	1.2	1.5	1.8	2.1	2.3	2.6
16.6	2.8	5.5	8.3	11.1	13.8	.6	0.2	0.4	0.7	1.0	1.3	1.5	1.8	2.1	2.4	2.6
16.7	2.8	5.6	8.4	11.2	13.9	.7	0.2	0.5	0.7	1.0	1.3	1.6	1.8	2.1	2.4	2.7
16.8	2.8	5.6	8.4	11.2	14.0	.8	0.2	0.5	0.8	1.0	1.3	1.6	1.9	2.1	2.4	2.7
16.9	2.9	5.7	8.5	11.3	14.1	.9	0.2	0.5	0.8	1.1	1.3	1.6	1.9	2.2	2.4	2.7
17.0	2.8	5.6	8.5	11.3	14.1	.0	0.0	0.3	0.6	0.9	1.2	1.5	1.7	2.0	2.3	2.6
17.1	2.8	5.7	8.5	11.4	14.2	.1	0.0	0.3	0.6	0.9	1.2	1.5	1.8	2.1	2.4	2.7
17.2	2.8	5.7	8.6	11.4	14.3	.2	0.1	0.3	0.6	0.9	1.2	1.5	1.8	2.1	2.4	2.7
17.3	2.9	5.8	8.6	11.5	14.4	.3	0.1	0.4	0.7	1.0	1.3	1.5	1.8	2.1	2.4	2.7
17.4	2.9	5.8	8.7	11.6	14.5	.4	0.1	0.4	0.7	1.0	1.3	1.6	1.9	2.2	2.4	2.7
17.5	2.9	5.8	8.8	11.7	14.6	.5	0.1	0.4	0.7	1.0	1.3	1.6	1.9	2.2	2.5	2.8
17.6	2.9	5.9	8.8	11.7	14.7	.6	0.2	0.5	0.8	1.0	1.3	1.6	1.9	2.2	2.5	2.8
17.7	3.0	5.9	8.9	11.8	14.8	.7	0.2	0.5	0.8	1.1	1.4	1.7	2.0	2.2	2.5	2.8
17.8	3.0	6.0	8.9	11.9	14.9	.8	0.2	0.5	0.8	1.1	1.4	1.7	2.0	2.3	2.6	2.9
17.9	3.0	6.0	9.0	12.0	15.0	.9	0.3	0.6	0.8	1.1	1.4	1.7	2.0	2.3	2.6	2.9
18.0	3.0	6.0	9.0	12.0	15.0	.0	0.0	0.3	0.6	0.9	1.2	1.5	1.8	2.2	2.5	2.8
18.1	3.0	6.0	9.0	12.0	15.1	.1	0.0	0.3	0.6	1.0	1.3	1.6	1.9	2.2	2.5	2.8
18.2	3.0	6.0	9.1	12.1	15.1	.2	0.1	0.4	0.7	1.0	1.3	1.6	1.9	2.2	2.5	2.8
18.3	3.0	6.1	9.1	12.2	15.2	.3	0.1	0.4	0.7	1.0	1.3	1.6	1.9	2.3	2.6	2.9
18.4	3.1	6.1	9.2	12.3	15.3	.4	0.1	0.4	0.7	1.0	1.4	1.7	2.0	2.3	2.6	2.9
18.5	3.1	6.2	9.3	12.3	15.4	.5	0.2	0.5	0.8	1.1	1.4	1.7	2.0	2.3	2.6	2.9
18.6	3.1	6.2	9.3	12.4	15.5	.6	0.2	0.5	0.8	1.1	1.4	1.7	2.0	2.3	2.7	3.0
18.7	3.1	6.3	9.4	12.5	15.6	.7	0.2	0.5	0.8	1.1	1.4	1.8	2.1	2.4	2.7	3.0
18.8	3.2	6.3	9.4	12.6	15.7	.8	0.2	0.6	0.9	1.2	1.5	1.8	2.1	2.4	2.7	3.0
18.9	3.2	6.3	9.5	12.6	15.8	.9	0.3	0.6	0.9	1.2	1.5	1.8	2.1	2.4	2.7	3.1
19.0	3.1	6.3	9.5	12.6	15.8	.0	0.0	0.3	0.6	1.0	1.3	1.6	1.9	2.3	2.6	2.9
19.1	3.2	6.3	9.5	12.7	15.9	.1	0.0	0.4	0.7	1.0	1.3	1.7	2.0	2.3	2.6	3.0
19.2	3.2	6.4	9.6	12.8	16.0	.2	0.1	0.4	0.7	1.0	1.4	1.7	2.0	2.3	2.7	3.0
19.3	3.2	6.4	9.6	12.9	16.1	.3	0.1	0.4	0.7	1.1	1.4	1.7	2.0	2.4	2.7	3.0
19.4	3.2	6.5	9.7	12.9	16.2	.4	0.1	0.5	0.8	1.1	1.4	1.8	2.1	2.4	2.7	3.1
19.5	3.3	6.5	9.8	13.0	16.3	.5	0.2	0.5	0.8	1.1	1.5	1.8	2.1	2.4	2.8	3.1
19.6	3.3	6.5	9.8	13.1	16.3	.6	0.2	0.5	0.8	1.2	1.5	1.8	2.1	2.5	2.8	3.1
19.7	3.3	6.6	9.9	13.2	16.4	.7	0.2	0.6	0.9	1.2	1.5	1.9	2.2	2.5	2.8	3.2
19.8	3.3	6.6	9.9	13.2	16.5	.8	0.3	0.6	0.9	1.2	1.6	1.9	2.2	2.5	2.9	3.2
19.9	3.4	6.7	10.0	13.3	16.6	.9	0.3	0.6	0.9	1.3	1.6	1.9	2.2	2.6	2.9	3.2
20.0	3.3	6.6	10.0	13.3	16.6	.0	0.0	0.3	0.7	1.0	1.4	1.7	2.0	2.4	2.7	3.1
20.1	3.3	6.7	10.0	13.4	16.7	.1	0.0	0.4	0.7	1.1	1.4	1.7	2.1	2.4	2.8	3.1
20.2	3.3	6.7	10.1	13.4	16.8	.2	0.1	0.4	0.8	1.1	1.4	1.8	2.1	2.5	2.8	3.1
20.3	3.4	6.8	10.1	13.5	16.9	.3	0.1	0.4	0.8	1.1	1.5	1.8	2.2	2.5	2.8	3.2
20.4	3.4	6.8	10.2	13.6	17.0	.4	0.1	0.5	0.8	1.2	1.5	1.8	2.2	2.5	2.9	3.2
20.5	3.4	6.8	10.3	13.7	17.1	.5	0.2	0.5	0.9	1.2	1.5	1.9	2.2	2.6	2.9	3.2
20.6	3.4	6.9	10.3	13.7	17.2	.6	0.2	0.5	0.9	1.2	1.6	1.9	2.3	2.6	2.9	3.3
20.7	3.5	6.9	10.4	13.8	17.3	.7	0.2	0.6	0.9	1.3	1.6	1.9	2.3	2.6	3.0	3.3
20.8	3.5	7.0	10.4	13.9	17.4	.8	0.3	0.6	1.0	1.3	1.6	2.0	2.3	2.7	3.0	3.3
20.9	3.5	7.0	10.5	14.0	17.5	.9	0.3	0.6	1.0	1.3	1.7	2.0	2.4	2.7	3.0	3.4
21.0	3.5	7.0	10.5	14.0	17.5	.0	0.0	0.4	0.7	1.1	1.4	1.8	2.1	2.5	2.9	3.2
21.1	3.5	7.0	10.5	14.0	17.6	.1	0.0	0.4	0.8	1.1	1.5	1.8	2.2	2.5	2.9	3.3
21.2	3.5	7.0	10.6	14.1	17.6	.2	0.1	0.4	0.8	1.1	1.5	1.9	2.2	2.6	2.9	3.3
21.3	3.5	7.1	10.6	14.2	17.7	.3	0.1	0.5	0.8	1.2	1.5	1.9	2.3	2.6	3.0	3.3
21.4	3.6	7.1	10.7	14.3	17.8	.4	0.1	0.5	0.9	1.2	1.6	1.9	2.3	2.7	3.0	3.4
21.5	3.6	7.2	10.8	14.3	17.9	.5	0.2	0.5	0.9	1.3	1.6	2.0	2.3	2.7	3.0	3.4
21.6	3.6	7.2	10.8	14.4	18.0	.6	0.2	0.6	0.9	1.3	1.6	2.0	2.4	2.7	3.1	3.4
21.7	3.6	7.3	10.9	14.5	18.1	.7	0.3	0.6	1.0	1.3	1.7	2.0	2.4	2.8	3.1	3.5
21.8	3.7	7.3	10.9	14.6	18.2	.8	0.3	0.6	1.0	1.4	1.7	2.1	2.4	2.8	3.2	3.5
21.9	3.7	7.3	11.0	14.6	18.3	.9	0.3	0.7	1.0	1.4	1.8	2.1	2.5	2.8	3.2	3.5
22.0	3.6	7.3	11.0	14.6	18.3	.0	0.0	0.4	0.7	1.1	1.5	1.9	2.2	2.6	3.0	3.4
22.1	3.7	7.3	11.0	14.7	18.4	.1	0.0	0.4	0.8	1.2	1.5	1.9	2.3	2.7	3.0	3.4
22.2	3.7	7.4	11.1	14.8	18.5	.2	0.1	0.4	0.8	1.2	1.6	1.9	2.3	2.7	3.1	3.4
22.3	3.7	7.4	11.1	14.9	18.6	.3	0.1	0.5	0.9	1.2	1.6	2.0	2.4	2.7	3.1	3.5
22.4	3.7	7.5	11.2	14.9	18.7	.4	0.1	0.5	0.9	1.3	1.6	2.0	2.4	2.8	3.1	3.5
22.5	3.8	7.5	11.3	15.0	18.8	.5	0.2	0.6	0.9	1.3	1.7	2.1	2.4	2.8	3.2	3.6
22.6	3.8	7.5	11.3	15.1	18.8	.6	0.2	0.6	1.0	1.3	1.7	2.1	2.5	2.8	3.2	3.6
22.7	3.8	7.6	11.4	15.2	18.9	.7	0.3	0.6	1.0	1.4	1.8	2.1	2.5	2.9	3.3	3.6
22.8	3.8	7.6	11.4	15.2	19.0	.8	0.3	0.7	1.0	1.4	1.8	2.2	2.5	2.9	3.3	3.7
22.9	3.9	7.7	11.5	15.3	19.1	.9	0.3	0.7	1.1	1.5	1.8	2.2	2.6	3.0	3.3	3.7
23.0	3.8	7.6	11.5	15.3	19.1	.0	0.0	0.4	0.8	1.2	1.6	2.0	2.3	2.7	3.1	3.5
23.1	3.8	7.7	11.5	15.4	19.2	.1	0.0	0.4	0.8	1.2	1.6	2.0	2.4	2.8	3.2	3.6
23.2	3.8	7.7	11.6	15.4	19.3	.2	0.1	0.5	0.9	1.3	1.6	2.0	2.4	2.8	3.2	3.6
23.3	3.9	7.8	11.6	15.5	19.4	.3	0.1	0.5	0.9	1.3	1.7	2.1	2.5	2.9	3.3	3.6
23.4	3.9	7.8	11.7	15.6	19.5	.4	0.2	0.5	0.9	1.3	1.7	2.1	2.5	2.9	3.3	3.7
23.5	3.9	7.8	11.8	15.7	19.6	.5	0.2	0.6	1.0	1.4	1.8	2.2	2.5	2.9	3.3	3.7
23.6	3.9	7.9	11.8	15.7	19.7	.6	0.2	0.6	1.0	1.4	1.8	2.2	2.6	3.0	3.4	3.8
23.7	4.0	7.9	11.9	15.8	19.8	.7	0.3	0.7	1.1	1.4	1.8	2.2	2.6	3.0	3.4	3.8
23.8	4.0	8.0	11.9	15.9	19.9	.8	0.3	0.7	1.1	1.5	1.9	2.3	2.7	3.1	3.4	3.8
23.9	4.0	8.0	12.0	16.0	20.0	.9	0.4	0.7	1.1	1.5	1.9	2.3	2.7	3.1	3.5	3.9
	10'	20'	30'	40'	50'		0'	1'	2'	3'	4'	5'	6'	7'	8'	9'

Dec. Inc.	Double Second Diff.	Corr.
16.0–17.9	1.0, 3.0, 4.9, 6.9, 8.9, 10.8, 12.8, 14.8, 16.7, 18.7, 20.7, 22.7, 24.6, 26.6, 28.6, 30.5, 32.5, 34.5	0.1, 0.2, 0.3, 0.4, 0.5, 0.6, 0.7, 0.8, 0.9, 1.0, 1.1, 1.2, 1.3, 1.4, 1.5, 1.6, 1.7
18.0–19.9	0.9, 2.8, 4.6, 6.5, 8.3, 10.2, 12.0, 13.9, 15.7, 17.6, 19.4, 21.3, 23.1, 25.0, 26.8, 28.7, 30.5, 32.3, 34.2	0.1, 0.2, 0.3, 0.4, 0.5, 0.6, 0.7, 0.8, 0.9, 1.0, 1.1, 1.2, 1.3, 1.4, 1.5, 1.6, 1.7, 1.8
20.0–21.9	0.9, 2.6, 4.4, 6.2, 7.9, 9.7, 11.4, 13.2, 14.9, 16.7, 18.5, 20.2, 22.0, 23.7, 25.5, 27.3, 29.0, 30.8, 32.5, 34.3	0.1, 0.2, 0.3, 0.4, 0.5, 0.6, 0.7, 0.8, 0.9, 1.0, 1.1, 1.2, 1.3, 1.4, 1.5, 1.6, 1.7, 1.8, 1.9
22.0–23.9	0.8, 2.5, 4.2, 5.9, 7.6, 9.3, 11.0, 12.7, 14.4, 16.1, 17.8, 19.5, 21.2, 22.8, 24.5, 26.2, 27.9, 29.6, 31.3, 33.0, 34.7	0.1, 0.2, 0.3, 0.4, 0.5, 0.6, 0.7, 0.8, 0.9, 1.0, 1.1, 1.2, 1.3, 1.4, 1.5, 1.6, 1.7, 1.8, 1.9, 2.0

Dec. Inc.	Tens 10'	20'	30'	40'	50'	Decimals	Units 0'	1'	2'	3'	4'	5'	6'	7'	8'	9'
24.0	4.0	8.0	12.0	16.0	20.0	.0	0.0	0.4	0.8	1.2	1.6	2.0	2.4	2.9	3.3	3.7
24.1	4.0	8.0	12.0	16.0	20.1	.1	0.0	0.4	0.9	1.3	1.7	2.1	2.5	2.9	3.3	3.7
24.2	4.0	8.0	12.1	16.1	20.1	.2	0.1	0.5	0.9	1.3	1.7	2.1	2.5	2.9	3.3	3.8
24.3	4.0	8.1	12.1	16.2	20.2	.3	0.1	0.5	0.9	1.3	1.8	2.2	2.6	3.0	3.4	3.8
24.4	4.1	8.1	12.2	16.3	20.3	.4	0.2	0.6	1.0	1.4	1.8	2.2	2.6	3.0	3.4	3.8
24.5	4.1	8.2	12.3	16.3	20.4	.5	0.2	0.6	1.0	1.4	1.8	2.2	2.7	3.1	3.5	3.9
24.6	4.1	8.2	12.3	16.4	20.5	.6	0.2	0.7	1.1	1.5	1.9	2.3	2.7	3.1	3.5	3.9
24.7	4.1	8.3	12.4	16.5	20.6	.7	0.3	0.7	1.1	1.5	1.9	2.3	2.7	3.1	3.6	4.0
24.8	4.2	8.3	12.4	16.6	20.7	.8	0.3	0.7	1.1	1.6	2.0	2.4	2.8	3.2	3.6	4.0
24.9	4.2	8.3	12.5	16.6	20.8	.9	0.4	0.8	1.2	1.6	2.0	2.4	2.8	3.2	3.6	4.0
25.0	4.1	8.3	12.5	16.6	20.8	.0	0.0	0.4	0.8	1.3	1.7	2.1	2.5	3.0	3.4	3.8
25.1	4.2	8.3	12.5	16.7	20.9	.1	0.0	0.5	0.9	1.3	1.7	2.2	2.6	3.0	3.4	3.9
25.2	4.2	8.4	12.6	16.8	21.0	.2	0.1	0.5	0.9	1.4	1.8	2.2	2.6	3.1	3.5	3.9
25.3	4.2	8.4	12.6	16.9	21.1	.3	0.1	0.6	1.0	1.4	1.8	2.3	2.7	3.1	3.5	4.0
25.4	4.2	8.5	12.7	16.9	21.2	.4	0.2	0.6	1.0	1.4	1.9	2.3	2.7	3.1	3.6	4.0
25.5	4.3	8.5	12.8	17.0	21.3	.5	0.2	0.6	1.1	1.5	1.9	2.3	2.8	3.2	3.6	4.0
25.6	4.3	8.5	12.8	17.1	21.3	.6	0.3	0.7	1.1	1.5	2.0	2.4	2.8	3.2	3.7	4.1
25.7	4.3	8.6	12.9	17.2	21.4	.7	0.3	0.7	1.1	1.6	2.0	2.4	2.8	3.3	3.7	4.1
25.8	4.3	8.6	12.9	17.2	21.5	.8	0.3	0.8	1.2	1.6	2.0	2.5	2.9	3.3	3.7	4.2
25.9	4.4	8.7	13.0	17.3	21.6	.9	0.4	0.8	1.2	1.7	2.1	2.5	2.9	3.4	3.8	4.2
26.0	4.3	8.6	13.0	17.3	21.6	.0	0.0	0.4	0.9	1.3	1.8	2.2	2.6	3.1	3.5	4.0
26.1	4.3	8.7	13.0	17.4	21.7	.1	0.0	0.5	0.9	1.4	1.8	2.3	2.7	3.1	3.6	4.0
26.2	4.3	8.7	13.1	17.4	21.8	.2	0.1	0.5	1.0	1.4	1.9	2.3	2.7	3.2	3.6	4.1
26.3	4.4	8.8	13.1	17.5	21.9	.3	0.1	0.6	1.0	1.5	1.9	2.3	2.8	3.2	3.7	4.1
26.4	4.4	8.8	13.2	17.6	22.0	.4	0.2	0.6	1.1	1.5	1.9	2.4	2.8	3.3	3.7	4.2
26.5	4.4	8.8	13.3	17.7	22.1	.5	0.2	0.7	1.1	1.5	2.0	2.4	2.9	3.3	3.8	4.2
26.6	4.4	8.9	13.3	17.7	22.2	.6	0.3	0.7	1.1	1.6	2.0	2.5	2.9	3.4	3.8	4.2
26.7	4.5	8.9	13.4	17.8	22.3	.7	0.3	0.8	1.2	1.6	2.1	2.5	3.0	3.4	3.8	4.3
26.8	4.5	9.0	13.4	17.9	22.4	.8	0.4	0.8	1.2	1.7	2.1	2.6	3.0	3.4	3.9	4.3
26.9	4.5	9.0	13.5	18.0	22.5	.9	0.4	0.8	1.3	1.7	2.2	2.6	3.0	3.5	3.9	4.4
27.0	4.5	9.0	13.5	18.0	22.5	.0	0.0	0.5	0.9	1.4	1.8	2.3	2.7	3.2	3.7	4.1
27.1	4.5	9.0	13.5	18.0	22.6	.1	0.0	0.5	1.0	1.4	1.9	2.3	2.8	3.3	3.7	4.2
27.2	4.5	9.0	13.6	18.1	22.6	.2	0.1	0.5	1.0	1.5	1.9	2.4	2.8	3.3	3.8	4.2
27.3	4.5	9.1	13.6	18.2	22.7	.3	0.1	0.6	1.1	1.5	2.0	2.4	2.9	3.3	3.8	4.3
27.4	4.6	9.1	13.7	18.3	22.8	.4	0.2	0.6	1.1	1.6	2.0	2.5	2.9	3.4	3.8	4.3
27.5	4.6	9.2	13.8	18.3	22.9	.5	0.2	0.7	1.1	1.6	2.1	2.5	3.0	3.4	3.9	4.4
27.6	4.6	9.2	13.8	18.4	23.0	.6	0.3	0.7	1.2	1.6	2.1	2.6	3.0	3.5	3.9	4.4
27.7	4.6	9.3	13.9	18.5	23.1	.7	0.3	0.8	1.2	1.7	2.2	2.6	3.1	3.5	4.0	4.4
27.8	4.7	9.3	13.9	18.6	23.2	.8	0.4	0.8	1.3	1.7	2.2	2.7	3.1	3.6	4.0	4.5
27.9	4.7	9.3	14.0	18.6	23.3	.9	0.4	0.9	1.3	1.8	2.2	2.7	3.2	3.6	4.1	4.5
28.0	4.6	9.3	14.0	18.6	23.3	.0	0.0	0.5	0.9	1.4	1.9	2.4	2.8	3.3	3.8	4.3
28.1	4.7	9.3	14.0	18.7	23.4	.1	0.0	0.5	1.0	1.5	1.9	2.4	2.9	3.4	3.8	4.3
28.2	4.7	9.4	14.1	18.8	23.5	.2	0.1	0.6	1.0	1.5	2.0	2.5	2.9	3.4	3.9	4.4
28.3	4.7	9.4	14.1	18.9	23.6	.3	0.1	0.6	1.1	1.6	2.0	2.5	3.0	3.5	3.9	4.4
28.4	4.7	9.5	14.2	18.9	23.7	.4	0.2	0.7	1.1	1.6	2.1	2.6	3.0	3.5	4.0	4.5
28.5	4.8	9.5	14.3	19.0	23.8	.5	0.2	0.7	1.2	1.7	2.1	2.6	3.1	3.6	4.0	4.5
28.6	4.8	9.5	14.3	19.1	23.8	.6	0.3	0.8	1.2	1.7	2.2	2.7	3.1	3.6	4.1	4.6
28.7	4.8	9.6	14.4	19.2	23.9	.7	0.3	0.8	1.3	1.8	2.2	2.7	3.2	3.7	4.1	4.6
28.8	4.8	9.6	14.4	19.2	24.0	.8	0.4	0.9	1.3	1.8	2.3	2.8	3.2	3.7	4.2	4.7
28.9	4.9	9.7	14:5	19.3	24.1	.9	0.4	0.9	1.4	1.9	2.3	2.8	3.3	3.8	4.2	4.7
29.0	4.8	9.6	14.5	19.3	24.1	.0	0.0	0.5	1.0	1.5	2.0	2.5	2.9	3.4	3.9	4.4
29.1	4.8	9.7	14.5	19.4	24.2	.1	0.0	0.5	1.0	1.5	2.0	2.5	3.0	3.5	4.0	4.5
29.2	4.8	9.7	14.6	19.4	24.3	.2	0.1	0.6	1.1	1.6	2.1	2.6	3.0	3.5	4.0	4.5
29.3	4.9	9.8	14.6	19.5	24.4	.3	0.1	0.6	1.1	1.6	2.1	2.6	3.1	3.6	4.1	4.6
29.4	4.9	9.8	14.7	19.6	24.5	.4	0.2	0.7	1.2	1.7	2.2	2.7	3.1	3.6	4.1	4.6
29.5	4.9	9.8	14.8	19.7	24.6	.5	0.2	0.7	1.2	1.7	2.2	2.7	3.2	3.7	4.2	4.7
29.6	4.9	9.9	14.8	19.7	24.7	.6	0.3	0.8	1.3	1.8	2.3	2.8	3.2	3.7	4.2	4.7
29.7	5.0	9.9	14.9	19.8	24.8	.7	0.3	0.8	1.3	1.8	2.3	2.8	3.3	3.8	4.3	4.8
29.8	5.0	10.0	14.9	19.9	24.9	.8	0.4	0.9	1.4	1.9	2.4	2.9	3.3	3.8	4.3	4.8
29.9	5.0	10.0	15.0	20.0	25.0	.9	0.4	0.9	1.4	1.9	2.4	2.9	3.4	3.9	4.4	4.9
30.0	5.0	10.0	15.0	20.0	25.0	.0	0.0	0.5	1.0	1.5	2.0	2.5	3.0	3.6	4.1	4.6
30.1	5.0	10.0	15.0	20.0	25.1	.1	0.1	0.6	1.1	1.6	2.1	2.6	3.1	3.6	4.1	4.6
30.2	5.0	10.0	15.1	20.1	25.1	.2	0.1	0.6	1.1	1.6	2.1	2.6	3.2	3.7	4.2	4.7
30.3	5.0	10.1	15.1	20.2	25.2	.3	0.2	0.7	1.2	1.7	2.2	2.7	3.2	3.7	4.2	4.7
30.4	5.1	10.1	15.2	20.3	25.3	.4	0.2	0.7	1.2	1.7	2.2	2.7	3.3	3.8	4.3	4.8
30.5	5.1	10.2	15.3	20.3	25.4	.5	0.3	0.8	1.3	1.8	2.3	2.8	3.3	3.8	4.3	4.8
30.6	5.1	10.2	15.3	20.4	25.5	.6	0.3	0.8	1.3	1.8	2.3	2.8	3.4	3.9	4.4	4.9
30.7	5.1	10.3	15.4	20.5	25.6	.7	0.4	0.9	1.4	1.9	2.4	2.9	3.4	3.9	4.4	4.9
30.8	5.2	10.3	15.4	20.6	25.7	.8	0.4	0.9	1.4	1.9	2.4	2.9	3.5	4.0	4.5	5.0
30.9	5.2	10.3	15.5	20.6	25.8	.9	0.5	1.0	1.5	2.0	2.5	3.0	3.5	4.0	4.5	5.0
31.0	5.1	10.3	15.5	20.6	25.8	.0	0.0	0.5	1.0	1.6	2.1	2.6	3.1	3.7	4.2	4.7
31.1	5.2	10.3	15.5	20.7	25.9	.1	0.1	0.6	1.1	1.6	2.2	2.7	3.2	3.7	4.3	4.8
31.2	5.2	10.4	15.6	20.8	26.0	.2	0.1	0.6	1.2	1.7	2.2	2.7	3.3	3.8	4.3	4.8
31.3	5.2	10.4	15.6	20.9	26.1	.3	0.2	0.7	1.2	1.7	2.3	2.8	3.3	3.8	4.4	4.9
31.4	5.2	10.5	15.7	20.9	26.2	.4	0.2	0.7	1.3	1.8	2.3	2.8	3.4	3.9	4.4	4.9
31.5	5.3	10.5	15.8	21.0	26.3	.5	0.3	0.8	1.3	1.8	2.4	2.9	3.4	3.9	4.5	5.0
31.6	5.3	10.5	15.8	21.1	26.3	.6	0.3	0.8	1.4	1.9	2.4	2.9	3.5	4.0	4.5	5.0
31.7	5.3	10.6	15.9	21.2	26.4	.7	0.4	0.9	1.4	1.9	2.5	3.0	3.5	4.0	4.6	5.1
31.8	5.3	10.6	15.9	21.2	26.5	.8	0.4	0.9	1.5	2.0	2.5	3.0	3.6	4.1	4.6	5.1
31.9	5.4	10.7	16.0	21.3	26.6	.9	0.5	1.0	1.5	2.0	2.6	3.1	3.6	4.1	4.7	5.2
	10'	20'	30'	40'	50'		0'	1'	2'	3'	4'	5'	6'	7'	8'	9'

Dec. Inc.	Double Second Diff.	Corr.
24.0–25.9	0.8, 2.5, 4.1, 5.8, 7.4, 9.1, 10.7, 12.3, 14.0, 15.6, 17.3, 18.9, 20.6, 22.2, 23.9, 25.5, 27.2, 28.8, 30.4, 32.1, 33.7, 35.4	[illegible]
26.0–27.9	0.8, 2.4, 4.0, 5.7, 7.3, 8.9, 10.5, 12.1, 13.7, 15.4, 17.0, 18.6, 20.2, 21.8, 23.4, 25.1, 26.7, 28.3, 29.9, 31.5, 33.1, 34.7	[illegible]
28.0–29.9	0.8, 2.4, 4.0, 5.6, 7.2, 8.8, 10.4, 12.0, 13.6, 15.2, 16.8, 18.4, 20.0, 21.6, 23.2, 24.8, 26.4, 28.0, 29.6, 31.2, 32.8, 34.4	[illegible]
30.0–31.9	0.8, 2.4, 4.0, 5.6, 7.2, 8.8, 10.4, 12.0, 13.6, 15.2, 16.8, 18.4, 20.0, 21.6, 23.2, 24.8, 26.4, 28.0, 29.6, 31.2, 32.8, 34.4	[illegible]

The Double-Second-Difference correction (Corr.) is always to be added to the tabulated altitude.

Table 42E

INTERPOLATION TABLE

Dec. Inc.	Tens 10′	20′	30′	40′	50′	Decimals	Units 0′	1′	2′	3′	4′	5′	6′	7′	8′	9′
28.0	4.6	9.3	14.0	18.6	23.3	.0	0.0	0.5	0.9	1.4	1.9	2.4	2.8	3.3	3.8	4.3
28.1	4.7	9.3	14.0	18.7	23.4	.1	0.0	0.5	1.0	1.5	1.9	2.4	2.9	3.4	3.8	4.3
28.2	4.7	9.4	14.1	18.8	23.5	.2	0.1	0.6	1.0	1.5	2.0	2.5	2.9	3.4	3.9	4.4
28.3	4.7	9.4	14.1	18.9	23.6	.3	0.1	0.6	1.1	1.6	2.0	2.5	3.0	3.5	3.9	4.4
28.4	4.7	9.5	14.2	18.9	23.7	.4	0.2	0.7	1.1	1.6	2.1	2.6	3.0	3.5	4.0	4.5
28.5	4.8	9.5	14.3	19.0	23.8	.5	0.2	0.7	1.2	1.7	2.1	2.6	3.1	3.6	4.0	4.5
28.6	4.8	9.5	14.3	19.1	23.8	.6	0.3	0.8	1.2	1.7	2.2	2.7	3.1	3.6	4.1	4.6
28.7	4.8	9.6	14.4	19.2	23.9	.7	0.3	0.8	1.3	1.8	2.2	2.7	3.2	3.7	4.1	4.6
28.8	4.8	9.6	14.4	19.2	24.0	.8	0.4	0.9	1.3	1.8	2.3	2.8	3.2	3.7	4.2	4.7
28.9	4.9	9.7	14.5	19.3	24.1	.9	0.4	0.9	1.4	1.9	2.3	2.8	3.3	3.8	4.2	4.7
29.0	4.8	9.6	14.5	19.3	24.1	.0	0.0	0.5	1.0	1.5	2.0	2.5	2.9	3.4	3.9	4.4
29.1	4.8	9.7	14.5	19.4	24.2	.1	0.0	0.5	1.0	1.5	2.0	2.5	3.0	3.5	4.0	4.5
29.2	4.8	9.7	14.6	19.4	24.3	.2	0.1	0.6	1.1	1.6	2.1	2.6	3.0	3.5	4.0	4.5
29.3	4.9	9.8	14.6	19.5	24.4	.3	0.1	0.6	1.1	1.6	2.1	2.6	3.1	3.6	4.1	4.6
29.4	4.9	9.8	14.7	19.6	24.5	.4	0.2	0.7	1.2	1.7	2.2	2.7	3.1	3.6	4.1	4.6
29.5	4.9	9.8	14.8	19.7	24.6	.5	0.2	0.7	1.2	1.7	2.2	2.7	3.2	3.7	4.2	4.7
29.6	4.9	9.9	14.8	19.7	24.7	.6	0.3	0.8	1.3	1.8	2.3	2.8	3.2	3.7	4.2	4.7
29.7	5.0	9.9	14.9	19.8	24.8	.7	0.3	0.8	1.3	1.8	2.3	2.8	3.3	3.8	4.3	4.8
29.8	5.0	10.0	14.9	19.9	24.9	.8	0.4	0.9	1.4	1.9	2.4	2.9	3.3	3.8	4.3	4.8
29.9	5.0	10.0	15.0	20.0	25.0	.9	0.4	0.9	1.4	1.9	2.4	2.9	3.4	3.9	4.4	4.9
30.0	5.0	10.0	15.0	20.0	25.0	.0	0.0	0.5	1.0	1.5	2.0	2.5	3.0	3.6	4.1	4.6
30.1	5.0	10.0	15.0	20.0	25.1	.1	0.1	0.6	1.1	1.6	2.1	2.6	3.1	3.6	4.1	4.6
30.2	5.0	10.0	15.1	20.1	25.1	.2	0.1	0.6	1.1	1.6	2.1	2.6	3.2	3.7	4.2	4.7
30.3	5.0	10.1	15.1	20.2	25.2	.3	0.2	0.7	1.2	1.7	2.2	2.7	3.2	3.7	4.2	4.7
30.4	5.1	10.1	15.2	20.3	25.3	.4	0.2	0.7	1.2	1.7	2.2	2.7	3.3	3.8	4.3	4.8
30.5	5.1	10.2	15.3	20.3	25.4	.5	0.3	0.8	1.3	1.8	2.3	2.8	3.3	3.8	4.3	4.8
30.6	5.1	10.2	15.3	20.4	25.5	.6	0.3	0.8	1.3	1.8	2.3	2.8	3.4	3.9	4.4	4.9
30.7	5.1	10.3	15.4	20.5	25.6	.7	0.4	0.9	1.4	1.9	2.4	2.9	3.4	3.9	4.4	4.9
30.8	5.2	10.3	15.4	20.6	25.7	.8	0.4	0.9	1.4	1.9	2.4	2.9	3.5	4.0	4.5	5.0
30.9	5.2	10.3	15.5	20.6	25.8	.9	0.5	1.0	1.5	2.0	2.5	3.0	3.5	4.0	4.5	5.0
31.0	5.1	10.3	15.5	20.6	25.8	.0	0.0	0.5	1.0	1.6	2.1	2.6	3.1	3.7	4.2	4.7
31.1	5.2	10.3	15.5	20.7	25.9	.1	0.1	0.6	1.1	1.6	2.2	2.7	3.2	3.7	4.3	4.8
31.2	5.2	10.4	15.6	20.8	26.0	.2	0.1	0.6	1.2	1.7	2.2	2.7	3.3	3.8	4.3	4.8
31.3	5.2	10.4	15.6	20.9	26.1	.3	0.2	0.7	1.2	1.7	2.3	2.8	3.3	3.8	4.4	4.9
31.4	5.2	10.5	15.7	20.9	26.2	.4	0.2	0.7	1.3	1.8	2.3	2.8	3.4	3.9	4.4	4.9
31.5	5.3	10.5	15.8	21.0	26.3	.5	0.3	0.8	1.3	1.8	2.4	2.9	3.4	3.9	4.5	5.0
31.6	5.3	10.5	15.8	21.1	26.3	.6	0.3	0.8	1.4	1.9	2.4	2.9	3.5	4.0	4.5	5.0
31.7	5.3	10.6	15.9	21.2	26.4	.7	0.4	0.9	1.4	1.9	2.5	3.0	3.5	4.0	4.6	5.1
31.8	5.3	10.6	15.9	21.2	26.5	.8	0.4	0.9	1.5	2.0	2.5	3.0	3.6	4.1	4.6	5.1
31.9	5.4	10.7	16.0	21.3	26.6	.9	0.5	1.0	1.5	2.0	2.6	3.1	3.6	4.1	4.7	5.2
32.0	5.3	10.6	16.0	21.3	26.6	.0	0.0	0.5	1.1	1.6	2.2	2.7	3.2	3.8	4.3	4.9
32.1	5.3	10.7	16.0	21.4	26.7	.1	0.1	0.6	1.1	1.7	2.2	2.8	3.3	3.8	4.4	4.9
32.2	5.3	10.7	16.1	21.4	26.8	.2	0.1	0.6	1.2	1.7	2.3	2.8	3.4	3.9	4.4	5.0
32.3	5.4	10.8	16.1	21.5	26.9	.3	0.2	0.7	1.2	1.8	2.3	2.9	3.4	4.0	4.5	5.0
32.4	5.4	10.8	16.2	21.6	27.0	.4	0.2	0.8	1.3	1.8	2.4	2.9	3.5	4.0	4.5	5.1
32.5	5.4	10.8	16.3	21.7	27.1	.5	0.3	0.8	1.4	1.9	2.4	3.0	3.5	4.1	4.6	5.1
32.6	5.4	10.9	16.3	21.7	27.2	.6	0.3	0.9	1.4	1.9	2.5	3.0	3.6	4.1	4.7	5.2
32.7	5.5	10.9	16.4	21.8	27.3	.7	0.4	0.9	1.5	2.0	2.5	3.1	3.6	4.2	4.7	5.3
32.8	5.5	11.0	16.4	21.9	27.4	.8	0.4	1.0	1.5	2.1	2.6	3.1	3.7	4.2	4.8	5.3
32.9	5.5	11.0	16.5	22.0	27.5	.9	0.5	1.0	1.6	2.1	2.7	3.2	3.7	4.3	4.8	5.4
33.0	5.5	11.0	16.5	22.0	27.5	.0	0.0	0.6	1.1	1.7	2.2	2.8	3.3	3.9	4.5	5.0
33.1	5.5	11.0	16.5	22.0	27.6	.1	0.1	0.6	1.2	1.7	2.3	2.8	3.4	4.0	4.5	5.1
33.2	5.5	11.0	16.6	22.1	27.6	.2	0.1	0.7	1.2	1.8	2.3	2.9	3.5	4.0	4.6	5.1
33.3	5.5	11.1	16.6	22.2	27.7	.3	0.2	0.7	1.3	1.8	2.4	3.0	3.5	4.1	4.6	5.2
33.4	5.6	11.1	16.7	22.3	27.8	.4	0.2	0.8	1.3	1.9	2.5	3.0	3.6	4.1	4.7	5.2
33.5	5.6	11.2	16.8	22.3	27.9	.5	0.3	0.8	1.4	2.0	2.5	3.1	3.6	4.2	4.7	5.3
33.6	5.6	11.2	16.8	22.4	28.0	.6	0.3	0.9	1.5	2.0	2.6	3.1	3.7	4.2	4.8	5.4
33.7	5.6	11.3	16.9	22.5	28.1	.7	0.4	0.9	1.5	2.1	2.6	3.2	3.7	4.3	4.9	5.4
33.8	5.7	11.3	16.9	22.6	28.2	.8	0.4	1.0	1.6	2.1	2.7	3.2	3.8	4.4	4.9	5.5
33.9	5.7	11.3	17.0	22.6	28.3	.9	0.5	1.1	1.6	2.2	2.7	3.3	3.9	4.4	5.0	5.5
34.0	5.6	11.3	17.0	22.6	28.3	.0	0.0	0.6	1.1	1.7	2.3	2.9	3.4	4.0	4.6	5.2
34.1	5.7	11.3	17.0	22.7	28.4	.1	0.1	0.6	1.2	1.8	2.4	2.9	3.5	4.1	4.7	5.2
34.2	5.7	11.4	17.1	22.8	28.5	.2	0.1	0.7	1.3	1.8	2.4	3.0	3.6	4.1	4.7	5.3
34.3	5.7	11.4	17.1	22.9	28.6	.3	0.2	0.7	1.3	1.9	2.5	3.0	3.6	4.2	4.8	5.3
34.4	5.7	11.5	17.2	22.9	28.7	.4	0.2	0.8	1.4	2.0	2.5	3.1	3.7	4.3	4.8	5.4
34.5	5.8	11.5	17.3	23.0	28.8	.5	0.3	0.9	1.4	2.0	2.6	3.2	3.7	4.3	4.9	5.5
34.6	5.8	11.5	17.3	23.1	28.8	.6	0.3	0.9	1.5	2.1	2.6	3.2	3.8	4.4	4.9	5.5
34.7	5.8	11.6	17.4	23.2	28.9	.7	0.4	1.0	1.6	2.1	2.7	3.3	3.9	4.4	5.0	5.6
34.8	5.8	11.6	17.4	23.2	29.0	.8	0.5	1.0	1.6	2.2	2.8	3.3	3.9	4.5	5.1	5.6
34.9	5.9	11.7	17.5	23.3	29.1	.9	0.5	1.1	1.7	2.2	2.8	3.4	4.0	4.5	5.1	5.7
35.0	5.8	11.6	17.5	23.3	29.1	.0	0.0	0.6	1.2	1.8	2.4	3.0	3.5	4.1	4.7	5.3
35.1	5.8	11.7	17.5	23.4	29.2	.1	0.1	0.7	1.2	1.8	2.4	3.0	3.6	4.2	4.8	5.4
35.2	5.8	11.7	17.6	23.4	29.3	.2	0.1	0.7	1.3	1.9	2.5	3.1	3.7	4.3	4.9	5.4
35.3	5.9	11.8	17.6	23.5	29.4	.3	0.2	0.8	1.4	2.0	2.5	3.1	3.7	4.3	4.9	5.5
35.4	5.9	11.8	17.7	23.6	29.5	.4	0.2	0.8	1.4	2.0	2.6	3.2	3.8	4.4	5.0	5.6
35.5	5.9	11.8	17.8	23.7	29.6	.5	0.3	0.9	1.5	2.1	2.7	3.3	3.8	4.4	5.0	5.6
35.6	5.9	11.9	17.8	23.7	29.7	.6	0.4	0.9	1.5	2.1	2.7	3.3	3.9	4.5	5.1	5.7
35.7	6.0	11.9	17.9	23.8	29.8	.7	0.4	1.0	1.6	2.2	2.8	3.4	4.0	4.6	5.1	5.7
35.8	6.0	12.0	17.9	23.9	29.9	.8	0.5	1.1	1.7	2.2	2.8	3.4	4.0	4.6	5.2	5.8
35.9	6.0	12.0	18.0	24.0	30.0	.9	0.5	1.1	1.7	2.3	2.9	3.5	4.1	4.7	5.3	5.9

Double Second Diff. and Corr. (Dec. Inc. 28.0–29.9)

Double Second Diff.	Corr.
0.8	0.1
2.4	0.2
4.0	0.3
5.6	0.4
7.2	0.5
8.8	0.6
10.4	0.7
12.0	0.8
13.6	0.9
15.2	1.0
16.8	1.1
18.4	1.2
20.0	1.3
21.6	1.4
23.2	1.5
24.8	1.6
26.4	1.7
28.0	1.8
29.6	1.9
31.2	2.0
32.8	2.1
34.4	

Double Second Diff. and Corr. (Dec. Inc. 30.0–31.9)

Double Second Diff.	Corr.
0.8	0.1
2.4	0.2
4.0	0.3
5.6	0.4
7.2	0.5
8.8	0.6
10.4	0.7
12.0	0.8
13.6	0.9
15.2	1.0
16.8	1.1
18.4	1.2
20.0	1.3
21.6	1.4
23.2	1.5
24.8	1.6
26.4	1.7
28.0	1.8
29.6	1.9
31.2	2.0
32.8	2.1
34.4	

Double Second Diff. and Corr. (Dec. Inc. 32.0–33.9)

Double Second Diff.	Corr.
0.8	0.1
2.4	0.2
4.0	0.3
5.7	0.4
7.3	0.5
8.9	0.6
10.5	0.7
12.1	0.8
13.7	0.9
15.4	1.0
17.0	1.1
18.6	1.2
20.2	1.3
21.8	1.4
23.4	1.5
25.1	1.6
26.7	1.7
28.3	1.8
29.9	1.9
31.5	2.0
33.1	2.1
34.7	

Double Second Diff. and Corr. (Dec. Inc. 34.0–35.9)

Double Second Diff.	Corr.
0.8	0.1
2.5	0.2
4.1	0.3
5.8	0.4
7.4	0.5
9.1	0.6
10.7	0.7
12.3	0.8
14.0	0.9
15.6	1.0
17.3	1.1
18.9	1.2
20.6	1.3
22.2	1.4
23.9	1.5
25.5	1.6
27.2	1.7
28.8	1.8
30.4	1.9
32.1	2.0
33.7	2.1
35.4	

Dec. Inc.	Tens 10′	20′	30′	40′	50′	Decimals	Units 0′	1′	2′	3′	4′	5′	6′	7′	8′	9′
36.0	6.0	12.0	18.0	24.0	30.0	.0	0.0	0.6	1.2	1.8	2.4	3.0	3.6	4.3	4.9	5.5
36.1	6.0	12.0	18.0	24.0	30.1	.1	0.1	0.7	1.3	1.9	2.5	3.1	3.7	4.3	4.9	5.5
36.2	6.0	12.0	18.1	24.1	30.1	.2	0.1	0.7	1.3	1.9	2.6	3.2	3.8	4.4	5.0	5.6
36.3	6.0	12.1	18.1	24.2	30.2	.3	0.2	0.8	1.4	2.0	2.6	3.2	3.8	4.4	5.0	5.7
36.4	6.1	12.1	18.2	24.3	30.3	.4	0.2	0.9	1.5	2.1	2.7	3.3	3.9	4.5	5.1	5.7
36.5	6.1	12.2	18.3	24.3	30.4	.5	0.3	0.9	1.5	2.1	2.7	3.3	4.0	4.6	5.2	5.8
36.6	6.1	12.2	18.3	24.4	30.5	.6	0.4	1.0	1.6	2.2	2.8	3.4	4.0	4.6	5.2	5.8
36.7	6.1	12.3	18.4	24.5	30.6	.7	0.4	1.0	1.6	2.3	2.9	3.5	4.1	4.7	5.3	5.9
36.8	6.2	12.3	18.4	24.6	30.7	.8	0.5	1.1	1.7	2.3	2.9	3.5	4.1	4.7	5.4	6.0
36.9	6.2	12.3	18.5	24.6	30.8	.9	0.5	1.2	1.8	2.4	3.0	3.6	4.2	4.8	5.4	6.0
37.0	6.1	12.3	18.5	24.6	30.8	.0	0.0	0.6	1.2	1.9	2.5	3.1	3.7	4.4	5.0	5.6
37.1	6.2	12.3	18.5	24.7	30.9	.1	0.1	0.7	1.3	1.9	2.6	3.2	3.8	4.4	5.1	5.7
37.2	6.2	12.4	18.6	24.8	31.0	.2	0.1	0.7	1.4	2.0	2.6	3.2	3.9	4.5	5.1	5.7
37.3	6.2	12.4	18.6	24.9	31.1	.3	0.2	0.8	1.4	2.1	2.7	3.3	3.9	4.6	5.2	5.8
37.4	6.2	12.5	18.7	24.9	31.2	.4	0.2	0.9	1.5	2.1	2.7	3.4	4.0	4.6	5.2	5.9
37.5	6.3	12.5	18.8	25.0	31.3	.5	0.3	0.9	1.6	2.2	2.8	3.4	4.1	4.7	5.3	5.9
37.6	6.3	12.5	18.8	25.1	31.3	.6	0.4	1.0	1.6	2.2	2.9	3.5	4.1	4.7	5.4	6.0
37.7	6.3	12.6	18.9	25.2	31.4	.7	0.4	1.1	1.7	2.3	2.9	3.6	4.2	4.8	5.4	6.1
37.8	6.3	12.6	18.9	25.2	31.5	.8	0.5	1.1	1.7	2.4	3.0	3.6	4.2	4.9	5.5	6.1
37.9	6.4	12.7	19.0	25.3	31.6	.9	0.6	1.2	1.8	2.4	3.1	3.7	4.3	4.9	5.6	6.2
38.0	6.3	12.6	19.0	25.3	31.6	.0	0.0	0.6	1.3	1.9	2.6	3.2	3.8	4.5	5.1	5.8
38.1	6.3	12.7	19.0	25.4	31.7	.1	0.1	0.7	1.3	2.0	2.6	3.3	3.9	4.6	5.2	5.8
38.2	6.3	12.7	19.1	25.4	31.8	.2	0.1	0.8	1.4	2.1	2.7	3.3	4.0	4.6	5.3	5.9
38.3	6.4	12.8	19.1	25.5	31.9	.3	0.2	0.8	1.5	2.1	2.8	3.4	4.0	4.7	5.3	6.0
38.4	6.4	12.8	19.2	25.6	32.0	.4	0.3	0.9	1.5	2.2	2.8	3.5	4.1	4.7	5.4	6.0
38.5	6.4	12.8	19.3	25.7	32.1	.5	0.3	1.0	1.6	2.2	2.9	3.5	4.2	4.8	5.5	6.1
38.6	6.4	12.9	19.3	25.7	32.2	.6	0.4	1.0	1.7	2.3	3.0	3.6	4.2	4.9	5.5	6.2
38.7	6.5	12.9	19.4	25.8	32.3	.7	0.4	1.1	1.7	2.4	3.0	3.7	4.3	4.9	5.6	6.2
38.8	6.5	13.0	19.4	25.9	32.4	.8	0.5	1.2	1.8	2.4	3.1	3.7	4.4	5.0	5.6	6.3
38.9	6.5	13.0	19.5	26.0	32.5	.9	0.6	1.2	1.9	2.5	3.1	3.8	4.4	5.1	5.7	6.4
39.0	6.5	13.0	19.5	26.0	32.5	.0	0.0	0.7	1.3	2.0	2.6	3.3	3.9	4.6	5.3	5.9
39.1	6.5	13.0	19.5	26.0	32.6	.1	0.1	0.7	1.4	2.0	2.7	3.4	4.0	4.7	5.3	6.0
39.2	6.5	13.0	19.6	26.1	32.6	.2	0.1	0.8	1.4	2.1	2.8	3.4	4.1	4.7	5.4	6.1
39.3	6.5	13.1	19.6	26.2	32.7	.3	0.2	0.9	1.5	2.2	2.8	3.5	4.1	4.8	5.5	6.1
39.4	6.6	13.1	19.7	26.3	32.8	.4	0.3	0.9	1.6	2.2	2.9	3.6	4.2	4.9	5.5	6.2
39.5	6.6	13.2	19.8	26.3	32.9	.5	0.3	1.0	1.6	2.3	3.0	3.6	4.3	4.9	5.6	6.3
39.6	6.6	13.2	19.8	26.4	33.0	.6	0.4	1.1	1.7	2.4	3.0	3.7	4.3	5.0	5.7	6.3
39.7	6.6	13.3	19.9	26.5	33.1	.7	0.5	1.1	1.8	2.4	3.1	3.8	4.4	5.1	5.7	6.4
39.8	6.7	13.3	19.9	26.6	33.2	.8	0.5	1.2	1.8	2.5	3.2	3.8	4.5	5.1	5.8	6.5
39.9	6.7	13.3	20.0	26.6	33.3	.9	0.6	1.3	1.9	2.6	3.2	3.9	4.5	5.2	5.9	6.5
40.0	6.6	13.3	20.0	26.6	33.3	.0	0.0	0.7	1.3	2.0	2.7	3.4	4.0	4.7	5.4	6.1
40.1	6.7	13.3	20.0	26.7	33.4	.1	0.1	0.7	1.4	2.1	2.8	3.4	4.1	4.8	5.5	6.1
40.2	6.7	13.4	20.1	26.8	33.5	.2	0.1	0.8	1.5	2.2	2.8	3.5	4.2	4.9	5.5	6.2
40.3	6.7	13.4	20.1	26.9	33.6	.3	0.2	0.9	1.6	2.2	2.9	3.6	4.3	4.9	5.6	6.3
40.4	6.7	13.5	20.2	26.9	33.7	.4	0.3	0.9	1.6	2.3	3.0	3.6	4.3	5.0	5.7	6.3
40.5	6.8	13.5	20.3	27.0	33.8	.5	0.3	1.0	1.7	2.4	3.0	3.7	4.4	5.1	5.7	6.4
40.6	6.8	13.5	20.3	27.1	33.8	.6	0.4	1.1	1.8	2.4	3.1	3.8	4.5	5.1	5.8	6.5
40.7	6.8	13.6	20.4	27.2	33.9	.7	0.5	1.1	1.8	2.5	3.2	3.8	4.5	5.2	5.9	6.5
40.8	6.8	13.6	20.4	27.2	34.0	.8	0.5	1.2	1.9	2.6	3.2	3.9	4.6	5.3	5.9	6.6
40.9	6.9	13.7	20.5	27.3	34.1	.9	0.6	1.3	2.0	2.6	3.3	4.0	4.7	5.3	6.0	6.7
41.0	6.8	13.6	20.5	27.3	34.1	.0	0.0	0.7	1.4	2.1	2.8	3.5	4.1	4.8	5.5	6.2
41.1	6.8	13.7	20.5	27.4	34.2	.1	0.1	0.8	1.5	2.1	2.8	3.5	4.2	4.9	5.6	6.3
41.2	6.8	13.7	20.6	27.4	34.3	.2	0.1	0.8	1.5	2.2	2.9	3.6	4.3	5.0	5.7	6.4
41.3	6.9	13.8	20.6	27.5	34.4	.3	0.2	0.9	1.6	2.3	3.0	3.7	4.4	5.0	5.7	6.4
41.4	6.9	13.8	20.7	27.6	34.5	.4	0.3	1.0	1.7	2.4	3.0	3.7	4.4	5.1	5.8	6.5
41.5	6.9	13.8	20.8	27.7	34.6	.5	0.3	1.0	1.7	2.4	3.1	3.8	4.5	5.2	5.9	6.6
41.6	6.9	13.9	20.8	27.7	34.7	.6	0.4	1.1	1.8	2.5	3.2	3.9	4.6	5.3	5.9	6.6
41.7	7.0	13.9	20.9	27.8	34.8	.7	0.5	1.2	1.9	2.6	3.3	3.9	4.6	5.3	6.0	6.7
41.8	7.0	14.0	20.9	27.9	34.9	.8	0.6	1.2	1.9	2.6	3.3	4.0	4.7	5.4	6.1	6.8
41.9	7.0	14.0	21.0	28.0	35.0	.9	0.6	1.3	2.0	2.7	3.4	4.1	4.8	5.5	6.2	6.8
42.0	7.0	14.0	21.0	28.0	35.0	.0	0.0	0.7	1.4	2.1	2.8	3.5	4.2	5.0	5.7	6.4
42.1	7.0	14.0	21.0	28.0	35.1	.1	0.1	0.8	1.5	2.2	2.9	3.6	4.3	5.0	5.7	6.4
42.2	7.0	14.0	21.1	28.1	35.1	.2	0.1	0.8	1.6	2.3	3.0	3.7	4.4	5.1	5.8	6.5
42.3	7.0	14.1	21.1	28.2	35.2	.3	0.2	0.9	1.6	2.3	3.0	3.8	4.5	5.2	5.9	6.6
42.4	7.1	14.1	21.2	28.3	35.3	.4	0.3	1.0	1.7	2.4	3.1	3.8	4.5	5.2	5.9	6.7
42.5	7.1	14.2	21.3	28.3	35.4	.5	0.4	1.1	1.8	2.5	3.2	3.9	4.6	5.3	6.0	6.7
42.6	7.1	14.2	21.3	28.4	35.5	.6	0.4	1.1	1.8	2.5	3.3	4.0	4.7	5.4	6.1	6.8
42.7	7.1	14.3	21.4	28.5	35.6	.7	0.5	1.2	1.9	2.6	3.3	4.0	4.7	5.5	6.2	6.9
42.8	7.2	14.3	21.4	28.6	35.7	.8	0.6	1.3	2.0	2.7	3.4	4.1	4.8	5.5	6.2	6.9
42.9	7.2	14.3	21.5	28.6	35.8	.9	0.6	1.3	2.1	2.8	3.5	4.2	4.9	5.6	6.3	7.0
43.0	7.1	14.3	21.5	28.6	35.8	.0	0.0	0.7	1.4	2.2	2.9	3.6	4.3	5.1	5.8	6.5
43.1	7.2	14.3	21.5	28.7	35.9	.1	0.1	0.8	1.5	2.2	3.0	3.7	4.4	5.1	5.9	6.6
43.2	7.2	14.4	21.6	28.8	36.0	.2	0.1	0.9	1.6	2.3	3.0	3.8	4.5	5.2	5.9	6.7
43.3	7.2	14.4	21.6	28.9	36.1	.3	0.2	0.9	1.7	2.4	3.1	3.8	4.6	5.3	6.0	6.7
43.4	7.2	14.5	21.7	28.9	36.2	.4	0.3	1.0	1.7	2.5	3.2	3.9	4.6	5.4	6.1	6.8
43.5	7.3	14.5	21.8	29.0	36.3	.5	0.4	1.1	1.8	2.5	3.3	4.0	4.7	5.4	6.2	6.9
43.6	7.3	14.5	21.8	29.1	36.3	.6	0.4	1.2	1.9	2.6	3.3	4.1	4.8	5.5	6.2	7.0
43.7	7.3	14.6	21.9	29.2	36.4	.7	0.5	1.2	2.0	2.7	3.4	4.1	4.9	5.6	6.3	7.0
43.8	7.3	14.6	21.9	29.2	36.5	.8	0.6	1.3	2.0	2.8	3.5	4.2	4.9	5.7	6.4	7.1
43.9	7.4	14.7	22.0	29.3	36.6	.9	0.7	1.4	2.1	2.8	3.6	4.3	5.0	5.7	6.5	7.2

Double Second Diff. and Corr. (Dec. Inc. 36.0–37.9)

Double Second Diff.	Corr.
0.8	0.1
2.5	0.2
4.2	0.3
5.9	0.4
7.6	0.5
9.3	0.6
11.0	0.7
12.7	0.8
14.4	0.9
16.1	1.0
17.8	1.1
19.5	1.2
21.2	1.3
22.8	1.4
24.5	1.5
26.2	1.6
27.9	1.7
29.6	1.8
31.3	1.9
33.0	2.0
34.7	

Double Second Diff. and Corr. (Dec. Inc. 38.0–39.9)

Double Second Diff.	Corr.
0.9	0.1
2.6	0.2
4.4	0.3
6.2	0.4
7.9	0.5
9.7	0.6
11.4	0.7
13.2	0.8
14.9	0.9
16.7	1.0
18.5	1.1
20.2	1.2
22.0	1.3
23.7	1.4
25.5	1.5
27.3	1.6
29.0	1.7
30.8	1.8
32.5	1.9
34.3	

Double Second Diff. and Corr. (Dec. Inc. 40.0–41.9)

Double Second Diff.	Corr.
0.9	0.1
2.8	0.2
4.6	0.3
6.5	0.4
8.3	0.5
10.2	0.6
12.0	0.7
13.9	0.8
15.7	0.9
17.6	1.0
19.4	1.1
21.3	1.2
23.1	1.3
25.0	1.4
26.8	1.5
28.7	1.6
30.5	.7
32.3	1.8
34.2	

Double Second Diff. and Corr. (Dec. Inc. 42.0–43.9)

Double Second Diff.	Corr.
1.0	0.1
3.0	0.2
4.9	0.3
6.9	0.4
8.9	0.5
10.8	0.6
12.8	0.7
14.8	0.8
16.7	0.9
18.7	1.0
20.7	1.1
22.7	1.2
24.6	1.3
26.6	1.4
28.6	1.5
30.5	1.6
32.5	1.7
34.5	

The Double-Second-Difference correction (Corr.) is always to be added to the tabulated altitude.

Table 42F

MOON (v = 10.4′) d = 12.9′

GHA 15 Hr.	= 315°52′	S 2°21′	
30 Min. 0 Sec.	= 7°10′	+07′	
v (10.4′) Corr.	= 05′	S 2°28′	
GHA	= 322°67′		
GHA	= 323°07′		
Assumed Long. W.	= 22°07′	Assumed Lat. = 45° N	
LHA	= 301°		

Hc	= 19°50′	d	= −46′	Z	= 114°
Corr.	= − 21	Dec. Inc.	= 28′	Zn	= 114°
Hc	= 19°29′	Corr.	= −21′		
Ho	= 19°00′				
	29′ Away				

SUN

GHA 15 Hr.	= 44°52′	Dec. = 23°21′N
30 Min. 0 Sec.	= 7°30′	
GHA	= 51°82′	
GHA	= 52°22′	
Assumed Long. W.	= 22°22′	Assumed Lat. = 45° N
LHA	= 30°	

		d	= +45′	360°
Hc	= 57°08′	Dec. Inc.	= 21′	Z= 122°
	+16	Corr.	= +16′	
Hc	= 57°24′			
Ho	= 57°30′			
	6′ Toward			

6-16-75 1630 Hours GMT (Guess)

Ho Moon = 19°00′
Ho Sun = 57°30′
Latitude = 45° N

MOON (v = 10.4′) d = 12.8′

GHA 1630 Hr.	=	330°21′	2°34′ S
30 Min. 0 Sec.	=	7°10′	+ 07′
v	=	05′	2°41′ S
GHA	=	337°36′	
Assumed Long. W.	=	37°36′	
LHA	=	300°	

Hc	=	18°71′	d	=	−45′	Z	=	114°
		−31′	Dec. Inc.	=	41′	Zn	=	114°
Hc	=	18°40′	Corr.	=	−31′			
Ho	=	18°60′						
		20′	Toward					

SUN

GHA 1630 Hr.	=	59°52′	Dec. = 23°21′ N
30 Min. 0 Sec.	=	7°30′	
GHA	=	66°82′	
GHA	=	67°22′	
Assumed Long. W.	=	37°22′	
LHA	=	30°	

Hc	=	57°08′	d	=	+45′			360°
		+16′	Dec. Inc.	=	21′	Z	=	122°
Hc	=	57°24′	Corr.	=	+16′	Zn	=	238°
Ho	=	57°30′						
		6′	Toward					

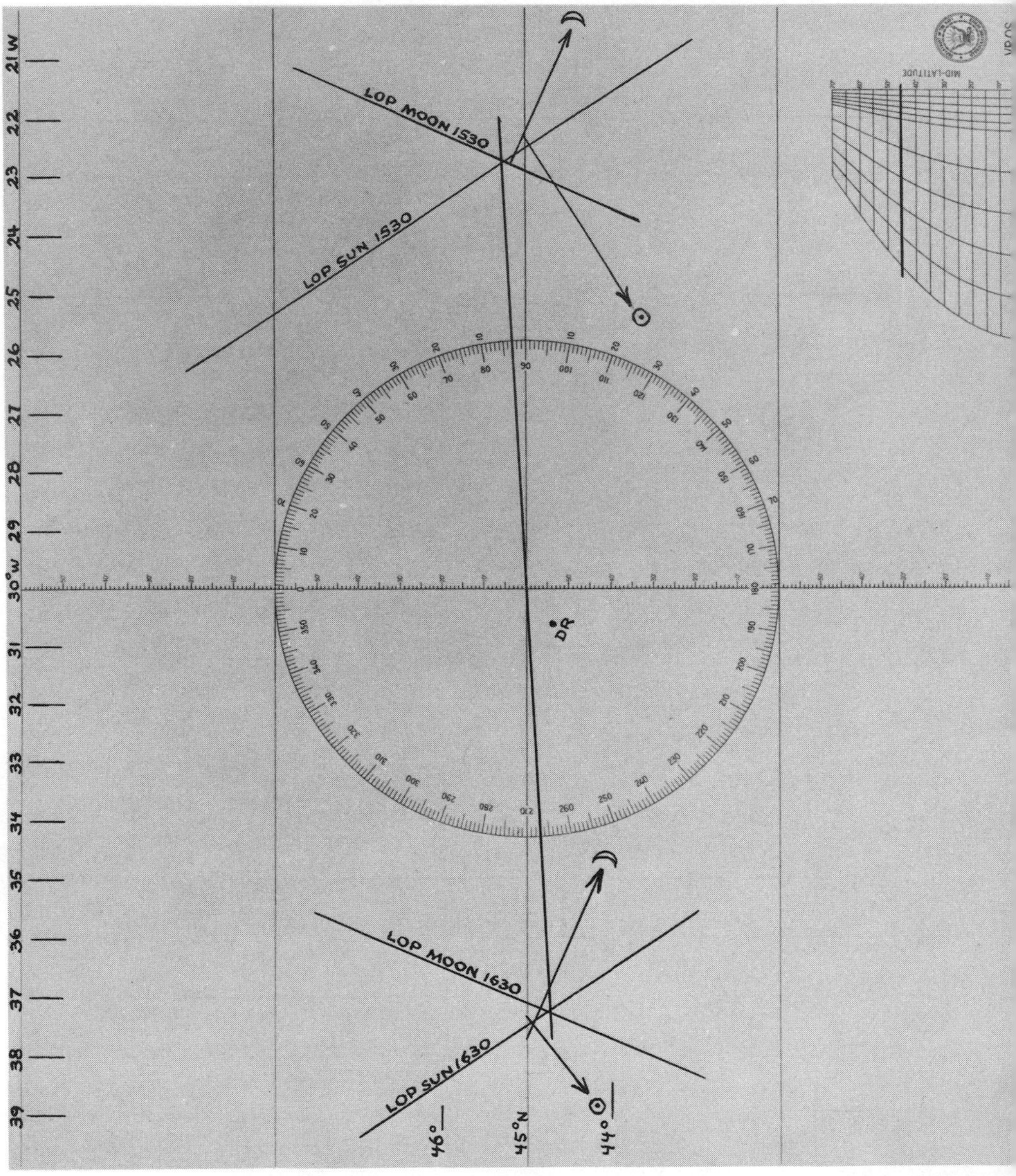
LOP MOON 1530
LOP SUN 1530
LOP MOON 1630
LOP SUN 1630
DR
46°
45°N
44°
21 W
30°W
MID-LATITUDE

Fig. 14.6

You can see from the plotting sheet (Figure 14.6) that the line between fixes has a shallow slope, so the intersection is broad. I read it as 30°20′ W, and since this example was based on a known position of 30° W, 45° N, you can see that this method could come in handy some day. Without time, you might be very glad to be able to get your longitude within 20 miles or so.

15
Solving the Navigational Triangle Directly

This chapter might also be called "Getting Back to Basics," since it goes back to the formulas that are used (in computers) to make the sight reduction tables. This chapter also includes a perpetual almanac of the sun, so with just this book, a sextant, and a watch, you can navigate by advancing sun lines and the noon sight.

Here are the basic formulas:

	I		II		III
(1)	$\sin Hc$	=	$\sin L \sin d$	+	$\cos L \cos d \cos t$
(2)	$\sin Z$	=	$\dfrac{\cos d \sin t}{\cos Hc}$		

Sin is short for *sine* and is pronounced "sign"; *cos* is short for *cosine* and is pronounced "cosign." The expression "sin L sin d" means sine of latitude multiplied by sine of declination.

Some general comments before you actually use these formulas: First, don't worry about solving the formulas—all you have to do is simple addition and subtraction. Second, there is no more accurate way to perform a sight reduction than this. Third, these formulas allow you to work directly from your DR position and thus *save* a few steps and *simplify* the plotting.

The first thing you need to know is that these formulas do not use LHA. Instead, they use what is called the *meridian angle*, called t in the right-hand parts of the equations. Remember that LHA is always measured westward from you to the celestial body and therefore can be larger than 180°. This is a convention adopted to keep azimuth simple. In the formulas, on the other hand, the meridian angle is the angle between your longitude line (meridian) and the longitude line of the celestial body, and it is always less than 180° (Figure 15.1). As you will see, there is a common-sense way around the slight problem of properly naming the azimuth (Zn).

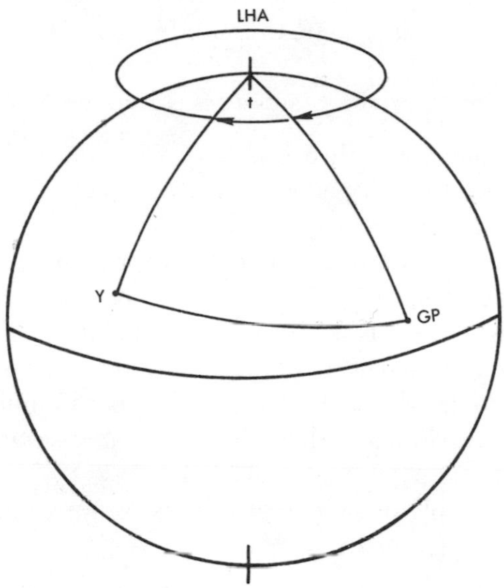

Fig. 15.1

You must know some rules in order to work with these formulas, so let's take them one by one.

First, if the declination of the body and your latitude are of *contrary* name, part II of the first equation is negative.

Second, if meridian angle (t) is larger than 90°, Part III of the first equation is negative.

Therefore, you can set up the following categories, using the terminology of the sight reduction tables:

A. Declination *same* name as latitude, meridian angle (t) *less* than 90°:

$$\sin Hc = \sin L \sin d + \cos L \cos d \cos t$$

B. Declination *same* name as latitude, meridian angle (t) *greater* than 90°:

$$\sin Hc = \sin L \sin d - \cos L \cos d \cos t$$

C. Declination *contrary* name to latitude, meridian angle (t) *less* than 90°:

$$\sin Hc = -\sin L \sin d + \cos L \cos d \cos t$$

D. Declination *contrary* name to latitude, meridian angle (t) *greater* than 90°:

$$\sin Hc = -\sin L \sin d - \cos L \cos d \cos t$$

Most of the time you will use *A* or *C*.

You do not have to worry about negative cases with the formula for azimuth angle (Z). It is always positive.

The table of sines and cosines (Table 45) begins on page 167. As you can see, all angles from 0° to 180° degrees are included with one line for each minute, so there is no need for interpolation. The middle two columns contain the logarithms of all these sines and cosines. The only thing you need to know about logarithms is that to multiply numbers you *add* their logarithms and to divide numbers you *subtract* their logarithms.

Here is an example of case *A*, declination *same* name as latitude, meridian angle (t) *less* than 90°:

Assume:		DR	= 15° N, 68 W
Latitude	= 15° N	GHA	= 123°
Declination	= 22° N		
t	= 55°		

Using these numbers, the formula is:

$\sin Hc = \sin 15° \times \sin 22° + \cos 15° \times \cos 22° \times \cos 55°$.

The sine of 15° is 0.25882 (see Table 45), and the sine of 22° is 0.37461. If you are good at multiplying long numbers, you could multiply these two, then add the result to the product of 0.96593 (cos 15°) times 0.92718 (cos 22°) times 0.57358 (cos 55°). The result is 0.61065. The formula indicates that sin *Hc* equals 0.61065. Therefore, look in the table to find out what angle has a sine of 0.61065. From the table, the angle is 37°38′. You can see that because the table is set up for each minute of angle, it is not necessary to interpolate; you just take the number closest to your result—in this case the number in the table was 0.61061. There is no way you can be off by more than 1 mile.

If you have a pocket calculator with a floating decimal and a capacity of over 5 digits, you can use it to do the arithmetic for you. If, however, you don't want to bother with a calculator and want to have a record of what you have done, you can use the logarithms given in the table:

Log of sine of 15°	=	9.41300
Log of sine of 22°	=	9.57358
		8.98658

In adding logarithms, you can forget about the carry-over in the numbers to the left of the decimal. In the addition here, the number is actually 18.98658. Just drop the 1 or 2 or 3 or whatever it happens to be.

Now you have the logarithm of the product of sin 15° and sin 22°. Look up this logarithm (8.98658) in the table. It is on the 5° page about halfway down. The closest logarithm in the table is 8.98549. You want the number that corresponds to this logarithm, and the rule is that it is always in the closest column—that is, the number is 0.09671 and not 0.99531, which is on the right-hand side across another logarithm column.

Now do the second set of multiplying:

Log cos 15°	=	9.98494
Log cos 22°	=	9.96717
Log cos 55°	=	9.75859
		(2)9.71070

The product of this multiplication is 0.51379, found on the 30° page in the left-hand column.

A note on using the table: The sine of 55° is read from the bottom up (see the column headings and the arrows at the bottom of the page). Note also that, on the other side of the page, the bottom left-hand number is the cosine of 124°60′ (that is, 125°) and that the cosine of 124°0′ is at the top left. So when you are using the left-hand column, observe carefully.

Now that you have the two numbers resulting from the multiplication, add them.

$$\begin{array}{rcl} & & 0.51379 \\ & & \underline{0.09671} \\ \sin Hc & = & 0.61050 \\ Hc & = & 37°38' \end{array}$$

Now solve for azimuth angle (Z).

$$\sin Z = \frac{\cos d \sin t}{\cos Hc}$$

Translated, this means that sine *Z* is equal to cosine *d times* sine *t divided by* cos *Hc*.

$$\sin Z = \frac{\cos 22° \sin 55°}{\cos 37°38'}$$

Using logarithms, you would compute *Z* like this:

$$\begin{array}{lcl} \text{Log cos } 22° & = & 9.96717 \\ \text{Log sin } 55° & = & \underline{9.91336} \\ & & (1)9.88053 \\ \text{Log cos } 37°38' & = & 9.89869 \text{ (subtract)} \\ \hline \text{Log sin } Z & = & 9.98184 \\ \sin Z & = & 0.95907 \\ Z & = & 73°33' \\ Z & = & 74° \end{array}$$

The azimuth angle (Z) that you have figured out from the formula is measured from north toward the east or west; or it is measured from south toward the east or west. In this table, *Z* is always less than 90°—i.e., 74° not 106°. If the body is north of you, the angle is measured from north toward east or west; if the body is south of you, the angle is measured from the south toward east or west. The practical way to determine how to apply the azimuth angle is to take a bearing when you are through taking sights. The only time you are apt to have difficulty is when Z is close to 90°. If it is so close that a bearing won't tell you whether the body is a little north or south of west or a little north or south of east, plot it as though it were due east or west. The same thing holds when Z is near 0°; the body is near meridian passage (noon), so make it north or south. Or, instead of fooling around with long, skinny triangles, take a noon sight.

In this example, the sun is west of you, so the angle is 74° toward the west from north (Figure 15.2).

To do all these computations efficiently, set up the form as shown in Figure 15.3. Examples of cases *B*, *C*, and *D* are worked out as examples. The assumed positions in the examples are the DR positions. So to plot them, draw the azimuth line through the DR toward the sun (or other body) and plot the LOP toward or away in the usual manner.

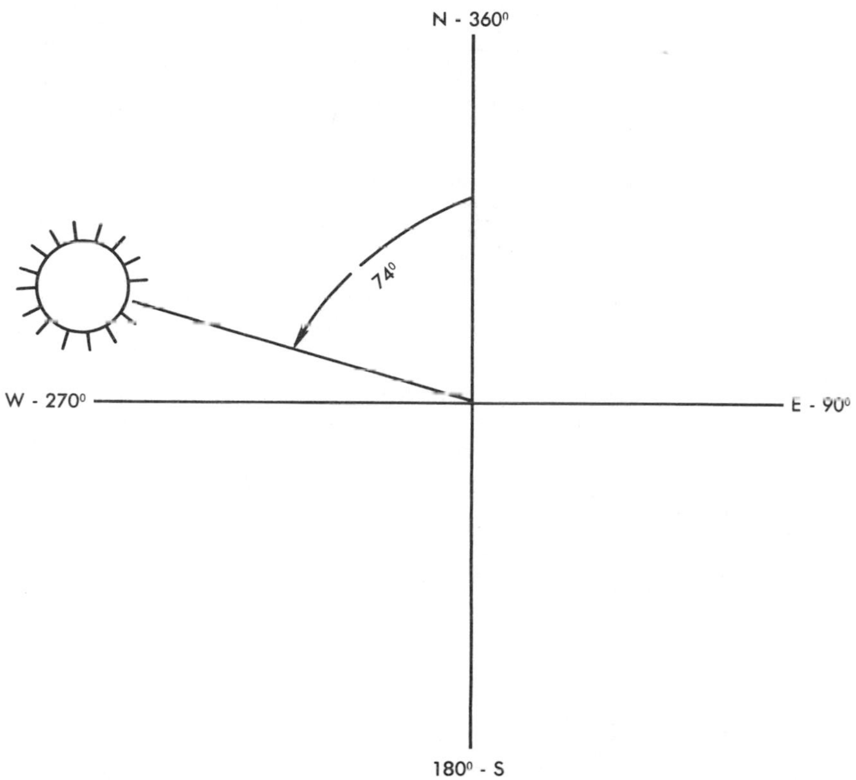

Figure 15.2. Bearing (Zn) = 360° (north) minus 74°. This is the basis of the rules for azimuth printed at the top of each page in H. O. 249.

Case *B*: Declination *same* name as latitude, meridian angle (t) *greater* than 90°

Lat = 25°S   Long = 110°E   GHA = 155°   t = 95°   dec = 9° S   Ho = −0°33′

Formulas: $\sin Hc = \sin L \sin d - \cos L \cos d \cos t$ $\text{Sin } Z = \dfrac{\cos d \sin t}{\cos Hc}$

Log sin L	= 9.62595	Log cos L	= 9.95728		
		Log cos d	= 9.99462	Log cos d	= 9.99462
Log sin d	= 9.19433	Log cos t	= 8.94030	Log sin t	= 9.99834
Log Product	= 8.82028	Log Product	= 8.89220	Log Product	= (1)9.99296
Product	= 0.06598	Product	= 0.07788		

Combine products per formula: −0.07788
    0.06598

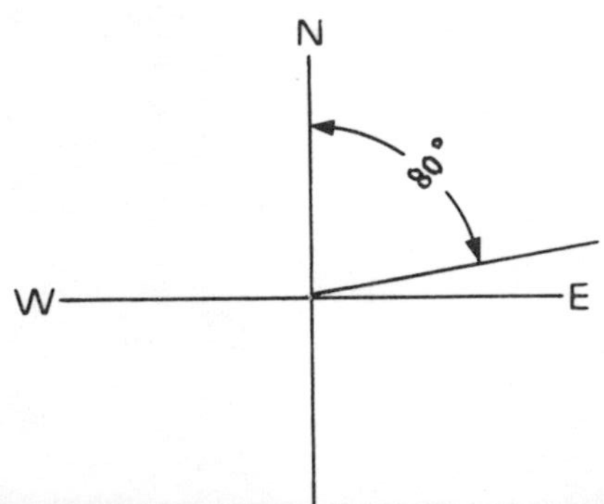

Sin Hc = −0.01190
Hc = −0°41′
Ho = −0°33′
8′ Toward

(1)9.99296
Log cos Hc = 9.99997 (subtract)
Log sin Z = 9.99299
Z = 79°44′
Z = 80°
Zn = 80°

Since the body is north and east of you, the Zn is called N 80° E, which corresponds to true direction of 80°.

Case C: Declination *contrary* name to latitude, meridian angle (t) *less* than 90°

Lat = 11° N Long = 160° W GHA = 124° t = 34° dec = 9° S Ho = 50°33′

Formulas: $\sin Hc = -\sin L \sin d + \cos L \cos d \cos t$ $\sin Z = \dfrac{\cos d \sin t}{\cos Hc}$

Log sin L	=	9.28060	Log cos L	=	9.99195			
			Log cos d	=	9.99462	Log cos d	=	9.99462
Log sin d	=	9.19433	Log cos t	=	9.91857	Log sin t	=	9.74756
Log Product	=	8.47493	Log Product	=	9.90514	Log Product	=	(1)9.74218
Product	=	0.02996	Product	=	0.80386			

Combine products per formula:

		0.80386				(1)9.74218
		−0.02996		Log cos Hc	=	9.80151 (subtract)
Sin Hc	=	.77390		Log sin Z	=	9.94067
Hc	=	50°43′		Z	=	60°43′
Ho	=	50°33′		Z	=	61°
		10′	Away			
				Zn	=	180° (south)
						−61
				Zn	=	119°

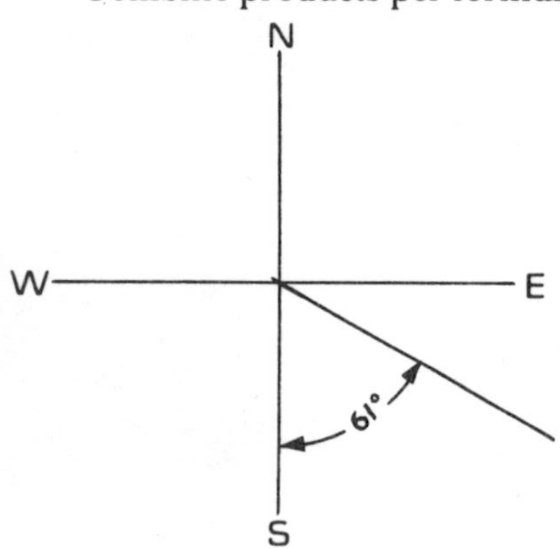

Body is south and east of you; Z is, therefore, S 61° E.

Case D: Declination *contrary* name to latitude, meridian angle (t) *greater* than 90°

Lat = 2° N   Long = 20° W   GHA = 110° t = 91°   dec = 1° S   Ho = −0°53′

Formulas: $\sin Hc = -\sin L \sin d - \cos L \cos d \cos t$ $\sin Z = \dfrac{\cos d \sin t}{\cos Hc}$

Log sin L = 8.54282		Log cos L = 9.99974			
		Log cos d = 9.99993		Log cos d = 9.99993	
Log sin d = 8.24186		Log cos t = 8.24186		Log sin t = 9.99993	
Log Product = 6.78468		Log Product = 8.24153		Log Product = (1)9.99986	
Product = −0.00058		Product = −0.01745			

Combine products per formula:

$$\begin{array}{rl} & -0.00058 \\ & \underline{-0.01745} \\ \sin Hc = & -0.01803 \\ Hc = & -1°02' \\ Ho = & \underline{-0°53'} \\ & 9' \text{ Toward} \end{array}$$

$$\begin{array}{rl} & (1)9.99986 \\ \text{Log } \cos Hc = & \underline{9.99993} \text{ (subtract)} \\ \text{Log } \sin Z = & 9.99993 \\ Z = & 89° \\ & 180° \text{ (south)} \\ Z = & \underline{89°} \\ Zn = & 269° \end{array}$$

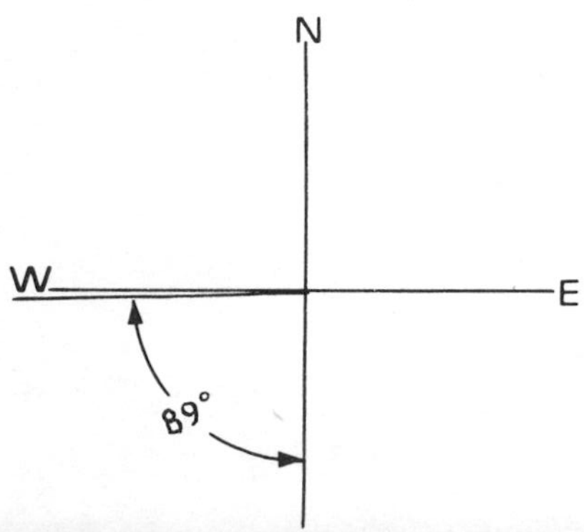

Body is south and west of you. Z is S 89° W. Here is a case where a compass bearing on the body would be helpful.

I have already suggested that these formulas can be solved by a pocket calculator. You look up the sines and cosines in the table and do the arithmetic with the calculator; then you look up in the table the angle whose sine you have calculated.

An even easier (and, naturally, more expensive) method is to use a pocket computer such as the Hewlett-Packard HP-45 or Texas Instruments SR-50 or Sears Roebuck 5878. These computers have the sines and cosines stored in a memory bank; all you have to do is enter the angle, and the device will supply the sine or cosine as required. When you have the sine on the display, you punch a button and the computer displays the corresponding angle.

The Hewlett-Packard computer has the convenient feature of being able to handle input of degrees and minutes. With the other two, you must deal with degrees and tenths of degrees. Thus, with the H-P, you could enter 36°30′, whereas with the others you would enter 36.5°. Since 6′ is one-tenth of a degree it is easy enough to use the less expensive computers. Still, even the least expensive (Texas Instruments) costs $150.

The major concern is how well one of these instruments will stand up under the conditions on a boat at sea. After all, you can use the tables even if they are wet, but will a wet computer function?

On the subject of electronic gadgets, the ultimate celestial navigation toy has arrived. It is a bread-box-size computer with its own program to solve any celestial navigation problem. If you want to reduce a star sight, you punch a button labeled "Star" and the computer then asks *you* for the required information—sextant reading, GHA, DR, and so on and displays the answer. All you have to do is take sights and times and thumb through the almanac. If this kind of thing is worth $1,200 to you, write to Micro Instrument Corporation (see Chapter 16).

I have suggested the foregoing methods of solving the navigational triangle as a way of reducing the *bulk* of equipment necessary. The sight reduction tables are bulky, and I know that I get tired of lugging them around from boat to boat and plane to plane. The problem is that they are *fast.* And this makes them hard to give up.

Another way to reduce the bulk without the risk of malfunction inherent in things electronic is to solve the triangle by means of the formulas and a slide rule. Many slide rules have sine scales—usually on the reverse side—and most are marked with cosines, since sines and cosines are complementary (that is, the sine of 10° is the same as the cosine of 80°; the angles always add up to 90°).

The mechanics of using a slide rule are fairly simple and are based on the principle of logarithms—to multiply you add; to divide you subtract. Every good slide rule, such as the Keuffel and Esser, comes with a thorough instruction manual.

The principal shortcoming of a standard slide rule for navigational purposes is that the scales are too short to allow reading to a fine enough degree. Thus, on a 10-inch rule, the scale from 80° to 90° covers one-eighth of an inch. If, however, you go to a large-diameter circular slide rule, you can get usable results. This is because the scale can be much longer. I have a Pickett "Atlas" circular slide rule that cost about the same as a set of sight reduction tables. It is 8 inches in diameter, and the multiplication scale, which is put on as a spiral, is about 17 feet long. Using it on Example *A*, I get an Hc of 37.65° or (multiplying 0.65 times 60′) 37°39′. You can't ask for much more than that! It's tedious but very compact, a metal disk 8 inches in diameter and less than one-eighth of an inch thick.

Turn now to the long-term almanac of the sun, which is printed as Table 43. As you see, it is divided into four columns with subheads for GHA and declination. Each month of the year is listed, and values are given usually for every three days, with a "quadrennial correction" value for GHA and declination. The base year of this table is 1956, so presume that you want the sun's GHA and declination for July 23, 1975, at 10-34-20 GMT:

1975
1956
———
19 = Years Since Table Base Year
19 ÷ 4 = 4 = Number of Quadrennial Units
3 Remaining = Use Column 3 in Table

Looking at the 4-year (quadrennial) corrections for GHA, you can see that multiplying them by 4 is not going to change things drastically (4 × −0.17′ for GHA on July 22 is less than 1′), so ignore the corrections for GHA.

APPENDIX X: LONG-TERM ALMANAC

SUN										
0		Quad. GHA Corr.	1		Date	2		Quad. Dec. Corr.	3	
GHA	Dec.		GHA	Dec.		GHA	Dec.		GHA	Dec.
				JANUARY						
° ′	° ′	′	° ′	° ′		° ′	° ′	′	° ′	° ′
179 14.9	23 06.1 S	−0.03	179 09.2	23 02.5 S	1	179 10.9	23 03.6 S	−0.15	179 12.9	23 04.7 S
178 53.7	22 50.8 S	−0.03	178 48.1	22 46.2 S	4	178 49.9	22 47.7 S	−0.20	178 51.8	22 49.1 S
178 33.4	22 31.5 S	−0.02	178 28.0	22 25.9 S	7	178 29.8	22 27.7 S	−0.22	178 31.5	22 29.4 S
178 14.0	22 08.2 S	−0.01	178 09.0	22 01.6 S	10	178 10.8	22 03.7 S	−0.26	178 12.2	22 05.8 S
177 55.9	21 41.0 S	0.00	177 51.3	21 33.4 S	13	177 52.9	21 35.9 S	−0.29	177 54.2	21 38.2 S
177 39.1	21 10.0 S	+0.01	177 35.0	21 01.5 S	16	177 36.4	21 04.2 S	−0.34	177 37.6	21 06.9 S
177 23.8	20 35.3 S	+0.04	177 20.2	20 26.0 S	19	177 21.5	20 28.9 S	−0.36	177 22.5	20 31.9 S
177 10.2	19 57.2 S	+0.05	177 07.1	19 47.0 S	22	177 08.1	19 50.2 S	−0.40	177 09.1	19 53.4 S
176 58.4	19 15.7 S	+0.08	176 55.7	19 04.6 S	25	176 56.4	19 08.1 S	−0.42	176 57.5	19 11.7 S
176 48.3	18 31.0 S	+0.10	176 46.0	18 19.1 S	28	176 46.6	18 22.9 S	−0.45	176 47.6	18 26.7 S
				FEBRUARY						
176 37.7	17 26.7 S	+0.12	176 35.9	17 13.9 S	1	176 36.4	17 18.0 S	−0.48	176 37.2	17 22.1 S
176 31.9	16 35.3 S	+0.13	176 30.4	16 21.8 S	4	176 31.0	16 26.1 S	−0.51	176 31.5	16 30.4 S
176 27.9	15 41.9 S	+0.15	176 26.9	15 27.1 S	7	176 27.4	15 31.7 S	−0.54	176 27.6	15 36.1 S
176 25.6	14 44.7 S	+0.15	176 25.1	14 30.0 S	10	176 25.5	14 34.8 S	−0.55	176 25.5	14 39.4 S
176 25.1	13 46.0 S	+0.17	176 25.2	13 30.8 S	13	176 25.4	13 35.8 S	−0.58	176 25.2	13 40.6 S
176 26.3	12 45.3 S	+0.18	176 27.0	12 29.6 S	16	176 26.9	12 34.7 S	−0.61	176 26.6	12 39.7 S
176 29.2	11 42.7 S	+0.20	176 30.3	11 26.6 S	19	176 30.0	11 31.8 S	−0.62	176 29.6	11 36.9 S
176 33.7	10 38.4 S	+0.22	176 35.2	10 22.0 S	22	176 34.7	10 27.3 S	−0.63	176 34.3	10 32.6 S
176 39.6	9 32.7 S	+0.23	176 41.4	9 15.9 S	25	176 40.8	9 21.3 S	−0.64	176 40.4	9 26.7 S
176 47.0	8 25.7 S	+0.24	176 49.0	8 08.0 S	28	176 48.3	8 14.1 S	−0.66	176 47.8	8 19.6 S
				MARCH						
176 52.5	7 40.4 S	+0.24	176 51.8	7 45.9 S	1	176 51.1	7 51.4 S	−0.66	176 50.6	7 57.0 S
177 01.8	6 31.7 S	+0.24	177 00.9	6 37.2 S	4	177 00.2	6 42.8 S	−0.67	176 59.6	6 48.5 S
177 12.1	5 22.1 S	+0.24	177 11.0	5 27.6 S	7	177 10.4	5 33.4 S	−0.68	177 09.5	5 39.1 S
177 23.2	4 11.8 S	+0.24	177 22.1	4 17.5 S	10	177 21.5	4 23.3 S	−0.69	177 20.4	4 28.9 S
177 35.1	3 01.1 S	+0.24	177 34.0	3 06.8 S	13	177 33.3	3 12.6 S	−0.71	177 32.1	3 18.3 S
177 47.6	1 50.0 S	+0.24	177 46.6	1 55.8 S	16	177 45.8	2 01.7 S	−0.71	177 44.5	2 07.4 S
178 00.6	0 38.9 S	+0.23	177 59.7	0 44.7 S	19	177 58.7	0 50.5 S	−0.70	177 57.5	0 56.2 S
178 14.0	0 32.2 N	+0.23	178 13.1	0 26.4 N	22	178 12.0	0 20.6 N	+0.70	178 10.8	0 14.9 N
178 27.7	1 43.1 N	+0.23	178 26.8	1 37.3 N	25	178 25.5	1 31.6 N	+0.70	178 24.5	1 25.8 N
178 41.4	2 53.7 N	+0.22	178 40.4	2 47.9 N	28	[illegible]	[illegible]	+0.70	178 38.2	2 36.5 N
				APRIL						
178 59.6	4 26.9 N	+0.22	178 58.5	4 21.3 N	1	178 57.4	4 15.6 N	+0.70	178 56.4	4 09.9 N
179 13.0	5 36.0 N	+0.19	179 11.9	5 30.5 N	4	179 10.9	5 24.8 N	+0.70	179 09.8	5 19.2 N
179 26.0	6 44.3 N	+0.18	179 24.9	6 38.8 N	7	179 24.0	6 33.2 N	+0.68	179 22.9	6 27.7 N
179 38.4	7 51.5 N	+0.17	179 37.4	7 46.1 N	10	179 36.6	7 40.6 N	+0.68	179 35.5	7 35.2 N
179 50.3	8 57.5 N	+0.14	179 49.5	8 52.2 N	13	179 48.6	8 46.8 N	+0.67	179 47.5	8 41.5 N
180 01.5	10 02.2 N	+0.12	180 00.8	9 56.9 N	16	179 59.9	9 51.7 N	+0.65	179 58.8	9 46.6 N
180 11.8	11 05.4 N	+0.10	180 11.3	11 00.2 N	19	180 10.4	10 55.2 N	+0.64	180 09.4	10 50.1 N
180 21.3	12 06.9 N	+0.09	180 20.9	12 01.9 N	22	180 19.9	11 57.0 N	+0.62	180 19.2	11 52.0 N
180 29.9	13 06.7 N	+0.08	180 29.4	13 01.8 N	25	180 28.5	12 57.1 N	+0.60	180 27.9	12 52.2 N
180 37.3	14 04.5 N	+0.06	180 36.8	13 59.8 N	28	180 36.0	13 55.2 N	+0.59	180 35.6	13 50.5 N
				MAY						
180 43.6	15 00.2 N	+0.04	180 43.1	14 55.7 N	1	180 42.5	14 51.3 N	+0.56	180 42.1	14 46.7 N
180 48.6	15 53.7 N	0.00	180 48.2	15 49.4 N	4	180 47.8	15 45.1 N	+0.54	180 47.4	15 40.8 N
180 52.4	16 44.8 N	−0.02	180 52.1	16 40.7 N	7	180 51.8	16 36.6 N	+0.52	180 51.4	16 32.5 N
180 54.8	17 33.4 N	−0.04	180 54.7	17 29.5 N	10	180 54.6	17 25.6 N	+0.50	180 54.2	17 21.8 N
180 56.0	18 19.4 N	−0.08	180 56.1	18 15.7 N	13	180 56.0	18 12.0 N	+0.47	180 55.7	18 08.4 N
180 55.9	19 02.6 N	−0.10	180 56.2	18 59.1 N	16	180 56.1	18 55.7 N	+0.44	180 55.9	18 52.3 N
180 54.5	19 42.9 N	−0.12	180 55.0	19 39.7 N	19	180 54.8	19 36.5 N	+0.40	180 54.9	19 33.3 N
180 52.0	20 20.2 N	−0.14	180 52.6	20 17.2 N	22	180 52.4	20 14.3 N	+0.37	180 52.7	20 11.3 N
180 48.3	20 54.4 N	−0.14	180 48.9	20 51.6 N	25	180 48.8	20 49.0 N	+0.34	180 49.2	20 46.3 N
180 43.5	21 25.3 N	−0.16	180 44.1	21 22.8 N	28	180 44.1	21 20.4 N	+0.30	180 44.7	21 18.0 N
				JUNE						
180 35.5	22 01.3 N	−0.18	180 36.0	21 59.2 N	1	180 36.3	21 57.2 N	+0.25	180 36.9	21 55.1 N
180 28.3	22 24.3 N	−0.22	180 28.9	22 22.5 N	4	180 29.3	22 20.8 N	+0.21	180 29.9	22 18.9 N
180 20.3	22 43.8 N	−0.24	180 21.0	22 42.3 N	7	180 21.5	22 40.8 N	+0.18	180 22.0	22 39.3 N
180 11.7	22 59.7 N	−0.25	180 12.5	22 58.5 N	10	180 13.0	22 57.3 N	+0.14	180 13.5	22 56.1 N
180 02.5	23 11.9 N	−0.26	180 03.5	23 11.0 N	13	180 04.0	23 10.1 N	+0.10	180 04.5	23 09.2 N
179 53.0	23 20.5 N	−0.27	179 54.1	23 19.9 N	16	179 54.5	23 19.3 N	+0.06	179 55.1	23 18.7 N
179 43.3	23 25.4 N	−0.27	179 44.4	23 25.1 N	19	179 44.7	23 24.8 N	+0.01	179 45.4	23 24.5 N
179 33.6	23 26.6 N	−0.27	179 34.7	23 26.6 N	22	179 34.9	23 26.5 N	−0.02	179 35.7	23 26.5 N
179 24.0	23 24.0 N	−0.26	179 24.9	23 24.3 N	25	179 25.1	23 24.6 N	−0.07	179 26.1	23 24.9 N
179 14.6	23 17.8 N	−0.26	179 15.4	23 18.3 N	28	179 15.7	23 18.9 N	−0.12	179 16.6	23 19.5 N

Table 43

APPENDIX X: LONG-TERM ALMANAC

SUN										
0		Quad. GHA Corr.	1		Date	2		Quad. Dec. Corr.	3	
GHA	Dec.		GHA	Dec.		GHA	Dec.		GHA	Dec.
					JULY					
° ′	° ′	′	° ′	° ′		° ′	° ′	′	° ′	° ′
179 05.7	23 07.8 N	−0.26	179 06.3	23 08.7 N	1	179 06.7	23 09.6 N	−0.15	179 07.5	23 10.4 N
178 57.2	22 54.2 N	−0.26	178 57.8	22 55.4 N	4	178 58.2	22 56.6 N	−0.19	178 58.8	22 57.7 N
178 49.4	22 37.0 N	−0.26	178 50.0	22 38.5 N	7	178 50.4	22 40.0 N	−0.24	178 50.8	22 41.4 N
178 42.4	22 16.3 N	−0.25	178 43.1	22 18.1 N	10	178 43.3	22 19.8 N	−0.27	178 43.7	22 21.5 N
178 36.4	21 52.2 N	−0.23	178 37.1	21 54.3 N	13	178 37.2	21 56.2 N	−0.31	178 37.4	21 58.2 N
178 31.5	21 24.7 N	−0.23	178 32.2	21 27.0 N	16	178 32.0	21 29.2 N	−0.35	178 32.3	21 31.5 N
178 27.8	20 53.9 N	−0.20	178 28.4	20 56.5 N	19	178 28.0	20 58.9 N	−0.39	178 28.3	21 01.5 N
178 25.3	20 19.9 N	−0.17	178 25.7	20 22.8 N	22	178 25.3	20 25.5 N	−0.41	178 25.6	20 28.3 N
178 24.1	19 42.9 N	−0.15	178 24.3	19 46.0 N	25	178 23.8	19 48.9 N	−0.44	178 24.1	19 52.0 N
178 24.3	19 02.9 N	−0.13	178 24.2	19 06.2 N	28	178 23.7	19 09.4 N	−0.47	178 23.9	19 12.8 N
					AUGUST					
178 26.5	18 05.2 N	−0.11	178 26.2	18 08.8 N	1	178 25.7	18 12.4 N	−0.52	178 25.6	18 16.0 N
178 29.7	17 18.8 N	−0.11	178 29.3	17 22.7 N	4	178 28.8	17 26.4 N	−0.56	178 28.5	17 30.2 N
178 34.2	16 29.9 N	−0.09	178 33.9	16 34.0 N	7	178 33.2	16 37.9 N	−0.58	178 32.7	16 41.9 N
178 40.1	15 38.6 N	−0.07	178 39.7	15 42.9 N	10	178 39.0	15 47.0 N	−0.60	178 38.3	15 51.1 N
178 47.3	14 45.0 N	−0.05	178 46.9	14 49.5 N	13	178 45.9	14 53.8 N	−0.62	178 45.2	14 58.1 N
178 55.8	13 49.4 N	−0.03	178 55.3	13 54.0 N	16	178 54.1	13 58.4 N	−0.64	178 53.4	14 03.0 N
179 05.4	12 51.7 N	0.00	179 04.9	12 56.5 N	19	179 03.5	13 01.1 N	−0.66	179 02.9	13 05.8 N
179 16.2	11 52.3 N	+0.02	179 15.5	11 57.2 N	22	179 14.1	12 01.9 N	−0.68	179 13.4	12 06.8 N
179 28.1	10 51.1 N	+0.04	179 27.1	10 56.1 N	25	179 25.7	11 01.0 N	−0.69	179 25.0	11 06.0 N
179 40.8	9 48.4 N	+0.06	179 39.7	9 53.5 N	28	179 38.3	9 58.5 N	−0.70	179 37.5	10 03.7 N
					SEPTEMBER					
179 58.9	8 22.6 N	+0.07	179 57.7	8 27.8 N	1	179 56.4	8 33.0 N	−0.72	179 55.4	8 38.3 N
180 13.3	7 16.8 N	+0.07	180 12.1	7 22.1 N	4	180 10.8	7 27.4 N	−0.73	180 09.6	7 32.8 N
180 28.2	6 09.9 N	+0.08	180 27.1	6 15.4 N	7	180 25.7	6 20.8 N	−0.74	180 24.5	6 26.2 N
180 43.6	5 02.1 N	+0.10	180 42.6	5 07.7 N	10	180 41.1	5 13.1 N	−0.75	180 39.8	5 18.7 N
180 59.4	3 53.6 N	+0.11	180 58.3	3 59.3 N	13	180 56.8	4 04.7 N	−0.75	180 55.5	4 10.3 N
181 15.3	2 44.5 N	+0.12	181 14.3	2 50.2 N	16	181 12.6	2 55.6 N	−0.75	181 11.5	3 01.3 N
181 31.4	1 34.8 N	+0.13	181 30.3	1 40.6 N	19	181 28.6	1 46.1 N	−0.75	181 27.5	1 51.8 N
181 47.3	0 24.9 N	+0.13	181 46.1	0 30.6 N	22	181 44.5	0 36.2 N	−0.75	181 43.5	0 42.0 N
182 03.0	0 45.2 S	+0.14	182 01.7	0 39.5 S	25	182 00.3	0 33.9 S	+0.74	181 59.3	0 28.1 S
182 18.3	1 55.3 S	+0.14	182 17.0	1 49.7 S	28	182 15.7	1 44.0 S	+0.74	182 14.7	1 38.3 S
					OCTOBER					
182 33.1	3 05.4 S	+0.14	182 31.9	2 59.7 S	1	182 30.7	2 54.1 S	+0.74	182 29.6	2 48.3 S
182 47.2	4 15.1 S	+0.13	182 46.1	4 09.5 S	4	182 45.0	4 03.8 S	+0.73	182 43.9	3 58.1 S
183 00.6	5 24.4 S	+0.12	182 59.6	5 18.7 S	7	182 58.5	5 13.1 S	+0.72	182 57.4	5 07.5 S
183 13.0	6 33.0 S	+0.12	183 12.2	6 27.4 S	10	183 11.1	6 21.9 S	+0.71	183 10.2	6 16.3 S
183 24.4	7 40.9 S	+0.12	183 23.8	7 35.3 S	13	183 22.7	7 29.9 S	+0.70	183 21.9	7 24.4 S
183 34.7	8 47.8 S	+0.11	183 34.2	8 42.3 S	16	183 33.1	8 37.0 S	+0.68	183 32.5	8 31.5 S
183 43.8	9 53.5 S	+0.11	183 43.3	9 48.2 S	19	183 42.3	9 42.9 S	+0.66	183 41.9	9 37.5 S
183 51.4	10 58.0 S	+0.10	183 50.9	10 52.8 S	22	183 50.1	10 47.6 S	+0.64	183 49.9	10 42.3 S
183 57.6	12 01.0 S	+0.10	183 57.1	11 55.9 S	25	183 56.5	11 50.9 S	+0.63	183 56.4	11 45.7 S
184 02.0	13 02.3 S	+0.08	184 01.6	12 57.4 S	28	184 01.3	12 52.5 S	+0.62	184 01.2	12 47.4 S
					NOVEMBER					
184 05.3	14 21.3 S	+0.06	184 05.1	14 16.6 S	1	184 05.1	14 11.8 S	+0.57	184 05.0	14 07.0 S
184 05.6	15 18.1 S	+0.04	184 05.7	15 13.5 S	4	184 05.8	15 08.9 S	+0.55	184 05.8	15 04.3 S
184 04.1	16 12.6 S	+0.03	184 04.4	16 08.2 S	7	184 04.7	16 03.8 S	+0.53	184 04.8	15 59.4 S
184 00.8	17 04.7 S	+0.02	184 01.3	17 00.5 S	10	184 01.6	16 56.3 S	+0.50	184 01.9	16 52.1 S
183 55.5	17 54.2 S	0.00	183 56.3	17 50.2 S	13	183 56.6	17 46.2 S	+0.46	183 57.2	17 42.2 S
183 48.4	18 40.8 S	−0.01	183 49.3	18 37.1 S	16	183 49.7	18 33.4 S	+0.42	183 50.6	18 29.6 S
183 39.5	19 24.5 S	−0.01	183 40.4	19 21.0 S	19	183 41.0	19 17.6 S	+0.40	183 42.1	19 14.0 S
183 28.7	20 05.0 S	−0.02	183 29.6	20 01.8 S	22	183 30.5	19 58.6 S	+0.36	183 31.7	19 55.3 S
183 16.0	20 42.3 S	−0.03	183 17.0	20 39.3 S	25	183 18.1	20 36.4 S	+0.33	183 19.5	20 33.4 S
183 01.6	21 16.0 S	−0.06	183 02.7	21 13.4 S	28	183 04.1	21 10.7 S	+0.29	183 05.4	21 08.0 S
					DECEMBER					
182 45.6	21 46.2 S	−0.08	182 46.9	21 43.8 S	1	182 48.5	21 41.4 S	+0.29	182 49.8	21 39.0 S
182 28.1	22 12.6 S	−0.08	182 29.6	22 10.5 S	4	182 31.3	22 08.4 S	+0.22	182 32.6	22 06.4 S
182 09.3	22 35.1 S	−0.09	182 11.1	22 33.4 S	7	182 12.8	22 31.6 S	+0.18	182 14.2	22 29.9 S
181 49.5	22 53.7 S	−0.09	181 51.4	22 52.3 S	10	181 53.0	22 50.8 S	+0.15	181 54.6	22 49.4 S
181 28.7	23 08.1 S	−0.08	181 30.7	23 07.1 S	13	181 32.3	23 06.0 S	+0.11	181 34.0	23 04.9 S
181 07.2	23 18.5 S	−0.08	181 09.2	23 17.8 S	16	181 10.8	23 17.0 S	+0.06	181 12.7	23 16.3 S
180 45.2	23 24.7 S	−0.08	180 47.1	23 24.3 S	19	180 48.8	23 23.9 S	+0.02	180 50.8	23 23.4 S
180 22.9	23 26.6 S	−0.08	180 24.7	23 26.5 S	22	180 26.5	23 26.5 S	−0.01	180 28.5	23 26.4 S
180 00.5	23 24.3 S	−0.08	180 02.2	23 24.6 S	25	180 04.2	23 24.9 S	−0.05	180 06.1	23 25.2 S
179 38.3	23 17.8 S	−0.07	179 39.9	23 18.4 S	28	179 41.9	23 19.1 S	−0.09	179 43.7	23 19.6 S

Table 43, continued

In this case, the declination correction amounts to about −2′, so proceed this way:

20°28′N	7-22	4 × −0.41 = −2′
−2′		
20°26′N	= 7-22 Declination	
19°52′N	= 7-25	
−2′		
19°50′N	= 7-25 Declination	

178°26′	= 7-22 GHA	20° 26′ N	
178°24′	= 7-25 GHA	19° 50′ N	
−2′	= 3-Day Change	− 36′	= 3-Day Change
−0.67′	= 1-Day Change	− 12′	= 1-Day Change
178°25′	= 7-23 GHA	20° 14′ N	= 7-23 Declination

Now this is for 0000 hours. You want an additional 10 hours, 34 minutes, and 20 seconds. Using the table "Conversion of Arc to Time" (Table 44), you can compute the following:

10 Hr. = 150°
34 Min. = 8.5° = 8°30′
20 Sec. = 5′
Adding: 178°25′
150°
8°30′
5′

336°60′ = 337° = GHA Sun 7-23-75 at 10-34-20 Hours

From the *Nautical Almanac* for 1975, GHA sun for this date and time is 336°59′. Close enough!

Now back to declination. At 0000 hours on 7-23-75 it was 20°14′ north. You found that the daily change was −12′, so the hourly change is −0.5′. Multiplying −0.5′ times 10, you find that the total change is −5′. The declination is therefore 20°09′ north (20°14′ − 5′). From the almanac, the declination is 20°09.6′ north. Again, close enough!

CONVERSION OF ARC TO TIME

0°–59°		60°–119°		120°–179°		180°–239°		240°–299°		300°–359°			0′·00	0′·25	0′·50	0′·75
°	h m	°	h m	°	h m	°	h m	°	h m	°	h m	′	m s	m s	m s	m s
0	0 00	60	4 00	120	8 00	180	12 00	240	16 00	300	20 00	0	0 00	0 01	0 02	0 03
1	0 04	61	4 04	121	8 04	181	12 04	241	16 04	301	20 04	1	0 04	0 05	0 06	0 07
2	0 08	62	4 08	122	8 08	182	12 08	242	16 08	302	20 08	2	0 08	0 09	0 10	0 11
3	0 12	63	4 12	123	8 12	183	12 12	243	16 12	303	20 12	3	0 12	0 13	0 14	0 15
4	0 16	64	4 16	124	8 16	184	12 16	244	16 16	304	20 16	4	0 16	0 17	0 18	0 19
5	0 20	65	4 20	125	8 20	185	12 20	245	16 20	305	20 20	5	0 20	0 21	0 22	0 23
6	0 24	66	4 24	126	8 24	186	12 24	246	16 24	306	20 24	6	0 24	0 25	0 26	0 27
7	0 28	67	4 28	127	8 28	187	12 28	247	16 28	307	20 28	7	0 28	0 29	0 30	0 31
8	0 32	68	4 32	128	8 32	188	12 32	248	16 32	308	20 32	8	0 32	0 33	0 34	0 35
9	0 36	69	4 36	129	8 36	189	12 36	249	16 36	309	20 36	9	0 36	0 37	0 38	0 39
10	0 40	70	4 40	130	8 40	190	12 40	250	16 40	310	20 40	10	0 40	0 41	0 42	0 43
11	0 44	71	4 44	131	8 44	191	12 44	251	16 44	311	20 44	11	0 44	0 45	0 46	0 47
12	0 48	72	4 48	132	8 48	192	12 48	252	16 48	312	20 48	12	0 48	0 49	0 50	0 51
13	0 52	73	4 52	133	8 52	193	12 52	253	16 52	313	20 52	13	0 52	0 53	0 54	0 55
14	0 56	74	4 56	134	8 56	194	12 56	254	16 56	314	20 56	14	0 56	0 57	0 58	0 59
15	1 00	75	5 00	135	9 00	195	13 00	255	17 00	315	21 00	15	1 00	1 01	1 02	1 03
16	1 04	76	5 04	136	9 04	196	13 04	256	17 04	316	21 04	16	1 04	1 05	1 06	1 07
17	1 08	77	5 08	137	9 08	197	13 08	257	17 08	317	21 08	17	1 08	1 09	1 10	1 11
18	1 12	78	5 12	138	9 12	198	13 12	258	17 12	318	21 12	18	1 12	1 13	1 14	1 15
19	1 16	79	5 16	139	9 16	199	13 16	259	17 16	319	21 16	19	1 16	1 17	1 18	1 19
20	1 20	80	5 20	140	9 20	200	13 20	260	17 20	320	21 20	20	1 20	1 21	1 22	1 23
21	1 24	81	5 24	141	9 24	201	13 24	261	17 24	321	21 24	21	1 24	1 25	1 26	1 27
22	1 28	82	5 28	142	9 28	202	13 28	262	17 28	322	21 28	22	1 28	1 29	1 30	1 31
23	1 32	83	5 32	143	9 32	203	13 32	263	17 32	323	21 32	23	1 32	1 33	1 34	1 35
24	1 36	84	5 36	144	9 36	204	13 36	264	17 36	324	21 36	24	1 36	1 37	1 38	1 39
25	1 40	85	5 40	145	9 40	205	13 40	265	17 40	325	21 40	25	1 40	1 41	1 42	1 43
26	1 44	86	5 44	146	9 44	206	13 44	266	17 44	326	21 44	26	1 44	1 45	1 46	1 47
27	1 48	87	5 48	147	9 48	207	13 48	267	17 48	327	21 48	27	1 48	1 49	1 50	1 51
28	1 52	88	5 52	148	9 52	208	13 52	268	17 52	328	21 52	28	1 52	1 53	1 54	1 55
29	1 56	89	5 56	149	9 56	209	13 56	269	17 56	329	21 56	29	1 56	1 57	1 58	1 59
30	2 00	90	6 00	150	10 00	210	14 00	270	18 00	330	22 00	30	2 00	2 01	2 02	2 03
31	2 04	91	6 04	151	10 04	211	14 04	271	18 04	331	22 04	31	2 04	2 05	2 06	2 07
32	2 08	92	6 08	152	10 08	212	14 08	272	18 08	332	22 08	32	2 08	2 09	2 10	2 11
33	2 12	93	6 12	153	10 12	213	14 12	273	18 12	333	22 12	33	2 12	2 13	2 14	2 15
34	2 16	94	6 16	154	10 16	214	14 16	274	18 16	334	22 16	34	2 16	2 17	2 18	2 19
35	2 20	95	6 20	155	10 20	215	14 20	275	18 20	335	22 20	35	2 20	2 21	2 22	2 23
36	2 24	96	6 24	156	10 24	216	14 24	276	18 24	336	22 24	36	2 24	2 25	2 26	2 27
37	2 28	97	6 28	157	10 28	217	14 28	277	18 28	337	22 28	37	2 28	2 29	2 30	2 31
38	2 32	98	6 32	158	10 32	218	14 32	278	18 32	338	22 32	38	2 32	2 33	2 34	2 35
39	2 36	99	6 36	159	10 36	219	14 36	279	18 36	339	22 36	39	2 36	2 37	2 38	2 39
40	2 40	100	6 40	160	10 40	220	14 40	280	18 40	340	22 40	40	2 40	2 41	2 42	2 43
41	2 44	101	6 44	161	10 44	221	14 44	281	18 44	341	22 44	41	2 44	2 45	2 46	2 47
42	2 48	102	6 48	162	10 48	222	14 48	282	18 48	342	22 48	42	2 48	2 49	2 50	2 51
43	2 52	103	6 52	163	10 52	223	14 52	283	18 52	343	22 52	43	2 52	2 53	2 54	2 55
44	2 56	104	6 56	164	10 56	224	14 56	284	18 56	344	22 56	44	2 56	2 57	2 58	2 59
45	3 00	105	7 00	165	11 00	225	15 00	285	19 00	345	23 00	45	3 00	3 01	3 02	3 03
46	3 04	106	7 04	166	11 04	226	15 04	286	19 04	346	23 04	46	3 04	3 05	3 06	3 07
47	3 08	107	7 08	167	11 08	227	15 08	287	19 08	347	23 08	47	3 08	3 09	3 10	3 11
48	3 12	108	7 12	168	11 12	228	15 12	288	19 12	348	23 12	48	3 12	3 13	3 14	3 15
49	3 16	109	7 16	169	11 16	229	15 16	289	19 16	349	23 16	49	3 16	3 17	3 18	3 19
50	3 20	110	7 20	170	11 20	230	15 20	290	19 20	350	23 20	50	3 20	3 21	3 22	3 23
51	3 24	111	7 24	171	11 24	231	15 24	291	19 24	351	23 24	51	3 24	3 25	3 26	3 27
52	3 28	112	7 28	172	11 28	232	15 28	292	19 28	352	23 28	52	3 28	3 29	3 30	3 31
53	3 32	113	7 32	173	11 32	233	15 32	293	19 32	353	23 32	53	3 32	3 33	3 34	3 35
54	3 36	114	7 36	174	11 36	234	15 36	294	19 36	354	23 36	54	3 36	3 37	3 38	3 39
55	3 40	115	7 40	175	11 40	235	15 40	295	19 40	355	23 40	55	3 40	3 41	3 42	3 43
56	3 44	116	7 44	176	11 44	236	15 44	296	19 44	356	23 44	56	3 44	3 45	3 46	3 47
57	3 48	117	7 48	177	11 48	237	15 48	297	19 48	357	23 48	57	3 48	3 49	3 50	3 51
58	3 52	118	7 52	178	11 52	238	15 52	298	19 52	358	23 52	58	3 52	3 53	3 54	3 55
59	3 56	119	7 56	179	11 56	239	15 56	299	19 56	359	23 56	59	3 56	3 57	3 58	3 59

The above table is for converting expressions in arc to their equivalent in time ; its main use in this Almanac is for the conversion of longitude for application to L.M.T. (*added* if *west*, *subtracted* if *east*) to give G.M.T. or vice versa, particularly in the case of sunrise, sunset, etc.

Table 44

This chapter has contained a good deal of information, so perhaps a short summary is in order. The two basic purposes of the chapter were (1) to present ways of solving the navigational triangle directly as a means of escaping the bulk and weight of the sight reduction tables, and (2) to make this book, by means of the Long-Term Almanac and the Sine-Cosine Tables, a complete, self-contained celestial navigation system to which you need to add only a sextant and watch.

What this means is that your initial investment in celestial navigation can consist solely of a relatively cheap sextant and this book.

LOGARITHMS

0° → ↓ ′	sin	Diff. 1′	sin	Diff. 1′	Diff. 1′	cos	cos	Diff. 1′	← 179° ↓ ′
0	0.00000	29	∞	—	0	10.00000	1.00000	0	60
1	.00029	29	6.46373	30103	0	.00000	.00000	0	59
2	.00058	29	.76476	17609	0	.00000	.00000	0	58
3	.00087	29	6.94085	12494	0	.00000	.00000	0	57
4	.00116	29	7.06579	9691	0	.00000	.00000	0	56
5	0.00145	30	7.16270	7918	0	10.00000	1.00000	0	55
6	.00175	29	.24188	6694	0	.00000	.00000	0	54
7	.00204	29	.30882	5800	0	.00000	.00000	0	53
8	.00233	29	.36682	5115	0	.00000	.00000	0	52
9	.00262	29	.41797	4576	0	.00000	.00000	0	51
10	0.00291	29	7.46373	4139	0	10.00000	1.00000	1	50
11	.00320	29	.50512	3779	0	.00000	0.99999	0	49
12	.00349	29	.54291	3476	0	.00000	.99999	0	48
13	.00378	29	.57767	3218	0	.00000	.99999	0	47
14	.00407	29	.60985	2997	0	.00000	.99999	0	46
15	0.00436	29	7.63982	2802	0	10.00000	0.99999	0	45
16	.00465	30	.66784	2633	1	10.00000	.99999	0	44
17	.00495	29	.69417	2483	0	9.99999	.99999	0	43
18	.00524	29	.71900	2348	0	.99999	.99999	1	42
19	.00553	29	.74248	2227	0	.99999	.99998	0	41
20	0.00582	29	7.76475	2119	0	9.99999	0.99998	0	40
21	.00611	29	.78594	2021	0	.99999	.99998	0	39
22	.00640	29	.80615	1930	0	.99999	.99998	0	38
23	.00669	29	.82545	1848	0	.99999	.99998	0	37
24	.00698	29	.84393	1773	0	.99999	.99998	1	36
25	0.00727	29	7.86166	1704	0	9.99999	0.99997	0	35
26	.00756	29	.87870	1639	0	.99999	.99997	0	34
27	.00785	29	.89509	1579	0	.99999	.99997	0	33
28	.00814	30	.91088	1524	1	.99999	.99997	1	32
29	.00844	29	.92612	1472	0	.99998	.99996	0	31
30	0.00873	29	7.94084	1424	0	9.99998	0.99996	0	30
31	.00902	29	.95508	1379	0	.99998	.99996	0	29
32	.00931	29	.96887	1336	0	.99998	.99996	1	28
33	.00960	29	.98223	1297	0	.99998	.99995	0	27
34	.00989	29	7.99520	1259	0	.99998	.99995	0	26
35	0.01018	29	8.00779	1223	0	9.99998	0.99995	0	25
36	.01047	29	.02002	1190	1	.99998	.99995	1	24
37	.01076	29	.03192	1158	0	.99997	.99994	0	23
38	.01105	29	.04350	1128	0	.99997	.99994	0	22
39	.01134	30	.05478	1100	0	.99997	.99994	1	21
40	0.01164	29	8.06578	1072	0	9.99997	0.99993	0	20
41	.01193	29	.07650	1046	0	.99997	.99993	0	19
42	.01222	29	.08696	1022	0	.99997	.99993	1	18
43	.01251	29	.09718	999	1	.99997	.99992	0	17
44	.01280	29	.10717	976	0	.99996	.99992	1	16
45	0.01309	29	8.11693	954	0	9.99996	0.99991	0	15
46	.01338	29	.12647	934	0	.99996	.99991	0	14
47	.01367	29	.13581	914	0	.99996	.99991	1	13
48	.01396	29	.14495	896	0	.99996	.99990	0	12
49	.01425	29	.15391	877	1	.99996	.99990	1	11
50	0.01454	29	8.16268	860	0	9.99995	0.99989	0	10
51	.01483	30	.17128	843	0	.99995	.99989	0	9
52	.01513	29	.17971	827	0	.99995	.99989	1	8
53	.01542	29	.18798	812	0	.99995	.99988	0	7
54	.01571	29	.19610	797	1	.99995	.99988	1	6
55	0.01600	29	8.20407	782	0	9.99994	0.99987	0	5
56	.01629	29	.21189	769	0	.99994	.99987	1	4
57	.01658	29	.21958	755	0	.99994	.99986	0	3
58	.01687	29	.22713	743	0	.99994	.99986	1	2
59	.01716	29	.23456	730	1	.99994	.99985	0	1
60	0.01745		8.24186			9.99993	0.99985		0
↑ 90° →	cos	Diff. 1′	cos	Diff. 1′	Diff. 1′	sin	sin	Diff. 1′	← 89° ↑

1° → ↓ ′	sin	Diff. 1′	sin	Diff. 1′	Diff. 1′	cos ←	cos	Diff. 1′	← 178° ↓ ′
0	0.01745	29	8.24186	717	0	9.99993	0.99985	1	60
1	.01774	29	.24903	706	0	.99993	.99984	0	59
2	.01803	29	.25609	695	0	.99993	.99984	1	58
3	.01832	30	.26304	684	1	.99993	.99983	0	57
4	.01862	29	.26988	673	0	.99992	.99983	1	56
5	0.01891	29	8.27661	663	0	9.99992	0.99982	0	55
6	.01920	29	.28324	653	0	.99992	.99982	1	54
7	.01949	29	.28977	644	0	.99992	.99981	1	53
8	.01978	29	.29621	634	1	.99992	.99980	0	52
9	.02007	29	.30255	624	0	.99991	.99980	1	51
10	0.02036	29	8.30879	616	0	9.99991	0.99979	0	50
11	.02065	29	.31495	608	1	.99991	.99979	1	49
12	.02094	29	.32103	599	0	.99990	.99978	1	48
13	.02123	29	.32702	590	0	.99990	.99977	0	47
14	.02152	29	.33292	583	0	.99990	.99977	1	46
15	0.02181	30	8.33875	575	1	9.99990	0.99976	0	45
16	.02211	29	.34450	568	0	.99989	.99976	1	44
17	.02240	29	.35018	560	0	.99989	.99975	1	43
18	.02269	29	.35578	553	0	.99989	.99974	0	42
19	.02298	29	.36131	547	1	.99989	.99974	1	41
20	0.02327	29	8.36678	539	0	9.99988	0.99973	1	40
21	.02356	29	.37217	533	0	.99988	.99972	0	39
22	.02385	29	.37750	526	1	.99988	.99972	1	38
23	.02414	29	.38276	520	0	.99987	.99971	1	37
24	.02443	29	.38796	514	0	.99987	.99970	1	36
25	0.02472	29	8.39310	508	1	9.99987	0.99969	0	35
26	.02501	29	.39818	502	0	.99986	.99969	1	34
27	.02530	30	.40320	496	0	.99986	.99968	1	33
28	.02560	29	.40816	491	1	.99986	.99967	1	32
29	.02589	29	.41307	485	0	.99985	.99966	0	31
30	0.02618	29	8.41792	480	0	9.99985	0.99966	1	30
31	.02647	29	.42272	474	1	.99985	.99965	1	29
32	.02676	29	.42746	470	0	.99984	.99964	1	28
33	.02705	29	.43216	464	0	.99984	.99963	0	27
34	.02734	29	.43680	459	1	.99984	.99963	1	26
35	0.02763	29	8.44139	455	0	9.99983	0.99962	1	25
36	.02792	29	.44594	450	0	.99983	.99961	1	24
37	.02821	29	.45044	445	1	.99983	.99960	1	23
38	.02850	29	.45489	441	0	.99982	.99959	0	22
39	.02879	29	.45930	436	0	.99982	.99959	1	21
40	0.02908	30	8.46366	433	1	9.99982	0.99958	1	20
41	.02938	29	.46799	427	0	.99981	.99957	1	19
42	.02967	29	.47226	424	0	.99981	.99956	1	18
43	.02996	29	.47650	419	1	.99981	.99955	1	17
44	.03025	29	.48069	416	0	.99980	.99954	1	16
45	0.03054	29	8.48485	411	1	9.99980	0.99953	1	15
46	.03083	29	.48896	408	0	.99979	.99952	0	14
47	.03112	29	.49304	404	0	.99979	.99952	1	13
48	.03141	29	.49708	400	1	.99979	.99951	1	12
49	.03170	29	.50108	396	0	.99978	.99950	1	11
50	0.03199	29	8.50504	393	1	9.99978	0.99949	1	10
51	.03228	29	.50897	390	0	.99977	.99948	1	9
52	.03257	29	.51287	386	0	.99977	.99947	1	8
53	.03286	30	.51673	382	1	.99977	.99946	1	7
54	.03316	29	.52055	379	0	.99976	.99945	1	6
55	0.03345	29	8.52434	376	1	9.99976	0.99944	1	5
56	.03374	29	.52810	373	0	.99975	.99943	1	4
57	.03403	29	.53183	369	1	.99975	.99942	1	3
58	.03432	29	.53552	367	0	.99974	.99941	1	2
59	.03461	29	.53919	363	0	.99974	.99940	1	1
60	0.03490		8.54282			9.99974	0.99939		0
↑ 91° →	cos	Diff. 1′	cos	Diff. 1′	Diff. 1′	sin	sin	Diff. 1′	↑ ← 88°

LOGARITHMS

2°→ ↓ ′	sin	Diff. 1′	sin	Diff. 1′	Diff. 1′	cos ←	cos	Diff. 1′	←177° ↓ ′
0	0. 03490	29	8. 54282	360	1	9. 99974	0. 99939	1	60
1	. 03519	29	. 54642	357	0	. 99973	. 99938	1	59
2	. 03548	29	. 54999	355	1	. 99973	. 99937	1	58
3	. 03577	29	. 55354	351	0	. 99972	. 99936	1	57
4	. 03606	29	. 55705	349	1	. 99972	. 99935	1	56
5	0. 03635	29	8. 56054	346	0	9. 99971	0. 99934	1	55
6	. 03664	29	. 56400	343	1	. 99971	. 99933	1	54
7	. 03693	30	. 56743	341	0	. 99970	. 99932	1	53
8	. 03723	29	. 57084	337	1	. 99970	. 99931	1	52
9	. 03752	29	. 57421	336	0	. 99969	. 99930	1	51
10	0. 03781	29	8. 57757	332	1	9. 99969	0. 99929	2	50
11	. 03810	29	. 58089	330	0	. 99968	. 99927	1	49
12	. 03839	29	. 58419	328	1	. 99968	. 99926	1	48
13	. 03868	29	. 58747	325	0	. 99967	. 99925	1	47
14	. 03897	29	. 59072	323	0	. 99967	. 99924	1	46
15	0. 03926	29	8. 59395	320	1	9. 99967	0. 99923	1	45
16	. 03955	29	. 59715	318	0	. 99966	. 99922	1	44
17	. 03984	29	. 60033	316	1	. 99966	. 99921	2	43
18	. 04013	29	. 60349	313	1	. 99965	. 99919	1	42
19	. 04042	29	. 60662	311	0	. 99964	. 99918	1	41
20	0. 04071	29	8. 60973	309	1	9. 99964	0. 99917	1	40
21	. 04100	29	. 61282	307	0	. 99963	. 99916	1	39
22	. 04129	30	. 61589	305	1	. 99963	. 99915	2	38
23	. 04159	29	. 61894	302	0	. 99962	. 99913	1	37
24	. 04188	29	. 62196	301	1	. 99962	. 99912	1	36
25	0. 04217	29	8. 62497	298	0	9. 99961	0. 99911	1	35
26	. 04246	29	. 62795	296	1	. 99961	. 99910	1	34
27	. 04275	29	. 63091	294	0	. 99960	. 99909	2	33
28	. 04304	29	. 63385	293	1	. 99960	. 99907	1	32
29	. 04333	29	. 63678	290	0	. 99959	. 99906	1	31
30	0. 04362	29	8. 63968	288	1	9. 99959	0. 99905	1	30
31	. 04391	29	. 64256	287	0	. 99958	. 99904	2	29
32	. 04420	29	. 64543	284	1	. 99958	. 99902	1	28
33	. 04449	29	. 64827	283	1	. 99957	. 99901	1	27
34	. 04478	29	. 65110	281	0	. 99956	. 99900	2	26
35	0. 04507	29	8. 65391	279	1	9. 99956	0. 99898	1	25
36	. 04536	29	. 65670	277	0	. 99955	. 99897	1	24
37	. 04565	29	. 65947	276	1	. 99955	. 99896	2	23
38	. 04594	29	. 66223	274	0	. 99954	. 99894	1	22
39	. 04623	30	. 66497	272	1	. 99954	. 99893	1	21
40	0. 04653	29	8. 66769	270	1	9. 99953	0. 99892	2	20
41	. 04682	29	. 67039	269	0	. 99952	. 99890	1	19
42	. 04711	29	. 67308	267	1	. 99952	. 99889	1	18
43	. 04740	29	. 67575	266	0	. 99951	. 99888	2	17
44	. 04769	29	. 67841	263	1	. 99951	. 99886	1	16
45	0. 04798	29	8. 68104	263	1	9. 99950	0. 99885	2	15
46	. 04827	29	. 68367	260	0	. 99949	. 99883	1	14
47	. 04856	29	. 68627	259	1	. 99949	. 99882	1	13
48	. 04885	29	. 68886	258	0	. 99948	. 99881	2	12
49	. 04914	29	. 69144	256	1	. 99948	. 99879	1	11
50	0. 04943	29	8. 69400	254	1	9. 99947	0. 99878	2	10
51	. 04972	29	. 69654	253	0	. 99946	. 99876	1	9
52	. 05001	29	. 69907	252	1	. 99946	. 99875	2	8
53	. 05030	29	. 70159	250	1	. 99945	. 99873	1	7
54	. 05059	29	. 70409	249	0	. 99944	. 99872	2	6
55	0. 05088	29	8. 70658	247	1	9. 99944	0. 99870	1	5
56	. 05117	29	. 70905	246	1	. 99943	. 99869	2	4
57	. 05146	29	. 71151	244	0	. 99942	. 99867	1	3
58	. 05175	30	. 71395	243	1	. 99942	. 99866	2	2
59	. 05205	29	. 71638	242	1	. 99941	. 99864	1	1
60	0. 05234		8. 71880			9. 99940	0. 99863		0
↑ 92°→	cos	Diff. 1′	cos	Diff. 1′	Diff. 1′	sin	sin	Diff. 1′	←87° ↑

3°→ ↓ ′	sin	Diff. 1′	sin	Diff. 1′	Diff. 1′	cos ←	cos	Diff. 1′	←176° ↓ ′
0	0. 05234	29	8. 71880	240	0	9. 99940	0. 99863	2	60
1	. 05263	29	. 72120	239	1	. 99940	. 99861	1	59
2	. 05292	29	. 72359	238	1	. 99939	. 99860	2	58
3	. 05321	29	. 72597	237	0	. 99938	. 99858	1	57
4	. 05350	29	. 72834	235	1	. 99938	. 99857	2	56
5	0. 05379	29	8. 73069	234	1	9. 99937	0. 99855	1	55
6	. 05408	29	. 73303	232	0	. 99936	. 99854	2	54
7	. 05437	29	. 73535	232	1	. 99936	. 99852	1	53
8	. 05466	29	. 73767	230	1	. 99935	. 99851	2	52
9	. 05495	29	. 73997	229	0	. 99934	. 99849	2	51
10	0. 05524	29	8. 74226	228	1	9. 99934	0. 99847	1	50
11	. 05553	29	. 74454	226	1	. 99933	. 99846	2	49
12	. 05582	29	. 74680	226	0	. 99932	. 99844	2	48
13	. 05611	29	. 74906	224	1	. 99932	. 99842	1	47
14	. 05640	29	. 75130	223	1	. 99931	. 99841	2	46
15	0. 05669	29	8. 75353	222	1	9. 99930	0. 99839	1	45
16	. 05698	29	. 75575	220	0	. 99929	. 99838	2	44
17	. 05727	29	. 75795	220	1	. 99929	. 99836	2	43
18	. 05756	29	. 76015	219	1	. 99928	. 99834	1	42
19	. 05785	29	. 76234	217	1	. 99927	. 99833	2	41
20	0. 05814	30	8. 76451	216	0	9. 99926	0. 99831	2	40
21	. 05844	29	. 76667	216	1	. 99926	. 99829	2	39
22	. 05873	29	. 76883	214	1	. 99925	. 99827	1	38
23	. 05902	29	. 77097	213	1	. 99924	. 99826	2	37
24	. 05931	29	. 77310	212	0	. 99923	. 99824	2	36
25	0. 05960	29	8. 77522	211	1	9. 99923	0. 99822	1	35
26	. 05989	29	. 77733	210	1	. 99922	. 99821	2	34
27	. 06018	29	. 77943	209	1	. 99921	. 99819	2	33
28	. 06047	29	. 78152	208	0	. 99920	. 99817	2	32
29	. 06076	29	. 78360	208	1	. 99920	. 99815	2	31
30	0. 06105	29	8. 78568	206	1	9. 99919	0. 99813	1	30
31	. 06134	29	. 78774	205	1	. 99918	. 99812	2	29
32	. 06163	29	. 78979	204	0	. 99917	. 99810	2	28
33	. 06192	29	. 79183	203	1	. 99917	. 99808	2	27
34	. 06221	29	. 79386	202	1	. 99916	. 99806	2	26
35	0. 06250	29	8. 79588	201	1	9. 99915	0. 99804	1	25
36	. 06279	29	. 79789	201	1	. 99914	. 99803	2	24
37	. 06308	29	. 79990	199	0	. 99913	. 99801	2	23
38	. 06337	29	. 80189	199	1	. 99913	. 99799	2	22
39	. 06366	29	. 80388	197	1	. 99912	. 99797	2	21
40	0. 06395	29	8. 80585	197	1	9. 99911	0. 99795	2	20
41	. 06424	29	. 80782	196	1	. 99910	. 99793	1	19
42	. 06453	29	. 80978	195	0	. 99909	. 99792	2	18
43	. 06482	29	. 81173	194	1	. 99909	. 99790	2	17
44	. 06511	29	. 81367	193	1	. 99908	. 99788	2	16
45	0. 06540	29	8. 81560	192	1	9. 99907	0. 99786	2	15
46	. 06569	29	. 81752	192	1	. 99906	. 99784	2	14
47	. 06598	29	. 81944	190	1	. 99905	. 99782	2	13
48	. 06627	29	. 82134	190	0	. 99904	. 99780	2	12
49	. 06656	29	. 82324	189	1	. 99904	. 99778	2	11
50	0. 06685	29	8. 82513	188	1	9. 99903	0. 99776	2	10
51	. 06714	29	. 82701	187	1	. 99902	. 99774	2	9
52	. 06743	30	. 82888	187	1	. 99901	. 99772	2	8
53	. 06773	29	. 83075	186	1	. 99900	. 99770	2	7
54	. 06802	29	. 83261	185	1	. 99899	. 99768	2	6
55	0. 06831	29	8. 83446	184	0	9. 99898	0. 99766	2	5
56	. 06860	29	. 83630	183	1	. 99898	. 99764	2	4
57	. 06889	29	. 83813	183	1	. 99897	. 99762	2	3
58	. 06918	29	. 83996	181	1	. 99896	. 99760	2	2
59	. 06947	29	. 84177	181	1	. 99895	. 99758	2	1
60	0. 06976		8. 84358			9. 99894	0. 99756		0
↑ 93°→	cos	Diff. 1′	cos	Diff. 1′	Diff. 1′	sin	sin	Diff. 1′	↑ ←86°

4° → ↓	sin	Diff. 1′	sin	Diff. 1′	Diff. 1′	cos ←	cos	Diff. 1′	← 175° ↓
′									′
0	0. 06976	29	8. 84358	181	1	9. 99894	0. 99756	2	60
1	. 07005	29	. 84539	179	1	. 99893	. 99754	2	59
2	. 07034	29	. 84718	179	1	. 99892	. 99752	2	58
3	. 07063	29	. 84897	178	0	. 99891	. 99750	2	57
4	. 07092	29	. 85075	177	1	. 99891	. 99748	2	56
5	0. 07121	29	8. 85252	177	1	9. 99890	0. 99746	2	55
6	. 07150	29	. 85429	176	1	. 99889	. 99744	2	54
7	. 07179	29	. 85605	175	1	. 99888	. 99742	2	53
8	. 07208	29	. 85780	175	1	. 99887	. 99740	2	52
9	. 07237	29	. 85955	173	1	. 99886	. 99738	2	51
10	0. 07266	29	8. 86128	173	1	9. 99885	0. 99736	2	50
11	. 07295	29	. 86301	173	1	. 99884	. 99734	3	49
12	. 07324	29	. 86474	171	1	. 99883	. 99731	2	48
13	. 07353	29	. 86645	171	1	. 99882	. 99729	2	47
14	. 07382	29	. 86816	171	1	. 99881	. 99727	2	46
15	0. 07411	29	8. 86987	169	1	9. 99880	0. 99725	2	45
16	. 07440	29	. 87156	169	0	. 99879	. 99723	2	44
17	. 07469	29	. 87325	169	1	. 99879	. 99721	2	43
18	. 07498	29	. 87494	167	1	. 99878	. 99719	3	42
19	. 07527	29	. 87661	168	1	. 99877	. 99716	2	41
20	0. 07556	29	8. 87829	166	1	9. 99876	0. 99714	2	40
21	. 07585	29	. 87995	166	1	. 99875	. 99712	2	39
22	. 07614	29	. 88161	165	1	. 99874	. 99710	2	38
23	. 07643	20	. 88326	164	1	. 99873	. 99708	3	37
24	. 07672	29	. 88490	164	1	. 99872	. 99705	2	36
25	0. 07701	29	8. 88654	163	1	9. 99871	0. 99703	2	35
26	. 07730	29	. 88817	163	1	. 99870	. 99701	2	34
27	. 07759	29	. 88980	162	1	. 99869	. 99699	3	33
28	. 07788	29	. 89142	162	1	. 99868	. 99696	2	32
29	. 07817	29	. 89304	160	1	. 99867	. 99694	2	31
30	0. 07846	29	8. 89464	161	1	9. 99866	0. 99692	3	30
31	. 07875	29	. 89625	159	1	. 99865	. 99689	2	29
32	. 07904	29	. 89784	159	1	. 99864	. 99687	2	28
33	. 07933	29	. 89943	159	1	. 99863	. 99685	2	27
34	. 07962	29	. 90102	158	1	. 99862	. 99683	3	26
35	0. 07991	29	8. 90260	157	1	9. 99861	0. 99680	2	25
36	. 08020	29	. 90417	157	1	. 99860	. 99678	2	24
37	. 08049	29	. 90574	156	1	. 99859	. 99676	3	23
38	. 08078	29	. 90730	155	1	. 99858	. 99673	2	22
39	. 08107	29	. 90885	155	1	. 99857	. 99671	3	21
40	0. 08136	29	8. 91040	155	1	9. 99856	0. 99668	2	20
41	. 08165	29	. 91195	154	1	. 99855	. 99666	2	19
42	. 08194	29	. 91349	153	1	. 99854	. 99664	3	18
43	. 08223	29	. 91502	153	1	. 99853	. 99661	2	17
44	. 08252	29	. 91655	152	1	. 99852	. 99659	2	16
45	0. 08281	29	8. 91807	152	1	9. 99851	0. 99657	3	15
46	. 08310	29	. 91959	151	2	. 99850	. 99654	2	14
47	. 08339	29	. 92110	151	1	. 99848	. 99652	3	13
48	. 08368	29	. 92261	150	1	. 99847	. 99649	2	12
49	. 08397	29	. 92411	150	1	. 99846	. 99647	3	11
50	0. 08426	29	8. 92561	149	1	9. 99845	0. 99644	2	10
51	. 08455	29	. 92710	149	1	. 99844	. 99642	3	9
52	. 08484	29	. 92859	148	1	. 99843	. 99639	2	8
53	. 08513	29	. 93007	147	1	. 99842	. 99637	2	7
54	. 08542	29	. 93154	147	1	. 99841	. 99635	3	6
55	0. 08571	29	8. 93301	147	1	9. 99840	0. 99632	2	5
56	. 08600	29	. 93448	146	1	. 99839	. 99630	3	4
57	. 08629	29	. 93594	146	1	. 99838	. 99627	2	3
58	. 08658	29	. 93740	145	1	. 99837	. 99625	3	2
59	. 08687	29	. 93885	145	2	. 99836	. 99622	3	1
60	0. 08716		8. 94030			9. 99834	0. 99619		0
↑ 94° →	cos	Diff. 1′	cos	Diff. 1′	Diff. 1′	sin	sin	Diff. 1′	← 85° ↑

LOGARITHMS

5° → ′ ↓	sin	Diff. 1′	sin	Diff. 1′	Diff. 1′	cos ←	cos	Diff. 1′	←174° ′ ↓
0	0. 08716	29	8. 94030	144	1	9. 99834	0. 99619	2	60
1	. 08745	29	. 94174	143	1	. 99833	. 99617	3	59
2	. 08774	29	. 94317	144	1	. 99832	. 99614	2	58
3	. 08803	28	. 94461	142	1	. 99831	. 99612	3	57
4	. 08831	29	. 94603	143	1	. 99830	. 99609	2	56
5	0. 08860	29	8. 94746	141	1	9. 99829	0. 99607	3	55
6	. 08889	29	. 94887	142	1	. 99828	. 99604	2	54
7	. 08918	29	. 95029	141	2	. 99827	. 99602	3	53
8	. 08947	29	. 95170	140	1	. 99825	. 99599	3	52
9	. 08976	29	. 95310	140	1	. 99824	. 99596	2	51
10	0. 09005	29	8 95450	139	1	9. 99823	0. 99594	3	50
11	. 09034	29	. 95589	139	1	. 99822	. 99591	3	49
12	. 09063	29	. 95728	139	1	. 99821	. 99588	2	48
13	. 09092	29	. 95867	138	1	. 99820	. 99586	3	47
14	. 09121	29	. 96005	138	2	. 99819	. 99583	3	46
15	0. 09150	29	8. 96143	137	1	9. 99817	0. 99580	2	45
16	. 09179	29	. 96280	137	1	. 99816	. 99578	3	44
17	. 09208	29	. 96417	136	1	. 99815	. 99575	3	43
18	. 09237	29	. 96553	136	1	. 99814	. 99572	2	42
19	. 09266	29	. 96689	136	1	. 99813	. 99570	3	41
20	0. 09295	29	8. 96825	135	2	9. 99812	0. 99567	3	40
21	. 09324	29	. 96960	135	1	. 99810	. 99564	2	39
22	. 09353	29	. 97095	134	1	. 99809	. 99562	3	38
23	. 09382	29	. 97229	134	1	. 99808	. 99559	3	37
24	. 09411	29	. 97363	133	1	. 99807	. 99556	3	36
25	0. 09440	29	8. 97496	133	2	9. 99806	0. 99553	2	35
26	. 09469	29	. 97629	133	1	. 99804	. 99551	3	34
27	. 09498	29	. 97762	132	1	. 99803	. 99548	3	33
28	. 09527	29	. 97894	132	1	. 99802	. 99545	3	32
29	. 09556	29	. 98026	131	1	. 99801	. 99542	2	31
30	0. 09585	29	8. 98157	131	2	9. 99800	0. 99540	3	30
31	. 09614	28	. 98288	131	1	. 99798	. 99537	3	29
32	. 09642	29	. 98419	130	1	. 99797	. 99534	3	28
33	. 09671	29	. 98549	130	1	. 99796	. 99531	3	27
34	. 09700	29	. 98679	129	2	. 99795	. 99528	2	26
35	0. 09729	29	8. 98808	129	1	9. 99793	0. 99526	3	25
36	. 09758	29	. 98937	129	1	. 99792	. 99523	3	24
37	. 09787	29	. 99066	128	1	. 99791	. 99520	3	23
38	. 09816	29	. 99194	128	2	. 99790	. 99517	3	22
39	. 09845	29	. 99322	128	1	. 99788	. 99514	3	21
40	0. 09874	29	8. 99450	127	1	9. 99787	0. 99511	3	20
41	. 09903	29	. 99577	127	1	. 99786	. 99508	2	19
42	. 09932	29	. 99704	126	2	. 99785	. 99506	3	18
43	. 09961	29	. 99830	126	1	. 99783	. 99503	3	17
44	. 09990	29	8. 99956	126	1	. 99782	. 99500	3	16
45	0. 10019	29	9. 00082	125	1	9. 99781	0. 99497	3	15
46	. 10048	29	. 00207	125	2	. 99780	. 99494	3	14
47	. 10077	29	. 00332	124	1	. 99778	. 99491	3	13
48	. 10106	29	. 00456	125	1	. 99777	. 99488	3	12
49	. 10135	29	. 00581	123	1	. 99776	. 99485	3	11
50	0. 10164	28	9. 00704	124	2	9. 99775	0. 99482	3	10
51	. 10192	29	. 00828	123	1	. 99773	. 99479	3	9
52	. 10221	29	. 00951	123	1	. 99772	. 99476	3	8
53	. 10250	29	. 01074	122	2	. 99771	. 99473	3	7
54	. 10279	29	. 01196	122	1	. 99769	. 99470	3	6
55	0. 10308	29	9. 01318	122	1	9. 99768	0. 99467	3	5
56	. 10337	29	. 01440	121	2	. 99767	. 99464	3	4
57	. 10366	29	. 01561	121	1	. 99765	. 99461	3	3
58	. 10395	29	. 01682	121	1	. 99764	. 99458	3	2
59	. 10424	29	. 01803	120	2	. 99763	. 99455	3	1
60	0. 10453		9. 01923			9. 99761	0. 99452		0
↑ 95° →	cos	Diff. 1′	cos	Diff. 1′	Diff. 1′	sin	sin	Diff. 1′	←84° ↑

LOGARITHMS

6°→ ↓ ′	sin	Diff. 1′	sin	Diff. 1′	Diff. 1′	cos ←	cos	Diff. 1′	←173° ↓ ′
0	0. 10453	29	9. 01923	120	1	9. 99761	0. 99452	3	60
1	. 10482	29	. 02043	120	1	. 99760	. 99449	3	59
2	. 10511	29	. 02163	120	2	. 99759	. 99446	3	58
3	. 10540	29	. 02283	119	1	. 99757	. 99443	3	57
4	. 10569	28	. 02402	118	1	. 99756	. 99440	3	56
5	0. 10597	29	9. 02520	119	2	9. 99755	0. 99437	3	55
6	. 10626	29	. 02639	118	1	. 99753	. 99434	3	54
7	. 10655	29	. 02757	117	1	. 99752	. 99431	3	53
8	. 10684	29	. 02874	118	2	. 99751	. 99428	4	52
9	. 10713	29	. 02992	117	1	. 99749	. 99424	3	51
10	0. 10742	29	9. 03109	117	1	9. 99748	0. 99421	3	50
11	. 10771	29	. 03226	116	2	. 99747	. 99418	3	49
12	. 10800	29	. 03342	116	1	. 99745	. 99415	3	48
13	. 10829	29	. 03458	116	2	. 99744	. 99412	3	47
14	. 10858	29	. 03574	116	1	. 99742	. 99409	3	46
15	0. 10887	29	9. 03690	115	1	9. 99741	0. 99406	4	45
16	. 10916	29	. 03805	115	2	. 99740	. 99402	3	44
17	. 10945	28	. 03920	114	1	. 99738	. 99399	3	43
18	. 10973	29	. 04034	115	1	. 99737	. 99396	3	42
19	. 11002	29	. 04149	113	2	. 99736	. 99393	3	41
20	0. 11031	29	9. 04262	114	1	9. 99734	0. 99390	4	40
21	. 11060	29	. 04376	114	2	. 99733	. 99386	3	39
22	. 11089	29	. 04490	113	1	. 99731	. 99383	3	38
23	. 11118	29	. 04603	112	2	. 99730	. 99380	3	37
24	. 11147	29	. 04715	113	1	. 99728	. 99377	3	36
25	0. 11176	29	9. 04828	112	1	9. 99727	0. 99374	4	35
26	. 11205	29	. 04940	112	2	. 99726	. 99370	3	34
27	. 11234	29	. 05052	112	1	. 99724	. 99367	3	33
28	. 11263	28	. 05164	111	2	. 99723	. 99364	4	32
29	. 11291	29	. 05275	111	1	. 99721	. 99360	3	31
30	0. 11320	29	9. 05386	111	2	9. 99720	0. 99357	3	30
31	. 11349	29	. 05497	110	1	. 99718	. 99354	3	29
32	. 11378	29	. 05607	110	1	. 99717	. 99351	4	28
33	. 11407	29	. 05717	110	2	. 99716	. 99347	3	27
34	. 11436	29	. 05827	110	1	. 99714	. 99344	3	26
35	0. 11465	29	9. 05937	109	2	0. 00713	0. 99341	4	25
36	. 11494	29	. 06046	109	1	. 99711	. 99337	3	24
37	. 11523	29	. 06155	109	2	. 99710	. 99334	3	23
38	. 11552	28	. 06264	108	1	. 99708	. 99331	4	22
39	. 11580	29	. 06372	109	2	. 99707	. 99327	3	21
40	0. 11609	29	9. 06481	108	1	9. 99705	0. 99324	4	20
41	. 11638	29	. 06589	107	2	. 99704	. 99320	3	19
42	. 11667	29	. 06696	108	1	. 99702	. 99317	3	18
43	. 11696	29	. 06804	107	2	. 99701	. 99314	4	17
44	. 11725	29	. 06911	107	1	. 99699	. 99310	3	16
45	0. 11754	29	9. 07018	106	2	9. 99698	0. 99307	4	15
46	. 11783	29	. 07124	107	1	. 99696	. 99303	3	14
47	. 11812	28	. 07231	106	2	. 99695	. 99300	3	13
48	. 11840	29	. 07337	105	1	. 99693	. 99297	4	12
49	. 11869	29	. 07442	106	2	. 99692	. 99293	3	11
50	0. 11898	29	9. 07548	105	1	9. 99690	0. 99290	4	10
51	. 11927	29	. 07653	105	2	. 99689	. 99286	3	9
52	. 11956	29	. 07758	105	1	. 99687	. 99283	4	8
53	. 11985	29	. 07863	105	2	. 99686	. 99279	3	7
54	. 12014	29	. 07968	104	1	. 99684	. 99276	4	6
55	0. 12043	28	9. 08072	104	2	9. 99683	0. 99272	3	5
56	. 12071	29	. 08176	104	1	. 99681	. 99269	4	4
57	. 12100	29	. 08280	103	2	. 99680	. 99265	3	3
58	. 12129	29	. 08383	103	1	. 99678	. 99262	4	2
59	. 12158	29	. 08486	103	2	. 99677	. 99258	3	1
60	0. 12187		9. 08589			9. 99675	0. 99255		0
↑ 96°→	cos	Diff. 1′	cos	Diff. 1′	Diff. 1′	sin	sin	Diff. 1′	←83° ↑

7° → ↓ ′	sin	Diff. 1′	sin	Diff. 1′	Diff. 1′	cos ←	cos	Diff. 1′	← 172° ↓ ′
0	0. 12187	29	9. 08589	103	1	9. 99675	0. 99255	4	60
1	. 12216	29	. 08692	103	2	. 99674	. 99251	3	59
2	. 12245	29	. 08795	102	2	. 99672	. 99248	4	58
3	. 12274	28	. 08897	102	1	. 99670	. 99244	4	57
4	. 12302	29	. 08999	102	2	. 99669	. 99240	3	56
5	0. 12331	29	9. 09101	101	1	9. 99667	0. 99237	4	55
6	. 12360	29	. 09202	102	2	. 99666	. 99233	3	54
7	. 12389	29	. 09304	101	1	. 99664	. 99230	4	53
8	. 12418	29	. 09405	101	2	. 99663	. 99226	4	52
9	. 12447	29	. 09506	100	2	. 99661	. 99222	3	51
10	0. 12476	28	9. 09606	101	1	9. 99659	0. 99219	4	50
11	. 12504	29	. 09707	100	2	. 99658	. 99215	4	49
12	. 12533	29	. 09807	100	1	. 99656	. 99211	3	48
13	. 12562	29	. 09907	99	2	. 99655	. 99208	4	47
14	. 12591	29	. 10006	100	2	. 99653	. 99204	4	46
15	0. 12620	29	9. 10106	99	1	9. 99651	0. 99200	3	45
16	. 12649	29	. 10205	99	2	. 99650	. 99197	4	44
17	. 12678	28	. 10304	98	1	. 99648	. 99193	4	43
18	. 12706	29	. 10402	99	2	. 99647	. 99189	3	42
19	. 12735	29	. 10501	98	2	. 99645	. 99186	4	41
20	0. 12764	29	9. 10599	98	1	9. 99643	0. 99182	4	40
21	. 12793	29	. 10697	98	2	. 99642	. 99178	3	39
22	. 12822	29	. 10795	98	2	. 99640	. 99175	4	38
23	. 12851	29	. 10893	97	1	. 99638	. 99171	4	37
24	. 12880	28	. 10990	97	2	. 99637	. 99167	4	36
25	0. 12908	29	9. 11087	97	2	9. 99635	0. 99163	3	35
26	. 12937	29	. 11184	97	1	. 99633	. 99160	4	34
27	. 12966	29	. 11281	96	2	. 99632	. 99156	4	33
28	. 12995	29	. 11377	97	1	. 99630	. 99152	4	32
29	. 13024	29	. 11474	96	2	. 99629	. 99148	4	31
30	0. 13053	28	9. 11570	96	2	9. 99627	0. 99144	3	30
31	. 13081	29	. 11666	95	1	. 99625	. 99141	4	29
32	. 13110	29	. 11761	96	2	. 99624	. 99137	4	28
33	. 13139	29	. 11857	95	2	. 99622	. 99133	4	27
34	. 13168	29	. 11952	95	2	. 99620	. 99129	4	26
35	0. 13197	29	9. 12047	95	1	9. 99618	0. 99125	3	25
36	. 13226	28	. 12142	94	2	. 99617	. 99122	4	24
37	. 13254	29	. 12236	95	2	. 99615	. 99118	4	23
38	. 13283	29	. 12331	94	1	. 99613	. 99114	4	22
39	. 13312	29	. 12425	94	2	. 99612	. 99110	4	21
40	0. 13341	29	9. 12519	93	2	9. 99610	0. 99106	4	20
41	. 13370	29	. 12612	94	1	. 99608	. 99102	4	19
42	. 13399	28	. 12706	93	2	. 99607	. 99098	4	18
43	. 13427	29	. 12799	93	2	. 99605	. 99094	3	17
44	. 13456	29	. 12892	93	2	. 99603	. 99091	4	16
45	0. 13485	29	9. 12985	93	1	9. 99601	0. 99087	4	15
46	. 13514	29	. 13078	93	2	. 99600	. 99083	4	14
47	. 13543	29	. 13171	92	2	. 99598	. 99079	4	13
48	. 13572	28	. 13263	92	1	. 99596	. 99075	4	12
49	. 13600	29	. 13355	92	2	. 99595	. 99071	4	11
50	0. 13629	29	9. 13447	92	2	9. 99593	0. 99067	4	10
51	. 13658	29	. 13539	91	2	. 99591	. 99063	4	9
52	. 13687	29	. 13630	92	1	. 99589	. 99059	4	8
53	. 13716	28	. 13722	91	2	. 99588	. 99055	4	7
54	. 13744	29	. 13813	91	2	. 99586	. 99051	4	6
55	0. 13773	29	9. 13904	90	2	9. 99584	0. 99047	4	5
56	. 13802	29	. 13994	91	1	. 99582	. 99043	4	4
57	. 13831	29	. 14085	90	2	. 99581	. 99039	4	3
58	. 13860	29	. 14175	91	2	. 99579	. 99035	4	2
59	. 13889	28	. 14266	90	2	. 99577	. 99031	4	1
60	0. 13917		9. 14356			9. 99575	0. 99027		0
↑ 97° →	cos	Diff. 1′	cos	Diff. 1′	Diff. 1′	sin	sin	Diff. 1′	← 82° ↑

LOGARITHMS

8°→ ↓ ′	sin	Diff. 1′	sin	Diff. 1′	Diff. 1′	cos ←	cos	Diff. 1′	←171° ↓ ′
0	0. 13917	29	9. 14356	89	1	9. 99575	0. 99027	4	60
1	. 13946	29	. 14445	90	2	. 99574	. 99023	4	59
2	. 13975	29	. 14535	89	2	. 99572	. 99019	4	58
3	. 14004	29	. 14624	90	2	. 99570	. 99015	4	57
4	. 14033	28	. 14714	89	2	. 99568	. 99011	5	56
5	0. 14061	29	9. 14803	88	1	9. 99566	0. 99006	4	55
6	. 14090	29	. 14891	89	2	. 99565	. 99002	4	54
7	. 14119	29	. 14980	89	2	. 99563	. 98998	4	53
8	. 14148	29	. 15069	88	2	. 99561	. 98994	4	52
9	. 14177	28	. 15157	88	2	. 99559	. 98990	4	51
10	0. 14205	29	9. 15245	88	1	9. 99557	0. 98986	4	50
11	. 14234	29	. 15333	88	2	. 99556	. 98982	4	49
12	. 14263	29	. 15421	87	2	. 99554	. 98978	5	48
13	. 14292	28	. 15508	88	2	. 99552	. 98973	4	47
14	. 14320	29	. 15596	87	2	. 99550	. 98969	4	46
15	0. 14349	29	9. 15683	87	2	9. 99548	0. 98965	4	45
16	. 14378	29	. 15770	87	1	. 99546	. 98961	4	44
17	. 14407	29	. 15857	87	2	. 99545	. 98957	4	43
18	. 14436	28	. 15944	86	2	. 99543	. 98953	5	42
19	. 14464	29	. 16030	86	2	. 99541	. 98948	4	41
20	0. 14493	29	9. 16116	87	2	9. 99539	0. 98944	4	40
21	. 14522	29	. 16203	86	2	. 99537	. 98940	4	39
22	. 14551	29	. 16289	85	2	. 99535	. 98936	5	38
23	. 14580	28	. 16374	86	1	. 99533	. 98931	4	37
24	. 14608	29	. 16460	85	2	. 99532	. 98927	4	36
25	0. 14637	29	9. 16545	86	2	9. 99530	0. 98923	4	35
26	. 14666	29	. 16631	85	2	. 99528	. 98919	5	34
27	. 14695	28	. 16716	85	2	. 99526	. 98914	4	33
28	. 14723	29	. 16801	85	2	. 99524	. 98910	4	32
29	. 14752	29	. 16886	84	2	. 99522	. 98906	4	31
30	0. 14781	29	9. 16970	85	2	9. 99520	0. 98902	5	30
31	. 14810	28	. 17055	84	1	. 99518	. 98897	4	29
32	. 14838	29	. 17139	84	2	. 99517	. 98893	4	28
33	. 14867	29	. 17223	84	2	. 99515	. 98889	5	27
34	. 14896	29	. 17307	84	2	. 99513	. 98884	4	26
35	0. 14925	29	9. 17391	83	2	9. 99511	0. 98880	4	25
36	. 14954	28	. 17474	84	2	. 99509	. 98876	5	24
37	. 14982	29	. 17558	83	2	. 99507	. 98871	4	23
38	. 15011	29	. 17641	83	2	. 99505	. 98867	4	22
39	. 15040	29	. 17724	83	2	. 99503	. 98863	5	21
40	0. 15069	28	9. 17807	83	2	9. 99501	0. 98858	4	20
41	. 15097	29	. 17890	83	2	. 99499	. 98854	5	19
42	. 15126	29	. 17973	82	2	. 99497	. 98849	4	18
43	. 15155	29	. 18055	82	1	. 99495	. 98845	4	17
44	. 15184	28	. 18137	83	2	. 99494	. 98841	5	16
45	0. 15212	29	9. 18220	82	2	9. 99492	0. 98836	4	15
46	. 15241	29	. 18302	81	2	. 99490	. 98832	5	14
47	. 15270	29	. 18383	82	2	. 99488	. 98827	4	13
48	. 15299	28	. 18465	82	2	. 99486	. 98823	5	12
49	. 15327	29	. 18547	81	2	. 99484	. 98818	4	11
50	0. 15356	29	9. 18628	81	2	9. 99482	0. 98814	5	10
51	. 15385	29	. 18709	81	2	. 99480	. 98809	4	9
52	. 15414	28	. 18790	81	2	. 99478	. 98805	5	8
53	. 15442	29	. 18871	81	2	. 99476	. 98800	4	7
54	. 15471	29	. 18952	81	2	. 99474	. 98796	5	6
55	0. 15500	29	9. 19033	80	2	9. 99472	0. 98791	4	5
56	. 15529	28	. 19113	80	2	. 99470	. 98787	5	4
57	. 15557	29	. 19193	80	2	. 99468	. 98782	4	3
58	. 15586	29	. 19273	80	2	. 99466	. 98778	5	2
59	. 15615	28	. 19353	80	2	. 99464	. 98773	4	1
60	0. 15643		9. 19433			9. 99462	0. 98769		0
↑ 98°→	cos	Diff. 1′	cos	Diff. 1′	Diff. 1′	sin	sin	Diff. 1′←	↑ 81°

LOGARITHMS

9°→ ↓ ′	sin	Diff. 1′	sin	Diff. 1′	Diff. 1′	cos ←	cos	Diff. 1′	←170° ↓ ′
0	0.15643	29	9.19433	80	2	9.99462	0.98769	5	60
1	.15672	29	.19513	79	2	.99460	.98764	4	59
2	.15701	29	.19592	80	2	.99458	.98760	5	58
3	.15730	28	.19672	79	2	.99456	.98755	4	57
4	.15758	29	.19751	79	2	.99454	.98751	5	56
5	0.15787	29	9.19830	79	2	9.99452	0.98746	5	55
6	.15816	29	.19909	79	2	.99450	.98741	4	54
7	.15845	28	.19988	79	2	.99448	.98737	5	53
8	.15873	29	.20067	78	2	.99446	.98732	4	52
9	.15902	29	.20145	78	2	.99444	.98728	5	51
10	0.15931	28	9.20223	79	2	9.99442	0.98723	5	50
11	.15959	29	.20302	78	2	.99440	.98718	4	49
12	.15988	29	.20380	78	2	.99438	.98714	5	48
13	.16017	29	.20458	77	2	.99436	.98709	5	47
14	.16046	28	.20535	78	2	.99434	.98704	4	46
15	0.16074	29	9.20613	78	3	9.99432	0.98700	5	45
16	.16103	29	.20691	77	2	.99429	.98695	5	44
17	.16132	28	.20768	77	2	.99427	.98690	4	43
18	.16160	29	.20845	77	2	.99425	.98686	5	42
19	.16189	29	.20922	77	2	.99423	.98681	5	41
20	0.16218	28	9.20999	77	2	9.99421	0.98676	5	40
21	.16246	29	.21076	77	2	.99419	.98671	4	39
22	.16275	29	.21153	76	2	.99417	.98667	5	38
23	.16304	29	.21229	77	2	.99415	.98662	5	37
24	.16333	28	.21306	76	2	.99413	.98657	5	36
25	0.16361	29	9.21382	76	2	9.99411	0.98652	4	35
26	.16390	29	.21458	76	2	.99409	.98648	5	34
27	.16419	28	.21534	76	3	.99407	.98643	5	33
28	.16447	29	.21610	75	2	.99404	.98638	5	32
29	.16476	29	.21685	76	2	.99402	.98633	4	31
30	0.16505	28	9.21761	75	2	9.99400	0.98629	5	30
31	.16533	29	.21836	76	2	.99398	.98624	5	29
32	.16562	29	.21912	75	2	.99396	.98619	5	28
33	.16591	29	.21987	75	2	.99394	.98614	5	27
34	.16620	28	.22062	75	2	.99392	.98609	5	26
35	0.16648	29	9.22137	74	2	9.99390	0.98604	4	25
36	.16677	29	.22211	75	3	.99388	.98600	5	24
37	.16706	28	.22286	75	2	.99385	.98595	5	23
38	.16734	29	.22361	74	2	.99383	.98590	5	22
39	.16763	29	.22435	74	2	.99381	.98585	5	21
40	0.16792	28	9.22509	74	2	9.99379	0.98580	5	20
41	.16820	29	.22583	74	2	.99377	.98575	5	19
42	.16849	29	.22657	74	3	.99375	.98570	5	18
43	.16878	28	.22731	74	2	.99372	.98565	4	17
44	.16906	29	.22805	73	2	.99370	.98561	5	16
45	0.16935	29	9.22878	74	2	9.99368	0.98556	5	15
46	.16964	28	.22952	73	2	.99366	.98551	5	14
47	.16992	29	.23025	73	2	.99364	.98546	5	13
48	.17021	29	.23098	73	3	.99362	.98541	5	12
49	.17050	28	.23171	73	2	.99359	.98536	5	11
50	0.17078	29	9.23244	73	2	9.99357	0.98531	5	10
51	.17107	29	.23317	73	2	.99355	.98526	5	9
52	.17136	28	.23390	72	2	.99353	.98521	5	8
53	.17164	29	.23462	73	3	.99351	.98516	5	7
54	.17193	29	.23535	72	2	.99348	.98511	5	6
55	0.17222	28	9.23607	72	2	9.99346	0.98506	5	5
56	.17250	29	.23679	73	2	.99344	.98501	5	4
57	.17279	29	.23752	71	2	.99342	.98496	5	3
58	.17308	28	.23823	72	3	.99340	.98491	5	2
59	.17336	29	.23895	72	2	.99337	.98486	5	1
60	0.17365		9.23967			9.99335	0.98481		0
↑ 99°→	cos	Diff. 1′	cos	Diff. 1′	Diff. 1′	sin	sin	Diff. 1′	←80° ↑

LOGARITHMS

10°→ ↓ ′	sin	Diff. 1′	sin	Diff. 1′	Diff. 1′	cos ←	cos	Diff. 1′	←169° ↓ ′
0	0. 17365	28	9. 23967	72	2	9. 99335	0. 98481	5	60
1	. 17393	29	. 24039	71	2	. 99333	. 98476	5	59
2	. 17422	29	. 24110	71	3	. 99331	. 98471	5	58
3	. 17451	28	. 24181	72	2	. 99328	. 98466	5	57
4	. 17479	29	. 24253	71	2	. 99326	. 98461	6	56
5	0. 17508	29	9. 24324	71	2	9. 99324	0. 98455	5	55
6	. 17537	28	. 24395	71	3	. 99322	. 98450	5	54
7	. 17565	29	. 24466	70	2	. 99319	. 98445	5	53
8	. 17594	29	. 24536	71	2	. 99317	. 98440	5	52
9	. 17623	28	. 24607	70	2	. 99315	. 98435	5	51
10	0. 17651	29	9. 24677	71	3	9. 99313	0. 98430	5	50
11	. 17680	28	. 24748	70	2	. 99310	. 98425	5	49
12	. 17708	29	. 24818	70	2	. 99308	. 98420	6	48
13	. 17737	29	. 24888	70	2	. 99306	. 98414	5	47
14	. 17766	28	. 24958	70	3	. 99304	. 98409	5	46
15	0. 17794	29	9. 25028	70	2	9. 99301	0. 98404	5	45
16	. 17823	29	. 25098	70	2	. 99299	. 98399	5	44
17	. 17852	28	. 25168	69	3	. 99297	. 98394	5	43
18	. 17880	29	. 25237	70	2	. 99294	. 98389	6	42
19	. 17909	28	. 25307	69	2	. 99292	. 98383	5	41
20	0. 17937	29	9. 25376	69	2	9. 99290	0. 98378	5	40
21	. 17966	29	. 25445	69	3	. 99288	. 98373	5	39
22	. 17995	28	. 25514	69	2	. 99285	. 98368	6	38
23	. 18023	20	. 25583	69	2	. 99283	. 98362	5	37
24	. 18052	29	. 25652	69	3	. 99281	. 98357	5	36
25	0. 18081	28	9. 25721	69	2	9. 99278	0. 98352	5	35
26	. 18109	29	. 25790	68	2	. 99276	. 98347	6	34
27	. 18138	28	. 25858	69	3	. 99274	. 98341	5	33
28	. 18166	29	. 25927	68	2	. 99271	. 98336	5	32
29	. 18195	29	. 25995	68	2	. 99269	. 98331	6	31
30	0. 18224	28	9. 26063	68	3	9. 99267	0. 98325	5	30
31	. 18252	29	. 26131	68	2	. 99264	. 98320	5	29
32	. 18281	28	. 26199	68	2	. 99262	. 98315	5	28
33	. 18309	29	. 26267	68	3	. 99260	. 98310	6	27
34	. 18338	29	. 26335	68	2	. 99257	. 98304	5	26
35	0. 18367	28	9. 26403	67	3	9. 99255	0. 98299	5	25
36	. 18395	29	. 26470	68	2	. 99252	. 98294	6	24
37	. 18424	28	. 26538	67	2	. 99250	. 98288	5	23
38	. 18452	29	. 26605	67	3	. 99248	. 98283	6	22
39	. 18481	28	. 26672	67	2	. 99245	. 98277	5	21
40	0. 18509	29	9. 26739	67	2	9. 99243	0. 98272	5	20
41	. 18538	29	. 26806	67	3	. 99241	. 98267	6	19
42	. 18567	28	. 26873	67	2	. 99238	. 98261	5	18
43	. 18595	29	. 26940	67	3	. 99236	. 98256	6	17
44	. 18624	28	. 27007	66	2	. 99233	. 98250	5	16
45	0. 18652	29	9. 27073	67	2	9. 99231	0. 98245	5	15
46	. 18681	29	. 27140	66	3	. 99229	. 98240	6	14
47	. 18710	28	. 27206	67	2	. 99226	. 98234	5	13
48	. 18738	29	. 27273	66	3	. 99224	. 98229	6	12
49	. 18767	28	. 27339	66	2	. 99221	. 98223	5	11
50	0. 18795	29	9. 27405	66	2	9. 99219	0. 98218	6	10
51	. 18824	28	. 27471	66	3	. 99217	. 98212	5	9
52	. 18852	29	. 27537	65	2	. 99214	. 98207	6	8
53	. 18881	29	. 27602	66	3	. 99212	. 98201	5	7
54	. 18910	28	. 27668	66	2	. 99209	. 98196	6	6
55	0. 18938	29	9. 27734	65	3	9. 99207	0. 98190	5	5
56	. 18967	28	. 27799	65	2	. 99204	. 98185	6	4
57	. 18995	29	. 27864	66	2	. 99202	. 98179	5	3
58	. 19024	28	. 27930	65	3	. 99200	. 98174	6	2
59	. 19052	29	. 27995	65	2	. 99197	. 98168	5	1
60	0. 19081		9. 28060			9. 99195	0. 98163		0
↑ 100°→	cos	Diff. 1′	→ cos	Diff. 1′	Diff. 1′	sin	sin	Diff. 1′	↑ ←79°

11°→ ↓ ′	sin	Diff. 1′	sin	Diff. 1′	Diff. 1′	cos ←	cos	Diff. 1′	←168° ↓ ′
0	0. 19081	28	9. 28060	65	3	9. 99195	0. 98163	6	60
1	. 19109	29	. 28125	65	2	. 99192	. 98157	5	59
2	. 19138	29	. 28190	64	3	. 99190	. 98152	6	58
3	. 19167	28	. 28254	65	2	. 99187	. 98146	6	57
4	. 19195	29	. 28319	65	3	. 99185	. 98140	5	56
5	0. 19224	28	9. 28384	64	2	9. 99182	0. 98135	6	55
6	. 19252	29	. 28448	64	3	. 99180	. 98129	5	54
7	. 19281	28	. 28512	65	2	. 99177	. 98124	6	53
8	. 19309	29	. 28577	64	3	. 99175	. 98118	6	52
9	. 19338	28	. 28641	64	2	. 99172	. 98112	5	51
10	0. 19366	29	9. 28705	64	3	9. 99170	0. 98107	6	50
11	. 19395	28	. 28769	64	2	. 99167	. 98101	5	49
12	. 19423	29	. 28833	63	3	. 99165	. 98096	6	48
13	. 19452	29	. 28896	64	2	. 99162	. 98090	6	47
14	. 19481	28	. 28960	64	3	. 99160	. 98084	5	46
15	0. 19509	29	9. 29024	63	2	9. 99157	0. 98079	6	45
16	. 19538	28	. 29087	63	3	. 99155	. 98073	6	44
17	. 19566	29	. 29150	64	2	. 99152	. 98067	6	43
18	. 19595	28	. 29214	63	3	. 99150	. 98061	5	42
19	. 19623	29	. 29277	63	2	. 99147	. 98056	6	41
20	0. 19652	28	9. 29340	63	3	9. 99145	0. 98050	6	40
21	. 19680	29	. 29403	63	2	. 99142	. 98044	5	39
22	. 19709	28	. 29466	63	3	. 99140	. 98039	6	38
23	. 19737	29	. 29529	62	2	. 99137	. 98033	6	37
24	. 19766	28	. 29591	63	3	. 99135	. 98027	6	36
25	0. 19794	29	9. 29654	62	2	9. 99132	0. 98021	5	35
26	. 19823	28	. 29716	63	3	. 99130	. 98016	6	34
27	. 19851	29	. 29779	62	3	. 99127	. 98010	6	33
28	. 19880	28	. 29841	62	2	. 99124	. 98004	6	32
29	. 19908	29	. 29903	63	3	. 99122	. 97998	6	31
30	0. 19937	28	9. 29966	62	2	9. 99119	0. 97992	5	30
31	. 19965	29	. 30028	62	3	. 99117	. 97987	6	29
32	. 19994	28	. 30090	61	2	. 99114	. 97981	6	28
33	. 20022	29	. 30151	62	3	. 99112	. 97975	6	27
34	. 20051	28	. 30213	62	3	. 99109	. 97969	6	26
35	0. 20079	29	9. 30275	61	2	9. 99106	0. 97963	5	25
36	. 20108	28	. 30336	62	3	. 99104	. 97958	6	24
37	. 20136	29	. 30398	61	2	. 99101	. 97952	6	23
38	. 20165	28	. 30459	62	3	. 99099	. 97946	6	22
39	. 20193	29	. 30521	61	3	. 99096	. 97940	6	21
40	0. 20222	28	9. 30582	61	2	9. 99093	0. 97934	6	20
41	. 20250	29	. 30643	61	3	. 99091	. 97928	6	19
42	. 20279	28	. 30704	61	2	. 99088	. 97922	6	18
43	. 20307	29	. 30765	61	3	. 99086	. 97916	6	17
44	. 20336	28	. 30826	61	3	. 99083	. 97910	5	16
45	0. 20364	29	9. 30887	60	2	9. 99080	0. 97905	6	15
46	. 20393	28	. 30947	61	3	. 99078	. 97899	6	14
47	. 20421	29	. 31008	60	3	. 99075	. 97893	6	13
48	. 20450	28	. 31068	61	2	. 99072	. 97887	6	12
49	. 20478	29	. 31129	60	3	. 99070	. 97881	6	11
50	0. 20507	28	9. 31189	61	3	9. 99067	0. 97875	6	10
51	. 20535	28	. 31250	60	2	. 99064	. 97869	6	9
52	. 20563	29	. 31310	60	3	. 99062	. 97863	6	8
53	. 20592	28	. 31370	60	3	. 99059	. 97857	6	7
54	. 20620	29	. 31430	60	2	. 99056	. 97851	6	6
55	0. 20649	28	9. 31490	59	3	9. 99054	0. 97845	6	5
56	. 20677	29	. 31549	60	3	. 99051	. 97839	6	4
57	. 20706	28	. 31609	60	2	. 99048	. 97833	6	3
58	. 20734	29	. 31669	59	3	. 99046	. 97827	6	2
59	. 20763	28	. 31728	60	3	. 99043	. 97821	6	1
60	0. 20791		9. 31788			9. 99040	0. 97815		0
↑ 101°→	cos	Diff. 1′	→ cos	Diff. 1′	Diff. 1′	sin	sin	Diff. 1′	←78° ↑

12°→ ↓ ′	sin	Diff. 1′	sin	Diff. 1′	Diff. 1′	cos ←	cos	←167° Diff. 1′	↓
0	0. 20791	29	9. 31788	59	2	9. 99040	0. 97815	6	60
1	. 20820	28	. 31847	60	3	. 99038	. 97809	6	59
2	. 20848	29	. 31907	59	3	. 99035	. 97803	6	58
3	. 20877	28	. 31966	59	2	. 99032	. 97797	6	57
4	. 20905	28	. 32025	59	3	. 99030	. 97791	7	56
5	0. 20933	29	9. 32084	59	3	9. 99027	0. 97784	6	55
6	. 20962	28	. 32143	59	2	. 99024	. 97778	6	54
7	. 20990	29	. 32202	59	3	. 99022	. 97772	6	53
8	. 21019	28	. 32261	58	3	. 99019	. 97766	6	52
9	. 21047	29	. 32319	59	3	. 99016	. 97760	6	51
10	0. 21076	28	9. 32378	59	2	9. 99013	0. 97754	6	50
11	. 21104	28	. 32437	58	3	. 99011	. 97748	6	49
12	. 21132	29	. 32495	58	3	. 99008	. 97742	7	48
13	. 21161	28	. 32553	59	3	. 99005	. 97735	6	47
14	. 21189	29	. 32612	58	2	. 99002	. 97729	6	46
15	0. 21218	28	9. 32670	58	3	9. 99000	0. 97723	6	45
16	. 21246	29	. 32728	58	3	. 98997	. 97717	6	44
17	. 21275	28	. 32786	58	3	. 98994	. 97711	6	43
18	. 21303	28	. 32844	58	2	. 98991	. 97705	7	42
19	. 21331	29	. 32902	58	3	. 98989	. 97698	6	41
20	0. 21360	28	9. 32960	58	3	9. 98986	0. 97692	6	40
21	. 21388	29	. 33018	57	3	. 98983	. 97686	6	39
22	. 21417	28	. 33075	58	2	. 98980	. 97680	7	38
23	. 21445	29	. 33133	57	3	. 98978	. 97673	6	37
24	. 21474	28	. 33190	58	3	. 98975	. 97667	6	36
25	0. 21502	28	9. 33248	57	3	9. 98972	0. 97661	6	35
26	. 21530	29	. 33305	57	2	. 98969	. 97655	7	34
27	. 21559	28	. 33362	58	3	. 98967	. 97648	6	33
28	. 21587	29	. 33420	57	3	. 98964	. 97642	6	32
29	. 21616	28	. 33477	57	3	. 98961	. 97636	6	31
30	0. 21644	28	9. 33534	57	3	9. 98958	0. 97630	7	30
31	. 21672	29	. 33591	56	2	. 98955	. 97623	6	29
32	. 21701	28	. 33647	57	3	. 98953	. 97617	6	28
33	. 21729	29	. 33704	57	3	. 98950	. 97611	7	27
34	. 21758	28	. 33761	57	3	. 98947	. 97604	6	26
35	0. 21786	28	9. 33818	56	3	9. 98944	0. 97598	6	25
36	. 21814	29	. 33874	57	3	. 98941	. 97592	7	24
37	. 21843	28	. 33931	56	2	. 98938	. 97585	6	23
38	. 21871	28	. 33987	56	3	. 98936	. 97579	6	22
39	. 21899	29	. 34043	57	3	. 98933	. 97573	7	21
40	0. 21928	28	9. 34100	56	3	9. 98930	0. 97566	6	20
41	. 21956	29	. 34156	56	3	. 98927	. 97560	7	19
42	. 21985	28	. 34212	56	3	. 98924	. 97553	6	18
43	. 22013	28	. 34268	56	2	. 98921	. 97547	6	17
44	. 22041	29	. 34324	56	3	. 98919	. 97541	7	16
45	0. 22070	28	9. 34380	56	3	9. 98916	0. 97534	6	15
46	. 22098	28	. 34436	55	3	. 98913	. 97528	7	14
47	. 22126	29	. 34491	56	3	. 98910	. 97521	6	13
48	. 22155	28	. 34547	55	3	. 98907	. 97515	7	12
49	. 22183	29	. 34602	56	3	. 98904	. 97508	6	11
50	0. 22212	28	9. 34658	55	3	9. 98901	0. 97502	6	10
51	. 22240	28	. 34713	56	2	. 98898	. 97496	7	9
52	. 22268	29	. 34769	55	3	. 98896	. 97489	6	8
53	. 22297	28	. 34824	55	3	. 98893	. 97483	7	7
54	. 22325	28	. 34879	55	3	. 98890	. 97476	6	6
55	0. 22353	29	9. 34934	55	3	9. 98887	0. 97470	7	5
56	. 22382	28	. 34989	55	3	. 98884	. 97463	6	4
57	. 22410	28	. 35044	55	3	. 98881	. 97457	7	3
58	. 22438	29	. 35099	55	3	. 98878	. 97450	6	2
59	. 22467	28	. 35154	55	3	. 98875	. 97444	7	1
60	0. 22495		9. 35209			9. 98872	0. 97437		0
↑ 102°→	cos	Diff. 1′	→ cos	Diff. 1′	Diff. 1′	sin	sin	Diff. 1′←77°	↑

13°→ ↓ ′	sin	Diff. 1′	sin	Diff. 1′	Diff. 1′	cos ←	cos	Diff. 1′	←166° ↓ ′
0	0.22495	28	9.35209	54	3	9.98872	0.97437	7	60
1	.22523	29	.35263	55	2	.98869	.97430	6	59
2	.22552	28	.35318	55	3	.98867	.97424	7	58
3	.22580	28	.35373	54	3	.98864	.97417	6	57
4	.22608	29	.35427	54	3	.98861	.97411	7	56
5	0.22637	28	9.35481	55	3	9.98858	0.97404	6	55
6	.22665	28	.35536	54	3	.98855	.97398	7	54
7	.22693	29	.35590	54	3	.98852	.97391	7	53
8	.22722	28	.35644	54	3	.98849	.97384	6	52
9	.22750	28	.35698	54	3	.98846	.97378	7	51
10	0.22778	29	9.35752	54	3	9.98843	0.97371	6	50
11	.22807	28	.35806	54	3	.98840	.97365	7	49
12	.22835	28	.35860	54	3	.98837	.97358	7	48
13	.22863	29	.35914	54	3	.98834	.97351	6	47
14	.22892	28	.35968	54	3	.98831	.97345	7	46
15	0.22920	28	9.36022	53	3	9.98828	0.97338	7	45
16	.22948	29	.36075	54	3	.98825	.97331	6	44
17	.22977	28	.36129	53	3	.98822	.97325	7	43
18	.23005	28	.36182	54	3	.98819	.97318	7	42
19	.23033	29	.36236	53	3	.98816	.97311	7	41
20	0.23062	28	9.36289	53	3	9.98813	0.97304	6	40
21	.23090	28	.36342	53	3	.98810	.97298	7	39
22	.23118	28	.36395	54	3	.98807	.97291	7	38
23	.23146	29	.36449	53	3	.98804	.97284	6	37
24	.23175	28	.36502	53	3	.98801	.97278	7	36
25	0.23203	28	9.36555	53	3	9.98798	0.97271	7	35
26	.23231	29	.36608	52	3	.98795	.97264	7	34
27	.23260	28	.36660	53	3	.98792	.97257	6	33
28	.23288	28	.36713	53	3	.98789	.97251	7	32
29	.23316	29	.36766	53	3	.98786	.97244	7	31
30	0.23345	28	9.36819	52	3	9.98783	0.97237	7	30
31	.23373	28	.36871	53	3	.98780	.97230	7	29
32	.23401	28	.36924	52	3	.98777	.97223	6	28
33	.23429	29	.36976	52	3	.98774	.97217	7	27
34	.23458	28	.37028	53	3	.98771	.97210	7	26
35	0.23486	28	9.37081	52	3	9.98768	0.97203	7	25
36	.23514	28	.37133	52	3	.98765	.97196	7	24
37	.23542	29	.37185	52	3	.98762	.97189	7	23
38	.23571	28	.37237	52	3	.98759	.97182	6	22
39	.23599	28	.37289	52	3	.98756	.97176	7	21
40	0.23627	29	9.37341	52	3	9.98753	0.97169	7	20
41	.23656	28	.37393	52	4	.98750	.97162	7	19
42	.23684	28	.37445	52	3	.98746	.97155	7	18
43	.23712	28	.37497	52	3	.98743	.97148	7	17
44	.23740	29	.37549	51	3	.98740	.97141	7	16
45	0.23769	28	9.37600	52	3	9.98737	0.97134	7	15
46	.23797	28	.37652	51	3	.98734	.97127	7	14
47	.23825	28	.37703	52	3	.98731	.97120	7	13
48	.23853	29	.37755	51	3	.98728	.97113	7	12
49	.23882	28	.37806	52	3	.98725	.97106	6	11
50	0.23910	28	9.37858	51	3	9.98722	0.97100	7	10
51	.23938	28	.37909	51	4	.98719	.97093	7	9
52	.23966	29	.37960	51	3	.98715	.97086	7	8
53	.23995	28	.38011	51	3	.98712	.97079	7	7
54	.24023	28	.38062	51	3	.98709	.97072	7	6
55	0.24051	28	9.38113	51	3	9.98706	0.97065	7	5
56	.24079	29	.38164	51	3	.98703	.97058	7	4
57	.24108	28	.38215	51	3	.98700	.97051	7	3
58	.24136	28	.38266	51	3	.98697	.97044	7	2
59	.24164	28	.38317	51	4	.98694	.97037	7	1
60	0.24192		9.38368			9.98690	0.97030		0
↑ 103°→	cos	Diff. 1′	→ cos	Diff. 1′	Diff. 1′	sin	sin	Diff. 1′	↑ ←76°

LOGARITHMS

14°→ ↓ ′	sin	Diff. 1′	sin	Diff. 1′	Diff. 1′	cos ←	cos	Diff. 1′	←165° ↓ ′
0	0. 24192	28	9. 38368	50	3	9. 98690	0. 97030	7	60
1	. 24220	29	. 38418	51	3	. 98687	. 97023	8	59
2	. 24249	28	. 38469	50	3	. 98684	. 97015	7	58
3	. 24277	28	. 38519	51	3	. 98681	. 97008	7	57
4	. 24305	28	. 38570	50	3	. 98678	. 97001	7	56
5	0. 24333	29	9. 38620	50	4	9. 98675	0. 96994	7	55
6	. 24362	28	. 38670	51	3	. 98671	. 96987	7	54
7	. 24390	28	. 38721	50	3	. 98668	. 96980	7	53
8	. 24418	28	. 38771	50	3	. 98665	. 96973	7	52
9	. 24446	28	. 38821	50	3	. 98662	. 96966	7	51
10	0. 24474	29	9. 38871	50	3	9. 98659	0. 96959	7	50
11	. 24503	28	. 38921	50	4	. 98656	. 96952	7	49
12	. 24531	28	. 38971	50	3	. 98652	. 96945	8	48
13	. 24559	28	. 39021	50	3	. 98649	. 96937	7	47
14	. 24587	28	. 39071	50	3	. 98646	. 96930	7	46
15	0. 24615	29	9. 39121	49	3	9. 98643	0. 96923	7	45
16	. 24644	28	. 39170	50	4	. 98640	. 96916	7	44
17	. 24672	28	. 39220	50	3	. 98636	. 96909	7	43
18	. 24700	28	. 39270	49	3	. 98633	. 96902	8	42
19	. 24728	28	. 39319	50	3	. 98630	. 96894	7	41
20	0. 24756	28	9. 39369	49	4	9. 98627	0. 96887	7	40
21	. 24784	29	. 39418	49	3	. 98623	. 96880	7	39
22	. 24813	28	. 39467	50	3	. 98620	. 96873	7	38
23	. 24841	28	. 39517	49	3	. 98617	. 96866	8	37
24	. 24869	28	. 39566	49	4	. 98614	. 96858	7	36
25	0. 24897	28	9. 39615	49	3	9. 98610	0. 96851	7	35
26	. 24925	29	. 39664	49	3	. 98607	. 96844	7	34
27	. 24954	28	. 39713	49	3	. 98604	. 96837	8	33
28	. 24982	28	. 39762	49	4	. 98601	. 96829	7	32
29	. 25010	28	. 39811	49	3	. 98597	. 96822	7	31
30	0. 25038	28	9. 39860	49	3	9. 98594	0. 96815	8	30
31	. 25066	28	. 39909	49	3	. 98591	. 96807	7	29
32	. 25094	28	. 39958	48	4	. 98588	. 96800	7	28
33	. 25122	29	. 40006	49	3	. 98584	. 96793	7	27
34	. 25151	28	. 40055	48	3	. 98581	. 96786	8	26
35	0. 25179	28	9. 40103	49	4	9. 98578	0. 96778	7	25
36	. 25207	28	. 40152	48	3	. 98574	. 96771	7	24
37	. 25235	28	. 40200	49	3	. 98571	. 96764	8	23
38	. 25263	28	. 40249	48	3	. 98568	. 96756	7	22
39	. 25291	29	. 40297	49	4	. 98565	. 96749	7	21
40	0. 25320	28	9. 40346	48	3	9. 98561	0. 96742	8	20
41	. 25348	28	. 40394	48	3	. 98558	. 96734	7	19
42	. 25376	28	. 40442	48	4	. 98555	. 96727	8	18
43	. 25404	28	. 40490	48	3	. 98551	. 96719	7	17
44	. 25432	28	. 40538	48	3	. 98548	. 96712	7	16
45	0. 25460	28	9. 40586	48	4	9. 98545	0. 96705	8	15
46	. 25488	28	. 40634	48	3	. 98541	. 96697	7	14
47	. 25516	29	. 40682	48	3	. 98538	. 96690	8	13
48	. 25545	28	. 40730	48	4	. 98535	. 96682	7	12
49	. 25573	28	. 40778	47	3	. 98531	. 96675	8	11
50	0. 25601	28	9. 40825	48	3	9. 98528	0. 96667	7	10
51	. 25629	28	. 40873	48	4	. 98525	. 96660	7	9
52	. 25657	28	. 40921	47	3	. 98521	. 96653	8	8
53	. 25685	28	. 40968	48	3	. 98518	. 96645	7	7
54	. 25713	28	. 41016	47	4	. 98515	. 96638	8	6
55	0. 25741	28	9. 41063	48	3	9. 98511	0. 96630	7	5
56	. 25769	29	. 41111	47	3	. 98508	. 96623	8	4
57	. 25798	28	. 41158	47	4	. 98505	. 96615	7	3
58	. 25826	28	. 41205	47	3	. 98501	. 96608	8	2
59	. 25854	28	. 41252	48	4	. 98498	. 96600	7	1
60	0. 25882		9. 41300			9. 98494	0. 96593		0
↑ 104°→	cos	Diff. 1′	°→ cos	Diff. 1′	Diff. 1′	sin	sin	Diff. 1′←	↑ 75°

15° → ′	sin	Diff. 1′	sin	Diff. 1′	Diff. 1′	cos ←	cos	Diff. 1′	← 164° ′
0	0. 25882	28	9. 41300	47	3	9. 98494	0. 96593	8	60
1	. 25910	28	. 41347	47	3	. 98491	. 96585	7	59
2	. 25938	28	. 41394	47	4	. 98488	. 96578	8	58
3	. 25966	28	. 41441	47	3	. 98484	. 96570	8	57
4	. 25994	28	. 41488	47	4	. 98481	. 96562	7	56
5	0. 26022	28	9. 41535	47	3	9. 98477	0. 96555	8	55
6	. 26050	29	. 41582	46	3	. 98474	. 96547	7	54
7	. 26079	28	. 41628	47	4	. 98471	. 96540	8	53
8	. 26107	28	. 41675	47	3	. 98467	. 96532	8	52
9	. 26135	28	. 41722	46	4	. 98464	. 96524	7	51
10	0. 26163	28	9. 41768	47	3	9. 98460	0. 96517	8	50
11	. 26191	28	. 41815	46	4	. 98457	. 96509	7	49
12	. 26219	28	. 41861	47	3	. 98453	. 96502	8	48
13	. 26247	28	. 41908	46	3	. 98450	. 96494	8	47
14	. 26275	28	. 41954	47	4	. 98447	. 96486	7	46
15	0. 26303	28	9. 42001	46	3	9. 98443	0. 96479	8	45
16	. 26331	28	. 42047	46	4	. 98440	. 96471	8	44
17	. 26359	28	. 42093	47	3	. 98436	. 96463	7	43
18	. 26387	28	. 42140	46	4	. 98433	. 96456	8	42
19	. 26415	28	. 42186	46	3	. 98429	. 96448	8	41
20	0. 26443	28	9. 42232	46	4	9. 98426	0. 96440	7	40
21	. 26471	29	. 42278	46	3	. 98422	. 96433	8	39
22	. 26500	28	. 42324	46	4	. 98419	. 96425	8	38
23	. 26528	28	. 42370	46	3	. 98415	. 96417	7	37
24	. 26556	28	. 42416	45	3	. 98412	. 96410	8	36
25	0. 26584	28	9. 42461	46	4	9. 98409	0. 96402	8	35
26	. 26612	28	. 42507	46	3	. 98405	. 96394	8	34
27	. 26640	28	. 42553	46	4	. 98402	. 96386	7	33
28	. 26668	28	. 42599	45	3	. 98398	. 96379	8	32
29	. 26696	28	. 42644	46	4	. 98395	. 96371	8	31
30	0. 26724	28	9. 42690	45	3	9. 98391	0. 96363	8	30
31	. 26752	28	. 42735	46	4	. 98388	. 96355	8	29
32	. 26780	28	. 42781	45	3	. 98384	. 96347	7	28
33	. 26808	28	. 42826	46	4	. 98381	. 96340	8	27
34	. 26836	28	. 42872	45	4	. 98377	. 96332	8	26
35	0. 26864	28	9. 42917	45	3	9. 98373	0. 96324	8	25
36	. 26892	28	. 42962	46	4	. 98370	. 96316	8	24
37	. 26920	28	. 43008	45	3	. 98366	. 96308	7	23
38	. 26948	28	. 43053	45	4	. 98363	. 96301	8	22
39	. 26976	28	. 43098	45	3	. 98359	. 96293	8	21
40	0. 27004	28	9. 43143	45	4	9. 98356	0. 96285	8	20
41	. 27032	28	. 43188	45	3	. 98352	. 96277	8	19
42	. 27060	28	. 43233	45	4	. 98349	. 96269	8	18
43	. 27088	28	. 43278	45	3	. 98345	. 96261	8	17
44	. 27116	28	. 43323	44	4	. 98342	. 96253	7	16
45	0. 27144	28	9. 43367	45	4	9. 98338	0. 96246	8	15
46	. 27172	28	. 43412	45	3	. 98334	. 96238	8	14
47	. 27200	28	. 43457	45	4	. 98331	. 96230	8	13
48	. 27228	28	. 43502	44	3	. 98327	. 96222	8	12
49	. 27256	28	. 43546	45	4	. 98324	. 96214	8	11
50	0. 27284	28	9. 43591	44	3	9. 98320	0. 96206	8	10
51	. 27312	28	. 43635	45	4	. 98317	. 96198	8	9
52	. 27340	28	. 43680	44	4	. 98313	. 96190	8	8
53	. 27368	28	. 43724	45	3	. 98309	. 96182	8	7
54	. 27396	28	. 43769	44	4	. 98306	. 96174	8	6
55	0. 27424	28	9. 43813	44	3	9. 98302	0. 96166	8	5
56	. 27452	28	. 43857	44	4	. 98299	. 96158	8	4
57	. 27480	28	. 43901	45	4	. 98295	. 96150	8	3
58	. 27508	28	. 43946	44	3	. 98291	. 96142	8	2
59	. 27536	28	. 43990	44	4	. 98288	. 96134	8	1
60	0. 27564		9. 44034			9. 98284	0. 96126		0
105° →	cos	Diff. 1′	→ cos	Diff. 1′	Diff. 1′	sin	sin	Diff. 1′	← 74°

16°→ ↓ ′	sin	Diff. 1′	sin	Diff. 1′	Diff. 1′	cos ←	cos	Diff. 1′	←163° ↓ ′
0	0.27564	28	9.44034	44	3	9.98284	0.96126	8	60
1	.27592	28	.44078	44	4	.98281	.96118	8	59
2	.27620	28	.44122	44	4	.98277	.96110	8	58
3	.27648	28	.44166	44	3	.98273	.96102	8	57
4	.27676	28	.44210	43	4	.98270	.96094	8	56
5	0.27704	27	9.44253	44	4	9.98266	0.96086	8	55
6	.27731	28	.44297	44	3	.98262	.96078	8	54
7	.27759	28	.44341	44	4	.98259	.96070	8	53
8	.27787	28	.44385	43	4	.98255	.96062	8	52
9	.27815	28	.44428	44	3	.98251	.96054	8	51
10	0.27843	28	9.44472	44	4	9.98248	0.96046	9	50
11	.27871	28	.44516	43	4	.98244	.96037	8	49
12	.27899	28	.44559	43	3	.98240	.96029	8	48
13	.27927	28	.44602	44	4	.98237	.96021	8	47
14	.27955	28	.44646	43	4	.98233	.96013	8	46
15	0.27983	28	9.44689	44	3	9.98229	0.96005	8	45
16	.28011	28	.44733	43	4	.98226	.95997	8	44
17	.28039	28	.44776	43	4	.98222	.95989	8	43
18	.28067	28	.44819	43	3	.98218	.95981	9	42
19	.28095	28	.44862	43	4	.98215	.95972	8	41
20	0.28123	27	9.44905	43	4	9.98211	0.95964	8	40
21	.28150	28	.44948	44	3	.98207	.95956	8	39
22	.28178	28	.44992	43	4	.98204	.95948	8	38
23	.28206	28	.45035	42	4	.98200	.95940	9	37
24	.28234	28	.45077	43	4	.98196	.95931	8	36
25	0.28262	28	9.45120	43	3	9.98192	0.95923	8	35
26	.28290	28	.45163	43	4	.98189	.95915	8	34
27	.28318	28	.45206	43	4	.98185	.95907	9	33
28	.28346	28	.45249	43	4	.98181	.95898	8	32
29	.28374	28	.45292	42	3	.98177	.95890	8	31
30	0.28402	27	9.45334	43	4	9.98174	0.95882	8	30
31	.28429	28	.45377	42	4	.98170	.95874	9	29
32	.28457	28	.45419	43	4	.98166	.95865	8	28
33	.28485	28	.45462	42	3	.98162	.95857	8	27
34	.28513	28	.45504	43	4	.98159	.95849	8	26
35	0.28541	28	9.45547	42	4	9.98155	0.95841	9	25
36	.28569	28	.45589	43	4	.98151	.95832	8	24
37	.28597	28	.45632	42	3	.98147	.95824	8	23
38	.28625	27	.45674	42	4	.98144	.95816	9	22
39	.28652	28	.45716	42	4	.98140	.95807	8	21
40	0.28680	28	9.45758	43	4	9.98136	0.95799	8	20
41	.28708	28	.45801	42	3	.98132	.95791	9	19
42	.28736	28	.45843	42	4	.98129	.95782	8	18
43	.28764	28	.45885	42	4	.98125	.95774	8	17
44	.28792	28	.45927	42	4	.98121	.95766	9	16
45	0.28820	27	9.45969	42	4	9.98117	0.95757	8	15
46	.28847	28	.46011	42	3	.98113	.95749	9	14
47	.28875	28	.46053	42	4	.98110	.95740	8	13
48	.28903	28	.46095	41	4	.98106	.95732	8	12
49	.28931	28	.46136	42	4	.98102	.95724	9	11
50	0.28959	28	9.46178	42	4	9.98098	0.95715	8	10
51	.28987	28	.46220	42	4	.98094	.95707	9	9
52	.29015	27	.46262	41	3	.98090	.95698	8	8
53	.29042	28	.46303	42	4	.98087	.95690	9	7
54	.29070	28	.46345	41	4	.98083	.95681	8	6
55	0.29098	28	9.46386	42	4	9.98079	0.95673	9	5
56	.29126	28	.46428	41	4	.98075	.95664	8	4
57	.29154	28	.46469	42	4	.98071	.95656	9	3
58	.29182	27	.46511	41	4	.98067	.95647	8	2
59	.29209	28	.46552	42	3	.98063	.95639	9	1
60	0.29237		9.46594			9.98060	0.95630		0
↑ 106°→ cos		Diff. 1′	→ cos	Diff. 1′	Diff. 1′	sin	sin	Diff. 1′	↑ ←73°

17° → ↓ ′	sin	Diff. 1′	sin	Diff. 1′	Diff. 1′	cos ←	cos	Diff. 1′	← 162° ↓ ′
0	0.29237	28	9.46594	41	4	9.98060	0.95630	8	60
1	.29265	28	.46635	41	4	.98056	.95622	9	59
2	.29293	28	.46676	41	4	.98052	.95613	8	58
3	.29321	27	.46717	41	4	.98048	.95605	9	57
4	.29348	28	.46758	42	4	.98044	.95596	8	56
5	0.29376	28	9.46800	41	4	9.98040	0.95588	9	55
6	.29404	28	.46841	41	4	.98036	.95579	8	54
7	.29432	28	.46882	41	3	.98032	.95571	9	53
8	.29460	27	.46923	41	4	.98029	.95562	8	52
9	.29487	28	.46964	41	4	.98025	.95554	9	51
10	0.29515	28	9.47005	40	4	9.98021	0.95545	9	50
11	.29543	28	.47045	41	4	.98017	.95536	8	49
12	.29571	28	.47086	41	4	.98013	.95528	9	48
13	.29599	27	.47127	41	4	.98009	.95519	8	47
14	.29626	28	.47168	41	4	.98005	.95511	9	46
15	0.29654	28	9.47209	40	4	9.98001	0.95502	9	45
16	.29682	28	.47249	41	4	.97997	.95493	8	44
17	.29710	27	.47290	40	4	.97993	.95485	9	43
18	.29737	28	.47330	41	3	.97989	.95476	9	42
19	.29765	28	.47371	40	4	.97986	.95467	8	41
20	0.29793	28	9.47411	41	4	9.97982	0.95459	9	40
21	.29821	28	.47452	40	4	.97978	.95450	9	39
22	.29849	27	.47492	41	4	.97974	.95441	8	38
23	.29876	28	.47533	40	4	.97970	.95433	9	37
24	.29904	28	.47573	40	4	.97966	.95424	9	36
25	0.29932	28	9.47613	41	4	9.97962	0.95415	8	35
26	.29960	27	.47654	40	4	.97958	.95407	9	34
27	.29987	28	.47694	40	4	.97954	.95398	9	33
28	.30015	28	.47734	40	4	.97950	.95389	9	32
29	.30043	28	.47774	40	4	.97946	.95380	8	31
30	0.30071	27	9.47814	40	4	9.97942	0.95372	9	30
31	.30098	28	.47854	40	4	.97938	.95363	9	29
32	.30126	28	.47894	40	4	.97934	.95354	9	28
33	.30154	28	.47934	40	4	.97930	.95345	8	27
34	.30182	27	.47974	40	4	.97926	.95337	9	26
35	0.30209	28	9.48014	40	4	9.97922	0.95328	9	25
36	.30237	28	.48054	40	4	.97918	.95319	9	24
37	.30265	27	.48094	39	4	.97914	.95310	9	23
38	.30292	28	.48133	40	4	.97910	.95301	8	22
39	.30320	28	.48173	40	4	.97906	.95293	9	21
40	0.30348	28	9.48213	39	4	9.97902	0.95284	9	20
41	.30376	27	.48252	40	4	.97898	.95275	9	19
42	.30403	28	.48292	40	4	.97894	.95266	9	18
43	.30431	28	.48332	39	4	.97890	.95257	9	17
44	.30459	27	.48371	40	4	.97886	.95248	8	16
45	0.30486	28	9.48411	39	4	9.97882	0.95240	9	15
46	.30514	28	.48450	40	4	.97878	.95231	9	14
47	.30542	28	.48490	39	4	.97874	.95222	9	13
48	.30570	27	.48529	39	4	.97870	.95213	9	12
49	.30597	28	.48568	39	5	.97866	.95204	9	11
50	0.30625	28	9.48607	40	4	9.97861	0.95195	9	10
51	.30653	27	.48647	39	4	.97857	.95186	9	9
52	.30680	28	.48686	39	4	.97853	.95177	9	8
53	.30708	28	.48725	39	4	.97849	.95168	9	7
54	.30736	27	.48764	39	4	.97845	.95159	9	6
55	0.30763	28	9.48803	39	4	9.97841	0.95150	8	5
56	.30791	28	.48842	39	4	.97837	.95142	9	4
57	.30819	27	.48881	39	4	.97833	.95133	9	3
58	.30846	28	.48920	39	4	.97829	.95124	9	2
59	.30874	28	.48959	39	4	.97825	.95115	9	1
60	0.30902		9.48998			9.97821	0.95106		0
↑ 107° →	cos	Diff. 1′	→ cos	Diff. 1′	Diff. 1′	sin	sin	Diff. 1′	← 72° ↑

LOGARITHMS

18°→ ↓ ′	sin	Diff. 1′	sin	Diff. 1′	Diff. 1′	cos ←	cos	Diff. 1′	←161° ↓ ′
0	0. 30902	27	9. 48998	39	4	9. 97821	0. 95106	9	60
1	. 30929	28	. 49037	39	5	. 97817	. 95097	9	59
2	. 30957	28	. 49076	39	4	. 97812	. 95088	9	58
3	. 30985	27	. 49115	38	4	. 97808	. 95079	9	57
4	. 31012	28	. 49153	39	4	. 97804	. 95070	9	56
5	0. 31040	28	9. 49192	39	4	9. 97800	0. 95061	9	55
6	. 31068	27	. 49231	38	4	. 97796	. 95052	9	54
7	. 31095	28	. 49269	39	4	. 97792	. 95043	10	53
8	. 31123	28	. 49308	39	4	. 97788	. 95033	9	52
9	. 31151	27	. 49347	38	5	. 97784	. 95024	9	51
10	0. 31178	28	9. 49385	39	4	9. 97779	0. 95015	9	50
11	. 31206	27	. 49424	38	4	. 97775	. 95006	9	49
12	. 31233	28	. 49462	38	4	. 97771	. 94997	9	48
13	. 31261	28	. 49500	39	4	. 97767	. 94988	9	47
14	. 31289	27	. 49539	38	4	. 97763	. 94979	9	46
15	0. 31316	28	9. 49577	38	5	9. 97759	0. 94970	9	45
16	. 31344	28	. 49615	39	4	. 97754	. 94961	9	44
17	. 31372	27	. 49654	38	4	. 97750	. 94952	9	43
18	. 31399	28	. 49692	38	4	. 97746	. 94943	10	42
19	. 31427	27	. 49730	38	4	. 97742	. 94933	9	41
20	0. 31454	28	9. 49768	38	4	9. 97738	0. 94924	9	40
21	. 31482	28	. 49806	38	5	. 97734	. 94915	9	39
22	. 31510	27	. 49844	38	4	. 97729	. 94906	9	38
23	. 31537	28	. 49882	38	4	. 97725	. 94897	9	37
24	. 31565	28	. 40020	38	4	. 97721	. 94888	10	36
25	0. 31593	27	9. 49958	38	4	9. 97717	0. 94878	9	35
26	. 31620	28	. 40006	38	5	. 97713	. 94869	9	34
27	. 31648	27	. 50034	38	4	. 97708	. 94860	9	33
28	. 31675	28	. 50072	38	4	. 97704	. 94851	9	32
29	. 31703	27	. 50110	38	4	. 97700	. 94842	10	31
30	0. 31730	28	9. 50148	37	5	9. 97696	0. 94832	9	30
31	. 31758	28	. 50185	38	4	. 97691	. 94823	9	29
32	. 31786	27	. 50223	38	4	. 97687	. 94814	0	28
33	. 31813	28	. 50261	37	4	. 97683	. 94805	10	27
34	. 31841	27	. 50298	38	5	. 97679	. 94795	9	26
35	0. 31868	28	9. 50336	38	4	9. 97674	0. 94786	9	25
36	. 31896	27	. 50374	37	4	. 97670	. 94777	9	24
37	. 31923	28	. 50411	38	4	. 97666	. 94768	10	23
38	. 31951	28	. 50449	37	5	. 97662	. 94758	9	22
39	. 31979	27	. 50486	37	4	. 97657	. 94749	9	21
40	0. 32006	28	9. 50523	38	4	9. 97653	0. 94740	10	20
41	. 32034	27	. 50561	37	4	. 97649	. 94730	9	19
42	. 32061	28	. 50598	37	5	. 97645	. 94721	9	18
43	. 32089	27	. 50635	38	4	. 97640	. 94712	10	17
44	. 32116	28	. 50673	37	4	. 97636	. 94702	9	16
45	0. 32144	27	9. 50710	37	4	9. 97632	0. 94693	9	15
46	. 32171	28	. 50747	37	5	. 97628	. 94684	10	14
47	. 32199	28	. 50784	37	4	. 97623	. 94674	9	13
48	. 32227	27	. 50821	37	4	. 97619	. 94665	9	12
49	. 32254	28	. 50858	38	5	. 97615	. 94656	10	11
50	0. 32282	27	9. 50896	37	4	9. 97610	0. 94646	9	10
51	. 32309	28	. 50933	37	4	. 97606	. 94637	10	9
52	. 32337	27	. 50970	37	5	. 97602	. 94627	9	8
53	. 32364	28	. 51007	36	4	. 97597	. 94618	9	7
54	. 32392	27	. 51043	37	4	. 97593	. 94609	10	6
55	0. 32419	28	9. 51080	37	5	9. 97589	0. 94599	9	5
56	. 32447	27	. 51117	37	4	. 97584	. 94590	10	4
57	. 32474	28	. 51154	37	4	. 97580	. 94580	9	3
58	. 32502	27	. 51191	36	5	. 97576	. 94571	10	2
59	. 32529	28	. 51227	37	4	. 97571	. 94561	9	1
60	0. 32557		9. 51264			9. 97567	0. 94552		0
↑ 108°→	cos	Diff. 1′	→cos	Diff. 1′	Diff. 1′	sin	sin	Diff. 1′	←71° ↑

LOGARITHMS

19°→ ↓ ′	sin	Diff. 1′	sin	Diff. 1′	Diff. 1′	cos ←	cos	Diff. 1′	←160° ↓ ′
0	0. 32557	27	9. 51264	37	4	9. 97567	0. 94552	10	60
1	. 32584	28	. 51301	37	5	. 97563	. 94542	9	59
2	. 32612	27	. 51338	36	4	. 97558	. 94533	10	58
3	. 32639	28	. 51374	37	4	. 97554	. 94523	9	57
4	. 32667	27	. 51411	36	5	. 97550	. 94514	10	56
5	0. 32694	28	9. 51447	37	4	9. 97545	0. 94504	9	55
6	. 32722	27	. 51484	36	5	. 97541	. 94495	10	54
7	. 32749	28	. 51520	37	4	. 97536	. 94485	9	53
8	. 32777	27	. 51557	36	4	. 97532	. 94476	10	52
9	. 32804	28	. 51593	36	5	. 97528	. 94466	9	51
10	0. 32832	27	9. 51629	37	4	9. 97523	0. 94457	10	50
11	. 32859	28	. 51666	36	4	. 97519	. 94447	9	49
12	. 32887	27	. 51702	36	5	. 97515	. 94438	10	48
13	. 32914	28	. 51738	36	4	. 97510	. 94428	10	47
14	. 32942	27	. 51774	37	5	. 97506	. 94418	9	46
15	0. 32969	28	9. 51811	36	4	9. 97501	0. 94409	10	45
16	. 32997	27	. 51847	36	5	. 97497	. 94399	9	44
17	. 33024	27	. 51883	36	4	. 97492	. 94390	10	43
18	. 33051	28	. 51919	36	4	. 97488	. 94380	10	42
19	. 33079	27	. 51955	36	5	. 97484	. 94370	9	41
20	0. 33106	28	9. 51991	36	4	9. 97479	0. 94361	10	40
21	. 33134	27	. 52027	36	5	. 97475	. 94351	9	39
22	. 33161	28	. 52063	36	4	. 97470	. 94342	10	38
23	. 33189	27	. 52099	36	5	. 97466	. 94332	10	37
24	. 33216	28	. 52135	36	4	. 97461	. 94322	9	36
25	0. 33244	27	9. 52171	36	4	9. 97457	0. 94313	10	35
26	. 33271	27	. 52207	35	5	. 97453	. 94303	10	34
27	. 33298	28	. 52242	36	4	. 97448	. 94293	9	33
28	. 33326	27	. 52278	36	5	. 97444	. 94284	10	32
29	. 33353	28	. 52314	36	4	. 97439	. 94274	10	31
30	0. 33381	27	9. 52350	35	5	9. 97435	0. 94264	10	30
31	. 33408	28	. 52385	36	4	. 97430	. 94254	9	29
32	. 33436	27	. 52421	35	5	. 97426	. 94245	10	28
33	. 33463	27	. 52456	36	4	. 97421	. 94235	10	27
34	. 33490	28	. 52492	35	5	. 97417	. 94225	10	26
35	0. 33518	27	9. 52527	36	4	9. 97412	0. 94215	9	25
36	. 33545	28	. 52563	35	5	. 97408	. 94206	10	24
37	. 33573	27	. 52598	36	4	. 97403	. 94196	10	23
38	. 33600	27	. 52634	35	5	. 97399	. 94186	10	22
39	. 33627	28	. 52669	36	4	. 97394	. 94176	9	21
40	0. 33655	27	9. 52705	35	5	9. 97390	0. 94167	10	20
41	. 33682	28	. 52740	35	4	. 97385	. 94157	10	19
42	. 33710	27	. 52775	36	5	. 97381	. 94147	10	18
43	. 33737	27	. 52811	35	4	. 97376	. 94137	10	17
44	. 33764	28	. 52846	35	5	. 97372	. 94127	9	16
45	0. 33792	27	9. 52881	35	4	9. 97367	0. 94118	10	15
46	. 33819	27	. 52916	35	5	. 97363	. 94108	10	14
47	. 33846	28	. 52951	35	5	. 97358	. 94098	10	13
48	. 33874	27	. 52986	35	4	. 97353	. 94088	10	12
49	. 33901	28	. 53021	35	5	. 97349	. 94078	10	11
50	0. 33929	27	9. 53056	36	4	9. 97344	0. 94068	10	10
51	. 33956	27	. 53092	34	5	. 97340	. 94058	9	9
52	. 33983	28	. 53126	35	4	. 97335	. 94049	10	8
53	. 34011	27	. 53161	35	5	. 97331	. 94039	10	7
54	. 34038	27	. 53196	35	4	. 97326	. 94029	10	6
55	0. 34065	28	9. 53231	35	5	9. 97322	0. 94019	10	5
56	. 34093	27	. 53266	35	5	. 97317	. 94009	10	4
57	. 34120	27	. 53301	35	4	. 97312	. 93999	10	3
58	. 34147	28	. 53336	34	5	. 97308	. 93989	10	2
59	. 34175	27	. 53370	35	4	. 97303	. 93979	10	1
60	0. 34202		9. 53405			9. 97299	0. 93969		0
↑ 109°→	cos	Diff. 1′	→ cos	Diff. 1′	Diff. 1′	sin	sin	Diff. 1′	←70° ↑

LOGARITHMS

20°→ ↓	sin	Diff. 1′	sin	Diff. 1′	Diff. 1′	cos ←	cos	Diff. 1′	←159° ↓
′									′
0	0. 34202	27	9. 53405	35	5	9. 97299	0. 93969	10	60
1	. 34229	28	. 53440	35	5	. 97294	. 93959	10	59
2	. 34257	27	. 53475	34	4	. 97289	. 93949	10	58
3	. 34284	27	. 53509	35	5	. 97285	. 93939	10	57
4	. 34311	28	. 53544	34	4	. 97280	. 93929	10	56
5	0. 34339	27	9. 53578	35	5	9. 97276	0. 93919	10	55
6	. 34366	27	. 53613	34	5	. 97271	. 93909	10	54
7	. 34393	28	. 53647	35	4	. 97266	. 93899	10	53
8	. 34421	27	. 53682	34	5	. 97262	. 93889	10	52
9	. 34448	27	. 53716	35	5	. 97257	. 93879	10	51
10	0. 34475	28	9. 53751	34	4	9. 97252	0. 93869	10	50
11	. 34503	27	. 53785	34	5	. 97248	. 93859	10	49
12	. 34530	27	. 53819	35	5	. 97243	. 93849	10	48
13	. 34557	27	. 53854	34	4	. 97238	. 93839	10	47
14	. 34584	28	. 53888	34	5	. 97234	. 93829	10	46
15	0. 34612	27	9. 53922	35	5	9. 97229	0. 93819	10	45
16	. 34639	27	. 53957	34	4	. 97224	. 93809	10	44
17	. 34666	28	. 53991	34	5	. 97220	. 93799	10	43
18	. 34694	27	. 54025	34	5	. 97215	. 93789	10	42
19	. 34721	27	. 54059	34	4	. 97210	. 93779	10	41
20	0. 34748	27	9. 54093	34	5	9. 97206	0. 93769	10	40
21	. 34775	28	. 54127	34	5	. 97201	. 93759	11	39
22	. 34803	27	. 54161	34	4	. 97196	. 93748	10	38
23	. 34830	27	. 54195	34	5	. 97192	. 93738	10	37
24	. 34857	27	. 54229	34	5	. 97187	. 93728	10	36
25	0. 34884	28	9. 54263	34	4	9. 97182	0. 93718	10	35
26	. 34912	27	. 54297	34	5	. 97178	. 93708	10	34
27	. 34939	27	. 54331	34	5	. 97173	. 93698	10	33
28	. 34966	27	. 54365	34	5	. 97168	. 93688	11	32
29	. 34993	28	. 54399	34	4	. 97163	. 93677	10	31
30	0. 35021	27	9. 54433	33	5	9. 97159	0. 93667	10	30
31	. 35048	27	. 54466	34	5	. 97154	. 93657	10	29
32	. 35075	27	. 54500	34	4	. 97149	. 93647	10	28
33	. 35102	28	. 54534	33	5	. 97145	. 93637	11	27
34	. 35130	27	. 54567	34	5	. 97140	. 93626	10	26
35	0. 35157	27	9. 54601	34	5	9. 97135	0. 93616	10	25
36	. 35184	27	. 54635	33	4	. 97130	. 93606	10	24
37	. 35211	28	. 54668	34	5	. 97126	. 93596	11	23
38	. 35239	27	. 54702	33	5	. 97121	. 93585	10	22
39	. 35266	27	. 54735	34	5	. 97116	. 93575	10	21
40	0. 35293	27	9. 54769	33	4	9. 97111	0. 93565	10	20
41	. 35320	27	. 54802	34	5	. 97107	. 93555	11	19
42	. 35347	28	. 54836	33	5	. 97102	. 93544	10	18
43	. 35375	27	. 54869	34	5	. 97097	. 93534	10	17
44	. 35402	27	. 54903	33	5	. 97092	. 93524	10	16
45	0. 35429	27	9. 54936	33	4	9. 97087	0. 93514	11	15
46	. 35456	28	. 54969	34	5	. 97083	. 93503	10	14
47	. 35484	27	. 55003	33	5	. 97078	. 93493	10	13
48	. 35511	27	. 55036	33	5	. 97073	. 93483	11	12
49	. 35538	27	. 55069	33	5	. 97068	. 93472	10	11
50	0. 35565	27	9. 55102	34	4	9. 97063	0. 93462	10	10
51	. 35592	27	. 55136	33	5	. 97059	. 93452	11	9
52	. 35619	28	. 55169	33	5	. 97054	. 93441	10	8
53	. 35647	27	. 55202	33	5	. 97049	. 93431	11	7
54	. 35674	27	. 55235	33	5	. 97044	. 93420	10	6
55	0. 35701	27	9. 55268	33	4	9. 97039	0. 93410	10	5
56	. 35728	27	. 55301	33	5	. 97035	. 93400	11	4
57	. 35755	27	. 55334	33	5	. 97030	. 93389	10	3
58	. 35782	28	. 55367	33	5	. 97025	. 93379	11	2
59	. 35810	27	. 55400	33	5	. 97020	. 93368	10	1
60	0. 35837		9. 55433			9. 97015	0. 93358		0
↑ 110°→	cos	Diff. 1′	→ cos	Diff. 1′	Diff. 1′	sin	sin	Diff. 1′←69° ↑	

LOGARITHMS

21°→ ′ ↓	sin	Diff. 1′	sin	Diff. 1′	Diff. 1′	cos ←	cos	←158° Diff. 1′	↓ ′
0	0. 35837	27	9. 55433	33	5	9. 97015	0. 93358	10	60
1	. 35864	27	. 55466	33	5	. 97010	. 93348	11	59
2	. 35891	27	. 55499	33	4	. 97005	. 93337	10	58
3	. 35918	27	. 55532	32	5	. 97001	. 93327	11	57
4	. 35945	28	. 55564	33	5	. 96996	. 93316	10	56
5	0. 35973	27	9. 55597	33	5	9. 96991	0. 93306	11	55
6	. 36000	27	. 55630	33	5	. 96986	. 93295	10	54
7	. 36027	27	. 55663	32	5	. 96981	. 93285	11	53
8	. 36054	27	. 55695	33	5	. 96976	. 93274	10	52
9	. 36081	27	. 55728	33	5	. 96971	. 93264	11	51
10	0. 36108	27	9. 55761	32	4	9. 96966	0. 93253	10	50
11	. 36135	27	. 55793	33	5	. 96962	. 93243	11	49
12	. 36162	28	. 55826	32	5	. 96957	. 93232	10	48
13	. 36190	27	. 55858	33	5	. 96952	. 93222	11	47
14	. 36217	27	. 55891	32	5	. 96947	. 93211	10	46
15	0. 36244	27	9. 55923	33	5	9. 96942	0. 93201	11	45
16	. 36271	27	. 55956	32	5	. 96937	. 93190	10	44
17	. 36298	27	. 55988	33	5	. 96932	. 93180	11	43
18	. 36325	27	. 56021	32	5	. 96927	. 93169	10	42
19	. 36352	27	. 56053	32	5	. 96922	. 93159	11	41
20	0. 36379	27	9. 56085	33	5	9. 96917	0. 93148	11	40
21	. 36406	28	. 56118	32	5	. 96912	. 93137	10	39
22	. 36434	27	. 56150	32	4	. 96907	. 93127	11	38
23	. 36461	27	. 56182	33	5	. 96903	. 93116	10	37
24	. 36488	27	. 56215	32	5	. 96898	. 93106	11	36
25	0. 36515	27	9. 56247	32	5	9. 96893	0. 93095	11	35
26	. 36542	27	. 56279	32	5	. 96888	. 93084	10	34
27	. 36569	27	. 56311	32	5	. 96883	. 93074	11	33
28	. 36596	27	. 56343	32	5	. 96878	. 93063	11	32
29	. 36623	27	. 56375	33	5	. 96873	. 93052	10	31
30	0. 36650	27	9. 56408	32	5	9. 96868	0. 93042	11	30
31	. 36677	27	. 56440	32	5	. 96863	. 93031	11	29
32	. 36704	27	. 56472	32	5	. 96858	. 93020	10	28
33	. 36731	27	. 56504	32	5	. 96853	. 93010	11	27
34	. 36758	27	. 56536	32	5	. 96848	. 92999	11	26
35	0. 36785	27	9. 56568	31	5	9. 96843	0. 92988	10	25
36	. 36812	27	. 56599	32	5	. 96838	. 92978	11	24
37	. 36839	28	. 56631	32	5	. 96833	. 92967	11	23
38	. 36867	27	. 56663	32	5	. 96828	. 92956	11	22
39	. 36894	27	. 56695	32	5	. 96823	. 92945	10	21
40	0. 36921	27	9. 56727	32	5	9. 96818	0. 92935	11	20
41	. 36948	27	. 56759	31	5	. 96813	. 92924	11	19
42	. 36975	27	. 56790	32	5	. 96808	. 92913	11	18
43	. 37002	27	. 56822	32	5	. 96803	. 92902	10	17
44	. 37029	27	. 56854	32	5	. 96798	. 92892	11	16
45	0. 37056	27	9. 56886	31	5	9. 96793	0. 92881	11	15
46	. 37083	27	. 56917	32	5	. 96788	. 92870	11	14
47	. 37110	27	. 56949	31	5	. 96783	. 92859	10	13
48	. 37137	27	. 56980	32	6	. 96778	. 92849	11	12
49	. 37164	27	. 57012	32	5	. 96772	. 92838	11	11
50	0. 37191	27	9. 57044	31	5	9. 96767	0. 92827	11	10
51	. 37218	27	. 57075	32	5	. 96762	. 92816	11	9
52	. 37245	27	. 57107	31	5	. 96757	. 92805	11	8
53	. 37272	27	. 57138	31	5	. 96752	. 92794	10	7
54	. 37299	27	. 57169	32	5	. 96747	. 92784	11	6
55	0. 37326	27	9. 57201	31	5	9. 96742	0. 92773	11	5
56	. 37353	27	. 57232	32	5	. 96737	. 92762	11	4
57	. 37380	27	. 57264	31	5	. 96732	. 92751	11	3
58	. 37407	27	. 57295	31	5	. 96727	. 92740	11	2
59	. 37434	27	. 57326	32	5	. 96722	. 92729	11	1
60	0. 37461		9. 57358			9. 96717	0. 92718		0
↑ 111°→	cos	Diff. 1′	→ cos	Diff. 1′	Diff. 1′	sin	sin	Diff. 1′ ←68°	↑

LOGARITHMS

22° → ′ ↓	sin	Diff. 1′	sin	Diff. 1′	Diff. 1′	cos ←	cos	Diff. 1′	← 157° ′ ↓
0	0. 37461	27	9. 57358	31	6	9. 96717	0. 92718	11	60
1	. 37488	27	. 57389	31	5	. 96711	. 92707	10	59
2	. 37515	27	. 57420	31	5	. 96706	. 92697	11	58
3	. 37542	27	. 57451	31	5	. 96701	. 92686	11	57
4	. 37569	26	. 57482	32	5	. 96696	. 92675	11	56
5	0. 37595	27	9. 57514	31	5	9. 96691	0. 92664	11	55
6	. 37622	27	. 57545	31	5	. 96686	. 92653	11	54
7	. 37649	27	. 57576	31	5	. 96681	. 92642	11	53
8	. 37676	27	. 57607	31	6	. 96676	. 92631	11	52
9	. 37703	27	. 57638	31	5	. 96670	. 92620	11	51
10	0. 37730	27	9. 57669	31	5	9. 96665	0. 92609	11	50
11	. 37757	27	. 57700	31	5	. 96660	. 92598	11	49
12	. 37784	27	. 57731	31	5	. 96655	. 92587	11	48
13	. 37811	27	. 57762	31	5	. 96650	. 92576	11	47
14	. 37838	27	. 57793	31	5	. 96645	. 92565	11	46
15	0. 37865	27	9. 57824	31	6	9. 96640	0. 92554	11	45
16	. 37892	27	. 57855	30	5	. 96634	. 92543	11	44
17	. 37919	27	. 57885	31	5	. 96629	. 92532	11	43
18	. 37946	27	. 57916	31	5	. 96624	. 92521	11	42
19	. 37973	26	. 57947	31	5	. 96619	. 92510	11	41
20	0. 37999	27	9. 57978	30	6	9. 96614	0. 92499	11	40
21	. 38026	27	. 58008	31	5	. 96608	. 92488	11	39
22	. 38053	27	. 58039	31	5	. 96603	. 92477	11	38
23	. 38080	27	. 58070	31	5	. 96598	. 92466	11	37
24	. 38107	27	. 58101	30	5	. 96593	. 92455	11	36
25	0. 38134	27	9. 58131	31	6	9. 96588	0. 92444	12	35
26	. 38161	27	. 58162	30	5	. 96582	. 92432	11	34
27	. 38188	27	. 58192	31	5	. 96577	. 92421	11	33
28	. 38215	26	. 58223	30	5	. 96572	. 92410	11	32
29	. 38241	27	. 58253	31	5	. 96567	. 92399	11	31
30	0. 38268	27	9. 58284	30	6	9. 96562	0. 92388	11	30
31	. 38295	27	. 58314	31	5	. 96556	. 92377	11	29
32	. 38322	27	. 58345	30	5	. 96551	. 92366	11	28
33	. 38349	27	. 58375	31	5	. 96546	. 92355	12	27
34	. 38376	27	. 58406	30	6	. 96541	. 92343	11	26
35	0. 38403	27	9. 58436	31	5	9. 96535	0. 92332	11	25
36	. 38430	26	. 58467	30	5	. 96530	. 92321	11	24
37	. 38456	27	. 58497	30	5	. 96525	. 92310	11	23
38	. 38483	27	. 58527	30	6	. 96520	. 92299	12	22
39	. 38510	27	. 58557	31	5	. 96514	. 92287	11	21
40	0. 38537	27	9. 58588	30	5	9. 96509	0. 92276	11	20
41	. 38564	27	. 58618	30	6	. 96504	. 92265	11	19
42	. 38591	26	. 58648	30	5	. 96498	. 92254	11	18
43	. 38617	27	. 58678	31	5	. 96493	. 92243	12	17
44	. 38644	27	. 58709	30	5	. 96488	. 92231	11	16
45	0. 38671	27	9. 58739	30	6	9. 96483	0. 92220	11	15
46	. 38698	27	. 58769	30	5	. 96477	. 92209	11	14
47	. 38725	27	. 58799	30	5	. 96472	. 92198	12	13
48	. 38752	26	. 58829	30	6	. 96467	. 92186	11	12
49	. 38778	27	. 58859	30	5	. 96461	. 92175	11	11
50	0. 38805	27	9. 58889	30	5	9. 96456	0. 92164	12	10
51	. 38832	27	. 58919	30	6	. 96451	. 92152	11	9
52	. 38859	27	. 58949	30	5	. 96445	. 92141	11	8
53	. 38886	26	. 58979	30	5	. 96440	. 92130	11	7
54	. 38912	27	. 59009	30	6	. 96435	. 92119	12	6
55	0. 38939	27	9. 59039	30	5	9. 96429	0. 92107	11	5
56	. 38966	27	. 59069	29	5	. 96424	. 92096	11	4
57	. 38993	27	. 59098	30	6	. 96419	. 92085	12	3
58	. 39020	26	. 59128	30	5	. 96413	. 92073	11	2
59	. 39046	27	. 59158	30	5	. 96408	. 92062	12	1
60	0. 39073		9. 59188			9. 96403	0. 92050		0
↑ 112° → cos		Diff. 1′	→ cos	Diff. 1′	Diff. 1′	sin	sin	Diff. 1′ ←	↑ 67°

LOGARITHMS

23°→ ↓ ′	sin	Diff. 1′	sin	Diff. 1′	Diff. 1′	cos ←	cos	Diff. 1′	←156° ↓ ′
0	0. 39073	27	9. 59188	30	6	9. 96403	0. 92050	11	60
1	. 39100	27	. 59218	29	5	. 96397	. 92039	11	59
2	. 39127	26	. 59247	30	5	. 96392	. 92028	12	58
3	. 39153	27	. 59277	30	6	. 96387	. 92016	11	57
4	. 39180	27	. 59307	29	5	. 96381	. 92005	11	56
5	0. 39207	27	9. 59336	30	6	9. 96376	0. 91994	12	55
6	. 39234	26	. 59366	30	5	. 96370	. 91982	11	54
7	. 39260	27	. 59396	29	5	. 96365	. 91971	12	53
8	. 39287	27	. 59425	30	6	. 96360	. 91959	11	52
9	. 39314	27	. 59455	29	5	. 96354	. 91948	12	51
10	0. 39341	26	9. 59484	30	6	9. 96349	0. 91936	11	50
11	. 39367	27	. 59514	29	5	. 96343	. 91925	11	49
12	. 39394	27	. 59543	30	5	. 96338	. 91914	12	48
13	. 39421	27	. 59573	29	6	. 96333	. 91902	11	47
14	. 39448	26	. 59602	30	5	. 96327	. 91891	12	46
15	0. 39474	27	9. 59632	29	6	9. 96322	0. 91879	11	45
16	. 39501	27	. 59661	29	5	. 96316	. 91868	12	44
17	. 39528	27	. 59690	30	6	. 96311	. 91856	11	43
18	. 39555	26	. 59720	29	5	. 96305	. 91845	12	42
19	. 39581	27	. 59749	29	6	. 96300	. 91833	11	41
20	0. 39608	27	9. 59778	30	5	9. 96294	0. 91822	12	40
21	. 39635	26	. 59808	29	5	. 96289	. 91810	11	39
22	. 39661	27	. 59837	29	6	. 96284	. 91799	12	38
23	. 39688	27	. 59866	29	5	. 96278	. 91787	12	37
24	. 39715	26	. 59895	29	6	. 96273	. 91775	11	36
25	0. 39741	27	9. 59924	30	5	9. 96267	0. 91764	12	35
26	. 39768	27	. 59954	29	6	. 96262	. 91752	11	34
27	. 39795	27	. 59983	29	5	. 96256	. 91741	12	33
28	. 39822	26	. 60012	29	6	. 96251	. 91729	11	32
29	. 39848	27	. 60041	29	5	. 96245	. 91718	12	31
30	0. 39875	27	9. 60070	29	6	9. 96240	0. 91706	12	30
31	. 39902	26	. 60099	29	5	. 96234	. 91694	11	29
32	. 39928	27	. 60128	29	6	. 96229	. 91683	12	28
33	. 39955	27	. 60157	29	5	. 96223	. 91671	11	27
34	. 39982	26	. 60186	29	6	. 96218	. 91660	12	26
35	0. 40008	27	9. 60215	29	5	9. 96212	0. 91648	12	25
36	. 40035	27	. 60244	29	6	. 96207	. 91636	11	24
37	. 40062	26	. 60273	29	5	. 96201	. 91625	12	23
38	. 40088	27	. 60302	29	6	. 96196	. 91613	12	22
39	. 40115	26	. 60331	28	5	. 96190	. 91601	11	21
40	0. 40141	27	9. 60359	29	6	9. 96185	0. 91590	12	20
41	. 40168	27	. 60388	29	5	. 96179	. 91578	12	19
42	. 40195	26	. 60417	29	6	. 96174	. 91566	11	18
43	. 40221	27	. 60446	28	6	. 96168	. 91555	12	17
44	. 40248	27	. 60474	29	5	. 96162	. 91543	12	16
45	0. 40275	26	9. 60503	29	6	9. 96157	0. 91531	12	15
46	. 40301	27	. 60532	29	5	. 96151	. 91519	11	14
47	. 40328	27	. 60561	28	6	. 96146	. 91508	12	13
48	. 40355	26	. 60589	29	5	. 96140	. 91496	12	12
49	. 40381	27	. 60618	28	6	. 96135	. 91484	12	11
50	0. 40408	26	9. 60646	29	6	9. 96129	0. 91472	11	10
51	. 40434	27	. 60675	29	5	. 96123	. 91461	12	9
52	. 40461	27	. 60704	28	6	. 96118	. 91449	12	8
53	. 40488	26	. 60732	29	5	. 96112	. 91437	12	7
54	. 40514	27	. 60761	28	6	. 96107	. 91425	11	6
55	0. 40541	26	9. 60789	29	6	9. 96101	0. 91414	12	5
56	. 40567	27	. 60818	28	5	. 96095	. 91402	12	4
57	. 40594	27	. 60846	29	6	. 96090	. 91390	12	3
58	. 40621	26	. 60875	28	5	. 96084	. 91378	12	2
59	. 40647	27	. 60903	28	6	. 96079	. 91366	11	1
60	0. 40674		9. 60931			9. 96073	0. 91355		0
↑ 113°→	cos	Diff. 1′	→ cos	Diff. 1′	Diff. 1′	sin	sin	Diff. 1′	↑ ←66°

LOGARITHMS

24°→ ↓ ′	sin	Diff. 1′	sin	Diff. 1′	Diff. 1′	cos ←	cos	Diff. 1′	←155° ↓ ′
0	0. 40674	26	9. 60931	29	6	9. 96073	0. 91355	12	60
1	. 40700	27	. 60960	28	5	. 96067	. 91343	12	59
2	. 40727	26	. 60988	28	6	. 96062	. 91331	12	58
3	. 40753	27	. 61016	29	6	. 96056	. 91319	12	57
4	. 40780	26	. 61045	28	5	. 96050	. 91307	12	56
5	0. 40806	27	9. 61073	28	6	9. 96045	0. 91295	12	55
6	. 40833	27	. 61101	28	5	. 96039	. 91283	11	54
7	. 40860	26	. 61129	29	6	. 96034	. 91272	12	53
8	. 40886	27	. 61158	28	6	. 96028	. 91260	12	52
9	. 40913	26	. 61186	28	5	. 96022	. 91248	12	51
10	0. 40939	27	9. 61214	28	6	9. 96017	0. 91236	12	50
11	. 40966	26	. 61242	28	6	. 96011	. 91224	12	49
12	. 40992	27	. 61270	28	5	. 96005	. 91212	12	48
13	. 41019	26	. 61298	28	6	. 96000	. 91200	12	47
14	. 41045	27	. 61326	28	6	. 95994	. 91188	12	46
15	0. 41072	26	9. 61354	28	6	9. 95988	0. 91176	12	45
16	. 41098	27	. 61382	29	5	. 95982	. 91164	12	44
17	. 41125	26	. 61411	27	6	. 95977	. 91152	12	43
18	. 41151	27	. 61438	28	6	. 95971	. 91140	12	42
19	. 41178	26	. 61466	28	5	. 95965	. 91128	12	41
20	0. 41204	27	9. 61494	28	6	9. 95960	0. 91116	12	40
21	. 41231	26	. 61522	28	6	. 95954	. 91104	12	39
22	. 41257	27	. 61550	28	6	. 95948	. 91092	12	38
23	. 41284	26	. 61578	28	5	. 95942	. 91080	12	37
24	. 41310	27	. 61606	28	6	. 95937	. 91068	12	36
25	0. 41337	26	9. 61634	28	6	9. 95931	0. 91056	12	35
26	. 41363	27	. 61662	27	5	. 95925	. 91044	12	34
27	. 41390	26	. 61689	28	6	. 95920	. 91032	12	33
28	. 41416	27	. 61717	28	6	. 95914	. 91020	12	32
29	. 41443	26	. 61745	28	6	. 95908	. 91008	12	31
30	0. 41469	27	9. 61773	27	5	9. 95902	0. 90996	12	30
31	. 41496	26	. 61800	28	6	. 95897	. 90984	12	29
32	. 41522	27	. 61828	28	6	. 95891	. 90972	12	28
33	. 41549	26	. 61856	27	6	. 95885	. 90960	12	27
34	. 41575	27	. 61883	28	6	. 95879	. 90948	12	26
35	0. 41602	26	9. 61911	28	5	9. 95873	0. 90936	12	25
36	. 41628	27	. 61939	27	6	. 95868	. 90924	13	24
37	. 41655	26	. 61966	28	6	. 95862	. 90911	12	23
38	. 41681	26	. 61994	27	6	. 95856	. 90899	12	22
39	. 41707	27	. 62021	28	6	. 95850	. 90887	12	21
40	0. 41734	26	9. 62049	27	5	9. 95844	0. 90875	12	20
41	. 41760	27	. 62076	28	6	. 95839	. 90863	12	19
42	. 41787	26	. 62104	27	6	. 95833	. 90851	12	18
43	. 41813	27	. 62131	28	6	. 95827	. 90839	13	17
44	. 41840	26	. 62159	27	6	. 95821	. 90826	12	16
45	0. 41866	26	9. 62186	28	5	9. 95815	0. 90814	12	15
46	. 41892	27	. 62214	27	6	. 95810	. 90802	12	14
47	. 41919	26	. 62241	27	6	. 95804	. 90790	12	13
48	. 41945	27	. 62268	28	6	. 95798	. 90778	12	12
49	. 41972	26	. 62296	27	6	. 95792	. 90766	13	11
50	0. 41998	26	9. 62323	27	6	9. 95786	0. 90753	12	10
51	. 42024	27	. 62350	27	5	. 95780	. 90741	12	9
52	. 42051	26	. 62377	28	6	. 95775	. 90729	12	8
53	. 42077	27	. 62405	27	6	. 95769	. 90717	13	7
54	. 42104	26	. 62432	27	6	. 95763	. 90704	12	6
55	0. 42130	26	9. 62459	27	6	9. 95757	0. 90692	12	5
56	. 42156	27	. 62486	27	6	. 95751	. 90680	12	4
57	. 42183	26	. 62513	28	6	. 95745	. 90668	13	3
58	. 42209	26	. 62541	27	6	. 95739	. 90655	12	2
59	. 42235	27	. 62568	27	5	. 95733	. 90643	12	1
60	0. 42262		9. 62595			9. 95728	0. 90631		0
↑ 114°→ cos		Diff. 1′	→ cos	Diff. 1′	Diff. 1′	sin	sin	Diff. 1′	←65° ↑

25°→ ↓ ′	sin	Diff. 1′	sin	Diff. 1′	Diff. 1′	cos ←	cos	Diff. 1′	←154° ↓ ′
0	0. 42262	26	9. 62595	27	6	9. 95728	0. 90631	13	60
1	. 42288	27	. 62622	27	6	. 95722	. 90618	12	59
2	. 42315	26	. 62649	27	6	. 95716	. 90606	12	58
3	. 42341	26	. 62676	27	6	. 95710	. 90594	12	57
4	. 42367	27	. 62703	27	6	. 95704	. 90582	13	56
5	0. 42394	26	9. 62730	27	6	9. 95698	0. 90569	12	55
6	. 42420	26	. 62757	27	6	. 95692	. 90557	12	54
7	. 42446	27	. 62784	27	6	. 95686	. 90545	13	53
8	. 42473	26	. 62811	27	6	. 95680	. 90532	12	52
9	. 42499	26	. 62838	27	6	. 95674	. 90520	13	51
10	0. 42525	27	9. 62865	27	5	9. 95668	0. 90507	12	50
11	. 42552	26	. 62892	26	6	. 95663	. 90495	12	49
12	. 42578	26	. 62918	27	6	. 95657	. 90483	13	48
13	. 42604	27	. 62945	27	6	. 95651	. 90470	12	47
14	. 42631	26	. 62972	27	6	. 95645	. 90458	12	46
15	0. 42657	26	9. 62999	27	6	9. 95639	0. 90446	13	45
16	. 42683	26	. 63026	26	6	. 95633	. 90433	12	44
17	. 42709	27	. 63052	27	6	. 95627	. 90421	13	43
18	. 42736	26	. 63079	27	6	. 95621	. 90408	12	42
19	. 42762	26	. 63106	27	6	. 95615	. 90396	13	41
20	0. 42788	27	9. 63133	26	6	9. 95609	0. 90383	12	40
21	. 42815	26	. 63159	27	6	. 95603	. 90371	13	39
22	. 42841	26	. 63186	27	6	. 95597	. 90358	12	38
23	. 42867	27	. 63213	26	6	. 95591	. 90346	12	37
24	. 42894	26	. 63239	27	6	. 95585	. 90334	13	36
25	0. 42920	26	9. 63266	26	6	9. 95579	0. 90321	12	35
26	. 42946	26	. 63292	27	6	. 95573	. 90309	13	34
27	. 42972	27	. 63319	26	6	. 95567	. 90296	12	33
28	. 42999	26	. 63345	27	6	. 95561	. 90284	13	32
29	. 43025	26	. 63372	26	6	. 95555	. 90271	12	31
30	0. 43051	26	9. 63398	27	6	9. 95549	0. 90259	13	30
31	. 43077	27	. 63425	26	6	. 95543	. 90246	13	29
32	. 43104	26	. 63451	27	6	. 95537	. 90233	12	28
33	. 43130	26	. 63478	26	6	. 95531	. 90221	13	27
34	. 43156	26	. 63504	27	6	. 95525	. 90208	12	26
35	0. 43182	27	9. 63531	26	6	9. 95519	0. 90196	13	25
36	. 43209	26	. 63557	26	6	. 95513	. 90183	12	24
37	. 43235	26	. 63583	27	7	. 95507	. 90171	13	23
38	. 43261	26	. 63610	26	6	. 95500	. 90158	12	22
39	. 43287	26	. 63636	26	6	. 95494	. 90146	13	21
40	0. 43313	27	9. 63662	27	6	9. 95488	0. 90133	13	20
41	. 43340	26	. 63689	26	6	. 95482	. 90120	12	19
42	. 43366	26	. 63715	26	6	. 95476	. 90108	13	18
43	. 43392	26	. 63741	26	6	. 95470	. 90095	13	17
44	. 43418	27	. 63767	27	6	. 95464	. 90082	12	16
45	0. 43445	26	9. 63794	26	6	9. 95458	0. 90070	13	15
46	. 43471	26	. 63820	26	6	. 95452	. 90057	12	14
47	. 43497	26	. 63846	26	6	. 95446	. 90045	13	13
48	. 43523	26	. 63872	26	6	. 95440	. 90032	13	12
49	. 43549	26	. 63898	26	7	. 95434	. 90019	12	11
50	0. 43575	27	9. 63924	26	6	9. 95427	0. 90007	13	10
51	. 43602	26	. 63950	26	6	. 95421	. 89994	13	9
52	. 43628	26	. 63976	26	6	. 95415	. 89981	13	8
53	. 43654	26	. 64002	26	6	. 95409	. 89968	12	7
54	. 43680	26	. 64028	26	6	. 95403	. 89956	13	6
55	0. 43706	27	9. 64054	26	6	9. 95397	0. 89943	13	5
56	. 43733	26	. 64080	26	7	. 95391	. 89930	12	4
57	. 43759	26	. 64106	26	6	. 95384	. 89918	13	3
58	. 43785	26	. 64132	26	6	. 95378	. 89905	13	2
59	. 43811	26	. 64158	26	6	. 95372	. 89892	13	1
60	0. 43837		9. 64184			9. 95366	0. 89879		0
↑ 115°→	cos	Diff. 1′	→ cos	Diff. 1′	Diff. 1′	sin	sin	Diff. 1′	↑ ←64°

LOGARITHMS

26°→ ↓	sin	Diff. 1′	sin	Diff. 1′	Diff. 1′	cos ←	cos	Diff. 1′	←153° ↓
′									′
0	0. 43837	26	9. 64184	26	6	9. 95366	0. 89879	12	60
1	. 43863	26	. 64210	26	6	. 95360	. 89867	13	59
2	. 43889	27	. 64236	26	6	. 95354	. 89854	13	58
3	. 43916	26	. 64262	26	7	. 95348	. 89841	13	57
4	. 43942	26	. 64288	25	6	. 95341	. 89828	12	56
5	0. 43968	26	9. 64313	26	6	9. 95335	0. 89816	13	55
6	. 43994	26	. 64339	26	6	. 95329	. 89803	13	54
7	. 44020	26	. 64365	26	6	. 95323	. 89790	13	53
8	. 44046	26	. 64391	26	7	. 95317	. 89777	13	52
9	. 44072	26	. 64417	25	6	. 95310	. 89764	12	51
10	0. 44098	26	9. 64442	26	6	9. 95304	0. 89752	13	50
11	. 44124	27	. 64468	26	6	. 95298	. 89739	13	49
12	. 44151	26	. 64494	25	6	. 95292	. 89726	13	48
13	. 44177	26	. 64519	26	7	. 95286	. 89713	13	47
14	. 44203	26	. 64545	26	6	. 95279	. 89700	13	46
15	0. 44229	26	9. 64571	25	6	9. 95273	0. 89687	13	45
16	. 44255	26	. 64596	26	6	. 95267	. 89674	12	44
17	. 44281	26	. 64622	25	7	. 95261	. 89662	13	43
18	. 44307	26	. 64647	26	6	. 95254	. 89649	13	42
19	. 44333	26	. 64673	25	6	. 95248	. 89636	13	41
20	0. 44359	26	9. 64698	26	6	9. 95242	0. 89623	13	40
21	. 44385	26	. 64724	25	7	. 95236	. 89610	13	39
22	. 44411	26	. 64749	26	6	. 95229	. 89597	13	38
23	. 44437	27	. 64775	25	6	. 95223	. 89584	13	37
24	. 44464	26	. 64800	26	6	. 95217	. 89571	13	36
25	0. 44490	26	9. 64826	25	7	9. 95211	0. 89558	13	35
26	. 44516	26	. 64851	26	6	. 95204	. 89545	13	34
27	. 44542	26	. 64877	25	6	. 95198	. 89532	13	33
28	. 44568	26	. 64902	25	7	. 95192	. 89519	13	32
29	. 44594	26	. 64927	26	6	. 95185	. 89506	13	31
30	0. 44620	26	9. 64953	25	6	9. 95179	0. 89493	13	30
31	. 44646	26	. 64978	25	6	. 95173	. 89480	13	29
32	. 44072	26	. 65003	26	7	. 95167	. 89467	13	28
33	. 44698	26	. 65029	25	6	. 95160	. 89454	13	27
34	. 44724	26	. 65054	25	6	. 95154	. 89441	13	26
35	0. 44750	26	9. 65079	25	7	9. 95148	0. 80128	13	25
36	. 44776	26	. 65104	26	6	. 95141	. 89415	13	24
37	. 44802	26	. 65130	25	6	. 95135	. 89402	13	23
38	. 44828	26	. 65155	25	7	. 95129	. 89389	13	22
39	. 44854	26	. 65180	25	6	. 95122	. 89376	13	21
40	0. 44880	26	9. 65205	25	6	9. 95116	0. 89363	13	20
41	. 44906	26	. 65230	25	7	. 95110	. 89350	13	19
42	. 44932	26	. 65255	26	6	. 95103	. 89337	13	18
43	. 44958	26	. 65281	25	7	. 95097	. 89324	13	17
44	. 44984	26	. 65306	25	6	. 95090	. 89311	13	16
45	0. 45010	26	9. 65331	25	6	9. 95084	0. 89298	13	15
46	. 45036	26	. 65356	25	7	. 95078	. 89285	13	14
47	. 45062	26	. 65381	25	6	. 95071	. 89272	13	13
48	. 45088	26	. 65406	25	6	. 95065	. 89259	14	12
49	. 45114	26	. 65431	25	7	. 95059	. 89245	13	11
50	0. 45140	26	9. 65456	25	6	9. 95052	0. 89232	13	10
51	. 45166	26	. 65481	25	7	. 95046	. 89219	13	9
52	. 45192	26	. 65506	25	6	. 95039	. 89206	13	8
53	. 45218	25	. 65531	25	6	. 95033	. 89193	13	7
54	. 45243	26	. 65556	24	7	. 95027	. 89180	13	6
55	0. 45269	26	9. 65580	25	6	9. 95020	0. 89167	14	5
56	. 45295	26	. 65605	25	7	. 95014	. 89153	13	4
57	. 45321	26	. 65630	25	6	. 95007	. 89140	13	3
58	. 45347	26	. 65655	25	6	. 95001	. 89127	13	2
59	. 45373	26	. 65680	25	7	. 94995	. 89114	13	1
60	0. 45399		9. 65705			9. 94988	0. 89101		0
↑ 116°→	cos	Diff. 1′	→ cos	Diff. 1′	Diff. 1′	sin	sin	Diff. 1′	↑ 1′←63°

LOGARITHMS

27°→ ↓ ′	sin	Diff. 1′	sin	Diff. 1′	Diff. 1′	cos ←	cos	Diff. 1′	←152° ↓ ′
0	0. 45399	26	9. 65705	24	6	9. 94988	0. 89101	14	60
1	. 45425	26	. 65729	25	7	. 94982	. 89087	13	59
2	. 45451	26	. 65754	25	6	. 94975	. 89074	13	58
3	. 45477	26	. 65779	25	7	. 94969	. 89061	13	57
4	. 45503	26	. 65804	24	6	. 94962	. 89048	13	56
5	0. 45529	25	9. 65828	25	7	9. 94956	0. 89035	14	55
6	. 45554	26	. 65853	25	6	. 94949	. 89021	13	54
7	. 45580	26	. 65878	24	7	. 94943	. 89008	13	53
8	. 45606	26	. 65902	25	6	. 94936	. 88995	14	52
9	. 45632	26	. 65927	25	7	. 94930	. 88981	13	51
10	0. 45658	26	9. 65952	24	6	9. 94923	0. 88968	13	50
11	. 45684	26	. 65976	25	6	. 94917	. 88955	13	49
12	. 45710	26	. 66001	24	7	. 94911	. 88942	14	48
13	. 45736	26	. 66025	25	6	. 94904	. 88928	13	47
14	. 45762	25	. 66050	25	7	. 94898	. 88915	13	46
15	0. 45787	26	9. 66075	24	6	9. 94891	0. 88902	14	45
16	. 45813	26	. 66099	25	7	. 94885	. 88888	13	44
17	. 45839	26	. 66124	24	7	. 94878	. 88875	13	43
18	. 45865	26	. 66148	25	6	. 94871	. 88862	14	42
19	. 45891	26	. 66173	24	7	. 94865	. 88848	13	41
20	0. 45917	25	9. 66197	24	6	9. 94858	0. 88835	13	40
21	. 45942	26	. 66221	25	7	. 94852	. 88822	14	39
22	. 45968	26	. 66246	24	6	. 94845	. 88808	13	38
23	. 45994	26	. 66270	25	7	. 94839	. 88795	13	37
24	. 46020	26	. 66295	24	6	. 94832	. 88782	14	36
25	0. 46046	26	9. 66319	24	7	9. 94826	0. 88768	13	35
26	. 46072	25	. 66343	25	6	. 94819	. 88755	14	34
27	. 46097	26	. 66368	24	7	. 94813	. 88741	13	33
28	. 46123	26	. 66392	24	7	. 94806	. 88728	13	32
29	. 46149	26	. 66416	25	6	. 94799	. 88715	14	31
30	0. 46175	26	9. 66441	24	7	9. 94793	0. 88701	13	30
31	. 46201	25	. 66465	24	6	. 94786	. 88688	14	29
32	. 46226	26	. 66489	24	7	. 94780	. 88674	13	28
33	. 46252	26	. 66513	24	6	. 94773	. 88661	14	27
34	. 46278	26	. 66537	25	7	. 94767	. 88647	13	26
35	0. 46304	26	9. 66562	24	7	9. 94760	0. 88634	14	25
36	. 46330	25	. 66586	24	6	. 94753	. 88620	13	24
37	. 46355	26	. 66610	24	7	. 94747	. 88607	14	23
38	. 46381	26	. 66634	24	6	. 94740	. 88593	13	22
39	. 46407	26	. 66658	24	7	. 94734	. 88580	14	21
40	0. 46433	25	9. 66682	24	7	9. 94727	0. 88566	13	20
41	. 46458	26	. 66706	25	6	. 94720	. 88553	14	19
42	. 46484	26	. 66731	24	7	. 94714	. 88539	13	18
43	. 46510	26	. 66755	24	7	. 94707	. 88526	14	17
44	. 46536	25	. 66779	24	6	. 94700	. 88512	13	16
45	0. 46561	26	9. 66803	24	7	9. 94694	0. 88499	14	15
46	. 46587	26	. 66827	24	7	. 94687	. 88485	13	14
47	. 46613	26	. 66851	24	6	. 94680	. 88472	14	13
48	. 46639	25	. 66875	24	7	. 94674	. 88458	13	12
49	. 46664	26	. 66899	23	7	. 94667	. 88445	14	11
50	0. 46690	26	9. 66922	24	6	9. 94660	0. 88431	14	10
51	. 46716	26	. 66946	24	7	. 94654	. 88417	13	9
52	. 46742	25	. 66970	24	7	. 94647	. 88404	14	8
53	. 46767	26	. 66994	24	6	. 94640	. 88390	13	7
54	. 46793	26	. 67018	24	7	. 94634	. 88377	14	6
55	0. 46819	25	9. 67042	24	7	9. 94627	0. 88363	14	5
56	. 46844	26	. 67066	24	6	. 94620	. 88349	13	4
57	. 46870	26	. 67090	23	7	. 94614	. 88336	14	3
58	. 46896	25	. 67113	24	7	. 94607	. 88322	14	2
59	. 46921	26	. 67137	24	7	. 94600	. 88308	13	1
60	0. 46947		9. 67161			9. 94593	0. 88295		0
↑ 117°→	cos	Diff. 1′	→ cos	Diff. 1′	Diff. 1′	sin	sin	Diff. 1′	←62° ↑

LOGARITHMS

28°→ ↓ ′	sin	Diff. 1′	sin	Diff. 1′	Diff. 1′	cos ←	cos	Diff. 1′	←151° ↓ ′
0	0. 46947	26	9. 67161	24	6	9. 94593	0. 88295	14	60
1	. 46973	26	. 67185	23	7	. 94587	. 88281	14	59
2	. 46999	25	. 67208	24	7	. 94580	. 88267	13	58
3	. 47024	26	. 67232	24	6	. 94573	. 88254	14	57
4	. 47050	26	. 67256	24	7	. 94567	. 88240	14	56
5	0. 47076	25	9. 67280	23	7	9. 94560	0. 88226	13	55
6	. 47101	26	. 67303	24	7	. 94553	. 88213	14	54
7	. 47127	26	. 67327	23	6	. 94546	. 88199	14	53
8	. 47153	25	. 67350	24	7	. 94540	. 88185	13	52
9	. 47178	26	. 67374	24	7	. 94533	. 88172	14	51
10	0. 47204	25	9. 67398	23	7	9. 94526	0. 88158	14	50
11	. 47229	26	. 67421	24	6	. 94519	. 88144	14	49
12	. 47255	26	. 67445	23	7	. 94513	. 88130	13	48
13	. 47281	25	. 67468	24	7	. 94506	. 88117	14	47
14	. 47306	26	. 67492	23	7	. 94499	. 88103	14	46
15	0. 47332	26	9. 67515	24	7	9. 94492	0. 88089	14	45
16	. 47358	25	. 67539	23	6	. 94485	. 88075	13	44
17	. 47383	26	. 67562	24	7	. 94479	. 88062	14	43
18	. 47409	25	. 67586	23	7	. 94472	. 88048	14	42
19	. 47434	26	. 67609	24	7	. 94465	. 88034	14	41
20	0. 47460	26	9. 67633	23	7	9. 94458	0. 88020	14	40
21	. 47486	25	. 67656	24	6	. 94451	. 88006	13	39
22	. 47511	26	. 67680	23	7	. 94445	. 87993	14	38
23	. 47537	25	. 67703	23	7	. 94438	. 87979	14	37
24	. 47562	26	. 67726	24	7	. 94431	. 87965	14	36
25	0. 47588	20	9. 67750	23	7	9. 94424	0. 87951	14	35
26	. 47614	25	. 67773	23	7	. 94417	. 87937	14	34
27	. 47639	26	. 67796	24	6	. 94410	. 87923	14	33
28	. 47665	25	. 67820	23	7	. 94404	. 87909	13	32
29	. 47690	26	. 67843	23	7	. 94397	. 87896	14	31
30	0. 47716	25	9. 67866	24	7	9. 94390	0. 87882	14	30
31	. 47741	26	. 67890	23	7	. 94383	. 87868	14	29
32	. 47767	26	. 67913	23	7	. 01376	. 87854	14	28
33	. 47793	25	. 67936	23	7	. 94369	. 87840	14	27
34	. 47818	26	. 67959	23	7	. 94362	. 87826	14	26
35	0. 47844	25	9. 07982	24	6	9. 94355	0. 87812	14	25
36	. 47869	26	. 68006	23	7	. 94349	. 87798	14	24
37	. 47895	25	. 68029	23	7	. 94342	. 87784	14	23
38	. 47920	26	. 68052	23	7	. 94335	. 87770	14	22
39	. 47946	25	. 68075	23	7	. 94328	. 87756	13	21
40	0. 47971	26	9. 68098	23	7	9. 94321	0. 87743	14	20
41	. 47997	25	. 68121	23	7	. 94314	. 87729	14	19
42	. 48022	26	. 68144	23	7	. 94307	. 87715	14	18
43	. 48048	25	. 68167	23	7	. 94300	. 87701	14	17
44	. 48073	26	. 68190	23	7	. 94293	. 87687	14	16
45	0. 48099	25	9. 68213	24	7	9. 94286	0. 87673	14	15
46	. 48124	26	. 68237	23	6	. 94279	. 87659	14	14
47	. 48150	25	. 68260	23	7	. 94273	. 87645	14	13
48	. 48175	26	. 68283	22	7	. 94266	. 87631	14	12
49	. 48201	25	. 68305	23	7	. 94259	. 87617	14	11
50	0. 48226	26	9. 68328	23	7	9. 94252	0. 87603	14	10
51	. 48252	25	. 68351	23	7	. 94245	. 87589	14	9
52	. 48277	26	. 68374	23	7	. 94238	. 87575	14	8
53	. 48303	25	. 68397	23	7	. 94231	. 87561	15	7
54	. 48328	26	. 68420	23	7	. 94224	. 87546	14	6
55	0. 48354	25	9. 68443	23	7	9. 94217	0. 87532	14	5
56	. 48379	26	. 68466	23	7	. 94210	. 87518	14	4
57	. 48405	25	. 68489	23	7	. 94203	. 87504	14	3
58	. 48430	26	. 68512	22	7	. 94196	. 87490	14	2
59	. 48456	25	. 68534	23	7	. 94189	. 87476	14	1
60	0. 48481		9. 68557			9. 94182	0. 87462		0
↑ 118°→	cos	Diff. 1′	→ cos	Diff. 1′	Diff. 1′	sin	sin	Diff. 1′	←61° ↑

LOGARITHMS

29°→ ′ ↓	sin	Diff. 1′	sin	Diff. 1′	Diff. 1′	cos ←	cos	Diff. 1′	←150° ↓ ′
0	0. 48481	25	9. 68557	23	7	9. 94182	0. 87462	14	60
1	. 48506	26	. 68580	23	7	. 94175	. 87448	14	59
2	. 48532	25	. 68603	22	7	. 94168	. 87434	14	58
3	. 48557	26	. 68625	23	7	. 94161	. 87420	14	57
4	. 48583	25	. 68648	23	7	. 94154	. 87406	15	56
5	0. 48608	26	9. 68671	23	7	9. 94147	0. 87391	14	55
6	. 48634	25	. 68694	22	7	. 94140	. 87377	14	54
7	. 48659	25	. 68716	23	7	. 94133	. 87363	14	53
8	. 48684	26	. 68739	23	7	. 94126	. 87349	14	52
9	. 48710	25	. 68762	22	7	. 94119	. 87335	14	51
10	0. 48735	26	9. 68784	23	7	9. 94112	0. 87321	15	50
11	. 48761	25	. 68807	22	7	. 94105	. 87306	14	49
12	. 48786	25	. 68829	23	8	. 94098	. 87292	14	48
13	. 48811	26	. 68852	23	7	. 94090	. 87278	14	47
14	. 48837	25	. 68875	22	7	. 94083	. 87264	14	46
15	0. 48862	26	9. 68897	23	7	9. 94076	0. 87250	15	45
16	. 48888	25	. 68920	22	7	. 94069	. 87235	14	44
17	. 48913	25	. 68942	23	7	. 94062	. 87221	14	43
18	. 48938	26	. 68965	22	7	. 94055	. 87207	14	42
19	. 48964	25	. 68987	23	7	. 94048	. 87193	15	41
20	0. 48989	25	9. 69010	22	7	9. 94041	0. 87178	14	40
21	. 49014	26	. 69032	23	7	. 94034	. 87164	14	39
22	. 49040	25	. 69055	22	7	. 94027	. 87150	14	38
23	. 49065	25	. 69077	23	8	. 94020	. 87136	15	37
24	. 49090	26	. 69100	22	7	. 94012	. 87121	14	36
25	0. 49116	25	9. 69122	22	7	9. 94005	0. 87107	14	35
26	. 49141	25	. 69144	23	7	. 93998	. 87093	14	34
27	. 49166	26	. 69167	22	7	. 93991	. 87079	15	33
28	. 49192	25	. 69189	23	7	. 93984	. 87064	14	32
29	. 49217	25	. 69212	22	7	. 93977	. 87050	14	31
30	0. 49242	26	9. 69234	22	7	9. 93970	0. 87036	15	30
31	. 49268	25	. 69256	23	8	. 93963	. 87021	14	29
32	. 49293	25	. 69279	22	7	. 93955	. 87007	14	28
33	. 49318	26	. 69301	22	7	. 93948	. 86993	15	27
34	. 49344	25	. 69323	22	7	. 93941	. 86978	14	26
35	0. 49369	25	9. 69345	23	7	9. 93934	0. 86964	15	25
36	. 49394	25	. 69368	22	7	. 93927	. 86949	14	24
37	. 49419	26	. 69390	22	8	. 93920	. 86935	14	23
38	. 49445	25	. 69412	22	7	. 93912	. 86921	15	22
39	. 49470	25	. 69434	22	7	. 93905	. 86906	14	21
40	0. 49495	26	9. 69456	23	7	9. 93898	0. 86892	14	20
41	. 49521	25	. 69479	22	7	. 93891	. 86878	15	19
42	. 49546	25	. 69501	22	8	. 93884	. 86863	14	18
43	. 49571	25	. 69523	22	7	. 93876	. 86849	15	17
44	. 49596	26	. 69545	22	7	. 93869	. 86834	14	16
45	0. 49622	25	9. 69567	22	7	9. 93862	0. 86820	15	15
46	. 49647	25	. 69589	22	8	. 93855	. 86805	14	14
47	. 49672	25	. 69611	22	7	. 93847	. 86791	14	13
48	. 49697	26	. 69633	22	7	. 93840	. 86777	15	12
49	. 49723	25	. 69655	22	7	. 93833	. 86762	14	11
50	0. 49748	25	9. 69677	22	7	9. 93826	0. 86748	15	10
51	. 49773	25	. 69699	22	8	. 93819	. 86733	14	9
52	. 49798	26	. 69721	22	7	. 93811	. 86719	15	8
53	. 49824	25	. 69743	22	7	. 93804	. 86704	14	7
54	. 49849	25	. 69765	22	8	. 93797	. 86690	15	6
55	0. 49874	25	9. 69787	22	7	9. 93789	0. 86675	14	5
56	. 49899	25	. 69809	22	7	. 93782	. 86661	15	4
57	. 49924	26	. 69831	22	7	. 93775	. 86646	14	3
58	. 49950	25	. 69853	22	8	. 93768	. 86632	15	2
59	. 49975	25	. 69875	22	7	. 93760	. 86617	14	1
60	0. 50000		9. 69897			9. 93753	0. 86603		0
↑ 119°→	cos	Diff. 1′	→ cos	Diff. 1′	Diff. 1′	sin	sin	Diff. 1′	←60° ↑

LOGARITHMS

30°→ ↓ ′	sin	Diff. 1′	sin	Diff. 1′	Diff. 1′	cos ←	cos	←149° Diff. 1′ ↓	′
0	0. 50000	25	9. 69897	22	7	9. 93753	0. 86603	15	60
1	. 50025	25	. 69919	22	8	. 93746	. 86588	15	59
2	. 50050	26	. 69941	22	7	. 93738	. 86573	14	58
3	. 50076	25	. 69963	21	7	. 93731	. 86559	15	57
4	. 50101	25	. 69984	22	7	. 93724	. 86544	14	56
5	0. 50126	25	9. 70006	22	8	9. 93717	0. 86530	15	55
6	. 50151	25	. 70028	22	7	. 93709	. 86515	14	54
7	. 50176	25	. 70050	22	7	. 93702	. 86501	15	53
8	. 50201	26	. 70072	21	8	. 93695	. 86486	15	52
9	. 50227	25	. 70093	22	7	. 93687	. 86471	14	51
10	0. 50252	25	9. 70115	22	7	9. 93680	0. 86457	15	50
11	. 50277	25	. 70137	22	8	. 93673	. 86442	15	49
12	. 50302	25	. 70159	21	7	. 93665	. 86427	14	48
13	. 50327	25	. 70180	22	8	. 93658	. 86413	15	47
14	. 50352	25	. 70202	22	7	. 93650	. 86398	14	46
15	0. 50377	26	9. 70224	21	7	9. 93643	0. 86384	15	45
16	. 50403	25	. 70245	22	8	. 93636	. 86369	15	44
17	. 50428	25	. 70267	21	7	. 93628	. 86354	14	43
18	. 50453	25	. 70288	22	7	. 93621	. 86340	15	42
19	. 50478	25	. 70310	22	8	. 93614	. 86325	15	41
20	0. 50503	25	9. 70332	21	7	9. 93606	0. 86310	15	40
21	. 50528	25	. 70353	22	8	. 93599	. 86295	14	39
22	. 50553	25	. 70375	21	7	. 93591	. 86281	15	38
23	. 50578	25	. 70396	22	7	. 93584	. 86266	15	37
24	. 50603	25	. 70418	21	8	. 93577	. 80251	14	36
25	0. 50628	26	9. 70439	22	7	9. 93569	0. 86237	15	35
26	. 50654	25	. 70461	21	8	. 93562	. 86222	15	34
27	. 50679	25	. 70482	22	7	. 93554	. 86207	15	33
28	. 50704	25	. 70504	21	8	. 93547	. 86192	14	32
29	. 50729	25	. 70525	22	7	. 93539	. 86178	15	31
30	0. 50754	25	9. 70547	21	7	9. 93532	0. 86163	15	30
31	. 50779	25	. 70568	22	8	. 93525	. 86148	15	29
32	. 50804	25	. 70590	21	7	. 93517	. 86133	14	28
33	. 50829	25	. 70611	22	8	. 93510	. 86119	15	27
34	. 50854	25	. 70633	21	7	. 93502	. 86104	15	26
35	0. 50879	25	0. 70654	21	8	0. 93495	0. 86089	15	25
36	. 50904	25	. 70675	22	7	. 93487	. 86074	15	24
37	. 50929	25	. 70697	21	8	. 93480	. 86059	14	23
38	. 50954	25	. 70718	21	7	. 93472	. 86045	15	22
39	. 50979	25	. 70739	22	8	. 93465	. 86030	15	21
40	0. 51004	25	9. 70761	21	7	9. 93457	0. 86015	15	20
41	. 51029	25	. 70782	21	8	. 93450	. 86000	15	19
42	. 51054	25	. 70803	21	7	. 93442	. 85985	15	18
43	. 51079	25	. 70824	22	8	. 93435	. 85970	14	17
44	. 51104	25	. 70846	21	7	. 93427	. 85956	15	16
45	0. 51129	25	9. 70867	21	8	9. 93420	0. 85941	15	15
46	. 51154	25	. 70888	21	7	. 93412	. 85926	15	14
47	. 51179	25	. 70909	22	8	. 93405	. 85911	15	13
48	. 51204	25	. 70931	21	7	. 93397	. 85896	15	12
49	. 51229	25	. 70952	21	8	. 93390	. 85881	15	11
50	0. 51254	25	9. 70973	21	7	9. 93382	0. 85866	15	10
51	. 51279	25	. 70994	21	8	. 93375	. 85851	15	9
52	. 51304	25	. 71015	21	7	. 93367	. 85836	15	8
53	. 51329	25	. 71036	22	8	. 93360	. 85821	15	7
54	. 51354	25	. 71058	21	8	. 93352	. 85806	14	6
55	0. 51379	25	9. 71079	21	7	9. 93344	0. 85792	15	5
56	. 51404	25	. 71100	21	8	. 93337	. 85777	15	4
57	. 51429	25	. 71121	21	7	. 93329	. 85762	15	3
58	. 51454	25	. 71142	21	8	. 93322	. 85747	15	2
59	. 51479	25	. 71163	21	7	. 93314	. 85732	15	1
60	0. 51504		9. 71184			9. 93307	0. 85717		0
↑ 120°→	cos	Diff. 1′	→ cos	Diff. 1′	Diff. 1′	sin	sin	Diff. 1′←59° ↑	

LOGARITHMS

31° ↓ ′	sin	Diff. 1′	sin	Diff. 1′	Diff. 1′	cos ←	cos	Diff. 1′	←148° ↓ ′
0	0.51504	25	9.71184	21	8	9.93307	0.85717	15	60
1	.51529	25	.71205	21	8	.93299	.85702	15	59
2	.51554	25	.71226	21	7	.93291	.85687	15	58
3	.51579	25	.71247	21	8	.93284	.85672	15	57
4	.51604	24	.71268	21	7	.93276	.85657	15	56
5	0.51628	25	9.71289	21	8	9.93269	0.85642	15	55
6	.51653	25	.71310	21	8	.93261	.85627	15	54
7	.51678	25	.71331	21	7	.93253	.85612	15	53
8	.51703	25	.71352	21	8	.93246	.85597	15	52
9	.51728	25	.71373	20	8	.93238	.85582	15	51
10	0.51753	25	9.71393	21	7	9.93230	0.85567	16	50
11	.51778	25	.71414	21	8	.93223	.85551	15	49
12	.51803	25	.71435	21	8	.93215	.85536	15	48
13	.51828	24	.71456	21	7	.93207	.85521	15	47
14	.51852	25	.71477	21	8	.93200	.85506	15	46
15	0.51877	25	9.71498	21	8	9.93192	0.85491	15	45
16	.51902	25	.71519	20	7	.93184	.85476	15	44
17	.51927	25	.71539	21	8	.93177	.85461	15	43
18	.51952	25	.71560	21	8	.93169	.85446	15	42
19	.51977	25	.71581	21	7	.93161	.85431	15	41
20	0.52002	24	9.71602	20	8	9.93154	0.85416	15	40
21	.52026	25	.71622	21	8	.93146	.85401	16	39
22	.52051	25	.71643	21	7	.93138	.85385	15	38
23	.52076	25	.71664	21	8	.93131	.85370	15	37
24	.52101	25	.71685	20	8	.93123	.85355	15	36
25	0.52126	25	9.71705	21	7	9.93115	0.85340	15	35
26	.52151	24	.71726	21	8	.93108	.85325	15	34
27	.52175	25	.71747	20	8	.93100	.85310	16	33
28	.52200	25	.71767	21	8	.93092	.85294	15	32
29	.52225	25	.71788	21	7	.93084	.85279	15	31
30	0.52250	25	9.71809	20	8	9.93077	0.85264	15	30
31	.52275	24	.71829	21	8	.93069	.85249	15	29
32	.52299	25	.71850	20	8	.93061	.85234	16	28
33	.52324	25	.71870	21	7	.93053	.85218	15	27
34	.52349	25	.71891	20	8	.93046	.85203	15	26
35	0.52374	25	9.71911	21	8	9.93038	0.85188	15	25
36	.52399	24	.71932	20	8	.93030	.85173	16	24
37	.52423	25	.71952	21	8	.93022	.85157	15	23
38	.52448	25	.71973	21	7	.93014	.85142	15	22
39	.52473	25	.71994	20	8	.93007	.85127	15	21
40	0.52498	24	9.72014	20	8	9.92999	0.85112	16	20
41	.52522	25	.72034	21	8	.92991	.85096	15	19
42	.52547	25	.72055	20	7	.92983	.85081	15	18
43	.52572	25	.72075	21	8	.92976	.85066	15	17
44	.52597	24	.72096	20	8	.92968	.85051	16	16
45	0.52621	25	9.72116	21	8	9.92960	0.85035	15	15
46	.52646	25	.72137	20	8	.92952	.85020	15	14
47	.52671	25	.72157	20	8	.92944	.85005	16	13
48	.52696	24	.72177	21	7	.92936	.84989	15	12
49	.52720	25	.72198	20	8	.92929	.84974	15	11
50	0.52745	25	9.72218	20	8	9.92921	0.84959	16	10
51	.52770	24	.72238	21	8	.92913	.84943	15	9
52	.52794	25	.72259	20	8	.92905	.84928	15	8
53	.52819	25	.72279	20	8	.92897	.84913	16	7
54	.52844	25	.72299	21	8	.92889	.84897	15	6
55	0.52869	24	9.72320	20	7	9.92881	0.84882	16	5
56	.52893	25	.72340	20	8	.92874	.84866	15	4
57	.52918	25	.72360	21	8	.92866	.84851	15	3
58	.52943	24	.72381	20	8	.92858	.84836	16	2
59	.52967	25	.72401	20	8	.92850	.84820	15	1
60	0.52992		9.72421			9.92842	0.84805		0
↑ 121°→	cos	Diff. 1′	→ cos	Diff. 1′	Diff. 1′	sin	sin	Diff. 1′	↑ ←58°

LOGARITHMS

32°→ ↓ ′	sin	Diff. 1′	sin	Diff. 1′	Diff. 1′	cos ←	cos	Diff. 1′	←147° ↓ ′
0	0. 52992	25	9. 72421	20	8	9. 92842	0. 84805	16	60
1	. 53017	24	. 72441	20	8	. 92834	. 84789	15	59
2	. 53041	25	. 72461	21	8	. 92826	. 84774	15	58
3	. 53066	25	. 72482	20	8	. 92818	. 84759	16	57
4	. 53091	24	. 72502	20	7	. 92810	. 84743	15	56
5	0. 53115	25	9. 72522	20	8	9. 92803	0. 84728	16	55
6	. 53140	24	. 72542	20	8	. 92795	. 84712	15	54
7	. 53164	25	. 72562	20	8	. 92787	. 84697	16	53
8	. 53189	25	. 72582	20	8	. 92779	. 84681	15	52
9	. 53214	24	. 72602	20	8	. 92771	. 84666	16	51
10	0. 53238	25	9. 72622	21	8	9. 92763	0. 84650	15	50
11	. 53263	25	. 72643	20	8	. 92755	. 84635	16	49
12	. 53288	24	. 72663	20	8	. 92747	. 84619	15	48
13	. 53312	25	. 72683	20	8	. 92739	. 84604	16	47
14	. 53337	24	. 72703	20	8	. 92731	. 84588	15	46
15	0. 53361	25	9. 72723	20	8	9. 92723	0. 84573	16	45
16	. 53386	25	. 72743	20	8	. 92715	. 84557	15	44
17	. 53411	24	. 72763	20	8	. 92707	. 84542	16	43
18	. 53435	25	. 72783	20	8	. 92699	. 84526	15	42
19	. 53460	24	. 72803	20	8	. 92691	. 84511	16	41
20	0. 53484	25	9. 72823	20	8	9. 92683	0. 84495	15	40
21	. 53509	25	. 72843	20	8	. 92675	. 84480	16	39
22	. 53534	24	. 72863	20	8	. 92667	. 84464	16	38
23	. 53558	25	. 72883	19	8	. 92659	. 84448	15	37
24	. 53583	24	. 72902	20	8	. 92651	. 84433	16	36
25	0. 53607	25	9. 72922	20	8	9. 92643	0. 84417	15	35
26	. 53632	24	. 72942	20	8	. 92635	. 84402	16	34
27	. 53656	25	. 72962	20	8	. 92627	. 84386	16	33
28	. 53681	24	. 72982	20	8	. 92619	. 84370	15	32
29	. 53705	25	. 73002	20	8	. 92611	. 84355	16	31
30	0. 53730	24	9. 73022	19	8	9. 92603	0. 84339	15	30
31	. 53754	25	. 73041	20	8	. 92595	. 84324	16	29
32	. 53779	25	. 73061	20	8	. 92587	. 84308	16	28
33	. 53804	24	. 73081	20	8	. 92579	. 84292	15	27
34	. 53828	25	. 73101	20	8	. 92571	. 84277	16	26
35	0. 53853	24	9. 73121	19	8	9. 92563	0. 84261	16	25
36	. 53877	25	. 73140	20	9	. 92555	. 84245	15	24
37	. 53902	24	. 73160	20	8	. 92546	. 84230	16	23
38	. 53926	25	. 73180	20	8	. 92538	. 84214	16	22
39	. 53951	24	. 73200	19	8	. 92530	. 84198	16	21
40	0. 53975	25	9. 73219	20	8	9. 92522	0. 84182	15	20
41	. 54000	24	. 73239	20	8	. 92514	. 84167	16	19
42	. 54024	25	. 73259	19	8	. 92506	. 84151	16	18
43	. 54049	24	. 73278	20	8	. 92498	. 84135	15	17
44	. 54073	24	. 73298	20	8	. 92490	. 84120	16	16
45	0. 54097	25	9. 73318	19	9	9. 92482	0. 84104	16	15
46	. 54122	24	. 73337	20	8	. 92473	. 84088	16	14
47	. 54146	25	. 73357	20	8	. 92465	. 84072	15	13
48	. 54171	24	. 73377	19	8	. 92457	. 84057	16	12
49	. 54195	25	. 73396	20	8	. 92449	. 84041	16	11
50	0. 54220	24	9. 73416	19	8	9. 92441	0. 84025	16	10
51	. 54244	25	. 73435	20	8	. 92433	. 84009	15	9
52	. 54269	24	. 73455	19	9	. 92425	. 83994	16	8
53	. 54293	24	. 73474	20	8	. 92416	. 83978	16	7
54	. 54317	25	. 73494	19	8	. 92408	. 83962	16	6
55	0. 54342	24	9. 73513	20	8	9. 92400	0. 83946	16	5
56	. 54366	25	. 73533	19	8	. 92392	. 83930	15	4
57	. 54391	24	. 73552	20	8	. 92384	. 83915	16	3
58	. 54415	25	. 73572	19	9	. 92376	. 83899	16	2
59	. 54440	24	. 73591	20	8	. 92367	. 83883	16	1
60	0. 54464		9. 73611			9. 92359	0. 83867		0
↑ 122°→ cos		Diff. 1′	→ cos	Diff. 1′	Diff. 1′	sin	sin	Diff. 1′	↑ ←57°

LOGARITHMS

33°→ ↓ ′	sin	Diff. 1′	sin	Diff. 1′	Diff. 1′	cos ←	cos	Diff. 1′	←146° ↓ ′
0	0. 54464	24	9. 73611	19	8	9. 92359	0. 83867	16	60
1	. 54488	25	. 73630	20	8	. 92351	. 83851	16	59
2	. 54513	24	. 73650	19	8	. 92343	. 83835	16	58
3	. 54537	24	. 73669	20	9	. 92335	. 83819	15	57
4	. 54561	25	. 73689	19	8	. 92326	. 83804	16	56
5	0. 54586	24	9. 73708	19	8	9. 92318	0. 83788	16	55
6	. 54610	25	. 73727	20	8	. 92310	. 83772	16	54
7	. 54635	24	. 73747	19	9	. 92302	. 83756	16	53
8	. 54659	24	. 73766	19	8	. 92293	. 83740	16	52
9	. 54683	25	. 73785	20	8	. 92285	. 83724	16	51
10	0. 54708	24	9. 73805	19	8	9. 92277	0. 83708	16	50
11	. 54732	24	. 73824	19	9	. 92269	. 83692	16	49
12	. 54756	25	. 73843	20	8	. 92260	. 83676	16	48
13	. 54781	24	. 73863	19	8	. 92252	. 83660	15	47
14	. 54805	24	. 73882	19	9	. 92244	. 83645	16	46
15	0. 54829	25	9. 73901	20	8	9. 92235	0. 83629	16	45
16	. 54854	24	. 73921	19	8	. 92227	. 83613	16	44
17	. 54878	24	. 73940	19	8	. 92219	. 83597	16	43
18	. 54902	25	. 73959	19	9	. 92211	. 83581	16	42
19	. 54927	24	. 73978	19	8	. 92202	. 83565	16	41
20	0. 54951	24	9. 73997	20	8	9. 92194	0. 83549	16	40
21	. 54975	24	. 74017	19	9	. 92186	. 83533	16	39
22	. 54999	25	. 74036	19	8	. 92177	. 83517	16	38
23	. 55024	24	. 74055	19	8	. 92169	. 83501	16	37
24	. 55048	24	. 74074	19	9	. 92161	. 83485	16	36
25	0. 55072	25	9. 74093	20	8	9. 92152	0. 83469	16	35
26	. 55097	24	. 74113	19	8	. 92144	. 83453	16	34
27	. 55121	24	. 74132	19	9	. 92136	. 83437	16	33
28	. 55145	24	. 74151	19	8	. 92127	. 83421	16	32
29	. 55169	25	. 74170	19	8	. 92119	. 83405	16	31
30	0. 55194	24	9. 74189	19	9	9. 92111	0. 83389	16	30
31	. 55218	24	. 74208	19	8	. 92102	. 83373	17	29
32	. 55242	24	. 74227	19	8	. 92094	. 83356	16	28
33	. 55266	25	. 74246	19	9	. 92086	. 83340	16	27
34	. 55291	24	. 74265	19	8	. 92077	. 83324	16	26
35	0. 55315	24	9. 74284	19	9	9. 92069	0. 83308	16	25
36	. 55339	24	. 74303	19	8	. 92060	. 83292	16	24
37	. 55363	25	. 74322	19	8	. 92052	. 83276	16	23
38	. 55388	24	. 74341	19	9	. 92044	. 83260	16	22
39	. 55412	24	. 74360	19	8	. 92035	. 83244	16	21
40	0. 55436	24	9. 74379	19	9	9. 92027	0. 83228	16	20
41	. 55460	24	. 74398	19	8	. 92018	. 83212	17	19
42	. 55484	25	. 74417	19	8	. 92010	. 83195	16	18
43	. 55509	24	. 74436	19	9	. 92002	. 83179	16	17
44	. 55533	24	. 74455	19	8	. 91993	. 83163	16	16
45	0. 55557	24	9. 74474	19	9	9. 91985	0. 83147	16	15
46	. 55581	24	. 74493	19	8	. 91976	. 83131	16	14
47	. 55605	25	. 74512	19	9	. 91968	. 83115	17	13
48	. 55630	24	. 74531	18	8	. 91959	. 83098	16	12
49	. 55654	24	. 74549	19	9	. 91951	. 83082	16	11
50	0. 55678	24	9. 74568	19	8	9. 91942	0. 83066	16	10
51	. 55702	24	. 74587	19	9	. 91934	. 83050	16	9
52	. 55726	24	. 74606	19	8	. 91925	. 83034	17	8
53	. 55750	25	. 74625	19	9	. 91917	. 83017	16	7
54	. 55775	24	. 74644	18	8	. 91908	. 83001	16	6
55	0. 55799	24	9. 74662	19	9	9. 91900	0. 82985	16	5
56	. 55823	24	. 74681	19	8	. 91891	. 82969	16	4
57	. 55847	24	. 74700	19	9	. 91883	. 82953	17	3
58	. 55871	24	. 74719	18	8	. 91874	. 82936	16	2
59	. 55895	24	. 74737	19	9	. 91866	. 82920	16	1
60	0. 55919		9. 74756			9. 91857	0. 82904		0
↑ 123°→	cos	Diff. 1′	→ cos	Diff. 1′	Diff. 1′	sin	sin	Diff. 1′←	56° ↑

LOGARITHMS

34°→ ↓	sin	Diff. 1′	sin	Diff. 1′	Diff. 1′	cos ←	cos	←145° Diff. 1′	↓
′									′
0	0. 55919	24	9. 74756	19	8	9. 91857	0. 82904	17	60
1	. 55943	25	. 74775	19	9	. 91849	. 82887	16	59
2	. 55968	24	. 74794	18	8	. 91840	. 82871	16	58
3	. 55992	24	. 74812	19	9	. 91832	. 82855	16	57
4	. 56016	24	. 74831	19	8	. 91823	. 82839	17	56
5	0. 56040	24	9. 74850	18	9	9. 91815	0. 82822	16	55
6	. 56064	24	. 74868	19	8	. 91806	. 82806	16	54
7	. 56088	24	. 74887	19	9	. 91798	. 82790	17	53
8	. 56112	24	. 74906	18	8	. 91789	. 82773	16	52
9	. 56136	24	. 74924	19	9	. 91781	. 82757	16	51
10	0. 56160	24	9. 74943	18	9	9. 91772	0. 82741	17	50
11	. 56184	24	. 74961	19	8	. 91763	. 82724	16	49
12	. 56208	24	. 74980	19	9	. 91755	. 82708	16	48
13	. 56232	24	. 74999	18	8	. 91746	. 82692	17	47
14	. 56256	24	. 75017	19	9	. 91738	. 82675	16	46
15	0. 56280	25	9. 75036	18	9	9. 91729	0. 82659	16	45
16	. 56305	24	. 75054	19	8	. 91720	. 82643	17	44
17	. 56329	24	. 75073	18	9	. 91712	. 82626	16	43
18	. 56353	24	. 75091	19	8	. 91703	. 82610	17	42
19	. 56377	24	. 75110	18	9	. 91695	. 82593	16	41
20	0. 56401	24	9. 75128	19	9	9. 91686	0. 82577	16	40
21	. 56425	24	. 75147	18	8	. 91677	. 82561	17	39
22	. 56449	24	. 75165	19	9	. 91669	. 82544	16	38
23	. 56473	24	. 75184	18	9	. 91660	. 82528	17	37
24	. 56497	24	. 75202	19	8	. 91651	. 82511	16	36
25	0. 56521	24	9. 75221	18	9	9. 91643	0. 82495	17	35
26	. 56545	24	. 75239	19	9	. 91634	. 82478	16	34
27	. 56569	24	. 75258	18	8	. 91625	. 82462	16	33
28	. 56593	24	. 75276	18	9	. 91617	. 82446	17	32
29	. 56617	24	. 75294	19	9	. 91608	. 82429	16	31
30	0. 56641	24	9. 75313	18	8	9. 91599	0. 82413	17	30
31	. 56665	24	. 75331	19	9	. 91591	. 82396	16	29
32	. 56689	24	. 75350	18	9	. 91582	. 82380	17	28
33	. 56713	23	. 75368	18	8	. 91573	. 82363	16	27
34	. 56736	24	. 75386	19	9	. 91565	. 82347	17	26
35	0. 56760	24	9. 75405	18	9	9. 91556	0. 82330	16	25
36	. 56784	24	. 75423	18	9	. 91547	. 82314	17	24
37	. 56808	24	. 75441	18	8	. 91538	. 82297	16	23
38	. 56832	24	. 75459	19	9	. 91530	. 82281	17	22
39	. 56856	24	. 75478	18	9	. 91521	. 82264	16	21
40	0. 56880	24	9. 75496	18	8	9. 91512	0. 82248	17	20
41	. 56904	24	. 75514	19	9	. 91504	. 82231	17	19
42	. 56928	24	. 75533	18	9	. 91495	. 82214	16	18
43	. 56952	24	. 75551	18	9	. 91486	. 82198	17	17
44	. 56976	24	. 75569	18	8	. 91477	. 82181	16	16
45	0. 57000	24	9. 75587	18	9	9. 91469	0. 82165	17	15
46	. 57024	23	. 75605	19	9	. 91460	. 82148	16	14
47	. 57047	24	. 75624	18	9	. 91451	. 82132	17	13
48	. 57071	24	. 75642	18	9	. 91442	. 82115	17	12
49	. 57095	24	. 75660	18	8	. 91433	. 82098	16	11
50	0. 57119	24	9. 75678	18	9	9. 91425	0. 82082	17	10
51	. 57143	24	. 75696	18	9	. 91416	. 82065	17	9
52	. 57167	24	. 75714	19	9	. 91407	. 82048	16	8
53	. 57191	24	. 75733	18	9	. 91398	. 82032	17	7
54	. 57215	23	. 75751	18	8	. 91389	. 82015	16	6
55	0. 57238	24	9. 75769	18	9	9. 91381	0. 81999	17	5
56	. 57262	24	. 75787	18	9	. 91372	. 81982	17	4
57	. 57286	24	. 75805	18	9	. 91363	. 81965	16	3
58	. 57310	24	. 75823	18	9	. 91354	. 81949	17	2
59	. 57334	24	. 75841	18	9	. 91345	. 81932	17	1
60	0. 57358		9. 75859			9. 91336	0. 81915		0
↑ 124°→	cos	Diff. 1′	→ cos	Diff. 1′	Diff. 1′	sin	sin	Diff. 1′←55°	↑

LOGARITHMS

35°→ ↓ ′	sin	Diff. 1′	sin	Diff. 1′	Diff. 1′	cos ←	cos	Diff. 1′	←144° ↓ ′
0	0. 57358	23	9. 75859	18	8	9. 91336	0. 81915	16	60
1	. 57381	24	. 75877	18	9	. 91328	. 81899	17	59
2	. 57405	24	. 75895	18	9	. 91319	. 81882	17	58
3	. 57429	24	. 75913	18	9	. 91310	. 81865	17	57
4	. 57453	24	. 75931	18	9	. 91301	. 81848	16	56
5	0. 57477	24	9. 75949	18	9	9. 91292	0. 81832	17	55
6	. 57501	23	. 75967	18	9	. 91283	. 81815	17	54
7	. 57524	24	. 75985	18	8	. 91274	. 81798	16	53
8	. 57548	24	. 76003	18	9	. 91266	. 81782	17	52
9	. 57572	24	. 76021	18	9	. 91257	. 81765	17	51
10	0. 57596	23	9. 76039	18	9	9. 91248	0. 81748	17	50
11	. 57619	24	. 76057	18	9	. 91239	. 81731	17	49
12	. 57643	24	. 76075	18	9	. 91230	. 81714	16	48
13	. 57667	24	. 76093	18	9	. 91221	. 81698	17	47
14	. 57691	24	. 76111	18	9	. 91212	. 81681	17	46
15	0. 57715	23	9. 76129	17	9	9. 91203	0. 81664	17	45
16	. 57738	24	. 76146	18	9	. 91194	. 81647	16	44
17	. 57762	24	. 76164	18	9	. 91185	. 81631	17	43
18	. 57786	24	. 76182	18	9	. 91176	. 81614	17	42
19	. 57810	23	. 76200	18	9	. 91167	. 81597	17	41
20	0. 57833	24	9. 76218	18	9	9. 91158	0. 81580	17	40
21	. 57857	24	. 76236	17	8	. 91149	. 81563	17	39
22	. 57881	23	. 76253	18	9	. 91141	. 81546	16	38
23	. 57904	24	. 76271	18	9	. 91132	. 81530	17	37
24	. 57928	24	. 76289	18	9	. 91123	. 81513	17	36
25	0. 57952	24	9. 76307	17	9	9. 91114	0. 81496	17	35
26	. 57976	23	. 76324	18	9	. 91105	. 81479	17	34
27	. 57999	24	. 76342	18	9	. 91096	. 81462	17	33
28	. 58023	24	. 76360	18	9	. 91087	. 81445	17	32
29	. 58047	23	. 76378	17	9	. 91078	. 81428	16	31
30	0. 58070	24	9. 76395	18	9	9. 91069	0. 81412	17	30
31	. 58094	24	. 76413	18	9	. 91060	. 81395	17	29
32	. 58118	23	. 76431	17	9	. 91051	. 81378	17	28
33	. 58141	24	. 76448	18	9	. 91042	. 81361	17	27
34	. 58165	24	. 76466	18	10	. 91033	. 81344	17	26
35	0. 58189	23	9. 76484	17	9	9. 91023	0. 81327	17	25
36	. 58212	24	. 76501	18	9	. 91014	. 81310	17	24
37	. 58236	24	. 76519	18	9	. 91005	. 81293	17	23
38	. 58260	23	. 76537	17	9	. 90996	. 81276	17	22
39	. 58283	24	. 76554	18	9	. 90987	. 81259	17	21
40	0. 58307	23	9. 76572	18	9	9. 90978	0. 81242	17	20
41	. 58330	24	. 76590	17	9	. 90969	. 81225	17	19
42	. 58354	24	. 76607	18	9	. 90960	. 81208	17	18
43	. 58378	23	. 76625	17	9	. 90951	. 81191	17	17
44	. 58401	24	. 76642	18	9	. 90942	. 81174	17	16
45	0. 58425	24	9. 76660	17	9	9. 90933	0. 81157	17	15
46	. 58449	23	. 76677	18	9	. 90924	. 81140	17	14
47	. 58472	24	. 76695	17	9	. 90915	. 81123	17	13
48	. 58496	23	. 76712	18	10	. 90906	. 81106	17	12
49	. 58519	24	. 76730	17	9	. 90896	. 81089	17	11
50	0. 58543	24	9. 76747	18	9	9. 90887	0. 81072	17	10
51	. 58567	23	. 76765	17	9	. 90878	. 81055	17	9
52	. 58590	24	. 76782	18	9	. 90869	. 81038	17	8
53	. 58614	23	. 76800	17	9	. 90860	. 81021	17	7
54	. 58637	24	. 76817	18	9	. 90851	. 81004	17	6
55	0. 58661	23	9. 76835	17	10	9. 90842	0. 80987	17	5
56	. 58684	24	. 76852	18	9	. 90832	. 80970	17	4
57	. 58708	23	. 76870	17	9	. 90823	. 80953	17	3
58	. 58731	24	. 76887	17	9	. 90814	. 80936	17	2
59	. 58755	24	. 76904	18	9	. 90805	. 80919	17	1
60	0. 58779		9. 76922			9. 90796	0. 80902		0
↑ 125°→	cos	Diff. 1′	→ cos	Diff. 1′	Diff. 1′	sin	sin	Diff. 1′	←54° ↑

LOGARITHMS

36°→ ↓ ′	sin	Diff. 1′	sin	Diff. 1′	Diff. 1′	cos ←	cos	Diff. 1′	←143° ↓ ′
0	0. 58779	23	9. 76922	17	9	9. 90796	0. 80902	17	60
1	. 58802	24	. 76939	18	10	. 90787	. 80885	18	59
2	. 58826	23	. 76957	17	9	. 90777	. 80867	17	58
3	. 58849	24	. 76974	17	9	. 90768	. 80850	17	57
4	. 58873	23	. 76991	18	9	. 90759	. 80833	17	56
5	0. 58896	24	9. 77009	17	9	9. 90750	0. 80816	17	55
6	. 58920	23	. 77026	17	10	. 90741	. 80799	17	54
7	. 58943	24	. 77043	18	9	. 90731	. 80782	17	53
8	. 58967	23	. 77061	17	9	. 90722	. 80765	17	52
9	. 58990	24	. 77078	17	9	. 90713	. 80748	18	51
10	0. 59014	23	9. 77095	17	10	9. 90704	0. 80730	17	50
11	. 59037	24	. 77112	18	9	. 90694	. 80713	17	49
12	. 59061	23	. 77130	17	9	. 90685	. 80696	17	48
13	. 59084	24	. 77147	17	9	. 90676	. 80679	17	47
14	. 59108	23	. 77164	17	10	. 90667	. 80662	18	46
15	0. 59131	23	9. 77181	18	9	9. 90657	0. 80644	17	45
16	. 59154	24	. 77199	17	9	. 90648	. 80627	17	44
17	. 59178	23	. 77216	17	9	. 90639	. 80610	17	43
18	. 59201	24	. 77233	17	10	. 90630	. 80593	17	42
19	. 59225	23	. 77250	18	9	. 90620	. 80576	18	41
20	0. 59248	24	9. 77268	17	9	9. 90611	0. 80558	17	40
21	. 59272	23	. 77285	17	10	. 90602	. 80541	17	39
22	. 59295	23	. 77302	17	9	. 90592	. 80524	17	38
23	. 59318	24	. 77319	17	9	. 90583	. 80507	18	37
24	. 59342	23	. 77336	17	9	. 90574	. 80489	17	36
25	0. 59365	24	9. 77353	17	10	9. 90565	0. 80472	17	35
26	. 59389	23	. 77370	17	9	. 90555	. 80455	17	34
27	. 59412	24	. 77387	18	9	. 90546	. 80438	18	33
28	. 59436	23	. 77405	17	10	. 90537	. 80420	17	32
29	. 59459	23	. 77422	17	9	. 90527	. 80403	17	31
30	0. 59482	24	9. 77439	17	9	9. 90518	0. 80386	18	30
31	. 59506	23	. 77456	17	10	. 90509	. 80368	17	29
32	. 59529	23	. 77473	17	9	. 90499	. 80351	17	28
33	. 59552	24	. 77490	17	10	. 90490	. 80334	18	27
34	. 59576	23	. 77507	17	9	. 90480	. 80316	17	26
35	0. 59599	23	9. 77524	17	9	9. 90471	0. 80299	17	25
36	. 59622	24	. 77541	17	10	. 90462	. 80282	18	24
37	. 59646	23	. 77558	17	9	. 90452	. 80264	17	23
38	. 59669	24	. 77575	17	9	. 90443	. 80247	17	22
39	. 59693	23	. 77592	17	10	. 90434	. 80230	18	21
40	0. 59716	23	9. 77609	17	9	9. 90424	0. 80212	17	20
41	. 59739	24	. 77626	17	10	. 90415	. 80195	17	19
42	. 59763	23	. 77643	17	9	. 90405	. 80178	18	18
43	. 59786	23	. 77660	17	10	. 90396	. 80160	17	17
44	. 59809	23	. 77677	17	9	. 90386	. 80143	18	16
45	0. 59832	24	9. 77694	17	9	9. 90377	0. 80125	17	15
46	. 59856	23	. 77711	17	10	. 90368	. 80108	17	14
47	. 59879	23	. 77728	16	9	. 90358	. 80091	18	13
48	. 59902	24	. 77744	17	10	. 90349	. 80073	17	12
49	. 59926	23	. 77761	17	9	. 90339	. 80056	18	11
50	0. 59949	23	9. 77778	17	10	9. 90330	0. 80038	17	10
51	. 59972	23	. 77795	17	9	. 90320	. 80021	18	9
52	. 59995	24	. 77812	17	10	. 90311	. 80003	17	8
53	. 60019	23	. 77829	17	9	. 90301	. 79986	18	7
54	. 60042	23	. 77846	16	10	. 90292	. 79968	17	6
55	0. 60065	24	9. 77862	17	9	9. 90282	0. 79951	17	5
56	. 60089	23	. 77879	17	10	. 90273	. 79934	18	4
57	. 60112	23	. 77896	17	9	. 90263	. 79916	17	3
58	. 60135	23	. 77913	17	10	. 90254	. 79899	18	2
59	. 60158	24	. 77930	16	9	. 90244	. 79881	17	1
60	0. 60182		9. 77946			9. 90235	0. 79864		0
↑ 126°→	cos	Diff. 1′	→ cos	Diff. 1′	Diff. 1′	sin	sin	Diff. 1′	↑ ←53°

LOGARITHMS

37° → ↓ ′	sin	Diff. 1′	sin	Diff. 1′	Diff. 1′	cos ←	cos	Diff. 1′	← 142° ↓ ′
0	0. 60182	23	9. 77946	17	10	9. 90235	0. 79864	18	60
1	. 60205	23	. 77963	17	9	. 90225	. 79846	17	59
2	. 60228	23	. 77980	17	10	. 90216	. 79829	18	58
3	. 60251	23	. 77997	16	9	. 90206	. 79811	18	57
4	. 60274	24	. 78013	17	10	. 90197	. 79793	17	56
5	0. 60298	23	9. 78030	17	9	9. 90187	0. 79776	18	55
6	. 60321	23	. 78047	16	10	. 90178	. 79758	17	54
7	. 60344	23	. 78063	17	9	. 90168	. 79741	18	53
8	. 60367	23	. 78080	17	10	. 90159	. 79723	17	52
9	. 60390	24	. 78097	16	10	. 90149	. 79706	18	51
10	0. 60414	23	9. 78113	17	9	9. 90139	0. 79688	17	50
11	. 60437	23	. 78130	17	10	. 90130	. 79671	18	49
12	. 60460	23	. 78147	16	9	. 90120	. 79653	18	48
13	. 60483	23	. 78163	17	10	. 90111	. 79635	17	47
14	. 60506	23	. 78180	17	10	. 90101	. 79618	18	46
15	0. 60529	24	9. 78197	16	9	9. 90091	0. 79600	17	45
16	. 60553	23	. 78213	17	10	. 90082	. 79583	18	44
17	. 60576	23	. 78230	16	9	. 90072	. 79565	18	43
18	. 60599	23	. 78246	17	10	. 90063	. 79547	17	42
19	. 60622	23	. 78263	17	10	. 90053	. 79530	18	41
20	0. 60645	23	9. 78280	16	9	9. 90043	0. 79512	18	40
21	. 60668	23	. 78296	17	10	. 90034	. 79494	17	39
22	. 60691	23	. 78313	16	10	. 90024	. 79477	18	38
23	. 60714	24	. 78329	17	9	. 90014	. 79459	18	37
24	. 60738	23	. 78346	16	10	. 90005	. 79441	17	36
25	0. 60761	23	9. 78362	17	10	9. 89995	0. 79424	18	35
26	. 60784	23	. 78379	16	9	. 89985	. 79406	18	34
27	. 60807	23	. 78395	17	10	. 89976	. 79388	17	33
28	. 60830	23	. 78412	16	10	. 89966	. 79371	18	32
29	. 60853	23	. 78428	17	9	. 89956	. 79353	18	31
30	0. 60876	23	9. 78445	16	10	9. 89947	0. 79335	17	30
31	. 60899	23	. 78461	17	10	. 89937	. 79318	18	29
32	. 60922	23	. 78478	16	9	. 89927	. 79300	18	28
33	. 60945	23	. 78494	16	10	. 89918	. 79282	18	27
34	. 60968	23	. 78510	17	10	. 89908	. 79264	17	26
35	0. 60991	24	9. 78527	16	10	9. 89898	0. 79247	18	25
36	. 61015	23	. 78543	17	9	. 89888	. 79229	18	24
37	. 61038	23	. 78560	16	10	. 89879	. 79211	18	23
38	. 61061	23	. 78576	16	10	. 89869	. 79193	17	22
39	. 61084	23	. 78592	17	10	. 89859	. 79176	18	21
40	0. 61107	23	9. 78609	16	9	9. 89849	0. 79158	18	20
41	. 61130	23	. 78625	17	10	. 89840	. 79140	18	19
42	. 61153	23	. 78642	16	10	. 89830	. 79122	17	18
43	. 61176	23	. 78658	16	10	. 89820	. 79105	18	17
44	. 61199	23	. 78674	17	9	. 89810	. 79087	18	16
45	0. 61222	23	9. 78691	16	10	9. 89801	0. 79069	18	15
46	. 61245	23	. 78707	16	10	. 89791	. 79051	18	14
47	. 61268	23	. 78723	16	10	. 89781	. 79033	17	13
48	. 61291	23	. 78739	17	10	. 89771	. 79016	18	12
49	. 61314	23	. 78756	16	9	. 89761	. 78998	18	11
50	0. 61337	23	9. 78772	16	10	9. 89752	0. 78980	18	10
51	. 61360	23	. 78788	17	10	. 89742	. 78962	18	9
52	. 61383	23	. 78805	16	10	. 89732	. 78944	18	8
53	. 61406	23	. 78821	16	10	. 89722	. 78926	18	7
54	. 61429	22	. 78837	16	10	. 89712	. 78908	17	6
55	0. 61451	23	9. 78853	16	9	9. 89702	0. 78891	18	5
56	. 61474	23	. 78869	17	10	. 89693	. 78873	18	4
57	. 61497	23	. 78886	16	10	. 89683	. 78855	18	3
58	. 61520	23	. 78902	16	10	. 89673	. 78837	18	2
59	. 61543	23	. 78918	16	10	. 89663	. 78819	18	1
60	0. 61566		9. 78934			9. 89653	0. 78801		0
↑ 127° →	cos	Diff. 1′	→ cos	Diff. 1′	Diff. 1′	sin	sin	Diff. 1′	← 52° ↑

LOGARITHMS

38°→ ↓	sin	Diff. 1′	sin	Diff. 1′	Diff. 1′	cos ←	cos	←141° Diff. 1′	↓
′									′
0	0.61566	23	9.78934	16	10	9.89653	0.78801	18	60
1	.61589	23	.78950	17	10	.89643	.78783	18	59
2	.61612	23	.78967	16	9	.89633	.78765	18	58
3	.61635	23	.78983	16	10	.89624	.78747	18	57
4	.61658	23	.78999	16	10	.89614	.78729	18	56
5	0.61681	23	9.79015	16	10	9.89604	0.78711	17	55
6	.61704	22	.79031	16	10	.89594	.78694	18	54
7	.61726	23	.79047	16	10	.89584	.78676	18	53
8	.61749	23	.79063	16	10	.89574	.78658	18	52
9	.61772	23	.79079	16	10	.89564	.78640	18	51
10	0.61795	23	9.79095	16	10	9.89554	0.78622	18	50
11	.61818	23	.79111	17	10	.89544	.78604	18	49
12	.61841	23	.79128	16	10	.89534	.78586	18	48
13	.61864	23	.79144	16	10	.89524	.78568	18	47
14	.61887	22	.79160	16	10	.89514	.78550	18	46
15	0.61909	23	9.79176	16	9	9.89504	0.78532	18	45
16	.61932	23	.79192	16	10	.89495	.78514	18	44
17	.61955	23	.79208	16	10	.89485	.78496	18	43
18	.61978	23	.79224	16	10	.89475	.78478	18	42
19	.62001	23	.79240	16	10	.89465	.78460	18	41
20	0.62024	22	9.79256	16	10	9.89455	0.78442	18	40
21	.62046	23	.79272	16	10	.89445	.78424	19	39
22	.62069	23	.79288	16	10	.89435	.78405	18	38
23	.62092	23	.79304	15	10	.89425	.78387	18	37
24	.62115	23	.79319	16	10	.89415	.78369	18	36
25	0.62138	22	9.79335	16	10	9.89405	0.78351	18	35
26	.62160	23	.79351	16	10	.89395	.78333	18	34
27	.62183	23	.79367	16	10	.89385	.78315	18	33
28	.62206	23	.79383	16	11	.89375	.78297	18	32
29	.62229	22	.79399	16	10	.89364	.78279	18	31
30	0.62251	23	9.79415	16	10	9.89354	0.78261	18	30
31	.62274	23	.79431	16	10	.89344	.78243	18	29
32	.62297	23	.79447	16	10	.89334	.78225	19	28
33	.62320	22	.79463	15	10	.89324	.78206	18	27
34	.62342	23	.79478	16	10	.89314	.78188	18	26
35	0.62365	23	9.79494	10	10	9.89304	0.78170	18	25
36	.62388	23	.79510	16	10	.89294	.78152	18	24
37	.62411	22	.79526	16	10	.89284	.78134	18	23
38	.62433	23	.79542	16	10	.89274	.78116	18	22
39	.62456	23	.79558	15	10	.89264	.78098	19	21
40	0.62479	23	9.79573	16	10	9.89254	0.78079	18	20
41	.62502	22	.79589	16	11	.89244	.78061	18	19
42	.62524	23	.79605	16	10	.89233	.78043	18	18
43	.62547	23	.79621	15	10	.89223	.78025	18	17
44	.62570	22	.79636	16	10	.89213	.78007	19	16
45	0.62592	23	9.79652	16	10	9.89203	0.77988	18	15
46	.62615	23	.79668	16	10	.89193	.77970	18	14
47	.62638	22	.79684	15	10	.89183	.77952	18	13
48	.62660	23	.79699	16	11	.89173	.77934	18	12
49	.62683	23	.79715	16	10	.89162	.77916	19	11
50	0.62706	22	9.79731	15	10	9.89152	0.77897	18	10
51	.62728	23	.79746	16	10	.89142	.77879	18	9
52	.62751	23	.79762	16	10	.89132	.77861	18	8
53	.62774	22	.79778	15	10	.89122	.77843	19	7
54	.62796	23	.79793	16	11	.89112	.77824	18	6
55	0.62819	23	9.79809	16	10	9.89101	0.77806	18	5
56	.62842	22	.79825	15	10	.89091	.77788	19	4
57	.62864	23	.79840	16	10	.89081	.77769	18	3
58	.62887	22	.79856	16	11	.89071	.77751	18	2
59	.62909	23	.79872	15	10	.89060	.77733	18	1
60	0.62932		9.79887			9.89050	0.77715		0
↑ 128°→	cos	Diff. 1′	→ cos	Diff. 1′	Diff. 1′	sin	sin	Diff. 1′←51°	↑

39°→ ↓ ′	sin	Diff. 1′	sin	Diff. 1′	Diff. 1′	cos ←	cos	Diff. 1′	←140° ↓ ′
0	0. 62932	23	9. 79887	16	10	9. 89050	0. 77715	19	60
1	. 62955	22	. 79903	15	10	. 89040	. 77696	18	59
2	. 62977	23	. 79918	16	10	. 89030	. 77678	18	58
3	. 63000	22	. 79934	16	11	. 89020	. 77660	19	57
4	. 63022	23	. 79950	15	10	. 89009	. 77641	18	56
5	0. 63045	23	9. 79965	16	10	9. 88999	0. 77623	18	55
6	. 63068	22	. 79981	15	11	. 88989	. 77605	19	54
7	. 63090	23	. 79996	16	10	. 88978	. 77586	18	53
8	. 63113	22	. 80012	15	10	. 88968	. 77568	18	52
9	. 63135	23	. 80027	16	10	. 88958	. 77550	19	51
10	0. 63158	22	9. 80043	15	11	9. 88948	0. 77531	18	50
11	. 63180	23	. 80058	16	10	. 88937	. 77513	19	49
12	. 63203	22	. 80074	15	10	. 88927	. 77494	18	48
13	. 63225	23	. 80089	16	11	. 88917	. 77476	18	47
14	. 63248	23	. 80105	15	10	. 88906	. 77458	19	46
15	0. 63271	22	9. 80120	16	10	9. 88896	0. 77439	18	45
16	. 63293	23	. 80136	15	11	. 88886	. 77421	19	44
17	. 63316	22	. 80151	15	10	. 88875	. 77402	18	43
18	. 63338	23	. 80166	16	10	. 88865	. 77384	18	42
19	. 63361	22	. 80182	15	11	. 88855	. 77366	19	41
20	0. 63383	23	9. 80197	16	10	9. 88844	0. 77347	18	40
21	. 63406	22	. 80213	15	10	. 88834	. 77329	19	39
22	. 63428	23	. 80228	16	11	. 88824	. 77310	18	38
23	. 63451	22	. 80244	15	10	. 88813	. 77292	19	37
24	. 63473	23	. 80259	15	10	. 88803	. 77273	18	36
25	0. 63496	22	9. 80274	16	11	9. 88793	0. 77255	19	35
26	. 63518	22	. 80290	15	10	. 88782	. 77236	18	34
27	. 63540	23	. 80305	15	11	. 88772	. 77218	19	33
28	. 63563	22	. 80320	16	10	. 88761	. 77199	18	32
29	. 63585	23	. 80336	15	10	. 88751	. 77181	19	31
30	0. 63608	22	9. 80351	15	11	9. 88741	0. 77162	18	30
31	. 63630	23	. 80366	16	10	. 88730	. 77144	19	29
32	. 63653	22	. 80382	15	11	. 88720	. 77125	18	28
33	. 63675	23	. 80397	15	10	. 88709	. 77107	19	27
34	. 63698	22	. 80412	16	11	. 88699	. 77088	18	26
35	0. 63720	22	9. 80428	15	10	9. 88688	0. 77070	19	25
36	. 63742	23	. 80443	15	10	. 88678	. 77051	18	24
37	. 63765	22	. 80458	15	11	. 88668	. 77033	19	23
38	. 63787	23	. 80473	16	10	. 88657	. 77014	18	22
39	. 63810	22	. 80489	15	11	. 88647	. 76996	19	21
40	0. 63832	22	9. 80504	15	10	9. 88636	0. 76977	18	20
41	. 63854	23	. 80519	15	11	. 88626	. 76959	19	19
42	. 63877	22	. 80534	16	10	. 88615	. 76940	19	18
43	. 63899	23	. 80550	15	11	. 88605	. 76921	18	17
44	. 63922	22	. 80565	15	10	. 88594	. 76903	19	16
45	0. 63944	22	9. 80580	15	11	9. 88584	0. 76884	18	15
46	. 63966	23	. 80595	15	10	. 88573	. 76866	19	14
47	. 63989	22	. 80610	15	11	. 88563	. 76847	19	13
48	. 64011	22	. 80625	16	10	. 88552	. 76828	18	12
49	. 64033	23	. 80641	15	11	. 88542	. 76810	19	11
50	0. 64056	22	9. 80656	15	10	9. 88531	0. 76791	19	10
51	. 64078	22	. 80671	15	11	. 88521	. 76772	18	9
52	. 64100	23	. 80686	15	11	. 88510	. 76754	19	8
53	. 64123	22	. 80701	15	10	. 88499	. 76735	18	7
54	. 64145	22	. 80716	15	11	. 88489	. 76717	19	6
55	0. 64167	23	9. 80731	15	10	9. 88478	0. 76698	19	5
56	. 64190	22	. 80746	16	11	. 88468	. 76679	18	4
57	. 64212	22	. 80762	15	10	. 88457	. 76661	19	3
58	. 64234	22	. 80777	15	11	. 88447	. 76642	19	2
59	. 64256	23	. 80792	15	11	. 88436	. 76623	19	1
60	0. 64279		9. 80807			9. 88425	0 76604		0
↑ 129°→	cos	Diff. 1′	→ cos	Diff. 1′	Diff. 1′	sin	sin	Diff. 1′	↑ ←50°

40°→ ↓ ′	sin	Diff. 1′	sin	Diff. 1′	Diff. 1′	cos ←	cos	Diff. 1′	←139° ↓ ′
0	0. 64279	22	9. 80807	15	10	9. 88425	0. 76604	18	60
1	. 64301	22	. 80822	15	11	. 88415	. 76586	19	59
2	. 64323	23	. 80837	15	10	. 88404	. 76567	19	58
3	. 64346	22	. 80852	15	11	. 88394	. 76548	18	57
4	. 64368	22	. 80867	15	11	. 88383	. 76530	19	56
5	0. 64390	22	9. 80882	15	10	9. 88372	0. 76511	19	55
6	. 64412	23	. 80897	15	11	. 88362	. 76492	19	54
7	. 64435	22	. 80912	15	11	. 88351	. 76473	18	53
8	. 64457	22	. 80927	15	10	. 88340	. 76455	19	52
9	. 64479	22	. 80942	15	11	. 88330	. 76436	19	51
10	0. 64501	23	9. 80957	15	11	9. 88319	0. 76417	19	50
11	. 64524	22	. 80972	15	10	. 88308	. 76398	18	49
12	. 64546	22	. 80987	15	11	. 88298	. 76380	19	48
13	. 64568	22	. 81002	15	11	. 88287	. 76361	19	47
14	. 64590	22	. 81017	15	10	. 88276	. 76342	19	46
15	0. 64612	23	9. 81032	15	11	9. 88266	0. 76323	19	45
16	. 64635	22	. 81047	14	11	. 88255	. 76304	18	44
17	. 64657	22	. 81061	15	10	. 88244	. 76286	19	43
18	. 64679	22	. 81076	15	11	. 88234	. 76267	19	42
19	. 64701	22	. 81091	15	11	. 88223	. 76248	19	41
20	0. 64723	23	9. 81106	15	11	9. 88212	0. 76229	19	40
21	. 64746	22	. 81121	15	10	. 88201	. 76210	18	39
22	. 64768	22	. 81136	15	11	. 88191	. 76192	19	38
23	. 64790	22	. 81151	15	11	. 88180	. 76173	19	37
24	. 64812	22	. 81166	14	11	. 88169	. 76154	19	36
25	0. 64834	22	9. 81180	15	10	9. 88158	0. 76135	19	35
26	. 64856	22	. 81195	15	11	. 88148	. 76116	19	34
27	. 64878	23	. 81210	15	11	. 88137	. 76097	19	33
28	. 64901	22	. 81225	15	11	. 88126	. 76078	19	32
29	. 64923	22	. 81240	14	10	. 88115	. 76059	18	31
30	0. 64945	22	9. 81254	15	11	9. 88105	0. 76041	19	30
31	. 64967	22	. 81269	15	11	. 88094	. 76022	19	29
32	. 64989	22	. 81284	15	11	. 88083	. 76003	19	28
33	. 65011	22	. 81299	15	11	. 88072	. 75984	19	27
34	. 65033	22	. 81314	14	10	. 88061	. 75965	19	26
35	0. 65055	22	9. 81328	15	11	9. 88051	0. 75946	19	25
36	. 65077	23	. 81343	15	11	. 88040	. 75927	19	24
37	. 65100	22	. 81358	14	11	. 88029	. 75908	19	23
38	. 65122	22	. 81372	15	11	. 88018	. 75889	19	22
39	. 65144	22	. 81387	15	11	. 88007	. 75870	19	21
40	0. 65166	22	9. 81402	15	11	9. 87996	0. 75851	19	20
41	. 65188	22	. 81417	14	10	. 87985	. 75832	19	19
42	. 65210	22	. 81431	15	11	. 87975	. 75813	19	18
43	. 65232	22	. 81446	15	11	. 87964	. 75794	19	17
44	. 65254	22	. 81461	14	11	. 87953	. 75775	19	16
45	0. 65276	22	9. 81475	15	11	9. 87942	0. 75756	18	15
46	. 65298	22	. 81490	15	11	. 87931	. 75738	19	14
47	. 65320	22	. 81505	14	11	. 87920	. 75719	19	13
48	. 65342	22	. 81519	15	11	. 87909	. 75700	20	12
49	. 65364	22	. 81534	15	11	. 87898	. 75680	19	11
50	0. 65386	22	9. 81549	14	10	9. 87887	0. 75661	19	10
51	. 65408	22	. 81563	15	11	. 87877	. 75642	19	9
52	. 65430	22	. 81578	14	11	. 87866	. 75623	19	8
53	. 65452	22	. 81592	15	11	. 87855	. 75604	19	7
54	. 65474	22	. 81607	15	11	. 87844	. 75585	19	6
55	0. 65496	22	9. 81622	14	11	9. 87833	0. 75566	19	5
56	. 65518	22	. 81636	15	11	. 87822	. 75547	19	4
57	. 65540	22	. 81651	14	11	. 87811	. 75528	19	3
58	. 65562	22	. 81665	15	11	. 87800	. 75509	19	2
59	. 65584	22	. 81680	14	11	. 87789	. 75490	19	1
60	0. 65606		9. 81694			9. 87778	0. 75471		0
↑ 130°→	cos	Diff. 1′	→ cos	Diff. 1′	Diff. 1′	sin	sin	Diff. 1′←	↑ 49°

41° ↓ ′	sin	Diff. 1′	sin	Diff. 1′	Diff. 1′	cos ←	cos	Diff. 1′	←138° ↓ ′
0	0. 65606	22	9. 81694	15	11	9. 87778	0. 75471	19	60
1	. 65628	22	. 81709	14	11	. 87767	. 75452	19	59
2	. 65650	22	. 81723	15	11	. 87756	. 75433	19	58
3	. 65672	22	. 81738	14	11	. 87745	. 75414	19	57
4	. 65694	22	. 81752	15	11	. 87734	. 75395	20	56
5	0. 65716	22	9. 81767	14	11	9. 87723	0. 75375	19	55
6	. 65738	21	. 81781	15	11	. 87712	. 75356	19	54
7	. 65759	22	. 81796	14	11	. 87701	. 75337	19	53
8	. 65781	22	. 81810	15	11	. 87690	. 75318	19	52
9	. 65803	22	. 81825	14	11	. 87679	. 75299	19	51
10	0. 65825	22	9. 81839	15	11	9. 87668	0. 75280	19	50
11	. 65847	22	. 81854	14	11	. 87657	. 75261	20	49
12	. 65869	22	. 81868	14	11	. 87646	. 75241	19	48
13	. 65891	22	. 81882	15	11	. 87635	. 75222	19	47
14	. 65913	22	. 81897	14	11	. 87624	. 75203	19	46
15	0. 65935	21	9. 81911	15	12	9. 87613	0. 75184	19	45
16	. 65956	22	. 81926	14	11	. 87601	. 75165	19	44
17	. 65978	22	. 81940	15	11	. 87590	. 75146	20	43
18	. 66000	22	. 81955	14	11	. 87579	. 75126	19	42
19	. 66022	22	. 81969	14	11	. 87568	. 75107	19	41
20	0. 66044	22	9. 81983	15	11	9. 87557	0. 75088	19	40
21	. 66066	22	. 81998	14	11	. 87546	. 75069	19	39
22	. 66088	21	. 82012	14	11	. 87535	. 75050	20	38
23	. 66109	22	. 82026	15	11	. 87524	. 75030	19	37
24	. 66131	22	. 82041	14	12	. 87513	. 75011	19	36
25	0. 66153	22	9. 82055	14	11	9. 87501	0. 74992	19	35
26	. 66175	22	. 82069	15	11	. 87490	. 74973	20	34
27	. 66197	21	. 82084	14	11	. 87479	. 74953	19	33
28	. 66218	22	. 82098	14	11	. 87468	. 74934	19	32
29	. 66240	22	. 82112	14	11	. 87457	. 74915	19	31
30	0. 66262	22	9. 82126	15	12	9. 87446	0. 74896	20	30
31	. 66284	22	. 82141	14	11	. 87434	. 74876	19	29
32	. 66306	21	. 82155	14	11	. 87423	. 74857	19	28
33	. 66327	22	. 82169	15	11	. 87412	. 74838	20	27
34	. 66349	22	. 82184	14	11	. 87401	. 74818	19	26
35	0. 66371	22	9. 82198	14	12	9. 87390	0. 74799	19	25
36	. 66393	21	. 82212	14	11	. 87378	. 74780	20	24
37	. 66414	22	. 82226	14	11	. 87367	. 74760	19	23
38	. 66436	22	. 82240	15	11	. 87356	. 74741	19	22
39	. 66458	22	. 82255	14	11	. 87345	. 74722	19	21
40	0. 66480	21	9. 82269	14	12	9. 87334	0. 74703	20	20
41	. 66501	22	. 82283	14	11	. 87322	. 74683	19	19
42	. 66523	22	. 82297	14	11	. 87311	. 74664	20	18
43	. 66545	21	. 82311	15	12	. 87300	. 74644	19	17
44	. 66566	22	. 82326	14	11	. 87288	. 74625	19	16
45	0. 66588	22	9. 82340	14	11	9. 87277	0. 74606	20	15
46	. 66610	22	. 82354	14	11	. 87266	. 74586	19	14
47	. 66632	21	. 82368	14	12	. 87255	. 74567	19	13
48	. 66653	22	. 82382	14	11	. 87243	. 74548	20	12
49	. 66675	22	. 82396	14	11	. 87232	. 74528	19	11
50	0. 66697	21	9. 82410	14	12	9. 87221	0. 74509	20	10
51	. 66718	22	. 82424	15	11	. 87209	. 74489	19	9
52	. 66740	22	. 82439	14	11	. 87198	. 74470	19	8
53	. 66762	21	. 82453	14	12	. 87187	. 74451	20	7
54	. 66783	22	. 82467	14	11	. 87175	. 74431	19	6
55	0. 66805	22	9. 82481	14	11	9. 87164	0. 74412	20	5
56	. 66827	21	. 82495	14	12	. 87153	. 74392	19	4
57	. 66848	22	. 82509	14	11	. 87141	. 74373	20	3
58	. 66870	21	. 82523	14	11	. 87130	. 74353	19	2
59	. 66891	22	. 82537	14	12	. 87119	. 74334	20	1
60	0. 66913		9. 82551			9. 87107	0. 74314		0
↑ 131° →	cos	Diff. 1′	→ cos	Diff. 1′	Diff. 1′	sin	sin	Diff. 1′	↑ ←48°

LOGARITHMS

42°→ ↓ ′	sin	Diff. 1′	sin	Diff. 1′	Diff. 1′	cos ←	cos	Diff. 1′	←137° ↓ ′
0	0. 66913	22	9. 82551	14	11	9. 87107	0. 74314	19	60
1	. 66935	21	. 82565	14	11	. 87096	. 74295	19	59
2	. 66956	22	. 82579	14	12	. 87085	. 74276	20	58
3	. 66978	21	. 82593	14	11	. 87073	. 74256	19	57
4	. 66999	22	. 82607	14	12	. 87062	. 74237	20	56
5	0. 67021	22	9. 82621	14	11	9. 87050	0. 74217	19	55
6	. 67043	21	. 82635	14	11	. 87039	. 74198	20	54
7	. 67064	22	. 82649	14	12	. 87028	. 74178	19	53
8	. 67086	21	. 82663	14	11	. 87016	. 74159	20	52
9	. 67107	22	. 82677	14	12	. 87005	. 74139	19	51
10	0. 67129	22	9. 82691	14	11	9. 86993	0. 74120	20	50
11	. 67151	21	. 82705	14	12	. 86982	. 74100	20	49
12	. 67172	22	. 82719	14	11	. 86970	. 74080	19	48
13	. 67194	21	. 82733	14	12	. 86959	. 74061	20	47
14	. 67215	22	. 82747	14	11	. 86947	. 74041	19	46
15	0. 67237	21	9. 82761	14	12	9. 86936	0. 74022	20	45
16	. 67258	22	. 82775	13	11	. 86924	. 74002	19	44
17	. 67280	21	. 82788	14	11	. 86913	. 73983	20	43
18	. 67301	22	. 82802	14	12	. 86902	. 73963	19	42
19	. 67323	21	. 82816	14	11	. 86890	. 73944	20	41
20	0. 67344	22	9. 82830	14	12	9. 86879	0. 73924	20	40
21	. 67366	21	. 82844	14	12	. 86867	. 73904	19	39
22	. 67387	22	. 82858	14	11	. 86855	. 73885	20	38
23	. 67409	21	. 82872	13	12	. 86844	. 73865	19	37
24	. 67430	22	. 82885	14	11	. 86832	. 73846	20	36
25	0. 67452	21	9. 82899	14	12	9. 86821	0. 73826	20	35
26	. 67473	22	. 82913	14	11	. 86809	. 73806	19	34
27	. 67495	21	. 82927	14	12	. 86798	. 73787	20	33
28	. 67516	22	. 82941	14	11	. 86786	. 73767	20	32
29	. 67538	21	. 82955	13	12	. 86775	. 73747	19	31
30	0. 67559	21	9. 82968	14	11	9. 86763	0. 73728	20	30
31	. 67580	22	. 82982	14	12	. 86752	. 73708	20	29
32	. 67602	21	. 82996	14	12	. 86740	. 73688	19	28
33	. 67623	22	. 83010	13	11	. 86728	. 73669	20	27
34	. 67645	21	. 83023	14	12	. 86717	. 73649	20	26
35	0. 67666	22	9. 83037	14	11	9. 86705	0. 73629	19	25
36	. 67688	21	. 83051	14	12	. 86694	. 73610	20	24
37	. 67709	21	. 83065	13	12	. 86682	. 73590	20	23
38	. 67730	22	. 83078	14	11	. 86670	. 73570	19	22
39	. 67752	21	. 83092	14	12	. 86659	. 73551	20	21
40	0. 67773	22	9. 83106	14	12	9. 86647	0. 73531	20	20
41	. 67795	21	. 83120	13	11	. 86635	. 73511	20	19
42	. 67816	21	. 83133	14	12	. 86624	. 73491	19	18
43	. 67837	22	. 83147	14	12	. 86612	. 73472	20	17
44	. 67859	21	. 83161	13	11	. 86600	. 73452	20	16
45	0. 67880	21	9. 83174	14	12	9. 86589	0. 73432	19	15
46	. 67901	22	. 83188	14	12	. 86577	. 73413	20	14
47	. 67923	21	. 83202	13	11	. 86565	. 73393	20	13
48	. 67944	21	. 83215	14	12	. 86554	. 73373	20	12
49	. 67965	22	. 83229	13	12	. 86542	. 73353	20	11
50	0. 67987	21	9. 83242	14	12	9. 86530	0. 73333	19	10
51	. 68008	21	. 83256	14	11	. 86518	. 73314	20	9
52	. 68029	22	. 83270	13	12	. 86507	. 73294	20	8
53	. 68051	21	. 83283	14	12	. 86495	. 73274	20	7
54	. 68072	21	. 83297	13	11	. 86483	. 73254	20	6
55	0. 68093	22	9. 83310	14	12	9. 86472	0. 73234	19	5
56	. 68115	21	. 83324	14	12	. 86460	. 73215	20	4
57	. 68136	21	. 83338	13	12	. 86448	. 73195	20	3
58	. 68157	22	. 83351	14	11	. 86436	. 73175	20	2
59	. 68179	21	. 83365	13	12	. 86425	. 73155	20	1
60	0. 68200		9. 83378			9. 86413	0. 73135		0
↑ 132°→	cos	Diff. 1′	→ cos	Diff. 1′	Diff. 1′	sin	sin	Diff. 1′	←47° ↑

43°→ ′	sin	Diff. 1′	sin	Diff. 1′	Diff. 1′	cos	cos	Diff. 1′	←136° ′
0	0. 68200	21	9. 83378	14	12	9. 86413	0. 73135	19	60
1	. 68221	21	. 83392	13	12	. 86401	. 73116	20	59
2	. 68242	22	. 83405	14	12	. 86389	. 73096	20	58
3	. 68264	21	. 83419	13	11	. 86377	. 73076	20	57
4	. 68285	21	. 83432	14	12	. 86366	. 73056	20	56
5	0. 68306	21	9. 83446	13	12	9. 86354	0. 73036	20	55
6	. 68327	22	. 83459	14	12	. 86342	. 73016	20	54
7	. 68349	21	. 83473	13	12	. 86330	. 72996	20	53
8	. 68370	21	. 83486	14	12	. 86318	. 72976	19	52
9	. 68391	21	. 83500	13	11	. 86306	. 72957	20	51
10	0. 68412	22	9. 83513	14	12	9. 86295	0. 72937	20	50
11	. 68434	21	. 83527	13	12	. 86283	. 72917	20	49
12	. 68455	21	. 83540	14	12	. 86271	. 72897	20	48
13	. 68476	21	. 83554	13	12	. 86259	. 72877	20	47
14	. 68497	21	. 83567	14	12	. 86247	. 72857	20	46
15	0. 68518	21	9. 83581	13	12	9. 86235	0. 72837	20	45
16	. 68539	22	. 83594	14	12	. 86223	. 72817	20	44
17	. 68561	21	. 83608	13	11	. 86211	. 72797	20	43
18	. 68582	21	. 83621	13	12	. 86200	. 72777	20	42
19	. 68603	21	. 83634	14	12	. 86188	. 72757	20	41
20	0. 68624	21	9. 83648	13	12	9. 86176	0. 72737	20	40
21	. 68645	21	. 83661	13	12	. 86164	. 72717	20	39
22	. 68666	22	. 83674	14	12	. 86152	. 72697	20	38
23	. 68688	21	. 83688	13	12	. 86140	. 72677	20	37
24	. 68709	21	. 83701	14	12	. 86128	. 72657	20	36
25	0. 68730	21	9. 83715	13	12	9. 86116	0. 72637	20	35
26	. 68751	21	. 83728	13	12	. 86104	. 72617	20	34
27	. 68772	21	. 83741	14	12	. 86092	. 72597	20	33
28	. 68793	21	. 83755	13	12	. 86080	. 72577	20	32
29	. 68814	21	. 83768	13	12	. 86068	. 72557	20	31
30	0. 68835	22	9. 83781	14	12	9. 86056	0. 72537	20	30
31	. 68857	21	. 83795	13	12	. 86044	. 72517	20	29
32	. 68878	21	. 83808	13	12	. 86032	. 72497	20	28
33	. 68899	21	. 83821	13	12	. 86020	. 72477	20	27
34	. 68920	21	. 83834	14	12	. 86008	. 72457	20	26
35	0. 68941	21	9. 83848	13	12	9. 85996	0. 72437	20	25
36	. 68962	21	. 83861	13	12	. 85984	. 72417	20	24
37	. 68983	21	. 83874	13	12	. 85972	. 72397	20	23
38	. 69004	21	. 83887	14	12	. 85960	. 72377	20	22
39	. 69025	21	. 83901	13	12	. 85948	. 72357	20	21
40	0. 69046	21	9. 83914	13	12	9. 85936	0. 72337	20	20
41	. 69067	21	. 83927	13	12	. 85924	. 72317	20	19
42	. 69088	21	. 83940	14	12	. 85912	. 72297	20	18
43	. 69109	21	. 83954	13	12	. 85900	. 72277	20	17
44	. 69130	21	. 83967	13	12	. 85888	. 72257	21	16
45	0. 69151	21	9. 83980	13	12	9. 85876	0. 72236	20	15
46	. 69172	21	. 83993	13	13	. 85864	. 72216	20	14
47	. 69193	21	. 84006	14	12	. 85851	. 72196	20	13
48	. 69214	21	. 84020	13	12	. 85839	. 72176	20	12
49	. 69235	21	. 84033	13	12	. 85827	. 72156	20	11
50	0. 69256	21	9. 84046	13	12	9. 85815	0. 72136	20	10
51	. 69277	21	. 84059	13	12	. 85803	. 72116	21	9
52	. 69298	21	. 84072	13	12	. 85791	. 72095	20	8
53	. 69319	21	. 84085	13	13	. 85779	. 72075	20	7
54	. 69340	21	. 84098	14	12	. 85766	. 72055	20	6
55	0. 69361	21	9. 84112	13	12	9. 85754	0. 72035	20	5
56	. 69382	21	. 84125	13	12	. 85742	. 72015	20	4
57	. 69403	21	. 84138	13	12	. 85730	. 71995	21	3
58	. 69424	21	. 84151	13	12	. 85718	. 71974	20	2
59	. 69445	21	. 84164	13	13	. 85706	. 71954	20	1
60	0. 69466		9. 84177			9. 85693	0. 71934		0
133°→	cos	Diff. 1′	→ cos	Diff. 1′	Diff. 1′	sin	sin	Diff. 1′	←46°

LOGARITHMS

44° → ↓ ′	sin	Diff. 1′	sin	Diff. 1′	Diff. 1′	cos ←	cos	← 135° Diff. 1′	↓ ′
0	0.69466	21	9.84177	13	12	9.85693	0.71934	20	60
1	.69487	21	.84190	13	12	.85681	.71914	20	59
2	.69508	21	.84203	13	12	.85669	.71894	21	58
3	.69529	20	.84216	13	12	.85657	.71873	20	57
4	.69549	21	.84229	13	13	.85645	.71853	20	56
5	0.69570	21	9.84242	13	12	9.85632	0.71833	20	55
6	.69591	21	.84255	14	12	.85620	.71813	21	54
7	.69612	21	.84269	13	12	.85608	.71792	20	53
8	.69633	21	.84282	13	13	.85596	.71772	20	52
9	.69654	21	.84295	13	12	.85583	.71752	20	51
10	0.69675	21	9.84308	13	12	9.85571	0.71732	21	50
11	.69696	21	.84321	13	12	.85559	.71711	20	49
12	.69717	20	.84334	13	13	.85547	.71691	20	48
13	.69737	21	.84347	13	12	.85534	.71671	21	47
14	.69758	21	.84360	13	12	.85522	.71650	20	46
15	0.69779	21	9.84373	12	13	9.85510	0.71630	20	45
16	.69800	21	.84385	13	12	.85497	.71610	20	44
17	.69821	21	.84398	13	12	.85485	.71590	21	43
18	.69842	20	.84411	13	13	.85473	.71569	20	42
19	.69862	21	.84424	13	12	.85460	.71549	20	41
20	0.69883	21	9.84437	13	12	9.85448	0.71529	21	40
21	.69904	21	.84450	13	13	.85436	.71508	20	39
22	.69925	21	.84463	13	12	.85423	.71488	20	38
23	.69946	20	.84476	13	12	.85411	.71468	21	37
24	.69966	21	.84489	13	13	.85399	.71447	20	36
25	0.69987	21	9.84502	13	12	9.85386	0.71427	20	35
26	.70008	21	.84515	13	13	.85374	.71407	21	34
27	.70029	20	.84528	12	12	.85361	.71386	20	33
28	.70049	21	.84540	13	12	.85349	.71366	21	32
29	.70070	21	.84553	13	13	.85337	.71345	20	31
30	0.70091	21	9.84566	13	12	9.85324	0.71325	20	30
31	.70112	20	.84579	13	13	.85312	.71305	21	29
32	.70132	21	.84592	13	12	.85299	.71284	20	28
33	.70153	21	.84605	13	13	.85287	.71264	21	27
34	.70174	21	.84618	12	12	.85274	.71243	20	26
35	0.70195	20	9.84630	13	12	9.85262	0.71223	20	25
36	.70215	21	.84643	13	13	.85250	.71203	21	24
37	.70236	21	.84656	13	12	.85237	.71182	20	23
38	.70257	20	.84669	13	13	.85225	.71162	21	22
39	.70277	21	.84682	12	12	.85212	.71141	20	21
40	0.70298	21	9.84694	13	13	9.85200	0.71121	21	20
41	.70319	20	.84707	13	12	.85187	.71100	20	19
42	.70339	21	.84720	13	13	.85175	.71080	21	18
43	.70360	21	.84733	12	12	.85162	.71059	20	17
44	.70381	20	.84745	13	13	.85150	.71039	20	16
45	0.70401	21	9.84758	13	12	9.85137	0.71019	21	15
46	.70422	21	.84771	13	13	.85125	.70998	20	14
47	.70443	20	.84784	12	12	.85112	.70978	21	13
48	.70463	21	.84796	13	13	.85100	.70957	20	12
49	.70484	21	.84809	13	13	.85087	.70937	21	11
50	0.70505	20	9.84822	13	12	9.85074	0.70916	20	10
51	.70525	21	.84835	12	13	.85062	.70896	21	9
52	.70546	21	.84847	13	12	.85049	.70875	20	8
53	.70567	20	.84860	13	13	.85037	.70855	21	7
54	.70587	21	.84873	12	12	.85024	.70834	21	6
55	0.70608	20	9.84885	13	13	9.85012	0.70813	20	5
56	.70628	21	.84898	13	13	.84999	.70793	21	4
57	.70649	21	.84911	12	12	.84986	.70772	20	3
58	.70670	20	.84923	13	13	.84974	.70752	21	2
59	.70690	21	.84936	13	12	.84961	.70731	20	1
60	0.70711		9.84949			9.84949	0.70711		0
↑ 134° →	cos	Diff. 1′	→ cos	Diff. 1′	Diff. 1′	sin	sin	Diff. 1′ ←	↑ 45°

16
Tools of the Trade

Now you have an idea of what you need to get started with celestial navigation. This chapter is devoted to comments on equipment and sources for obtaining what you want.

Sextants

You really only need to add a sextant to this book to be able to practice celestial navigation at home. This sextant could be as cheap as the plastic Davis MK I (about $15) or as expensive as the best Plath (about $550). There are many choices between these two figures and, all in all, I feel that the best one to start with is the EBBCO for about $60. The EBBCO, like the best modern sextants, is a micrometer-drum sextant with a telescope. After using it, you will be perfectly adapted to any of the fine "professional" sextants on the market. I can personally testify that it is a perfectly usable sextant at sea in actual practice. As a matter of fact, when it is blowing and the water is flying, I use the EBBCO by choice. It is very light and much easier to clean if it gets doused—just wash it off under the galley pump. If you should later decide to buy a professional sextant, your EBBCO is the perfect backup to have aboard.

Listed below are some other makes of sextants with comments.

Japanese Sextants. Beware; I find the ones I have seen crudely made. In his booklet, *Choosing a Marine Sextant*, Robert Kleid of Fairfield, Connecticut, lists three typical flaws of Japanese sextants. One is the manner in which the micrometer drum is secured to its shaft—it could lead to a disaster. The drum is not pinned to the shaft. There are much better sextants for about the same money.

German Sextants. The larger Plath and the Cassens Plath are really fine instruments in almost every respect. I have the lightweight model of the large Plath. It has two shortcomings, however. First, the 4-power scope is too powerful for small-boat use in anything but a calm sea. This can be corrected by purchasing a 2.5-power scope (secondhand, since they aren't made anymore). Second, the light alloy used tends to build up microscopic amounts of corrosion on the arc, which, due to the perfect machining of the instrument, causes the index arm to drag. I have overcome this problem with lithium grease. Source: C. Plath, GmbH, Hamburg, Germany.

English Sextants. English sextants are good, well-made instruments, and their prices seem relatively reasonable. I particularly like the direct lighting feature, as it can illuminate your watch as well as the arc and greatly facilitate dawn and dusk observations. Sources: EBBCO—East Berks Boat Company, Wergrove, Berkshire, England; Hughes—Kelvin-Hughes, Ltd., Minories, London, England; Heath—Heath Navigational, Ltd., 33 Avery Hill Road, New Eltham, London, England.

American Sextants. Davis Instuments in Oakland, California, manufactures two plastic sextants. The cheapest one—a vernier rather than a micrometer model—is the one I used when I first learned. It is acceptable, but strictly for practice. It has no scope, and it is very easy for the index arm to move after you have taken a sight. This same fault applies to the more expensive model (also a vernier type), which costs about $60. With a micrometer-drum sextant, the reading cannot change unless you turn the drum. With the other types, a slight joggle can alter the reading.

Time

A good short-wave radio will enable you to pick up time signals world wide. Many portable Radio Direction Finders (RDFs) have two or three spots in the right range—2.5, 5, 10, 15, or 25 Megahertz (MHz or MC) for WWV at Fort Collins, Colorado; and 3.3 and 7.7 MHz for CHU in Canada.

As I mentioned in the chapter on time, a neat, compact, time-only receiver (Time Kube) is made and sold by Radio Shack, Inc., of Fort Worth, Texas. I have one and I like it. It costs about $50. There is one model for WWV and another for CHU.

Chronometers are fascinating, appealing, and expensive! It is hard to find a quartz-crystal chronometer for less than $200, and I really think that on a small boat there are better uses for $200. One of the most frequent criticisms against quartz crystals is that they are too sensitive to temperature changes, causing the chronometer to slow down when the temperature drops and to run faster when the temperature rises. Although this sensitivity is negligible in the even environment of a large ship, it makes the quartz crystal chronometer less than ideal for the rough and tumble of a small boat. A quartz wristwatch, on the other hand, overcomes this problem to a large degree—the body heat from your arm helps keep the temperature within a tolerance of a few degrees at all times. Thus, if you do contemplate buying a chronometer, you should investigate quartz crystal wristwatches.

Computers and Calculators

Sources for details on the computers and calculators mentioned in this book are:

Texas Instruments, Inc.
P.O. Box 3640, M/5 84M
Dallas, Texas 75221

Hewlett-Packard
19310 Pruneridge Avenue
Cupertino, California 95014

Sears Roebuck and Company
(address the store in your locality)

Micro Instrument Company
12901 Crenshaw Boulevard
Hawthorne, California 90250

Plotting Sheets

I use the Universal Plotting Sheet published by the Hydrographic Office in Washington, D.C. Like all other government materials, it is available from the Superintendent of Documents, Washington, D.C. 20402. This plotting sheet is a convenient size for small-boat use—14 × 12 inches.

There doesn't seem to be a British equivalent of this sheet, but sheets are available for various latitudes with meridians already ruled in. The main source for all British navigational materials is the Hydrographic Department, Minister of Defence, Taunton, Somerset, TA1 2DN England.

Almanacs

There are only two almanacs in general use today, the *Nautical Almanac* and the *Air Almanac*. The British versions of these are referred to also by number: *Air Publication 1602* (A.P. 1602) and *Nautical Publication 314* (N.P. 314). The *Air Almanac* is also reproduced with minor changes of language in France and Spain. The *Nautical Almanac* is similarly reproduced in Brazil, Denmark, Greece, Indonesia, Italy, Mexico, Norway, and Sweden.

The advantages of the *Air Almanac* are the ease and speed with which data can be extracted and the excellent star chart and sky diagrams. The disadvantage is that it is published in three parts at $6.50 each. Not only is it three times as bulky as the *Nautical Almanac*; it is also three times as expensive. All in all for a small-boat owner, the *Nautical Almanac* is a better bet. Its interpolation for GHA is more accurate, and you can buy the information for a whole year at once. The *Air Almanac* is published quarterly, and it might not be convenient for you to pick it up.

Sight Reduction Tables

Two sight reduction tables are used in this book—primarily H.O. 249, because of its speed and the great simplification of star sights offered by Volume I. H.O. 249 is in three volumes: Volume I, *Selected Stars*; Volume II, *Latitudes 0-39°*; Volume III, *Latitudes 40°-89°*.

Like everything else in life, H.O. 249 does have its shortcomings. The principal one is that Volumes II and III cover only bodies whose declinations are less than 29° North or South. This range of declination, however, covers all the planets, the sun, the moon, and 43 of the 57 navigational stars; so you can see that this is not a serious shortcoming for a small-boat sailor. Volume I is republished every five years to catch up with the precession and nutation correction; Volumes II and III, of course, are good for as long as the laws of spherical trigonometry remain in effect.

H.O. 229, on the other hand, is totally complete. Its six volumes cover all latitudes and all declinations. They also weigh a ton and take up a foot of shelf space. The interpolation table is finicky and, while terribly accurate, is really more than is required on a small boat.

The English counterpart to H.O. 249 is A.P. 3270 and is printed on thinner paper than H.O. 249. As a result, it is less bulky. It is also hard bound instead of loose-leaf. These two considerations resulted in my ordering a set for myself.

The British equivalent of H.O. 229 is N.P. 401. It has been slightly reduced in size, and the paper is thinner, so it is not quite so bulky as the American. Otherwise, it is identical.

Other Sight Reduction Tables

Before the advent of H.O. 229, the official sight reduction table was H.O. 214, and it was taught at the United States Naval Academy. It has the characteristic of dealing with meridian angle (*t*) rather than LHA, so the rules for azimuth are the same as those outlined in Chapter 15. Although it is now out of print, it is a good table, consisting of nine relatively thin volumes. The only real shortcoming is that it does not provide solutions for observations (Ho) of less than 5°.

Also out of print now is H.O. 211, which is one of the most compact sight reduction methods available. The book is only 50 pages, measures a mere 9″ × 6″, and can solve any navigational triangle. My one cavil with it is that instead of using the two fundamental formulas (see Chapter 15) it solves the navigational triangle in another way that requires four. To me this is more tedious than simply going back to the two basic sine-cosine formulas.

If you really like celestial navigation, look up this method; the book is certainly the most compact available, and the method is quite elegant. My edition dates from 1940, when it cost 90 cents!

General Sources

Here is a list of various companies specializing in the sale of sextants, tables, plotting sheets, and so on.

California

G. E. Butler Company
356 California Street
San Francisco, CA 94100

Coast Navigation School
418 East Canon Perdido
Santa Barbara, CA 93102

Southwest Instrument Company
235 W. Seventh Street
San Pedro, CA 90731

Connecticut

Kleid Navigation, Inc.
24 Lee Drive
Fairfield, CT 06430

Illinois

Navigation Equipment Company
228 W. Chicago Avenue
Chicago, IL 60600

Louisiana

Baker, Lyman & Company
308 Magazine Street
New Orleans, LA 70100

Maryland

Maryland Nautical Sales
406 Water Street
Baltimore, MD 21200

Weems & Plath, Inc.
P.O. Box 1991
Annapolis, MD 21400

New York

M. Low Inc.
110 Hudson Street
New York, NY 10013

New York Nautical Instrument & Service
140 W. Broadway
New York, NY 10013

Pennsylvania

Victor Aguste Gustin
105 S. Second Street
Philadelphia, PA 19100

Texas

Baker, Lyman & Company
Cotton Exchange Building
Houston, TX 77000

R. H. John Chart Agency
515 Twenty-first Street
Galveston, TX 77550

Washington

"Captains"
1324 Second Avenue
Seattle, WA 98100

Further Reading

For readers who want to go further into celestial navigation, there are three major "standard" works:

Bowditch, Nathaniel. *American Practical Navigator.* Published by the Hydrographic Office as H.O. 9. This is truly an epic tome, including incredible amounts of information on practically every aspect of navigation. It is available from the Superintendent of Documents, Washington, D.C. 20402.

Dunlap, G. D., and Shufeldt, H. H. *Dutton's Navigation and Piloting.* Naval Institute Press, Annapolis, Md. This is the text for the Naval Academy—thorough, rigorous, with meticulously detailed drawings.

Mixter, George W. *Primer of Navigation.* Van Nostrand Reinhold, New York, N.Y. If your dad knows celestial navigation, this is probably the book he learned it from. It has been recently updated and has quizzes at the end of the chapters.

Sight Reduction Formats

To use the sight reduction formats that follow, simply take out the parts you need and set them up on your worksheet.

Thus, if you are going to take a sight for a sun line and are using H.O. 249 and the *Nautical Almanac*, you take 1, 2, 3, 4, 5, and 9 from the left-hand column.

If you are going to use the *Air Almanac* and H.O. 229, you use 1, 2, and 3 from the left-hand column, 4 and 5 from the middle column, and 9 from the right-hand column.

If you are going to take a multibody fix, set the necessary formats up in parallel rows on your worksheet. That way when you open the almanac or table, you can get all the data at once.

Nautical Almanac and H.O. 249			*Air Almanac*		H.O. 229
1. Date :					
Log Reads:					
2. WT	=				
F (−) S (+)	=				
WT to GMT	=				
GMT	=				
3. IC = ______ On	=	− Corr.			
= ______ Off	=	+ Corr.			
4. Hs	=		Hs	=	
Dip	=		Dip	=	
IC	=		IC	=	
ha	=		ha	=	
Main Corr.	=		Ref.	=	
L or U	=		Semi diam.	=	
Add'l. Corr.	=		HP Corr.	=	
Ho	=		Ho	=	
5. SUN			SUN		
Dec.	=		Dec.	=	
GHA Hr.	=		GHA Hr., Min.	=	
Min., Sec.	=		Min., Sec.	=	

GHA	=	
(+360?)	=	
Assumed Long. W.	=	(−)
Assumed Long. E.	=	(+)
LHA	=	
(−360?°)	=	
LHA	=	
Assumed Lat.	=	

6. MOON

HP	=	
Dec.	=	
Increasing?	=	
Decreasing?	=	
(d =) Corr. =	=	
Dec.	=	
GHA Hr.	=	
Min., Sec.	=	
(v =) Corr.	=	
GHA	=	
(+360?)	=	
Assumed Long. W.	=	(−)
Assumed Long. E.	=	(+)
LHA	=	
(−360?)	=	

GHA	=	
(+360?)	=	
Assumed Long. W.	=	(−)
Assumed Long. E.	=	(+)
LHA	=	
(−360?)	=	
LHA	=	
Assumed Lat.	=	

MOON

HP	=	
Dec.	=	
Increasing?	=	
Decreasing?	=	
(d =) Corr.	=	
Dec.	=	
GHA Hr., Min.	=	
Min., Sec.	=	
(v =) Corr.	=	
GHA	=	
(+360?)	=	
Assumed Long. W.	=	(−)
Assumed Long. E.	=	(+)
LHA	=	
(−360?)	=	

LHA	=	
Assumed Lat.	=	
7. PLANET		
Dec.	=	
GHA Hr.,	=	
Min., Sec.	=	
(v =) Corr.	=	
GHA	=	
(+360?)	=	
Assumed Long. W.	=	(−)
Assumed Long. E.	=	(+)
LHA	=	
(−360?)	=	
LHA	=	
Assumed Lat.	=	
8. STAR		
Dec.	=	
GHA Aries—Hr.	=	
Min., Sec.	=	
GHA	=	
(+360?)	=	
Assumed Long. W.	=	(−)
Assumed Long. E.	=	(+)

LHA	=	
Assumed Lat.	=	
PLANET		
Dec.	=	
GHA Hr., Min.	=	
Min., Sec.	=	
(v =) Corr.	=	
GHA	=	
(+360?)	=	
Assumed Long. W.	=	(−)
Assumed Long. E.	=	(+)
LHA	=	
(−360?)	=	
LHA	=	
Assumed Lat.	=	
STAR		
Dec.	=	
GHA Aries—Hr. Min.	=	
Min., Sec.	=	
SHA Star	=	
GHA Star	=	
(+360?)	=	
Assumed Long. W.	=	(−)
Assumed Long. E.	=	(+)

LHA Aries	=			LHA Star	=		
(−360?)	=			(−360?)	=		
LHA Aries	=			LHA Star	=		
Assumed Lat.	=			Assumed Lat.	=		
9. Hc	=	d	=	Hc	=	d	=
Corr.	=	Dec. Inc.	= x	Corr.	=	Dec. Inc.	=
Hc	=	Corr.	=	Hc	=	Tens	=
						U's, d's	=
Ho	=			Ho	=	DSD	=
						Corr.	=
Z	=	Zn	=	Z	=	Zn	=

Appendix
BBC Schedule of Time Broadcasts

BBC Time Signals.

NATURE OF BROADCAST: The B.B.C. time signal consists of an automatic transmission by the standard clock at the Greenwich Observatory of 6 dots, one for each second from the 55th to the 60th inclusive. The final dot is the time signal.

ERROR: Normally accurate to 0.1 second.

Radio One

HOURS OF TRANSMISSION:	FREQ.:
0600	
0900	1214 kHz, A3.
1100	
1830 (except Sunday)	

Radio Two

HOURS OF TRANSMISSION:	FREQ.:
0600	200 kHz, A3;
0900	89.1 MHz, Wrotham;
1100	90.0 MHz, Dover;
1300	90.1 MHz, Brighton.
1830	

Radio Four.

HOURS OF TRANSMISSION:	FREQ.:
0600 (except Sun.), 0700, 0800, 1200, 1700, 2200 (except Sat.)	North 692, Scotland 809, Wales 881, London 908, West 1092, Midland 1088, North 1151, N. Ireland 1340, South 1457, London (Ramsgate) 1484 kHz, A3. Brighton(West) 94.5, Dover 94.4, Wrotham 93.5 MHz, F3.

European Service.

HOURS OF TRANSMISSION:	METER BAND:
0400	49, 41, 31, 25.
0500	49, 41, 31, 25, 19.
0600	75, 49, 41, 31, 25, 19.
1300	31, 25, 19, 16, 13.
1600	31, 25, 19, 16.
1800	49, 41, 31.
1900	31, 25, 19, 16.
1930	31, 25, 19, 16.
2000	49, 41, 31, 25, 19.
2300	75, 49, 41, 31.

Index

T

U

V

W

Z